Introduction to Catastrophe Risk Modelling

Focusing on the physics of the catastrophe process and addressed directly to advanced students, this innovative textbook quantifies dozens of perils, both natural and human-made, and covers the latest developments in catastrophe modelling. Combining basic statistics, applied physics, natural and environmental sciences, civil engineering, and psychology, the text remains at an introductory level, focusing on fundamental concepts for a comprehensive understanding of catastrophe phenomenology and risk quantification. A broad spectrum of perils is covered, including geophysical, hydrological, meteorological, climatological, biological, extraterrestrial, technological, and socio-economic, as well as events caused by domino effects and global warming. Following industry standards, the text provides the necessary tools to develop a CAT model from hazard to loss assessment. Online resources include a CAT risk model starter kit and a CAT risk modelling 'sandbox' with a Python Jupyter tutorial. Every process, described by equations, (pseudo)codes, and illustrations, is fully reproducible, allowing students to solidify knowledge through practice.

Arnaud Mignan is a renowned expert in catastrophe risk assessment, who has worked at leading firms and institutes in London, Zurich, and Shenzhen over the past two decades. He developed innovative models while at Risk Management Solutions and catastrophic domino-effect models – featured in a national newspaper and Euronews – while at the Swiss Federal Institute of Technology. Later, as an Associate Professor at The Institute of Risk Analysis, Prediction and Management (Risks-X), established by the Southern University of Science and Technology (SUSTech), and as a Guest Professor at the ETH Risk Center in Zurich, he crafted the course *Introduction to Catastrophe Risk Modelling*. Since 2023, he has been CEO of Mignan Risk Analytics GmbH and acts as a leading specialist for the World Bank.

'Professor Mignan has the benefit of substantial insurance industry experience to complement his academic achievements. This is a timely book, which incorporates the important lessons from the major disasters of recent years, and should be a valuable guide for students of catastrophe modelling.'

Gordon Woo, Moody's/RMS

'This book is a definitive guide on catastrophe risk modelling, expertly covering all major perils and their interconnected cascade effects. Its remarkably clear pedagogical style, with numerous concrete examples and worked-out models, makes it an invaluable resource for understanding complex risks. It combines comprehensive insights with practical step-by-step instructions combining physical modelling, concrete engineering implementations and the relevant decision making and financial economic dimensions. I took great pleasure in exploring its extraordinary rich content and I am sure that it will become the key reference for professionals, academics, and students.'

Didier Sornette, Southern University of Science and Technology

Introduction to Catastrophe Risk Modelling

A Physics-Based Approach

ARNAUD MIGNAN
Mignan Risk Analytics GmbH, Greater Zurich Area, Switzerland
Institute of Risk Analysis, Prediction and Management (Risks-X), SUSTech, Shenzhen, China

CAMBRIDGE
UNIVERSITY PRESS

CAMBRIDGE
UNIVERSITY PRESS

Shaftesbury Road, Cambridge CB2 8EA, United Kingdom

One Liberty Plaza, 20th Floor, New York, NY 10006, USA

477 Williamstown Road, Port Melbourne, VIC 3207, Australia

314–321, 3rd Floor, Plot 3, Splendor Forum, Jasola District Centre, New Delhi – 110025, India

103 Penang Road, #05–06/07, Visioncrest Commercial, Singapore 238467

Cambridge University Press is part of Cambridge University Press & Assessment, a department of the University of Cambridge.

We share the University's mission to contribute to society through the pursuit of education, learning and research at the highest international levels of excellence.

www.cambridge.org
Information on this title: www.cambridge.org/highereducation/isbn/9781009437387

DOI: 10.1017/9781009437370

When citing this work, please include a reference to the DOI 10.1017/9781009437370

First published 2025

A catalogue record for this publication is available from the British Library

Library of Congress Cataloging-in-Publication Data
Names: Mignan, Arnaud, author.
Title: Introduction to catastrophe risk modelling : a physics-based approach / Arnaud Mignan, Mignan Risk Analytics Gmbh Institute of Risk Analysis, Prediction & Management (Risks-X), SUSTech, Shenzhen, China.
Description: Cambridge, United Kingdom ; New York, NY, USA : Cambridge University Press, 2025. | Includes bibliographical references and index.
Identifiers: LCCN 2024011641 | ISBN 9781009437387 (hardback) | ISBN 9781009437370 (ebook)
Subjects: LCSH: Emergency management – Mathematical models. | Risk – Mathematical models. | Risk management. | Hazard mitigation.
Classification: LCC HV551.2 .M55 2025 | DDC 363.34/8011–dc23/eng/20240612
LC record available at https://lccn.loc.gov/2024011641

ISBN 978-1-009-43738-7 Hardback
ISBN 978-1-009-43734-9 Paperback

Additional resources for this publication at www.cambridge.org/mignan

*To my daughters Cléore and Lilly-Belle,
with the hope that they can better navigate
the uncertainties of tomorrow's world.*

Contents

List of Plates page xi
List of Figures xii
List of Tables xv
List of Boxes xvi
Foreword xvii
Preface xix
Acknowledgements xxii
List of Abbreviations and Symbols xxiii

Part I Basics of CAT Modelling

1 Overview of Catastrophe Risks 3
1.1 Concepts and Terminology 4
1.1.1 Catastrophe, Disaster, and Peril 4
1.1.2 Hazard and Risk 5
1.1.3 Scale of Catastrophe Risks 7
1.1.4 Uncertainty, Randomness, and Probability 11
1.1.5 Catastrophe (CAT) Model 13
1.2 History of Catastrophe Science 16
1.2.1 From Ancient Myths to Medieval Texts 16
1.2.2 Mathematization 18
1.2.3 Computerization 19
1.3 Peril Taxonomy 21
1.3.1 Geophysical Perils 22
1.3.2 Hydrological Perils 27
1.3.3 Meteorological Perils 31
1.3.4 Climatological Perils 35
1.3.5 Biological Perils 39
1.3.6 Extraterrestrial Perils 41
1.3.7 Technological Perils 43
1.3.8 Socio-Economic Perils 50
1.4 Concluding Remarks 56
1.5 Exercises 58

2 Probabilistic Hazard Assessment 60
2.1 Basic Concepts 60

2.1.1 Stochastic Event Set 61
2.1.2 Hazard Intensity Footprint Catalogue 63
2.1.3 Hazard Metrics 64
2.1.4 Hazard as an Energy-Transfer Process 66
2.2 Event Source 67
2.2.1 Point Sources 69
2.2.2 Line Sources 73
2.2.3 Track Sources 81
2.2.4 Area Sources and Diffuse Sources 83
2.3 Event Size Distribution 86
2.3.1 Power-Law Distribution 87
2.3.2 Extreme Value Distributions 94
2.4 Event Intensity Footprint 97
2.4.1 Analytical Expressions of Static Spatial Diffusion 98
2.4.2 Threshold Models of Passive Event Emergence 104
2.4.3 Numerical Models of Dynamic Event Propagation 106
2.5 Application to Probabilistic Seismic Hazard Assessment 106
2.5.1 Stochastic Rupture Set (Event Table) 107
2.5.2 Earthquake Ground-Shaking Footprint Catalogue 112
2.5.3 Hazard Metrics 113
2.6 Concluding Remarks 114
2.7 Exercises 115

3 Probabilistic Risk Assessment 118
3.1 Exposure Module 119
3.1.1 Exposure Characteristics 120
3.1.2 Exposure Footprint 124
3.2 Vulnerability Module 129
3.2.1 Basic Concepts 129
3.2.2 Expert Judgement Approach 132
3.2.3 Empirical Approach 136
3.2.4 Engineering-Based Approach 137
3.2.5 Vulnerability Functions per Peril 140
3.3 Loss Module 145
3.3.1 Loss Tables 146
3.3.2 Risk Metrics 147
3.3.3 The CAT Risk Modelling Sandbox 151
3.4 Concluding Remarks 156
3.5 Exercises 156

Part II Advanced CAT Modelling

4 Dynamic Hazard Modelling 161
4.1 Cellular Automata and Agent-Based Models 162

4.1.1 Basic Concepts 162
4.1.2 Blackout (BO) Cellular Automaton 167
4.1.3 Flood (FF, PF, . . .) Cellular Automaton 168
4.1.4 Landslide (LS) Cellular Automaton 170
4.1.5 Wildfire (WF) Cellular Automaton 173
4.1.6 Stampede (St) Agent-Based Model 174
4.1.7 Social Unrest (SU) Agent-Based Model 176
4.2 System Dynamics 178
4.2.1 Concepts 178
4.2.2 Epidemic (Ep) Dynamics 179
4.2.3 Pest Infestation (PI) Dynamics 183
4.2.4 Armed Conflict (AC) Dynamics 188
4.3 Concluding Remarks 189
4.4 Exercises 190

5 Modelling of Cascading Effects 193
5.1 Interdependencies in CAT Models 194
5.1.1 Alternative to the Poisson Distribution 194
5.1.2 Chain-of-Events Encoding 197
5.1.3 Non-stationarity of CAT Model Parameters 203
5.2 Domino Effects 206
5.2.1 Domino Effects at Refineries (Ex, Fi, TS) 206
5.2.2 Cascading Failures in Physical Networks ($\cdots \rightarrow$ BO) 209
5.2.3 Generalization to Any Peril Interaction 211
5.3 Compound Events in the Complex Earth System 212
5.3.1 Global Warming 213
5.3.2 Ecosystem Collapse 221
5.3.3 Societal and Civilizational Collapse 224
5.4 Concluding Remarks 229
5.5 Exercises 230

Part III CAT Risk Management

6 CAT Risk Management 235
6.1 Risk Transfer and Risk Reduction Instruments 236
6.1.1 Risk Transfer via Catastrophe Insurance 236
6.1.2 Risk Transfer via CAT Bonds 240
6.1.3 Risk Reduction Instruments 242
6.2 Peril Forecasting 245
6.2.1 Tropical Cyclone (TC) Forecasting 246
6.2.2 Earthquake (EQ) Forecasting 248
6.3 Risk Governance: A Case Study 253
6.3.1 Case Study: Seismic Risk at a Geothermal Plant 255
6.3.2 Epistemic Uncertainties in CAT Models 258

6.3.3 Decision-Making under Uncertainties 263
6.4 Final General Conclusions 271
6.5 Exercises 272

References 276
Index 331

The plate section is to be found between pages 166 and 167

Plates

1.1 Annual damage losses due to disasters during the period 1950–2020
1.2 Typical scheme of a risk matrix
1.4 Ranking of catastrophic losses per peril category
1.8 Geographical distribution of some of the main natural perils on Earth
2.4 Examples of localized sources (points, lines, and tracks)
2.20 The 2003 European heatwave footprint defined with a threshold method
3.5 Damage footprint of the 2011 Joplin tornado and vulnerability curve fit
3.10 One instance of the GenMR digital template (default parameterization)
3.11 Intensity footprint and loss footprint in the GenMR digital template
4.6 Example of a fluvial flood footprint modelled with a simple cellular automaton
4.8 Examples of simulated landslides triggered by rainstorms
4.9 The forest-fire model
5.13 Maps of GMST changes observed from May 1953 to May 2023

Figures

1.1	Annual damage losses due to disasters during the period 1950–2020	*page* 3
1.2	Typical scheme of a risk matrix	6
1.3	Scale of typical catastrophes compared to other event types	7
1.4	Ranking of catastrophic losses per peril category	8
1.5	Comparing aleatory and epistemic uncertainties with the roll of a dice	12
1.6	The risk process in CAT modelling	14
1.7	Woodcuts from Lycosthenes' *Prodigiorum ac ostentorum chronicon*	17
1.8	Geographical distribution of some of the main natural perils on Earth	23
1.9	Illustrating the earthquake process and potential observed damage	24
1.10	Examples of historical volcanic eruptions	26
1.11	Examples of historical landslides	27
1.12	Illustrating the flooding process and potential observed extent	28
1.13	Illustrating the tsunami process and quantifying its causes	30
1.14	Illustrating the tropical cyclone process and structure	32
1.15	Dust storm during the Dust Bowl caused by a severe drought	37
1.16	The 2003 Cedar wildfire, Southern California	38
1.17	Illustrating the epidemic process and the structure of a virus	40
1.18	Illustrating the airburst process and potential observed damage	42
1.19	Examples of human-made explosions	46
1.20	The Vajont Dam, before and after the 1963 landslide	47
1.21	Maslow's hierarchy of needs in the context of catastrophe risks	51
1.22	Exercise 1.3: Ad hoc example illustrating the catastrophe risk process	59
2.1	Illustrating the main steps taken in the hazard module	61
2.2	Examples of hazard curves derived from simulated winter storms	65
2.3	Understanding the hazard process in terms of energy transfer	66
2.4	Examples of localized sources (points, lines, and tracks)	68
2.5	Relationship between catchment characteristics and water discharge	72
2.6	Fault rupture characteristics and relationship to earthquake magnitude	74
2.7	Storm surge as a line source and relationship to wind speed	76
2.8	Illustrating how a stochastic tornado line source can be generated	78
2.9	Schematic representation of tsunami source types	79
2.10	Pressure and wind speed profiles of a tropical cyclone	82
2.11	Specific case of the environment as a diffuse source	84
2.12	Landslide susceptibility defined from the factor of safety	85
2.13	Empirical size distribution of various perils	90
2.14	Size distributions of earthquakes and landslides	92

2.15 GEV application to the Potomac River peak discharge 97
2.16 Examples of earthquake ground-motion footprints 100
2.17 Comparing overpressure footprints of explosions and asteroid impacts 101
2.18 Example of tornado wind field with matching intensity footprint 102
2.19 Modelling of the intensity footprint of Hurricane Katrina 103
2.20 The 2003 European heatwave footprint defined with a threshold method 105
2.21 Map of crustal sources in the European Database of Seismogenic Faults 108
2.22 Earthquake ground-motion footprint catalogue 109
2.23 Examples of ground-motion fields for a given rupture 113
2.24 Hazard maps and hazard curves for our ad hoc PSHA 113
2.25 Exercise 2.2: Map and time series of past monster occurrences 116
3.1 Workflow of the CAT risk model framework 119
3.2 Exposure characteristics and relationship to different hazard loads 123
3.3 Exposure footprints for Malta obtained from different databases 127
3.4 Examples of fragility curves and of a vulnerability curve 129
3.5 Damage footprint of the 2011 Joplin tornado and vulnerability curve fit 137
3.6 Vulnerability curves fitted to the 1993 Meuse flood claims 138
3.7 Engineering-based approach to fragility curve modelling 139
3.8 Generic vulnerability curves for different hazard intensity loads 144
3.9 Example of exceedance probability (EP) curve 150
3.10 One instance of the GenMR digital template (default parameterization) 153
3.11 Intensity footprint and loss footprint in the GenMR digital template 154
4.1 Cellular automaton neighbourhoods 163
4.2 The sandpile model 165
4.3 Cellular automata in the CAT modelling framework 166
4.4 Example of a blackout sandpile model 168
4.5 Water transfer rules for a flood cellular automaton 169
4.6 Example of a fluvial flood footprint modelled with a simple cellular automaton 171
4.7 Mass transfer rules for a landslide cellular automaton 172
4.8 Examples of simulated landslides triggered by rainstorms 172
4.9 The forest-fire model 174
4.10 Crowd dynamics ABM for stampede simulation 175
4.11 Example of a civil-violence ABM for riot simulation 177
4.12 Stocks, flows, and feedback loops in system dynamics 179
4.13 Causal loop diagram of the SIR model 180
4.14 SIR model of the spread of an infectious disease 182
4.15 A SIR process simulated with a toy ABM 183
4.16 Prey–predator population dynamics following the Lotka–Volterra ODEs 185
4.17 Spruce budworm outbreak as a cusp catastrophe 187
4.18 Exercise 4.1a: Sandpile iterations 191
5.1 Statistical clustering modelled with a negative binomial distribution 195
5.2 Impact of clustering on different types of EP curves 198
5.3 Some concepts of graph theory to model chains of events 200
5.4 Illustrating the process of earthquake clustering 201

5.5 Implementation of chains of events in a YET 202
5.6 Long-term risk drivers in exposure and vulnerability 203
5.7 Combined roles of volcanic eruptions and earthquakes in building damage 204
5.8 Accelerated building fatigue from earthquake clustering 205
5.9 Domino effects at refineries, derived from ENSAD 208
5.10 Cascading failure in a power grid due to a natural event 210
5.11 General graph of peril interactions 212
5.12 Global warming and emergence of temperature extremes 214
5.13 Maps of GMST changes observed from May 1953 to May 2023 215
5.14 Concept of a compound event 216
5.15 Causal loop diagram of the ESCIMO global climate model 217
5.16 Examples of models describing ecosystem collapse due to a regime shift 224
5.17 Diagram of the World2 model interrelating the five variables 227
5.18 Results of the World2 model's standard run 228
5.19 Exercise 5.3c: Set-up of the critical infrastructure 231
6.1 Illustration of an insurance financial structure 239
6.2 Impact of earthquake reduction designs on building vulnerability 243
6.3 Hurricane track forecasting using a Markov model 247
6.4 Precursory pattern prior to the 1997 $M = 6$ Umbria and Marche earthquake 250
6.5 The emblematic aftershock sequence of the 1891 Great Nobi earthquake 252
6.6 CAT modelling in the general risk governance scheme 253
6.7 Risk governance case study: seismic risk at a geothermal plant 256
6.8 Epistemic uncertainties considered in the case study 262
6.9 EP-curve epistemic uncertainties visualized in a 'spaghetti' plot 263
6.10 Cumulative prospect theory parameterization 268
6.11 Exercise 6.1: The beta distribution 272
6.12 Exercise 6.3: Ad hoc forecast heat map and ROC curve (to add) 274

Tables

1.1	Peril taxonomy	*page* 21
1.2	The 10 deadliest earthquakes in CATDAT (1900–2023[a])	25
1.3	The 10 costliest weather and climate disasters in the US	32
1.4	The 10 most damaging windstorms in the period 1979–2012	33
1.5	The 10 deadliest epidemics in 1500–2023	40
1.6	Complexity of technological perils simplified	44
1.7	All 42 perils ordered alphabetically by their identifiers	56
2.1	An ad hoc event table	62
2.2	An ad hoc simulation table	63
2.3	Volcanic Explosivity Index (VEI)	73
2.4	Enhanced Fujita path-portion weight per declared EF tornado	77
2.5	Power-law size distribution parameters for different perils	89
2.6	Excerpt from the EDSF crustal fault source model	108
2.7	Excerpt from the generated event table	111
2.8	Exercise 2.2: Catalogue of past monster occurrences	115
3.1	Examples of consequence functions	132
3.2	Abbreviated Modified Mercalli Intensity scale	133
3.3	Abbreviated TORRO hailstorm intensity (H) scale	134
3.4	Abbreviated Saffir–Simpson scale	135
3.5	Abbreviated Enhanced Fujita (EF) scale	135
3.6	Generic damage scale for blasts	141
3.7	Generic damage scale for earthquakes	142
3.8	Generic vulnerability tables for floods	143
3.9	A fictitious event loss table (ELT)	147
3.10	Fictitious ELT with EP calculations	150
3.11	Example of a fictitious ELT with two columns to fill	157
6.1	Exercise 6.4: Aftershock probabilities to be completed	274

Boxes

1.1	A Note on Catastrophe Databases	*page* 9
1.2	Catastrophes in Literature	19
2.1	The Monte Carlo Method (with Inverse Transform Sampling)	80
2.2	The Giant-Catastrophe Metaphor	87
2.3	Basic Power-Law Statistics	88
2.4	Pseudocode to Plot a CCDF from Empirical Data	91
2.5	Stochastic Ruptures for Unknown Fault Locations	110
2.6	Runaway Ruptures: the 2023 *M*7.8 Türkiye–Syria Earthquake	111
3.1	Modelling Urban Land-Use Patterns	121
3.2	Modelling Road Networks	125
3.3	Modelling Built-Up Land Footprints	126
3.4	Downscaling	128
3.5	Generating YLTs from an ELT	148
3.6	EP Curve: Going from Eq. (3.20) to Eq. (3.21)	149
3.7	Build Your Own CAT Risk Model from Scratch (Hazard Part)	154
3.8	Build Your Own CAT Risk Model from Scratch (Risk Part)	155
4.1	Game of Life and Rule 30	163
4.2	Zombie Apocalypse Modelled with a Modified SIR Model	180
4.3	Logistic Growth	184
4.4	Catastrophe Theory	187
5.1	Square Matrix Definitions for Chains of Events	198
5.2	Trophic Cascades	221
5.3	The Legacy of J.W. Forrester	226
6.1	Portfolio Diversification	237
6.2	Measuring the Success of a Forecast: The 'Finley Affair'	251
6.3	Bayesian Inference	260
6.4	St. Petersburg Paradox	266

Foreword

Catastrophe science advances with the occurrence of major events, each of which provides an opportunity to learn. Every decade there is a need for a new text on catastrophe modelling, which reflects the lessons learned in previous decades. Arnaud Mignan has written the principal text for the 2020s, a decade in which the global concern over climate change overshadows the management of all weather risks.

Each decade is remembered by seismologists and earthquake engineers for key earthquake disasters. The East Anatolian Fault earthquake in Türkiye–Syria of 6 February 2023 is one of these. Five years earlier, Arnaud and I co-authored a paper on runaway earthquakes on the North Anatolian Fault, which anticipated this type of rare rupture sequence. Valuable insight into rare catastrophes with little or no historical precedent can be gained from Arnaud's book.

As a Risk Management Solutions (RMS) colleague prior to joining the Federal Institute of Technology (ETH) Zurich, Arnaud has gained a very broad knowledge of catastrophe modelling, founded on substantial practical experience, and is well qualified to write a textbook and teach a course on this subject, for which this book would be requisite reading.

A notable feature of this book is the inclusion of exercises for students. These will be a useful teaching aid for those lecturing on catastrophe modelling, in either an academic or a business environment.

Dr. Gordon Woo
Catastrophist, Moody's/RMS
July 2023

Preface

Catastrophe risk modelling is a relatively young profession which has gained from decades of research and development in the domains of natural science, engineering, actuarial science, and computer science. In 2010, I made a rather unconventional decision to return to academia after having worked several years in the catastrophe risk modelling industry, at Risk Management Solutions (RMS). My motivation was to study catastrophes from a broader and more theoretical perspective than what was typically demanded by the insurance industry. This set me on a decade-long journey at the Swiss Federal Institute of Technology (ETH) Zurich, into the fascinating field of multi-risk assessment, where various expertise – on earthquakes, floods, storms, severe accidents, and more – had to be merged into a unified framework. I would often peruse my copy of *Catastrophe Modeling: A New Approach to Managing Risk*, signed for me by my RMS colleague and editor Patricia Grossi, and which remains to this day one of the main textbooks on the subject matter. Simultaneously, I diligently accumulated a thousand or more scientific articles that, in some way or another, related to catastrophe science. When the opportunity arose in 2020 to accept an associate professorship at the Southern University of Science and Technology (SUSTech) in Shenzhen, I was presented with a unique chance to shape a course of my choosing. Naturally, I decided to offer an introductory course on catastrophe risk modelling. With this decision, the time had come to collate 15 years of combined experience in the industry and academia to describe the catastrophe concept as an entity in itself.

This textbook has been crafted to accompany the course *Introduction to Catastrophe Risk Modelling*, first taught at SUSTech during the spring semester of 2021 and later offered at ETH Zurich in the subsequent year. The decision to create a comprehensive textbook instead of simple lecture notes came from an important observation and some fortuitous event: previous textbooks were primarily addressed to insurers and practitioners in disaster risk reduction. Recognizing a notable gap in the literature, I aspired to fill it by writing the very first STEM book dedicated to the field of catastrophe modelling. The writing of this textbook was, in part, made possible by extended writing sessions during the challenging periods of quarantine, which coincided with my frequent travels between Switzerland and China. These travels occurred at the height of the COVID-19 pandemic, an event that will feature among the historical catastrophes discussed in the forthcoming pages.

The textbook has several key features that distinguish it from others in the field:

- **Comprehensive coverage:** It encompasses a broad spectrum of perils, ranging from natural disasters to human-made events, including earthquakes, volcanic eruptions, landslides, floods, storms, wildfires, epidemics, asteroid impacts, explosions, blackouts, and social unrest, among many others (42 in total).

- **Unified approach:** The textbook adopts a systematic and unifying approach to catastrophe (CAT) modelling by categorizing and harmonizing all perils through first principles and straightforward computational rules. This approach simplifies the complex landscape of CAT modelling.
- **Theoretical and practical insights:** It provides both theoretical foundations and practical insights into CAT modelling. Readers will gain a deep understanding of the subject while also acquiring the essential tools used in the industry. Additionally, the textbook offers step-by-step guidance for building a simplified CAT risk model from scratch.
- **Hands-on learning:** To facilitate direct application of the acquired knowledge, the textbook offers a suite of supplementary materials. This includes the CAT Risk Modelling Sandbox (version 1), featuring an interactive notebook and computer programs, allowing readers to experiment with various disaster scenarios. Furthermore, a CAT Risk Modelling Starter Kit is provided in different programming languages, giving individuals the chance to construct their own CAT risk models.

This textbook is designed for readers who ideally possess a scientific or engineering background and have a keen interest in exploring the fundamental, societal, and/or interdisciplinary aspects of catastrophe risk analysis. The material covered in this text is very broad in scope, encompassing principles of statistics, applied physics, environmental sciences, civil engineering, and psychology. However, it is important to note that this book maintains an introductory level of presentation, with a central focus on delivering fundamental concepts that enable a comprehensive understanding of catastrophe phenomenology and risk quantification. While the application of first principles is advocated throughout the chapters, interested readers will discover a wealth of references pointing to more in-depth and advanced studies if they wish to dig deeper into specific areas. To enhance comprehension and facilitate practical applications, any given risk process is explained from various angles with illustrations and equations complemented by in-text pseudocodes and exercises, as well as concrete code examples using various catastrophe scenarios in the supplementary materials. It therefore follows Marvin Minsky's dictum that 'you don't understand anything until you learn it more than one way'.

Catastrophe risk modelling deals with adverse events, which carry the potential for significant economic and human losses. In a world where zero risk does not exist, the catastrophes considered in this textbook, while rare in an individual's lifespan, are of immense concern for our society as a whole. The student of risk aims at solving the important problems faced by our world due to natural and anthropogenic perils. Such endeavour must start with a better understanding of our intertwined environmental and engineered world system. In an era overshadowed by the potential consequences of climate change, acquiring a better understanding of these events becomes not only imperative but also time-sensitive to save lives and mitigate damage.

This textbook is structured into three distinct parts to facilitate the reader's learning progress. Part I provides the foundation of CAT modelling (Chapters 1-3) and Part II explores more advanced facets of this discipline (Chapters 4 and 5 – optional, with some more advanced mathematics required for full comprehension). Finally, Part III introduces standard CAT risk management strategies (Chapter 6). Chapter 1 serves as a comprehensive introduction, offering readers a broad understanding of catastrophe risks. It covers essential concepts and terminology, traces the historical development of catastrophe risk analysis, and provides a *tour d'horizon* of various perils and infamous catastrophes. Chapters 2 and 3 are dedicated to the fundamental principles of probabilistic hazard and risk assessment methods, respectively. By the conclusion of Part I, readers will be well-equipped to build a basic CAT risk model from scratch, encompassing a variety of perils. Part II explores the dynamic aspects of catastrophes. Chapter 4 addresses individual events, while Chapter 5 dissects some of the main cascading effects and feedback loops observed in the complex Earth system, including those driven by global warming. Finally, Part III presents key techniques associated with risk transfer and risk reduction. Topics include catastrophe insurance and short-term peril forecasting, guided by CAT model results and an understanding of different behaviours under risk and uncertainty. To illustrate the concept of risk governance, this section also features a case study on anthropogenic earthquake risk at geothermal plants.

It is my sincere hope that this textbook will inspire and nurture new vocations in the captivating and socially significant field of catastrophe risk modelling.

Acknowledgements

I extend my deepest gratitude to my wife and two daughters for their understanding throughout the arduous process of writing this textbook. Their patience – during the countless evenings and weekends I spent dedicated to this endeavour – is remarkable.

A profound acknowledgment is also owed to the Institute of Risk Analysis, Prediction and Management (Risks-X) and to the Department of Earth and Space Sciences (ESS), SUSTech, which generously provided me with the opportunity to shape an entirely new course that led to this book. Without their support for academic freedom, this project would not have come to fruition. My thanks go particularly to Professor Xiaofei Chen (Risks-X Dean, ESS Head), Professor Didier Sornette (Risks-X co-Dean), and Professor Wei Zhang (ESS Deputy Head, Head of Teaching) who believed in me. The 'domino effects' section of this book was also partially funded by the National Natural Science Foundation of China, grant no. 42074054.

I am also indebted to the reviewers of the textbook, who provided some of their time to comment on the manuscript. Their insights have significantly enriched the quality of the material. In alphabetical order: Dr Laurentiu Danciu (ETHZ), Professor Yu Feng (Sun Yat-sen Uni.), Dr Ahoura Jafarimanesh (PwC Switzerland), again Professor Didier Sornette (ETHZ/SUSTech/Risks-X), Professor Zhan Tian and Professor Shaun Wang (SUSTech), Professor Ziqi Wang (Berkeley), and Dr Gordon Woo (RMS). I'd also like to thank Professor (emeritus) Ulrich Golüke (Business School of Lausanne) for his help in crafting an exercise based on the Earth System Climate Interpretable Model (ESCIMO) climate model.

Finally, I express my sincere appreciation to the mentors, colleagues, and friends who have played pivotal roles and made scientific life a pleasure. In chronological order: Professor Geoffrey King (IPG Paris) and Professor David Bowman (Cal State Fullerton) for memorable and fun discussions on earthquake forecasting through my PhD years; Dr Robert Muir-Wood (RMS), who gave me the liberty to innovate on extreme earthquake modelling and to work on some more exotic risks; Professor Domenico Giardini and Professor Stefan Wiemer (ETHZ), who entrusted me with a decade's worth of interesting risk projects; my companions in stimulating scientific walks around ETH Zurich, Marco Broccardo (now Professor, University of Trento) and again Laurentiu Danciu; Professor Bozidar Stojadinovic and Dr Hélène Schernberg (ETHZ), who gave me the opportunity to test my course at this institute while a guest professor at its Risk Center; and last but not least, Yuting Xie and Professor Ke Wu (SUSTech/Risks-X) for their continuous help in the past few years.

Abbreviations and Symbols

List of Abbreviations

AAL	average annual loss
ABM	agent-based model
AEP	aggregate exceedance probability
CAT	catastrophe
CCDF	complementary cumulative distribution function
CDF	cumulative distribution function
CPI	consumer price index
CPT	cumulative prospect theory
DS	damage state
eCDF	empirical cumulative distribution function
EDSF	European Database of Seismogenic Faults
EF	Enhanced Fujita (scale)
EGS	enhanced geothermal system
ELT	event loss table
EM-DAT	EMergency events DATabase
ENSAD	ENergy-related Severe Accident Database
EP	exceedance probability (curve)
EUT	expected utility theory
GDP	gross domestic product
GEM	Global Earthquake Model
GenMR	Generic Multi-Risk (digital template)
GEV	generalized extreme value (distribution)
GMPE	ground motion prediction equation
GMST	global mean surface temperature
GPD	generalized Pareto distribution
IBTrACS	International Best Track Archive for Climate Stewardship
ID	identifier (of an event or simulation)
IPCC	Intergovernmental Panel on Climate Change
IR	individual risk
LaMEVE	Large Magnitude Explosive Volcanic Eruptions (database)
LCOE	levelized cost of electricity
MDR	mean damage ratio
MMI	Modified Mercalli Intensity

NIMBY	not in my backyard
NOAA	National Oceanic and Atmospheric Association
ODE	ordinary differential equation
OEP	occurrence exceedance probability
PDF	probability density function
PGA	peak ground acceleration
PSHA	probabilistic seismic hazard assessment
SIR	susceptible–infectious–recovered (model)
SOC	self-organized criticality
TLS	traffic light system
TORRO	Tornado and Storm Research Organisation
TVaR	tail value at risk
USGS	United States Geological Survey
VaR	value at risk
VCE	vapour cloud explosion
VEI	volcanic explosivity index
YET	year event table
YLT	year loss table

Note: For peril identifiers (ID), see Table 1.7. Country codes adhere to the ISO 3166-1 alpha-2 standard, such as 'US' for the United States of America.

List of Main Symbols Used

Variable	Meaning	Dimension[a]
a	Event productivity	[ ]
A	Area	$[L^2]$
$\mathbf{A}$ (a_{ij})	Adjacency matrix (element)	[ ]
b	Event size ratio	[ ]
B	Holland's parameter (winds)	[ ]
$\mathcal{B}$	Beta distribution	[ ]
c	Generic constant, parameter[b]	[various]
C	Centrality measure	[ ]
d	Displacement, distance	[L]
D	Damage (ratio, probability)	[ ]
E	Energy	$[ML^2T^{-2}]$
f	Probability density function (PDF)	[ ]
f_*	Generic function (* a letter)	[various]
F	Cumulative distribution function (CDF)	[ ]
$\bar{F}$	Complementary CDF (CCDF)	[ ]
F_S	Factor of safety (landslides)	[ ]

g	Gravitational acceleration	[LT^{-2}]
$\mathcal{G}\ (\mathcal{E}, \mathcal{V})$	Graph (edge, vertex)	[ ]
h	Height, thickness	[L]
$\mathcal{H}$	Heaviside step function	[ ]
i, j, k	Natural numbers (indices)	[ ]
I	Intensity (of a hazard)	[various]
K	Carrying capacity	[ ]
l, W	Length (or level), width	[L]
L	Loss	[various]
m	Mass	[M]
M	Magnitude (earthquakes)	[ ]
n, N	Numbers, counts	[ ]
$\mathcal{N}$	Normal distribution	[ ]
$p\ (\mathbb{P})$	Probability (distribution)	[ ]
P	Pressure	[$ML^{-1}T^{-2}$]
$\mathbf{P}\ (p_{ij})$	Transition matrix (element)	[ ]
q	Quantile	[ ]
Q	Water discharge (floods)	[L^3T^{-1}]
$\mathbf{Q}\ (q_{ij})$	Transition-rate matrix (element)	[T^{-1}]
r	Radius, distance	[L]
R	Growth or spread rate	[T^{-1}]
R_0	Basic reproduction number (epidemics)	[ ]
S	Size (generic) of event	[various]
$\mathcal{S}\ (\mathcal{A},\mathcal{O})$	State (action, outcome)	[ ]
$t\ (\tau)$	Time (characteristic)	[T]
T	Temperature	[K]
T_*	Period (* a letter)	[T]
u	Random number	[ ]
$\mathcal{U}$	Uniform distribution	[ ]
v	Velocity, speed	[LT^{-1}]
V	Volume	[L^3]
w	Weight	[ ]
x, y	Coordinates	[L]
$X,Y(x, y)$	Random variables (values of)	[ ]
y_p, z_p	Return period, return level	[T],[var.]
z	Depth, altitude	[L]
α	Power exponent	[ ]
β	Transmission rate (epidemics)	[T^{-1}]
γ	Recovery rate (epidemics)	[T^{-1}]
δt	Time interval	[T]
ΔH_c	Heat of combustion (VCE)	[L^2T^{-2}]
ΔT	Return period, time shift	[T]
$\Delta\sigma$	Stress change (earthquakes)	[$ML^{-1}T^{-2}$]

$\Delta *$	Difference of $*$ (excluding T)	[various]
ϵ	Noise component	[various]
η	Efficiency, fraction	[ ]
θ, ϕ	Angles (including directions)	[ ]
Θ	Parameter set	[various]
κ_D	Vulnerability index	[ ]
λ (Λ)	Event rate (cumulative)	[T^{-1}]
$\mu, \mathbb{E}$	Mean (location), expectation	[various]
ν	Exposure (asset) value, cost	[various]
ξ	Shape (GEV, GPD)	[ ]
ρ	Density	[ML^{-3}]
σ,	Standard deviation (scale),	[various]
σ^2(Var)	Variance	
ϕ_d, ψ	Index of dispersion, dispersion statistic	[ ]

[a] [L] represents length, [M] mass, [T] time, [K] temperature, and [ ] dimensionless.

[b] If $*$ is the standard notation in a given discipline, c_* may be adopted in this textbook to avoid confusion otherwise arising from the use of similar symbols across different disciplines.

PART I

BASICS OF CAT MODELLING

1 Overview of Catastrophe Risks

We live in a complex and risky world. Natural calamities – including earthquakes, tsunamis, storms, floods, wildfires, epidemics, among others – can profoundly affect our society and economy. Similarly, human-made disasters, accidental or deliberate, such as armed conflicts, critical infrastructure failures, explosions, and toxic spills, can also lead to dire consequences.

The number of loss-generating disasters continues to exhibit an increasing trend (Munich Re, 2017; Guha-Sapir et al., 2022), as depicted in Figure 1.1. The primary cause of this trend is population growth coupled with the concentration of more people and assets in regions susceptible to hazards (Bilham, 2009; Gall et al., 2011; Merz et al., 2021). Notable catastrophes of the 2010s include the 2011 Tohoku Japanese earthquake, which triggered a massive tsunami leading to the Fukushima nuclear disaster (Norio et al., 2011); the Great 2011 Thailand floods that disrupted the global manufacturing supply chain (SwissRe sigma, 2012; Abe and Ye, 2013); Hurricane Harvey (2017), which caused unprecedented flooding in Texas due to torrential rainfall (e.g., Qina et al., 2020); and the COVID-19 pandemic, which brought most of the world to a halt (Guan et al., 2020). During the finalization

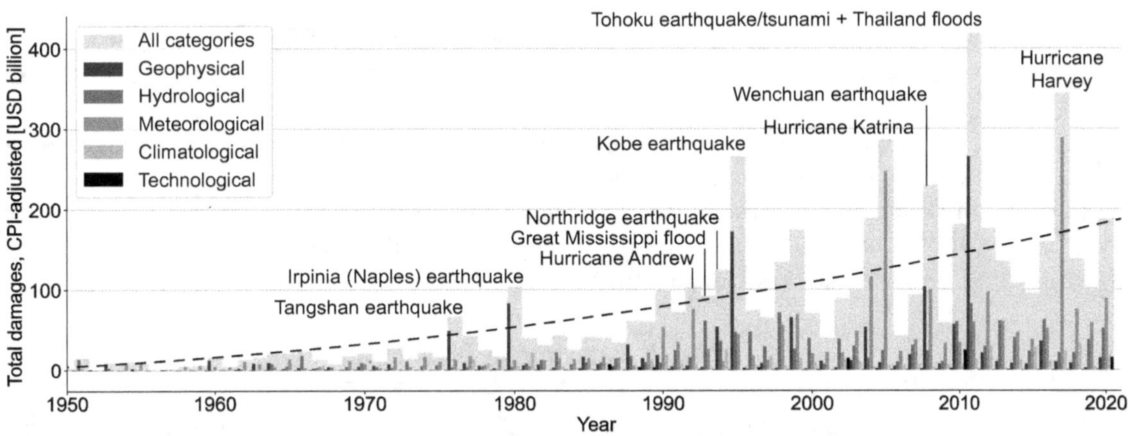

Fig. 1.1 Annual damage losses attributed to natural and human-made disasters during the period 1950–2020, as estimated from the international disaster database, or EMergency events DATabase (EM-DAT) (Guha-Sapir et al., 2022) (data used with permission from the Centre for Research on the Epidemiology of Disasters (CRED)). The year 2011, recording the highest losses, earned the epithet *annus horribilis* by The Geneva Association (Courbage and Stahel, 2012). Accelerating trend in economic losses highlighted by the dashed curve. A black and white version of this figure will appear in some formats. For the colour version, please refer to the plate section.

of this textbook, on 6 February 2023, the strongest earthquake in Türkiye in more than 80 years occurred; it was the fifth-deadliest earthquake of the twenty-first century (Dal Zilio and Ampuero, 2023). On 10 September, Storm Daniel, a rare Mediterranean hurricane, led to catastrophic flooding in Libya due to the collapse of dams caused by record-breaking rainfall. This marks the deadliest 'Medicane' on record (Wikipedia, 2023). Also during the writing of this textbook, major conflicts unfolded, including the war in Ukraine (Rawtani et al., 2022) and the Gaza–Israel asymmetric war (Steinbock, 2023). Global warming has become an exacerbating factor, with evidence showing an increase in the frequency and severity of droughts (Vicente-Serrano et al., 2014), wildfires (Westerling et al., 2006; Abatzoglou and Williams, 2016), tropical cyclones (Knutson et al., 2020), and other perils with rising temperature (van Aalst, 2006; IPCC, 2021). What are the physical characteristics of these catastrophes? How do experts assess their likelihood, intensity, and resulting damage? What is the conceivable range of catastrophe risks that individuals and society at large face today?

All of these questions will be addressed in detail throughout this textbook. The objective of this introductory chapter is to cover the main aspects of catastrophe risk from a qualitative perspective, offering an overview of what will be explored in quantitative terms in the subsequent chapters. Section 1.1 starts with the definition of fundamental terms and concepts. Section 1.2 then describes the historical development of catastrophe risk science, which was often influenced by the societal impact of some infamous catastrophes. Finally, Section 1.3 provides an exhaustive list of perils categorized by their physical origin, including geophysical, hydrological, meteorological, climatological, biological, extraterrestrial, technological, and socio-economic perils.

1.1 Concepts and Terminology

Whenever possible, official definitions published by the United Nations International Strategy for Disaster Reduction (UNISDR) will be employed, as UNISDR seeks to foster a shared comprehension and application of concepts related to disaster risk reduction (UNISDR, 2009). Some other catastrophe (CAT) model-related definitions originate from Lloyd's Market Association (2013).

1.1.1 Catastrophe, Disaster, and Peril

The term **catastrophe**, derived from the Ancient Greek verb meaning 'to bring down', originally denoted a rapid and calamitous event within the plot of a theatrical tragedy (Woo, 1999, p. 1). Over time, its meaning extended to include *sudden events, whether natural or human-made, with severe consequences*. Examples of natural catastrophes can be earthquakes, volcanic eruptions, storms, or floods; examples of human-made catastrophes include chemical plant explosions, nuclear meltdowns, dam failures, and more. The term **peril** is used to label these different types of events, or phenomena, which have the potential to cause damage (Foote et al., 2017, p. 301). An exhaustive list of perils will be provided in Section 1.3. The term **disaster** is often considered synonymous with catastrophe (Grossi and Kunreuther, 2005a, glossary), although there is a tendency to view a catastrophe as having broader impacts than a disaster. An illustrative sentence is: 'My offer was refused – and a disaster eventually turned into a catastrophe' (Trump, 1987). The word's etymology suggests a significant and sudden misfortune, attributed to the unfavourable position of certain planets in the old astrological interpretation of events. The official definition of a disaster, as proposed by the UNISDR (2009), is 'a serious disruption of the functioning of a community or a society involving widespread human, material, economic or environmental losses and impacts, which exceeds the ability of the affected community or society to cope using its own resources'. As there is no clear predefined threshold to distinguish between these two terms, catastrophe and disaster will be used interchangeably throughout this textbook.

What constitutes a disaster, or a catastrophe, can vary depending on the point of view. People involved in a car crash surely perceive it as a personal disaster, while society at large considers it a relatively minor incident. For quantitatively defining widespread losses, various metrics and thresholds are used in different contexts. In the case of severe human-made accidents, as exemplified by the ENergy-related Severe Accident Database (ENSAD), such events are often defined by specific criteria such as leading to five or more fatalities or economic losses exceeding five million dollars, among other factors (Burgherr and Hirschberg, 2008). In the insurance industry, a catastrophe is typically defined as an event that results in economic losses surpassing tens of millions of dollars or involves more than a dozen fatalities (Jones et al., 2017, p. 2). To be declared a disaster, EM-DAT requires an event to cause 10 or more deaths, affect 100 or more people, or trigger the declaration of a state of emergency or a call for international assistance (Guha-Sapir et al., 2022). Can global warming be considered a catastrophe? It is a medium-to-long-term trend and hence not a catastrophe per se, which is, by definition, a unique and well-delimited event. Global warming, however, can increase the frequency and severity of catastrophes.

1.1.2 Hazard and Risk

Hazard is any 'dangerous phenomenon, substance, human activity or condition that may cause loss of life, injury or other health impacts, property damage, loss of livelihoods and services, social and economic disruption, or environmental damage' (UNISDR, 2009). Any peril is therefore hazardous by nature. 'Risk may be defined as a compound measure of the probability and magnitude of adverse effect' (Lowrance, 1980, p. 5). UNISDR (2009) provides an almost identical definition for **risk**: '*the combination of the probability of an event and its negative consequences*'. Therefore, a statement about risk must include a quantitative, or at least semi-quantitative, description of the likelihood of negative-outcome events and of the losses that they are expected to generate. One of the simplest ways to visualize risk is via the concept of a **risk matrix** (Cox, 1998, 2008), two examples of which are shown in Figure 1.2. They are defined as 5×5 matrices where each cell represents a given likelihood and level of impact. **Likelihood**, in a semi-quantitative approach, is synonymous with frequency, chance, or probability (for a formal definition of probability,

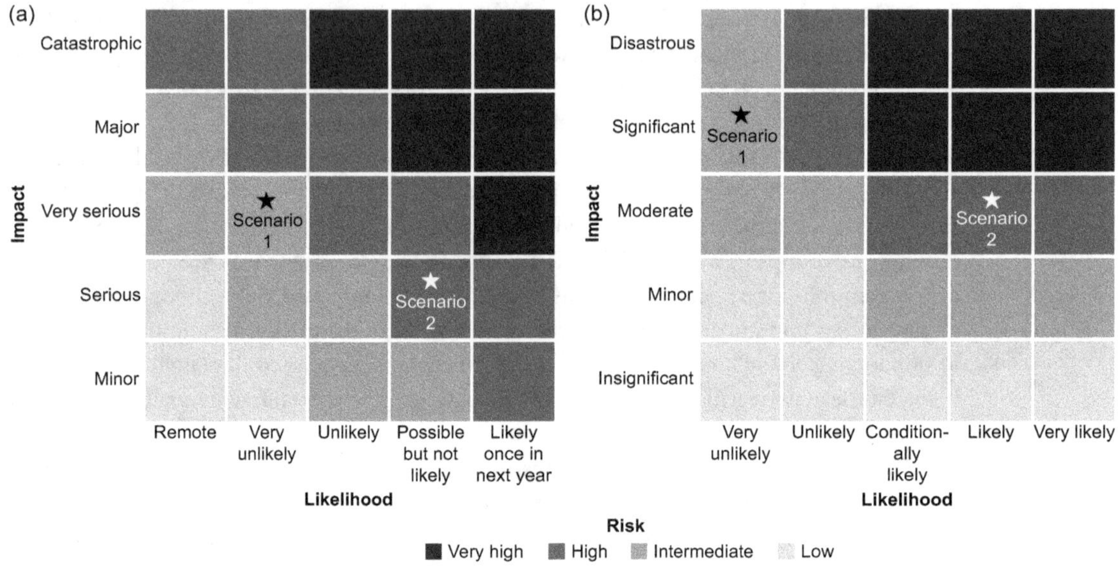

Fig. 1.2 Typical scheme of a risk matrix where scenarios are ranked as a function of likelihood and impact for comparative risk judgements. Risk categories according to (a) Cox (1998, fig. III.1.1) and (b) Komendantova et al. (2014, fig. 3). A black and white version of this figure will appear in some formats. For the colour version, please refer to the plate section.

see Section 1.1.4). Frequency levels commonly go from one event expected annually (very likely) to one event every 10,000 years (remote) (Cox, 1998). **Impact** can be synonymous with magnitude, intensity, or severity of the event (when defined at the hazard level) or with damage or loss (when defined at the risk level). These terms will be more precisely defined in the next two chapters. For now, risk can be defined by the simple equation:

$$\text{risk} = \text{likelihood} \times \text{impact}, \tag{1.1}$$

where $\times$ represents the complex interplay between likelihood and impact (product or convolution operator).

UNISDR (2009) defines **risk assessment** as 'a methodology to determine the nature and extent of risk by analysing potential hazards and evaluating existing conditions of vulnerability that together could potentially harm exposed people, property, services, livelihoods and the environment on which they depend'. Risk assessment, which can be deterministic or probabilistic, employs empirical evidence, expert judgement, and physical models to evaluate risks. Calculating risks is the main topic of Parts I and II of this textbook. **Risk management**, covered in Part III, is 'the systematic approach and practice of managing uncertainty to minimise potential harm and loss' (UNISDR, 2009). The two main processes of risk management are risk transfer and risk reduction. **Risk transfer** is 'the process of formally or informally shifting the financial consequences of particular risks from one party to another whereby a household, community, enterprise or state authority will obtain resources from the other party after a disaster occurs, in exchange for ongoing or compensatory social or financial benefits provided to that other party' (UNISDR, 2009). **Risk**

reduction is 'the concept and practice of reducing disaster risks through systematic efforts to analyse and manage the causal factors of disasters, including through reduced exposure to hazards, lessened vulnerability of people and property, wise management of land and the environment, and improved preparedness for adverse events' (UNISDR, 2009). Finally, **risk governance** encompasses all of the above, as well as the additional ways different stakeholders (individuals and institutions, public and private) cooperate for efficient risk management (e.g., Scolobig et al., 2017, table A1).

1.1.3 Scale of Catastrophe Risks

The catastrophes explored in this textbook differ significantly from events within the realm of common experience, making it useful to draw comparisons with other risky events. As previously introduced, the risk matrix offers a scale of risk. In Figure 1.3, this scale is simplified into a 3×3 matrix, categorizing events based on their likelihood (low, medium, or high) and consequences (low, medium, or high). This matrix enables us to distinguish between accidental events, typical catastrophes, and extreme catastrophes.

High-probability–low-consequences events are everyday-type **accidents** that typically affect isolated individuals or small groups of people. They commonly occur on the road, at home, at the workplace, or in places of entertainment. In addition to accidents, this category includes diseases (excluding infectious diseases that can spread to large groups) and violent crimes. It is important to note that the notion that the most catastrophic events are the riskiest is a misconception. Globally, when considering premature deaths

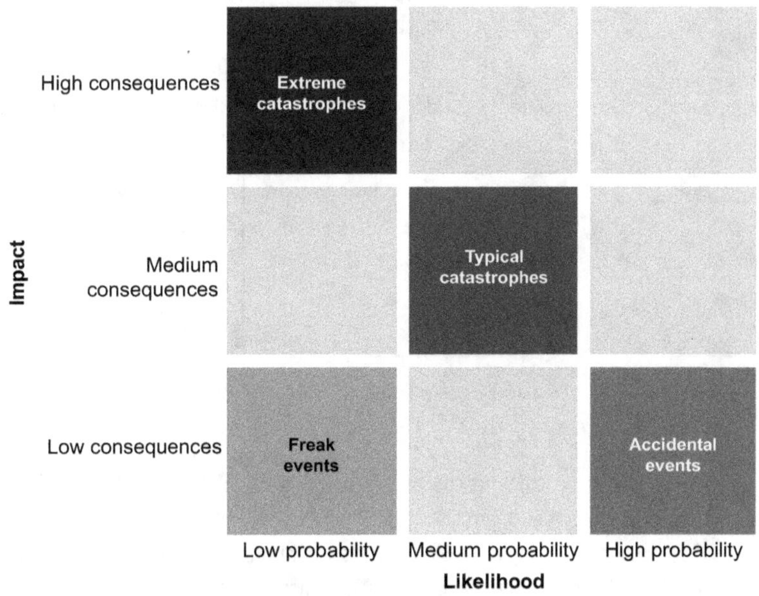

Fig. 1.3 Scale of typical catastrophes compared to other event types in a risk matrix. Typical catastrophes have 'medium' consequences relative to the much more extreme events that could occur.

per year, diseases (e.g., cardiovascular diseases, cancer) followed by common accidents (e.g., car crashes, drownings) far surpass natural disasters and terrorism-related fatalities (Ritchie et al., 2018). For instance, car accidents alone result in 1.2 million deaths worldwide per year, with more than 90% of individual events involving only a few casualties (Smil, 2008, p. 240). High-probability–low-consequences events are well understood because they can be statistically quantified. They are typically predictable, with little or no element of surprise. Additionally, they represent a relatively stable background of risk, although increasing trends in such events often raise concerns and warrant in-depth analysis. Infamous examples include the chromium-6 water supply pollution in a small town in California, portrayed in the movie *Erin Brockovich* (Banks, 2003), and the 2020s US fentanyl crisis (Friedman and Shover, 2023).

Medium-probability–medium-consequences events, as outlined in Figure 1.3, represent the typical catastrophes occurring globally every few years or decades, typically with local to regional impacts. These events constitute the central focus of discussion in this textbook. An analysis of EM-DAT (Guha-Sapir et al., 2022), summarized in Figure 1.4, reveals that during the early twenty-first century (2001–2020), **weather events**, which include storms and floods (i.e., hydrological and meteorological events combined) accounted for approximately 70% of damage losses, while geophysical events (specifically

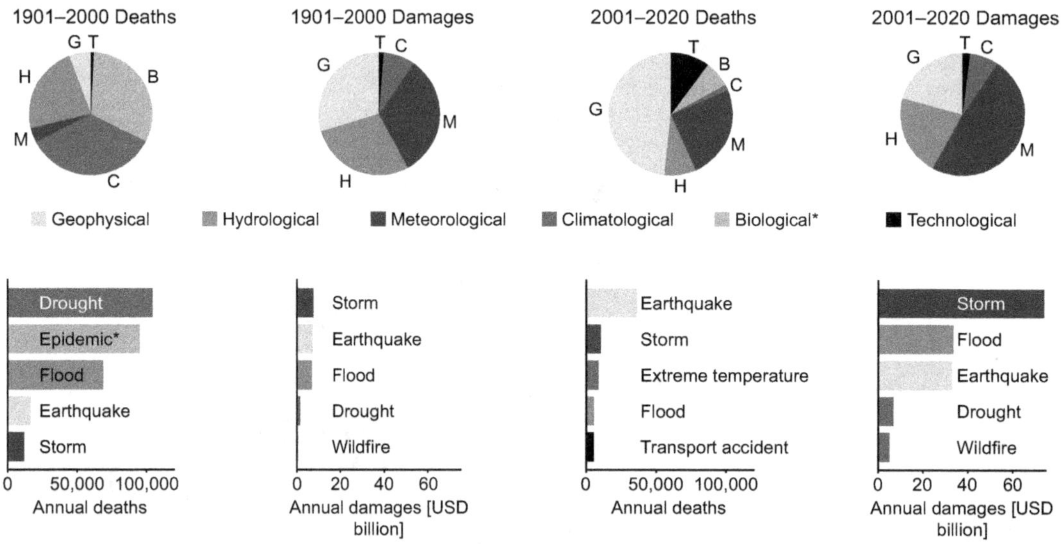

Fig. 1.4 Ranking of catastrophic losses per peril category in terms of deaths and damage for the twentieth and early twenty-first centuries, inferred from EM-DAT (Guha-Sapir et al., 2022) (data used with permission from the CRED). * Epidemic data incomplete in EM-DAT (pers. comm., R. Below, February 2023), with COVID-19 deaths not yet counted and the 1918 Spanish Flu, responsible for 32 million deaths (Marani et al., 2021a), missing due to early-twentieth-century data incompleteness. A black and white version of this figure will appear in some formats. For the colour version, please refer to the plate section.

earthquakes) were responsible for nearly half of all fatalities. These findings align with the results obtained from annual global compilations provided by Swiss Re and summarized in Smil (2008, p. 14). While similar rankings are observed for damage losses in the twentieth century, there are significant shifts in the causes of fatalities. During that period (1901–2000), most deaths were attributed to climatological events, followed by biological and hydrological events in EM-DAT. However, EM-DAT is incomplete for the first parts of the twentieth century, with the 1918 Spanish Flu missing, for example. If included, biological events would become the deadliest natural events. Compared to common accidents, catastrophes exhibit significant variability over the years (Ritchie et al., 2018), primarily due to their highly skewed size distribution (to be addressed in the next chapter). This means that *overall losses are predominantly influenced by the most extreme events*. In the twentieth century, those were the two world wars (Smil, 2008, fig. 2.19) (not included in EM-DAT), epidemics, droughts (leading to famines), and floods, which collectively claimed tens of millions of lives. The 2001–2020 period in EM-DAT does not include COVID-19 fatalities, which would again reconfigure the rankings of perils. There are other issues in addition to event volatility and data incompleteness in catastrophe databases (Box 1.1), with some caution always advised when attempting to interpret them.

Low-probability–high-consequences events are very rare catastrophes characterized by extreme consequences, ranging from regional to global in scale. They can be further divided into three main subcategories: tail events, meta-events, and speculative events, aligning with the classification of existential risks made by Smil (2008) – known catastrophic risks, plausible catastrophic risks, and entirely speculative risks, respectively. **Tail events** (or mega-events), which are a continuation of the typical catastrophes mentioned earlier, represent the most extreme instances that a peril can physically attain. They can be

Box 1.1 **A Note on Catastrophe Databases**

Catastrophe loss data exhibit significant heterogeneity and are often incomplete in terms of spatial or temporal coverage. Different catastrophe databases may list different sets of events or provide different estimates for the same event. For instance, Dixon et al. (2005) highlighted the impact of database biases on conflicting rankings for heat-related mortality compared to cold-related mortality. Similarly, Ashley and Gilson (2009) questioned the accuracy of lightning-related deaths by comparing several databases, revealing discrepancies. Furthermore, disparities of up to one order of magnitude in landslide-related deaths have been observed, with landslide impacts frequently being under-represented in catastrophe databases (Petley, 2012). Additional biases in natural hazard loss data come from variations in minimum recording thresholds, peril categorizations, levels of event aggregation, and methods for accounting losses (Gall et al., 2009). Similar causes of divergence are observed in databases of human-made disasters, such as of data breaches (cyberattacks), with added challenges linked to potential secrecy (e.g., Neto et al., 2021). In brief, biases are common when dealing with rare events data (King and Zeng, 2001). The catastrophe loss statistics presented in Figures 1.1 and 1.4, and later in this textbook, should be viewed as indicative in nature and should not be accepted at face value. All numbers are always referenced so that the reader can investigate the original sources for the matching metadata.

so severe that they are considered existential risks, such as mega-impacts, mega-tsunamis or mega-eruptions (Smil, 2008). Estimating their likelihood often involves extrapolating from the size distribution of smaller events. Notably, mega-droughts, with the potential for societal collapse, are not as rare as one might think, with known historical precedents (Cook et al., 2007). Additionally, the likelihood of a super-eruption (with an erupted volume greater than 450 km^3) is greater than that of a mega-impact by a 1-km diameter asteroid (Self, 2006). **Meta-events** encompass the simultaneous occurrence of multiple events, either coincidentally (referred to as the '**perfect storm**' concept; e.g., Paté-Cornell, 2012) or due to cascading effects. The combination of numerous dynamic processes often results in non-linear responses and risk amplification (Mignan et al., 2014). Estimating their likelihood can be challenging due to potential knowledge gaps regarding the number of variables involved. A thermonuclear war is a pertinent example, particularly if defined as the culmination of lower-scale conflicts. Similarly, the crossing of tipping points in the climate system, leading to a regime shift in the complex Earth system, is considered a low-probability–high-consequences event (Lenton, 2011). These extreme events will be studied in detail in Part II (Chapter 5). **Speculative events**, representing the final category, correspond to risks that so far remain within the realm of science fiction but might materialize, for example as emergent technological perils (Bostrom, 2002, 2019). Examples include artificial intelligence taking over the world, in the manner of *The Terminator* (Cameron, 1984), or an alien invasion (Neal, 2014).

Closer to Earth, environmental trends, such as global warming or resource depletion – also the subject matter of Chapter 5 – tend to elevate the likelihood and severity of catastrophes towards the realm of extreme events. This could mean that some typical catastrophes shift toward the high-probability–medium-consequences or medium-probability–high-consequences categories of Figure 1.3. Events in the high-probability–high-consequences category also become conceivable in the case of climatic regime shifts (Lenton, 2011).

Finally, low-probability–low-consequences freak events include all the marginal risk oddities. These events can range from casualties due to falling meteorites (La Paz, 1951) to casualties due to falling into a sinkhole (Van Den Eeckhaut et al., 2007b), and even ship collisions with rogue waves (Nikolkina and Didenkulova, 2011). The study of freak events may lead to bizarre analyses, such as the hazard assessment of bearded men in laboratories (Barbeito et al., 1967). This category is mentioned briefly for the sake of completeness, and no further examples will be provided in this textbook.

Additional scales can be employed to provide further characterization of risky events. Woo (1999) introduced a general spatial scale with three classes: microscale (local, one to hundreds of metres), mesoscale (local to regional, tens to hundreds of kilometres), and macroscale (continental to global, thousands of kilometres). Furthermore, the temporal scale can be defined in terms of event duration instead of return period (e.g., Gill and Malamud, 2014). Earthquakes generally occur within seconds while tropical cyclones may take up to a few days to cross a region. Pandemics and droughts can last from months to years. This illustrates that the term 'sudden' may vary across perils, but it is essential that the event duration is relatively short relative to its return period, maintaining consistency in the characterization of these events.

1.1.4 Uncertainty, Randomness, and Probability

Risk exists because of uncertainty. When the likelihood of rain tomorrow stands at 100%, it becomes a fact rather than a risk. It is certain that one should carry an umbrella to stay dry or, alternatively, remain at home. If it were certain that a major catastrophe would strike a specific city next week, immediate actions would be taken to evacuate everyone. As articulated by Clark (1980), 'at the centre of the risk problem are people and their fears. Fears of loss, fears of injury, and – most of all – fears of the unknown'. *Quantifying risk involves measuring uncertain negative outcomes in terms of probabilities.* An uncertain outcome, such as the occurrence of an event denoted as X, is governed by **randomness**, or in other words, is subject to chance (with $X \in 0, 1$, where $X = 1$ means 'occurs' while $X = 0$ means 'does not occur'). X is referred to as a **random variable**, described by a random phenomenon. The formalization of random processes will be covered in Chapter 2, along with event sampling methods. Considering an experiment where the number of occurrences of event X in N trials is $N(X = 1)$, the **probability** of X taking the value 1, denoted as $\mathbb{P}(X = 1)$ and satisfying the condition $0 \leq \mathbb{P}(X = 1) \leq 1$, signifies the value to which the ratio $N(X = 1)/N$ converges as N approaches infinity (Grimmett and Stirzaker, 2001, p. 4), or:

$$\mathbb{P}(X = 1) = \lim_{N \to \infty} \frac{N(X = 1)}{N}. \tag{1.2}$$

This is also referred to as the **objective probability**. In the case of catastrophes, as well as most other natural processes, probability estimates themselves carry uncertainty. This implies that multiple types of uncertainties need to be addressed for proper CAT risk assessment.

The first type of uncertainty is ***aleatory uncertainty****, which represents the inherent and irreducible randomness of a process*. It is also known as stochastic uncertainty, or process error (Mitchell-Wallace, 2017, p. 168). Classic examples encompass games of chance, such as coin tosses, dice rolls, or card draws. In the heyday of Newtonian physics, chance was initially perceived as a manifestation of imperfect knowledge in a deterministic universe (Woo, 1999, p. 71). However, Edward Lorenz, an early explorer of chaos through his work on weather prediction, demonstrated that natural processes exhibit instability (Lorenz, 1963). While the deterministic trajectory of an asteroid leading to an impact has virtually zero aleatory uncertainty (Halley, 1705), the same cannot be said for complex phenomena such as storms, earthquakes, and most other potential catastrophes which all occur with a true probability $\mathbb{P}(X = 1)$.

The second type of uncertainty, **epistemic uncertainty**, arises from a lack of perfect knowledge and can be mitigated through the acquisition of more information. This category encompasses various forms of uncertainty, including a lack of understanding of the process (model error, assumption error), inaccurate and/or imprecise data (data error, measurement error), or lack of data (undersampling, parameter uncertainty). Quantifying epistemic uncertainties involves providing the probability that a specific quantity, such as the true probability of an event, assumes a particular value. According to Eq. (1.2), when using a fair six-sided dice as an example where the true probability of landing on a specific face x is $\mathbb{P}(X = x) = p = 1/6$, the relative frequency $N(X = x)/N$ – known as the

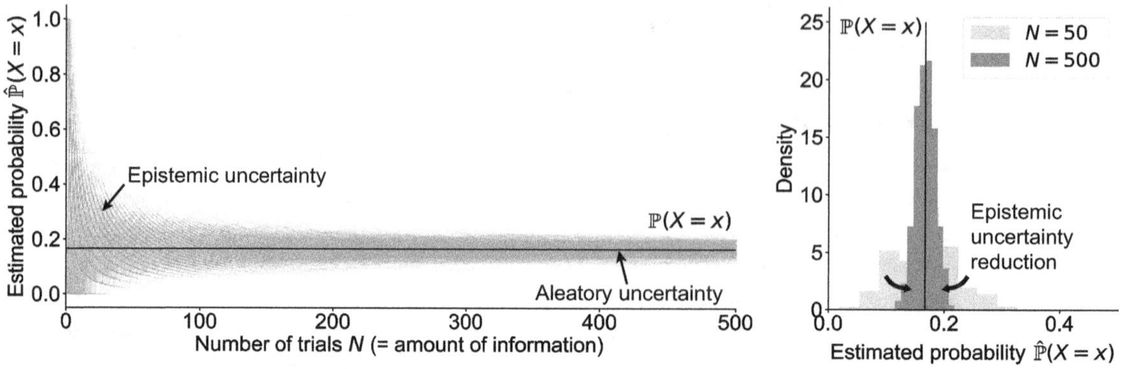

Fig. 1.5 Comparing aleatory and epistemic uncertainties with the roll of a dice. The true probability of landing on a specific face x, $\mathbb{P}(X = x) = p = 1/6$, is irreducible while its estimate $\hat{\mathbb{P}}(X = x)$ is uncertain and describable by a probability distribution. This epistemic uncertainty is reducible with more information, here by increasing the number of trials N.

empirical probability or **frequency-based probability** $\hat{\mathbb{P}}(X = x)$ – converges towards the true probability while the epistemic uncertainties tend to zero, as the number of trials N increases (Fig. 1.5). The objective is to glean insights from as much historical data as possible to arrive at an accurate and precise estimate of $\mathbb{P}(X = x)$. However, catastrophes are too rare to make any realistic attempt to estimate their true probability of occurrence (Exs 1.1 and 1.2).

Moreover, in contrast to simple games of chance, the assumption that natural events are independent and identically distributed (**i.i.d.** in statistics) is a significant oversimplification. Frank Knight, the economist, asserted in his influential 1921 dissertation, *Risk, Uncertainty and Profit*, that 'there is no possibility of forming in any way groups of instances of sufficient homogeneity to make possible a quantitative determination of true probability. [They] are far too unique, generally speaking, for any sort of statistical tabulation to have any value for guidance. The conception of an objective measurable probability or chance is simply inapplicable' (Knight, 1921, p. 231). Initially applied to economics, **Knightian uncertainty** is equally relevant to catastrophes (Woo, 2011, p. 109). A degree of belief can still be quantified using subjective probabilities, based on the (limited) evidence accessible to an individual. However, this approach necessitates a departure from the Frequentist method (i.e., Eq. (1.2)), opting instead for the Bayesian approach, which recognizes an inherent subjective element, grounded in prior knowledge. Bayesian inference will be introduced in the context of peril forecasting in Part III.

There is actually a third type, called **ontological uncertainty**, which pertains to unknown unknowns, and defies quantification (Mitchell-Wallace, 2017, p. 167). These unknown unknowns introduce a level of ambiguity that can blur the distinction between aleatory and epistemic uncertainties. Any new discovery has the potential to alter our perspective on what constitutes inherent randomness or missing information, further complicating the delineation between these two types of uncertainty.

The many subtleties in the types of uncertainties can be confusing, especially since not all terms have been introduced yet (such as primary and secondary uncertainties in CAT models; Section 1.1.5). For the sake of simplicity, only aleatory uncertainties will be explicitly addressed in Parts I and II of this textbook. Event sets will be generated, occurrence probabilities calculated, and hazard intensity fields and damage functions provided only for a limited number of models and data sets per peril. It means that the underlying epistemic uncertainties will remain implicit for a start. The specific role of epistemic uncertainties and of subjective probabilities on decision-making will be the subject matter of Part III (Chapter 6). It will be worth remembering throughout this textbook that any CAT risk model is subjective by construction (or a 'fiction'; Frigg, 2010), since it represents only a simplified version of the risk process believed to take place, and that the rarity of catastrophes renders impractical any objective interpretation of probability, which could require data beyond the lifespan of the universe.

1.1.5 Catastrophe (CAT) Model

Lloyd's Market Association (2013) defines a **catastrophe model** as 'a computerised system that generates a robust set of simulated events and estimates the magnitude, intensity, and location of the event to determine the amount of damage and calculate the insured loss as a result of a catastrophic event such as a hurricane or an earthquake'. The keyword in this sentence is 'simulated'. Since historical records are too short to be representative of the future in the context of rare catastrophes, plausible events must be generated based on statistical and physical models, which are first calibrated to past observations to then extrapolate the space of possibilities. Despite the inherent complexity of such a framework, a CAT model delivers outputs to the end user in the form of simple metrics, graphs, and maps. The main users are (re)insurance firms, civil protection agencies, and other risk management practitioners. These entities can create their own CAT models, purchase models from CAT model vendors, and use public or other open-source platforms such as HAZUS (Schneider and Schauer, 2006), the Central American Probabilistic Risk Assessment (CAPRA) platform (Cardona et al., 2012), CLIMADA (Aznar-Siguan and Bresch, 2019), or Oasis, among others (Chen et al., 2009; Schmidt et al., 2011). The four basic components of a catastrophe model are the hazard module, exposure data (or inventory) module, vulnerability module, and loss (or financial) module (Fig. 1.6a) (Grossi et al., 2005; Jones et al., 2017). A simplified representation of the full process is given by the **Risk Triangle** of Crichton (1999) where 'risk is the probability of a loss, and this depends on three elements, hazard, vulnerability and exposure'. The size of risk is represented by the triangle area, which depends on the size of the three other components represented by the lengths of the triangle's sides. As a consequence, if there is no hazard, no vulnerability or no exposure, there is no risk (Fig. 1.6b).

1.1.5.1 Hazard Module

The **hazard module** plays a crucial role in characterizing the sources and the other characteristics of events within a specified spatiotemporal window for a given peril. This includes

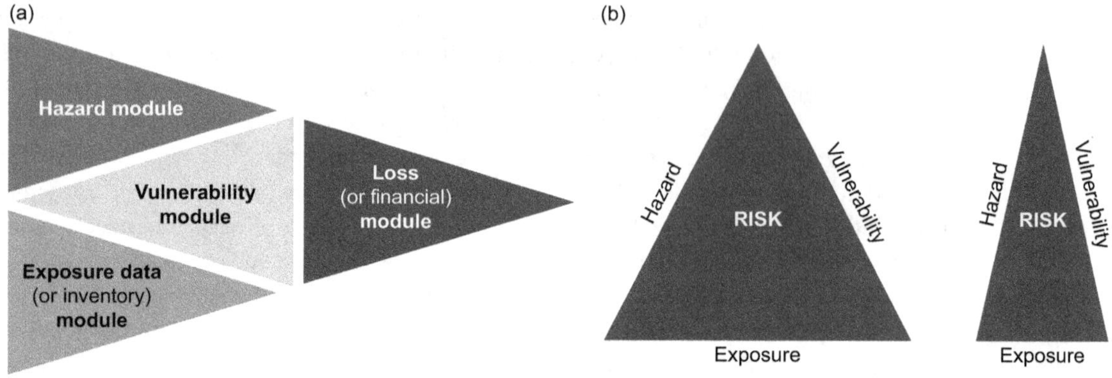

Fig. 1.6 The risk process in CAT modelling: (a) Standard structure of a CAT model with its four modules. A similar representation can be found in Jones et al. (2017, fig. 1.2); see also, for example, Grossi et al. (2005, fig. 2.2) or Aznar-Siguan and Bresch (2019, fig. 1). (b) The Risk Triangle of Crichton (1999).

an *assessment of the frequency of events in relation to their sizes, as well as their intensities*. To illustrate, in the context of different perils: for earthquakes, the source is a fault segment and the intensity is the level of ground-shaking; for storms, the source is the projected path and the intensity is the wind speed; for industrial explosions, the source is the hazardous material container and the intensity is the blast overpressure; etc. This module generates three main products: a set of events, a catalogue of matching spatial intensity footprints per event, and a suite of hazard metrics. The hazard module is highly peril-specific and requires domain expertise such as in geophysics, meteorology, hydrology, or civil engineering (for human-made perils). Chapter 2 will describe the computational framework of the hazard module for a wide variety of perils.

1.1.5.2 Exposure Data (or Inventory) Module

The **exposure data module** is responsible for characterizing the inventory of buildings and other structures that are at risk. In insurance terminology, this inventory is often referred to as a portfolio (see Part III). This module focuses on gathering and organizing essential parameters related to the exposed assets. These parameters include the geographic coordinates or address information, the monetary value or replacement cost, the occupancy type (e.g., residential, commercial, industrial), and the structure characteristics, such as building material, number of storeys or building height, year built. Depending on the specific risk assessment needs, this module can also include content value (for estimating content-related losses), number of inhabitants (for casualty estimation), and business revenue (for business interruption assessment). Peril-specific elements can also be included, such as the roof (under the impact of storms or volcanic eruptions), or protections against flooding or ground-shaking. Examples of both real and simulated inventories will be described in Chapter 3.

1.1.5.3 Vulnerability Module

The **vulnerability module** estimates the potential damage as the ratio of repair costs to replacement costs, based on functions that combine the outputs of the hazard and exposure modules. This necessitates expertise in engineering and access to a database of **damage functions** (also referred to as impact functions) that perform the *calculation of damage levels for hazard intensity values, conditional on the characteristics of the structures at risk*. The resulting output provides a direct means to calculate the losses. Consequently, this module serves as the interface connecting the other three modules (Fig. 1.6a). Chapter 3 will describe the various methods existing to assess damage and provide selected damage functions for specific perils, in the form of vulnerability and fragility curves, both of which will be defined in that chapter.

1.1.5.4 Loss (or Financial) Module

The **loss module** is responsible for evaluating and quantifying potential financial losses. The **ground-up losses**, or total losses, are calculated based on the extent of damage incurred, combined with the replacement value of the exposed assets. This calculation is performed before considering the impact of any insurance policies. From there, 'the financial module will typically accumulate through location-level terms, to policy and then programme level conditions, at each stage applying limits, deductibles, and special conditions that have been coded into the model. The resulting output is an event loss table that provides an assessment of the financial risk exposure to individual events' (Lloyd's Market Association, 2013). Basically, financial structures are applied to calculate the **insured losses** from the ground-up losses. The event loss table is the final outcome of the CAT model upon which risk metrics are applied. It is essentially the event set from the hazard module updated with event losses. Risk metrics include key measures such as the average annual loss, exceedance probability curves, and other actuarial indicators. Losses are typically defined in economic terms, representing monetary values. However, they can also be expressed in human terms, such as fatalities or injuries. Using standardized units of loss enables the comparison of risks across different perils on a consistent basis. Chapter 3 will introduce and define the main risk metrics, providing insights into how these metrics are used to assess the risk. Additionally, the insurance-related aspects of the loss module will be explored in Part III (Chapter 6).

All the elements necessary for the creation of a basic CAT model will be available to the reader by the end of Part I. Risk quantification will be based on the following modelling pipeline, which is an update of Eq. (1.1):

$$\text{risk} = \underbrace{\text{likelihood} \times \text{intensity}}_{\text{hazard}} \times \text{vulnerability} \times \text{exposure} \quad \text{with } \overbrace{\text{intensity} \times \text{vulnerability} \times \text{exposure}}^{\text{loss, or impact}}. \tag{1.3}$$

The Risk Triangle is here retrieved with risk = hazard × vulnerability × exposure, and where the probability of a loss originates from the likelihood of an event of a given size, leading to a certain intensity (Ex. 1.3). Both aleatory and epistemic uncertainties, as discussed in Section 1.1.4, are categorized as **primary uncertainty** when referring to the event set generation, and to **secondary uncertainty** for all the other steps of the CAT modelling pipeline (hazard intensity, damage, and loss levels; Foote et al., 2017, p. 371). For the purpose of this textbook, relatively simple models will be considered. Readers interested in exploring these models and understanding their integration within CAT models can proceed directly to Chapter 2. Alternatively, you may continue reading below to learn about the historical aspects of catastrophe science and gain insights into the various types of perils that will be modelled later in the text.

1.2 History of Catastrophe Science

The historical perspective outlined below describes the evolution of catastrophe science, tracing its development from the early days of human civilization to the contemporary CAT models employed today. Initially, disasters were documented as a means to alert future generations of potential threats. Early engineering solutions emerged to mitigate building collapses, and eventually, instruments were devised to predict and provide advance warnings about impending calamities. The advent of the scientific revolution led to a more profound comprehension of hazards and risks through the mathematical analysis of observed phenomena. In the twentieth century, the computerization of science marked the decisive milestone necessary for creating realistic models that could capture the interactions between natural processes, the built environment, and our socio-economic system. Throughout history, the profound impact of catastrophes on society has consistently driven the parallel advancement of risk management and innovation, as will become increasingly evident in this section.

1.2.1 From Ancient Myths to Medieval Texts

Prior to historical records and well before the development of the scientific method, **myths** were an effective way to propagate information over many generations about important events, such as past disasters. Some continued to prevail in some parts of the world until recent centuries. In these stories, catastrophes were often depicted as giant creatures, from the giant catfish *namazu* in Japan (Ludwin et al., 2007) to the Fenris Wolf that was trapped and chained deep under the mountains of Scandinavia (Mörner, 2007). Nowadays, lessons can be learned about the occurrence of mega-catastrophes in the distant past from the analysis of these tales, which – although cryptic – are the only records available. This is done for example via the recent discipline of geomythology (Piccardi and Masse, 2007). Myths of the sky falling down and of a great darkness can, for example, refer to past asteroid or comet impacts, as well as to giant volcanic eruptions (Masse and Masse, 2007). Maori and Aboriginal legends allude to significant cosmogenic events associated with giant waves and fires. Some of those great catastrophes may have occurred as recently as 1500 (Bryant et al., 2007). Since asteroid impacts leave virtually no trace when impacting the ocean or when exploding above the ground in airbursts (Mignan et al., 2011), the information contained in oral tradition and rock paintings gives some clues of the potential size and frequency of such elusive events.

Some other messages have been left for future generations as markers on carved stones. **Tsunami stones** are found along the coastline of Japan. One such tablet reads 'High dwellings are the peace and harmony of our descendants. Remember the calamity of the great tsunamis. Do not build any homes below this point' (Van Luik et al., 2013). Yet, the token of a past catastrophe half hidden on a forested hillside can easily be disregarded or forgotten by modern generations. **Hunger stones** are another example, marking places of extreme droughts and associated famines. They remain hidden from sight until they emerge at the bottom of a dried river. One infamous inscription reads: '*Wenn du mich siehst dann weine*' [if you see me, you will weep] (Elleder et al., 2020). It is sometimes said that it takes only three generations to forget. The concept of a passive and enduring stone marker represents an elegant solution to the vicissitudes of time and is nowadays employed to warn future generations of nuclear waste sites (Lomberg and Hora, 1997). Even more universal markers are envisioned in Lincoln Child's science fiction novel *Deep Storm* (Child, 2007).

Disasters form a common plot in Ancient Greek and Roman literature, such as in the great myths (e.g., Homer's Iliad) and religious texts (e.g., from the great flood in the Book of Genesis to the 10 plagues in the Book of Exodus). The description of the eruption of Mount Vesuvius in the year 79 by Pliny the Younger is a renowned example of Latin text on a specific catastrophe (Vittori et al., 2007). As for the superstition of the European Dark Ages, it is reflected in disasters being considered bad omens in the Medieval texts. Figure 1.7 reproduces some woodcut engravings from *Prodigiorum ac ostentorum chronicon* [Chronicles of Omens and Portents] (Lycosthenes, 1557), where earthquakes, storms, floods, pests, fires, and conflicts are depicted, among other calamities. Investigation of the

Fig. 1.7 Woodcuts from Lycosthenes' *Prodigiorum ac ostentorum chronicon* [Chronicles of Omens and Portents] symbolizing various types of perils (Lycosthenes, 1557). From the copy at the Bibliothèque interuniversitaire de santé (Paris) available at archive.org (public domain).

disaster-related literature provides insights into past catastrophes and their consequences on society. Such information has for long been overlooked (Sargiacomo et al., 2021). Manuscript databases, such as the Corpus *Latinorum Et Mediaevalium Naturae Scriptorum* for the Mediterranean region, now catalogue disasters from scattered sources (Vittori et al., 2007).

Not only were disasters recorded early on, but risk management strategies were also put into place. As early as the second millennium BC, the Babylonian code of Hammurabi covered the role of earthquakes, making it the earliest earthquake building code in the world as well as an early concept of insurance against building collapse (Woo, 1999, p. 260). In the year 132, during the Eastern Han Dynasty, scientist and engineer Zhang Heng invented the first earthquake ground-motion detector (Ben-Menahem, 1995), consisting of a large bronze vessel surrounded by dragons' heads holding unstable bronze balls.

1.2.2 Mathematization

The development of the scientific method led to the mathematization of the phenomena related to catastrophes. Often forgotten, it was the study of comets which confirmed Newton's laws, and it was the study of winds which led to the development of partial derivatives (Woo, 1999). In 1687, Sir Isaac Newton published his *Philosophiæ Naturalis Principia Mathematica*, in which he outlined the laws of gravity and motion. It was, however, his colleague, Edmond Halley, who first proved Newton's law of gravitation (Spencer Jones, 1947). From numerous comet observations, Halley concluded that one of the celestial objects had a 76-year return period (Halley, 1705). Halley's Comet returned as predicted in 1758. This represents a unique case of deterministic predictability of a potential natural hazard, explainable from classical Newtonian mechanics. Halley also produced the first **life tables**, relating mortality with age in a population (Halley, 1693). His life table later served as a reference to compute the price of life annuities (Bacaër, 2011, chapter 2). In 1746, a few years after Halley's death, a prize was awarded by the Berlin Academy of Sciences for the best paper on the laws governing winds. The winner was the French mathematician Jean d'Alembert, whose article is best remembered for introducing to calculus the concept of a partial derivative to calculate the dynamics of winds (d'Alembert, 1746), without which numerical storm forecasting would be impossible today (Woo, 1999, p. 2). Soon after, in 1760, mathematician Daniel Bernoulli presented a work on the dynamics of smallpox, which pioneered the field of epidemiology (Bernoulli, 1766). He used the concept of differential equations previously introduced by Newton and Leibniz, as well as Halley's life table, to apply his theory. Interestingly, d'Alembert criticized Bernoulli's assumptions about the probability of infection and of dying from smallpox being independent of age. He suggested a different solution that did not require these assumptions (Bacaër, 2011, chapter 4).

While the eighteenth century was paving the way toward CAT modelling, a catastrophe destroyed the city of Lisbon in 1755 (refer to Box 1.2 for its description in literature). The Great Lisbon earthquake, which possibly reached magnitude 9 on the Richter scale (Muir-Wood and Mignan, 2009), led to intense fires and a large tsunami. Appointed by the king, the Marquis de Pombal coordinated the reconstruction of the city with rational land

Box 1.2 Catastrophes in Literature

Great catastrophes have been recorded by famous novelists in works of journalism, from the worst British hurricane ever recorded, described in *The Storm* (Defoe, 1704) to the 1906 San Francisco earthquake in *The Story of an Eyewitness* (London, 1906). Other catastrophes have been described as allegories, one of the best examples being the 1755 Lisbon earthquake in *Poème sur le désastre de Lisbonne* (Voltaire, 1756) and *Candide* (Voltaire, 1759). In *The Magic Mountain* of Thomas Mann, two fictional characters talk about it: 'I have read that Goethe said to his servant, that night in his bedchamber _ ... you are confusing two catastrophes. You are thinking of the earthquake of Messina. I have in mind the one that visited Lisbon in the year 1755._ Pardon _ Well, Voltaire was outraged by it' (Mann, 1927). Other famous examples of allegories include John Steinbeck's *The Grapes of Wrath* on the impact of the severe drought and dust storms during the Dust Bowl (Steinbeck, 1939) and Albert Camus' *La Peste* on the consequences on an epidemic in the French Algerian city of Oran (Camus, 1947). Catastrophes naturally became a central theme of the science fiction genre. *The Last Man* by Mary Shelley of Frankenstein fame, published in 1826, describes a depopulated Earth in the late twenty-first century, ravaged by a new pandemic (Shelley, 1826). The apocalyptic story *The Conversation of Eiros and Charmion*, written by Edgar Allan Poe, depicts a comet impact in what could be the earliest modern fictional work about a celestial catastrophe (Poe, 1839). It was published after the 1832 Biela's Comet panic (Arago, 1832) and Halley's Comet return in 1835. By the end of the nineteenth century, 'popular culture had become saturated with an "apocalyptic imaginary"' of both human-made and natural disasters (Bulfin, 2015).

planning and modular construction. The 'mathematics of materials' here led to the first seismically engineered construction project, known as Pombaline buildings (Zack, 2015). Engineering went from practical (craftsmanship) to scientific (modern engineering) with Galileo and his *Discorsi* published in 1638. Mathematicians improved on Galileo's model in the following century. Robert Hooke and Christopher Wren presented mathematical thoughts about the strength of arches to rebuild London only a few years after it had been devastated by the 1666 Great Fire (Zack, 2015). Jacob Bernoulli, and later his nephew Daniel and Leonhard Euler, developed a theory of beam bending which would become widely used in the nineteenth century (the Euler–Bernoulli beam theory). Collapse fragility curves can still be based on their differential equations (e.g., Kumar and Gardoni, 2014).

1.2.3 Computerization

In the mid-nineteenth century, visionary Charles Babbage, father of the mechanical computer, used his machines to calculate the impact of volcanic heat on the changing elevation of a temple in Italy (Babbage, 1847). By the 1860s, insurance life tables were calculated with such types of mechanical difference machines to the great benefit of actuaries (Lewin et al., 1989). It was a period when both the size and the number of risks to be insured began outstripping the capacities of the insurance industry, leading to the foundation of the first reinsurance companies. For example, the 1842 Hamburg and 1861 Glarus Fires were instrumental in the creation of Cologne Re and Swiss Re, respectively (Jones et al., 2017, p. 3). New sensors were built, including the modern anemometer and seismograph. While the basic concept of the anemometer was developed as early as the fifteenth century, it was in 1846 that the design was improved so that the rotation rate was proportional to the average wind speed (Robinson, 1849). The first useful seismograph system, recording ground displacements, was constructed in Japan in 1880 but could record only local events. A modified version could record a Japanese earthquake in 1889 from Potsdam, in Germany, marking that year as the birth of instrumental seismology (von Rebeur-Paschwitz, 1889; Ben-Menahem, 1995). The accumulation of disaster-related data started and prefigured the future computational needs of the twentieth century.

Numerical computing was already in the making by the 1910s when Lewis Fry Richardson was developing weather forecasts by hand using an iteration method at a time when 'computers' were still referring to people doing the computations (Lynch, 2008). Numerical simulations (by a machine) date back to 1945 with research by John von Neuman and Stan Ulam at Los Alamos (US), where the **Monte Carlo method** (or repeated random sampling) was developed using the first electronic computer, the Electronic Numerical Integrator and Computer (ENIAC) (Metropolis, 1987). The computer grandly facilitated iterative work and led to a new wave of discoveries. The handwritten annuity tables laboriously calculated by the actuaries of the nineteenth century could be accomplished by a modern computer within a matter of minutes. It is no surprise then that the majority of actuary offices were computerizing during the 1960s (Lewin et al., 1989). His work requiring thousands of iterations, Edward Lorenz could not have been an early observer of chaos through his work on weather prediction without an electronic computer (Lorenz, 1963) (the so-called 'butterfly effect'). It was again the computer which allowed Benoit Mandelbrot, who coined the term fractal, to render their beauty on a screen for the first time in 1980 (Mandelbrot, 1986). Fractals, some of the most striking manifestations of chaos (Peitgen and Richter, 1986), can shape environmental phenomena such as terrains (Fournier et al., 1982) and cities (Batty, 1991), and control the size distribution of disasters (King, 1983). Storm forecasting, which requires the partial differential equations originally derived by d'Alembert, developed significantly in that period. By 1970, forecasting the path of a hurricane would take only a few seconds with the HURRAN program (Hope and Neumann, 1970). Using supercomputers, Wilhelmson et al. (1990) were some of the first to develop a four-dimensional storm model to simulate and visualize realistic thunderstorms and tornado-generating storms. In the case of floods, it was in the 1980s that access to digital elevation models allowed for more realistic modelling (Beven, 2020). As stated by Thurner et al. (2018, p. 29) regarding the modelling of complex phenomena, 'what we can do today would have been unthinkable only a few decades ago' due to lack of computing power.

Don Friedman pioneered CAT modelling with the computer simulations of hazards and risk that he developed throughout the 1970s (Friedman, 1984). Thanks to computer-based models and geographic information systems, CAT modelling companies (or CAT model vendors) were afterward founded: AIR Worldwide in 1987, Risk Management Solutions (RMS) in 1988, EQECAT in 1994. The public HAZUS platform from the Federal Emergency Management Agency (FEMA) followed in 1997 (Grossi et al., 2005, p. 24). At first, use of these proprietary models was not widespread but two major catastrophes changed this in 1989, Hurricane Hugo and the Loma Prieta earthquake, which sent a warning signal to the insurance industry. Prior to Hurricane Hugo in 1989, the insurance industry in the US had never suffered a loss of over USD 1 billion from a single disaster. Hurricane Andrew caused USD 15.5 billion in insured losses in 1992 while the 1994 Northridge earthquake amounted to USD 12.5 billion. Many insurance companies became insolvent following these events (Grossi and Kunreuther, 2005b), proving the importance of CAT modelling.

1.3 Peril Taxonomy

The numerous perils mentioned earlier have not yet received formal definitions and classifications. Perils encompass a diverse range of occurrences, both natural and human-made, and will now be systematically categorized based on their physical origins. The proposed classification is defined in Table 1.1 and includes geophysical, hydrological, meteorological, climatological, biological, extraterrestrial, technological, and socio-economic perils (42 perils in total). It should be noted that the proposed classification may slightly differ

Table 1.1 Peril taxonomy

Category	Perils (with their two-character identifiers)	Section
Geophysical	Earthquake (EQ), volcanic eruption (VE), landslide (LS)	1.3.1
Hydrological	Fluvial flood (FF), pluvial flood (PF), storm surge (SS), tsunami (Ts)	1.3.2
Meteorological (basic)	Windstorm (WS), rainstorm (RS), hailstorm (HS), lightning (Li), tornado (To)	1.3.3
Meteorological (complex)	Tropical cyclone (TC), extratropical cyclone (xC), convective storm (CS)	1.3.3
Climatological	Heatwave (HW), cold wave (CW), drought (Dr), wildfire (WF)	1.3.4
Biological	Epidemic (Ep), pest infestation (PI)	1.3.5
Extraterrestrial	Asteroid (and comet) impact (AI), geomagnetic storm (GS)	1.3.6
Technological	Blackout (BO), cyberattack (CA), dam flood (DF), energy supply failure (Ef), explosion (Ex), fire (Fi), information system failure (If), toxic spill (TS), transport system failure (Tf), water system failure (Wf)	1.3.7
Socio-economic	Armed conflict (AC), banking system failure (Bf), business interruption (BI), famine (Fa), financial crisis (FC), public service failure (Sf), stampede (St), social unrest (SU), terrorist attack (TA)	1.3.8

from others existing in the literature. For example, Munich Re (2017) includes tsunamis in geophysical events and mass movement in hydrological events. Gill and Malamud (2014), in their review, also include tsunamis in geophysical events and define wildfires as biophysical events. Meteorological and hydrological perils due to rainfall can also be grouped in so-called weather events (NOAA-NCEI, 2023). EM-DAT merges tsunamis with earthquakes and considers stampedes as a technological peril since the database has no socio-economic class (Guha-Sapir et al., 2022). All of these differences can often lead to confusion when investigating historical catastrophe data (Box 1.1). Efforts toward standardization have occurred and evolved throughout the years (UNDRR and ISC, 2020) (Ex. 1.4). The classification introduced in this textbook is designed to differentiate between peril processes and peril triggers. For instance, both tsunamis and storm surges are considered hydrological events since they initiate from a water mass movement and cause flooding. However, tsunamis are triggered by geophysical events whereas storm surges are triggered by storms.

This section will provide an outline, in very broad and qualitative terms, of the phenomenology of the main perils known to exist. Some examples of the largest and most damaging historical events will also be given. Methods to quantitatively assess the size, intensity, and damage potential of most of the perils described below will be provided in the following chapters. Some of the metrics and damage scales that will be mentioned, such as earthquake magnitude, Volcanic Explosivity Index, Saffir–Simpson scale, or Fujita scale, will be formally defined in Chapters 2 and 3.

Figure 1.8 provides a map of the global distribution of some of the main natural perils, showing that no region on Earth is safe from the powers of Nature. **Natural disasters** (geophysical, hydrological, meteorological, climatological, biological, and extraterrestrial) will be described first, followed by **human-made disasters** (technological and socio-economic). Human-made perils can be accidental or malicious. When accidental, they are primarily due to human error (Mignan et al., 2022a) but may also be triggered by another disaster.

1.3.1 Geophysical Perils

Geophysical perils correspond to the events originating from the solid Earth and due to geological or geomorphological processes. This includes earthquakes, volcanic eruptions, and landslides.

1.3.1.1 Earthquakes (EQ)

An **earthquake** describes the sudden movement along a fault rupture, as well as the resulting ground-shaking that is due to the release of elastic energy previously accumulated by plate tectonics on the fault and dispersed in seismic waves (Fig. 1.9). Ground-shaking is the primary cause of structural and non-structural damage during an earthquake. Secondary causes include **liquefaction** (i.e., soil temporarily behaving

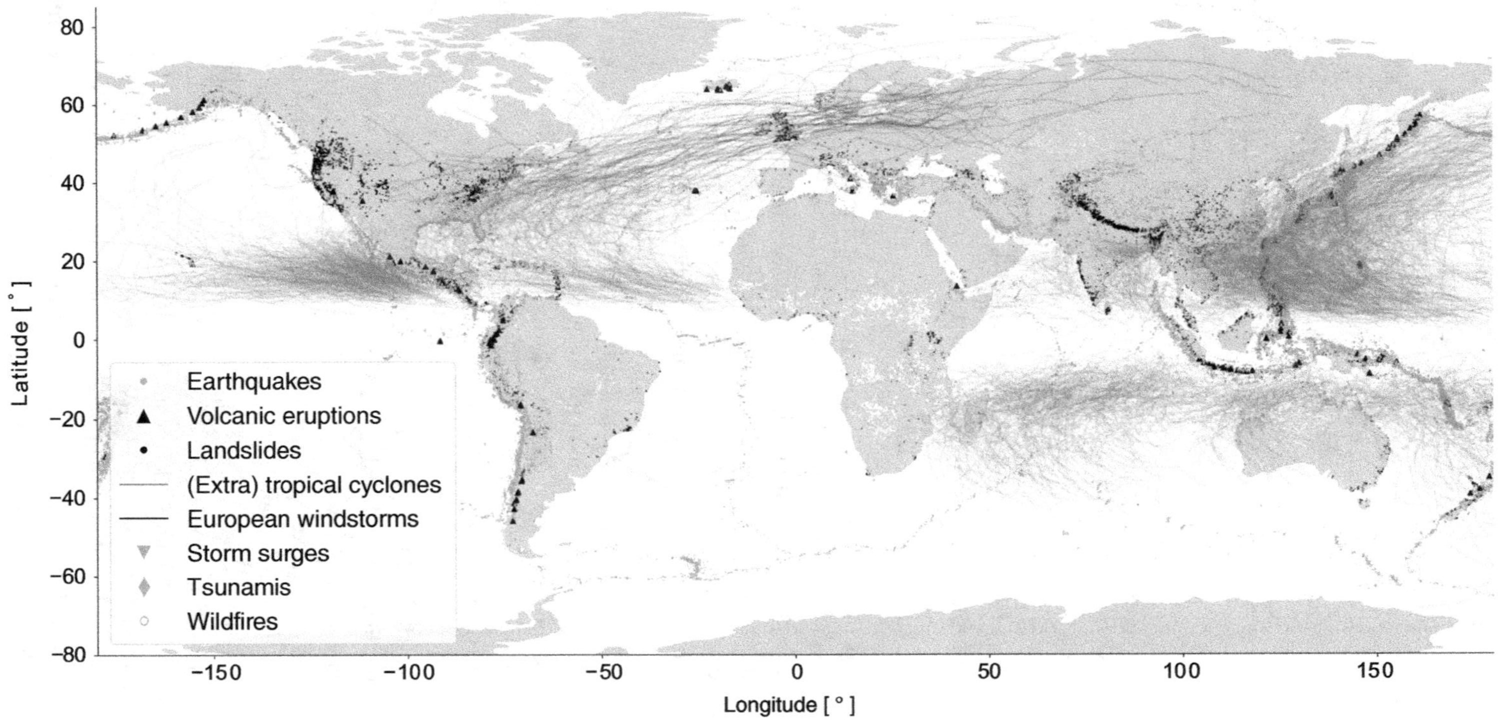

Fig. 1.8 Geographical distribution of some of the main natural perils on Earth. Earthquakes from the International Seismological Centre-Global Earthquake Model (ISC-GEM) Global Instrumental Earthquake Catalogue (ISC, 2023; Storchak et al., 2013, 2015; Di Giacomo et al., 2018) (CC-BY-SA, with permission); landslides (rain-induced) from the Global (rain-induced) Landslide Catalog (GLC) (Kirschbaum et al., 2010, 2015); storm surges from SURGEDAT (Needham and Keim, 2012); tropical and extratropical cyclones from the International Best Track Archive for Climate Stewardship (IBTrACS) (Knapp et al., 2010, 2018); tsunamis from the Global Historical Tsunami Database (NGDC/WDS, 2022); volcanic eruptions from the global database Large Magnitude Explosive Volcanic Eruptions (LaMEVE) (Crosweller et al., 2012); wildfires from the FRY global database of fire patches (Laurent et al., 2018a,b) (CC-BY 3.0); European windstorms from the eXtreme WindStorms (XWS) catalogue (Met Office, University of Reading and University of Exeter, 2023; Roberts et al., 2014) (CC-BY 4.0, https://creativecommons.org/licenses/by/4.0/). A black and white version of this figure will appear in some formats. For the colour version, please refer to the plate section.

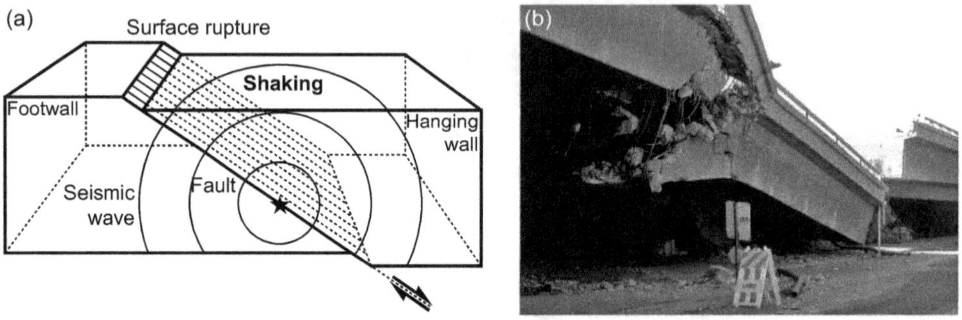

Fig. 1.9 Illustrating the earthquake process and potential observed damage. (a) The earthquake process, shown here for a normal (dip-slip) fault, where the block above the fault has moved downward relative to the block below (i.e., extensional tectonic regime). The star represents the earthquake **hypocentre**, which is the point where the rupture nucleates. The location directly above that point on the surface is called the **epicentre**. More details about possible fault mechanisms can be found in Chapter 2. (b) Photograph taken after the 1994 $M = 6.7$ Northridge, California, earthquake highlighting infrastructure damage (USGS, public domain).

like a liquid due to the shaking) and the surface fault rupture (e.g., Bird and Bommer, 2004). With the main process explaining earthquakes being tectonic stress buildup, the majority of earthquakes are observed at the boundary of tectonic plates (Fig. 1.8). Stress redistribution following a rupture leads to additional, mostly smaller, earthquakes called **aftershocks** (King et al., 1994), although doublets or triplets of large earthquakes can also cluster within hours to months of each other (Mignan et al., 2018; Dal Zilio and Ampuero, 2023). Volcanic eruptions and asteroid impacts have the potential to trigger large earthquakes, and earthquakes can trigger volcanic eruptions (Hill et al., 2002). Earthquakes are also induced by explosions, deep fluid stimulation and other human interactions with the underground (Foulger et al., 2018). The largest earthquakes ever recorded by instruments are the 1960 magnitude $M = 9.5$ Valdivia Great Chilean earthquake, the 1964 $M = 9.2$ Great Alaskan earthquake, the 2004 $M = 9.1$ Great Sumatra–Andaman Islands earthquake, and the 2011 $M = 9.1$ Great East Japan, Tohoku, earthquake (ISC, 2023). The 10 deadliest earthquakes that occurred between 1900 and 2023 are listed in Table 1.2, based on data from the CATDAT Damaging Earthquake Database (Daniell et al., 2011, table 2). Prior to the 2011 Tohoku earthquake, the costliest earthquake was the 1995 Kobe earthquake, remembered as the industrialized world's first experience of a catastrophic urban earthquake (Chang, 2000). Secondary effects of earthquakes, such as tsunamis, landslides, and fires, have been observed to account for approximately 25% of fatalities and economic losses on average, but in rare occasions, they can dominate the losses. This is the case for the tsunamis caused by the 2004 Sumatra and 2011 Tohoku earthquakes, the landslides caused by the 1949 Khait earthquake (Soviet Union, now Tajikistan), or the fires caused by the 1906 San Francisco and 1923 Kanto earthquakes (Daniell et al., 2017). The greatest single fire loss in US history was the one triggered by the 1906 San Francisco earthquake (Scawthorn et al., 2019).

Table 1.2 The 10 deadliest earthquakes in CATDAT (1900–2023[a])

Name	Date	Magnitude M	Fatalities[a]
Haiyuan, China	16/12/1920	7.8	273,465
Tangshan, China	27/07/1976	7.8	242,419
Great Sumatra–Andaman Islands	26/12/2004	9.1	228,194
Great Kanto, Japan	01/09/1923	7.9	142,831
Haiti	12/01/2010	7.0	137,000
Ashgabat, Turkmenistan	05/10/1948	7.3	122,000
Wenchuan, China	12/05/2008	7.9	88,287
Kashmir, Pakistan	08/10/2005	7.6	87,364
Messina, Italy	28/12/1908	7.1	85,926
Ancash, Peru	31/05/1970	7.9	66,794

[a] 1900–2010 numbers from Daniell et al. (2011, table 2) (CC-BY 3.0) with data from the CATDAT Damaging Earthquake Database. Median estimates given for fatalities due to both ground-shaking and secondary perils (tsunamis, landslides, and fires). Note that the Daniell et al. (2011) table remains correct as of December 2023 with the deadliest event in the period 2011–2023 being the earthquake doublet which occurred in Türkiye and Syria in 2023 with just under 60,000 fatalities (approximate value).

1.3.1.2 Volcanic Eruptions (VE)

Volcanic eruptions occur as magma (molten rock) reaches the Earth's surface. There are two types of volcanic eruptions in terms of activity, explosive and effusive eruptions. The majority of active volcanos are located along the Pacific Ring of Fire (Fig. 1.8). **Explosive eruptions** are characterized by gas-driven explosions that propel unconsolidated pyroclastic material outside the volcanic edifice (i.e., **tephra**), while **effusive eruptions** are characterized by the production of lava flows. Explosive eruptions are the most damaging. Their size is expressed by the volume (or mass) of magma released, measured by the Volcanic Explosivity Index (VEI) (Newhall and Self, 1982) (see Table 2.3). Prior to such an eruption, the pressure in the magma chamber rises, accumulating potential energy. Eventually, some magma moves into the conduit above the chamber and expands, leading to an upward accelerating flow. Most of the energy released during the eruption is thermal in nature (Pyle, 1995). Explosive eruptions are not only destructive, they can also be detrimental to the climate by their emissions. The 1815 Tambora eruption, which produced a tephra volume greater than 100 km^3 (VEI-7), was the largest eruption recorded in history and also the deadliest. It cooled the global atmosphere by almost 1°C, provoking crop failures and mass starvation in 1815–1816, ultimately leading to more than 50,000 deaths (e.g., Self, 2006). The deadliest volcanic event of the twentieth century was the 1902 Mont Pelée eruption in Martinique (Lesser Antilles), with pyroclastic flows responsible for more than 36,000 fatalities (Tilling and Lipman, 1993). Notable volcanic eruptions in the late twentieth century include the 1991 VEI-6 Mount Pinatubo, Philippines, eruption and 1980 VEI-5 Mount St. Helens eruption, the worst volcanic disaster in US history (Fig. 1.10a and b). These events killed 800 and 57 people, respectively (Tilling and Lipman, 1993).

Fig. 1.10 Examples of historical volcanic eruptions. (a) VEI-5 eruption of Mount St. Helens in 1980 (USGS, public domain). (b) VEI-6 eruption of Mount Pinatubo in 1991, taken from the east side of Clark Air Base (USGS, public domain). (c) VEI5-6 Hunga Tonga eruption in 2022, taken from the GOES-17 weather satellite (NOAA, public domain, via Wikimedia Commons).

The 2010 eruptions of Eyjafjallajökull, Iceland, although of relatively moderate size (VEI-4), caused enormous disruption to air travel (see transport system failures in Section 1.3.7). *Lahars* (i.e., mudflows or debris flows) are one of the principal secondary hazards resulting from explosive volcanic activity (Gudmundsson, 2015). Volcanic eruptions can also cause tsunamis, acid rain, and air pollution (Marshall et al., 2022). The 2022 underwater Hunga Tonga volcanic eruption, for example, produced the largest global perturbation of stratospheric aerosols since the Mount Pinatubo eruption (Sellitto et al., 2022) and the largest tsunami since the 1883 Krakatoa eruption (Lynett et al., 2022) (Fig. 1.10c).

1.3.1.3 Landslides (LS)

A **landslide**, or mass movement, is a downslope movement of soil or rocks occurring as a result of discrete shear failure under the action of gravity (Fig. 1.11). The main causes for initial instability are water saturation due to heavy rain (Iverson, 2000) and earthquake ground-shaking (Jibson, 1993). A particularly dangerous type of landslide is the **debris flow**, which consists of a mixture of solid particles and water (with less than 40% of the flow's volume made of water). It is called a ***lahar*** when occurring in a volcanic region with the solid particle component made of pyroclastic material. Primary *lahars* occur during eruptions and are caused by the bursting of the crater lake or the melting of ice; secondary *lahars* result from heavy rain in areas of abundant tephra (Gudmundsson, 2015). **Mudflows** and **avalanches** are mass movements involving mud particles and snow/ice, respectively. Anthropogenic factors, such as modification of the topography or of land use, can increase the likelihood of landslides (Jaboyedoff et al., 2016). Similarly, wildfires can be followed by landslides because they remove vegetation and ground cover, making the soil more unstable during the next rainstorm (Rengers et al., 2020). It is easily understood from this description why landslides are commonly regarded as secondary perils. Mountainous regions, where the terrain slope is high on average, are the most prone to landslides (Nadim et al., 2006). Losses due to isolated landslides are rarely of concern from a reinsurance

Fig. 1.11 Examples of historical landslides. (a) The 2005 La Conchita, California, landslide, which killed 10 people (USGS, public domain). (b) Example of damage by a debris flow in Qushan, Beichuan County, China, following the 2008 Wenchuan earthquake (USGS, public domain). (c) The 2014 Oso, Washington landslide, which killed 43 people (USGS, public domain).

industry perspective (Hillier, 2017, p. 246); however, their cumulative impact is significant. More than 2,500 non-seismic fatal landslides occurring worldwide in the period 2004–2010 caused a total of 32,322 recorded deaths (Petley, 2012). In the same period, the 2005 Kashmir and 2008 Wenchuan earthquakes alone led to 25,000 and over 20,000 landslide-related fatalities, respectively (Dunning et al., 2007; Yin et al., 2011). Rain-induced landslides can also be very deadly, an example being the debris flows generated during the 1999 Vargas rainstorm in Venezuela, which led to more than 30,000 deaths (Wieczorek et al., 2001). In 1985, 23,000 people were killed when the town of Armero, Columbia, was destroyed by a *lahar* (Gudmundsson, 2015). Landslides can also trigger tsunamis, as well as floods via so-called landslide lakes or landslide dams (Cui et al., 2009).

1.3.2 Hydrological Perils

Hydrological perils include all the natural events that cause **flooding**, meaning phenomena that lead to an accumulation of water on land where it is absent in normal circumstances. The three main types of floods are fluvial floods, pluvial floods, and coastal floods (Fig. 1.12a). The first two are primarily caused by inland **precipitation** (i.e., the fall of solid or liquid water over land), which is the major driver of the hydrologic cycle over land (Pagano and Sorooshian, 2002). **Coastal floods** can be storm surges (meteorological origin) or tsunamis (non-meteorological origin). They can lead to saline intrusion, causing damage to farmland. Note that tsunamis are generally not included in loss counts of flood events but are included in earthquake losses (Table 1.2). The most damaging flood events have a regional or even continental-scale footprint (e.g., Bates et al., 2021). Such a wide spatial extent can put a halt to a country's economic sectors. For example, the Great 2011 Thailand floods, which covered an area the size of Switzerland and were due to record monsoon rainfall, greatly affected the country's farmland and industrial parks. The insured

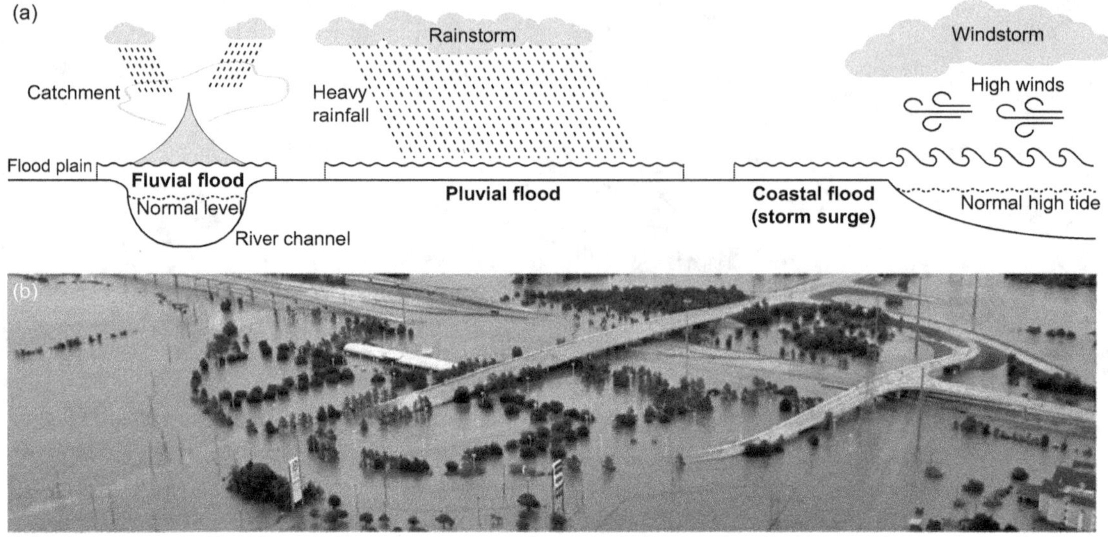

Fig. 1.12 Illustrating the flooding process and potential observed extent. (a) The three main flood processes: fluvial, pluvial, and coastal. (b) Extensive inland flooding in Texas associated with Hurricane Harvey in 2017 (USGS, public domain).

losses (USD 12 billion) ranked at the time as the highest ever worldwide from a freshwater flood disaster (total economic losses were USD 30 billion) (SwissRe sigma, 2012). Such extreme inland floods are often caused by unusual successions of similar rainstorm tracks (Merz et al., 2021) or by powerful storms stalling overland (e.g., 2017 Hurricane Harvey, Fig. 1.12b; Section 1.3.3). More than 175,000 people were killed by floods from 1975 to 2001 worldwide (Jonkman, 2005), and since 2011, floods have caused more than a third of natural catastrophe-related fatalities (Swiss Re Institute, 2022). Flooding can also lead to debris flows (e.g., Totschnig et al., 2011) and health problems via water contamination. Flooding by the overtopping or failure of artificial dams and levees can be highly destructive (see Section 1.3.7).

1.3.2.1 Fluvial Floods (FF)

Fluvial (or river) floods occur when a river overtops its banks and inundates the surrounding area. The main cause is heavy and/or sustained rainfall (but also snow or ice melt) that generates a high run-off in a catchment, causing the downstream river to overflow as it goes beyond its maximum discharge capacity. The run-off intensity depends on how much rainwater can infiltrate into the soil. Successive rainstorms can therefore increase the wetness of the catchment, and in turn increase the likelihood of an extreme river flood (Merz et al., 2021). Much rarer processes leading to fluvial flood include landslide dams (Cui et al., 2009), river reversal due to an earthquake, and glacier bursts by volcanic eruptions. When the rainfall spatial footprint is vast, flooding can be due to both direct precipitation (i.e., pluvial flood) and river discharge. For the case of the Great 2011 Thailand floods, Loc et al.

(2020) estimated that three-fourths of the excess water originated from direct precipitation. The 1931 Yangtze river floods were the deadliest fluvial floods in the last century with fatality estimates ranging from 400,000 to 3.7 million (Kundzewicz and Takeuchi, 1999).

1.3.2.2 Pluvial Floods (PF)

A **pluvial flood** results from heavy precipitation during an intense rainstorm, which creates a rapid overland inundation (Szewranski et al., 2018). Surface run-off happens when the rain intensity exceeds the infiltration capacity of the soil, leading to excess rainwater. It can be increased by land-use changes, such as urbanization and soil compaction (Merz et al., 2021). **Flash floods** occur during very short but extreme rainfalls, which lead to a quick water level rise and high velocity flow. These events result in the highest average mortality per event compared to fluvial floods (Jonkman, 2005). Surface water floods caused by the failure of the urban drainage system are not directly dangerous to life but may cause significant economic losses. A pluvial flood can happen at any location since it does not depend on the presence of a river system.

1.3.2.3 Storm Surges (SS)

Storm surges are generated by strong winds, with pre-landfall wind speeds likely to be the main driver for maximum surge heights. Typically, the part of the coastline that observes onshore winds (blowing from water to land) experiences greater surge levels, with water heights increasing closer to the path of the cyclone. In contrast, lower surge levels are expected on the offshore winds' side. The coastal flood severity can then vary depending on the dynamics of the storm, but also on the background tide level and the geometry of the continental shelf and coastline. The spatiotemporal complexity of the process means that a tropical cyclone less intense than another one may produce a greater surge (Needham and Keim, 2011). Tropical cyclones of size 5 on the Saffir–Simpson scale can generate a surge more than 5.5 m in height (Simpson, 1974), which can be devastating (see Table 3.4). For example, a 1970 cyclone in the Bay of Bengal produced a 9.1-m surge that killed 300,000 people in Bangladesh (the deadliest recorded tropical-cyclone-related event; Cerveny et al., 2017). Another one in 1991 killed 140,000 people in the same country. In the US, the 1900 Galveston Hurricane generated a 6.1-m surge that killed between 6,000 and 8,000 people making it the deadliest natural disaster in US history. More recently, in 2005, Hurricane Katrina led to an 8.5-m surge which claimed more than 1,800 lives (all numbers from the review of Needham and Keim, 2011). It also remains the costliest natural disaster in US history with a consumer price index (CPI)-adjusted loss of USD 190 billion (NOAA-NCEI, 2023). Extratropical cyclones can also generate coastal flooding by the combination of very strong winds and high tide. Storm Anatol for example generated a 5.9-m surge on the North Sea coasts back in 1999 (Munich Re, 2002). The deadliest storm surge caused by an extratropical cyclone in the past century occurred in 1953 in the North Sea; it killed over 2,000 people in coastal areas of the United Kingdom, Belgium, and the Netherlands (Jonkman and Kelman, 2005).

1.3.2.4 Tsunamis (Ts)

A **tsunami** (from Japanese *tsu* = harbour, *nami* = wave) is a sea wave generated by a non-meteorological source, such as a submarine earthquake, a volcanic eruption, a landslide, or an asteroid impact. The wave initiation requires an impulsive vertical displacement of the surface of the body of water. A volcanic eruption can generate a tsunami in various ways, through pyroclastic flows, explosions, flank collapse, and submarine eruptions. Tsunami waves can cross the oceans over very long distances without any major energy loss. As the water depth decreases closer to the coast, the tsunami waves decrease in speed; the wavelength is shortened and the energy within each wave is concentrated in less water, increasing their amplitude (Fig. 1.13a; Röbke and Vött, 2017). Due to the high velocity of the wave, damage can be caused by even small-amplitude events. Damage can also be caused by projectiles. According to the National Oceanic and Atmospheric Association (NOAA)/World Data Service (WDS) tsunami database (NGDC/WDS, 2022), tsunamis caused almost 300,000 deaths in the 1900–2022 period, with more than 70% of tsunamis being due to earthquakes (Fig. 1.13b). The 2004 Indian Ocean tsunami triggered by the Great Sumatra–Andaman Islands earthquake accounts for most of the deaths in that period, with 227,900 fatalities. The massive 2011 tsunami triggered by the Great Tohoku earthquake comes second, with approximately 18,400 deaths. The estimated economic losses for these two events are USD 10 billion and 220 billion, respectively (NGDC/WDS, 2022). At a similar order of magnitude, the tsunami caused by the 1755 Lisbon earthquake, in Portugal, reached the Caribbean islands and generated ***seiches*** (i.e., oscillating water in an enclosed body of water) as far north as Loch Ness in Scotland (e.g., Muir-Wood and Mignan, 2009). The maximum run-up height ever recorded was 524 m in Lituya Bay in Alaska in 1958. This mega-tsunami was caused by a landslide, itself triggered by an earthquake (Fritz et al., 2009). The hypothetical collapse of the Cumbre Vieja volcano at La Palma, Canary Islands, which would drop hundreds of cubic metres of rock into the ocean, could generate a tsunami reaching 25 m high along the American coastlines (Ward and Day, 2001).

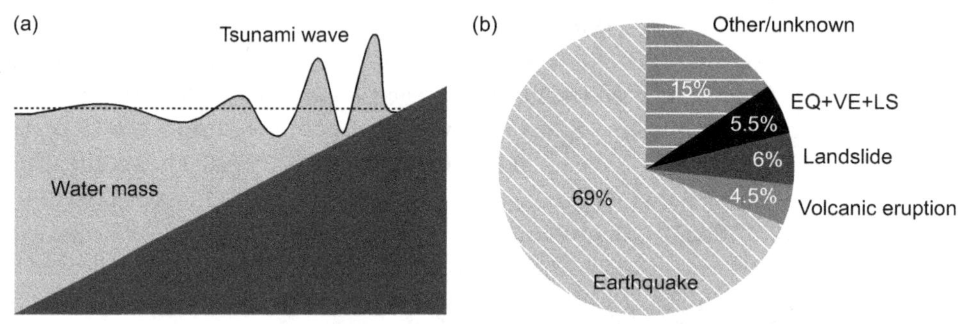

Fig. 1.13 Illustrating the tsunami process and quantifying its causes. (a) 'Cross section of a tsunami as its long waves move through the ocean and compress as they approach the coast', redrawn and simplified from NOAA (2023). (b) Cause of a tsunami, in percentage, derived from the NOAA/WDS tsunami database (NGDC/WDS, 2022).

1.3.3 Meteorological Perils

Meteorological perils are all the extreme atmospheric phenomena that can be dangerous. This includes wind (windstorm, WS), heavy rainfall (rainstorm, RS), hail (hailstorm, HS), and lightning (Li, i.e., thunderstorm). Weather catastrophes are often a combination of several perils. Damage during a tropical cyclone (TC) or an extratropical cyclone (xC) may be due to both wind and intense rainfall. Damage during a severe convective storm (CS) may be due to hail and/or tornados (To). The thunderstorms during a tropical cyclone may have the characteristics of severe convective storms. Note that rainstorms can be considered invisible perils in the sense that they lead to damage only indirectly via flood events (Mignan et al., 2014). Despite their specificities, all **storms** are motion systems that transform potential energy and latent heat into wind kinetic energy.

1.3.3.1 Tropical Cyclones (TC)

A **tropical cyclone** is a rotating weather system which tends to form between 5° and 30° of latitude (Fig. 1.8) where six main conditions are met for the process of tropical cyclogenesis. Conditions include sea surface temperatures exceeding 26°C and sufficient Coriolis force (due to the Earth's rotation), among others (Gray, 1998). Beyond these two latitudinal bands, cyclonic storms are called extratropical cyclones (see Section 1.3.3.2). A mature tropical cyclone is a symmetrical spiral structure constituted of rainbands of thunderstorms and centred on the storm eye (Fig. 1.14; Emanuel, 1986). **Tropical storms** correspond to winds between 17 m/s and 33 m/s (61–118 km/h). Following the Saffir–Simpson hurricane wind scale (Simpson, 1974), tropical storms with sustained wind speeds exceeding 33 m/s (118 km/h) become **hurricanes** (in North Atlantic and Central and East Pacific Oceans). They are called **typhoons** over the western North Pacific Ocean and **cyclones** over the Indian Ocean and South Pacific. In recent years, the term '**Medicane**' has been employed to describe Mediterranean Sea storms with tropical cyclone characteristics (e.g., Androulidakis et al., 2023). Event severity is described by categories 1 to 5 on the Saffir–Simpson scale (see Table 3.4). Tropical cyclones typically lose their energy after landing, due to the absence of heat generation from warm ocean waters. Consequently, the storm system gradually weakens, leading to its breakdown, although rainfall and winds may persist with diminished intensity. In rare circumstances, a reintensification of the storm can occur through the internal release of latent heat from moisture condensation, providing an additional, albeit rare, source of energy (Grossman and Rodenhuis, 1975). Hurricanes are the costliest natural disasters in the US. Of the 10 costliest weather or climate catastrophes in the country, the top eight are hurricanes (Table 1.3; NOAA-NCEI, 2023). The 1979 super-Typhoon Tip, which crossed Japan, was the largest in term of surface and the most intense tropical cyclone ever recorded, with wind gusts up to 85 m/s (306 km/h) (Dunnavan and Diercks, 1980). Storm surge (Section 1.3.2.3) is an important secondary peril of tropical cyclones, which causes almost 50% of the fatalities, while only 8% are attributed to wind (for Atlantic hurricanes; Rappaport, 2014). Inland flooding (via intense rainfall), tornadoes, hail, and lightning are other sub-perils potentially associated with tropical cyclones (e.g. Hillier, 2017, p. 200). Hurricane Ivan (2004) for example produced 117 tornadoes that

Table 1.3 The 10 costliest weather and climate disasters in the US

Name	Year	Economic losses[a] [USD billion]	Fatalities
Hurricane Katrina	2005	190	1,833
Hurricane Harvey	2017	151	89
Hurricane Ian	2022	113	152
Hurricane Maria	2017	109	2,981
Hurricane Sandy	2012	84	159
Hurricane Ida	2021	80	96
Hurricane Irma	2017	60	97
Hurricane Andrew	1992	57	61
US drought/heatwave	1988	51	454
Midwest Flooding	1993	44	48

[a] Economic losses (CPI-adjusted) and fatality numbers from NOAA-NCEI (2023) include both primary and secondary losses (e.g., flooding).

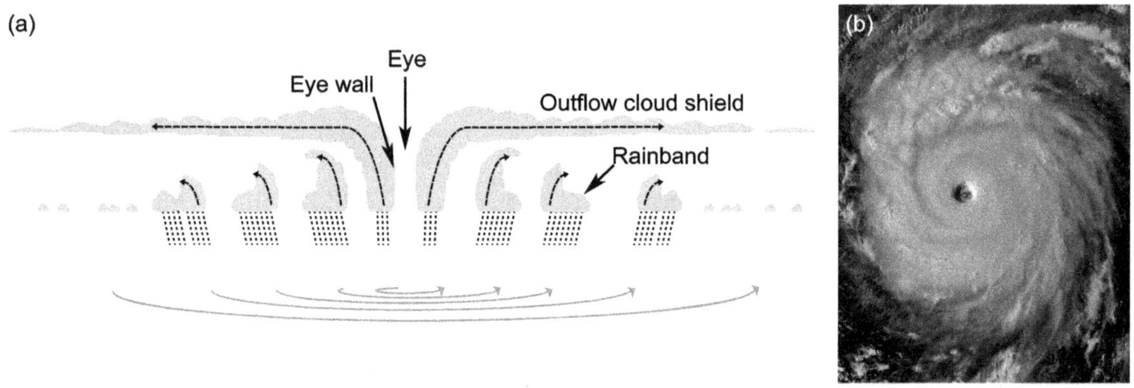

Fig. 1.14 Illustrating the tropical cyclone process and structure. (a) The tropical cyclone process. Bordering the eye of the storm is the **eye wall**, where the strongest winds and heaviest rainfall are found. Beyond the eye wall are **rainbands**, with thunderstorms and rain showers. Ascending air is shown by the dashed arrows (see Chapter 2 for the pressure and wind profiles of a tropical cyclone). (b) Hurricane Katrina on 28 August 2005 when it was at its maximum intensity as a Category 5 storm (NOAA, public domain).

led to USD 97 million in property damage (e.g., Belanger et al., 2009). In 2017, Hurricane Harvey's extreme rainfall resulted in unprecedented flooding in the Houston area, Texas (Fig. 1.12b; Qina et al., 2020).

1.3.3.2 Extratropical Cyclones (xC)

Cyclonic storms occurring in mid-latitudes (30°–80°) are termed **extratropical cyclones**. Within the Northern Atlantic, these storms often manifest as winter storms, representing distinctive features of the European climate and commonly known as **windstorms**. Their

Table 1.4 The 10 most damaging windstorms in the period 1979–2012

Name	Date	v_{max} [m/s]	Insured losses[a] [USD billion]
Daria	25 January 1990	37.9	8.2
Lothar	26 December 1999	36.7	8.0
Kyrill	18 January 2007	36.4	6.7
Great Storm of 87	16 October 1987	39.5	6.3
Vivian	26 February 1990	35.2	5.6
Klaus	24 January 2009	37.2	3.5
Martin	27 December 1999	37.2	3.3
Xynthia	27 February 2010	32.6	2.9
Anatol	03 December 1999	39.9	2.6
Erwin	08 January 2005	39.2	2.2

[a] Numbers from Roberts et al. (2014, table 1) (CC-BY 3.0) indexed to 2012 values.

origin lies in the instability associated with the polar front, which delineates polar air from tropical air (Friedman, 1999). This instability arises from the interaction between cold and warm air masses, resulting in the release of potential energy in the form of wind. Additionally, they may arise from the remnants of powerful tropical cyclones moving northward. Unlike tropical cyclones, which typically dissipate over land as they lose their primary heat source, windstorms can experience intensification post-landfall, depending on the presence or not of sharp atmospheric temperature gradients. Notably, these storms exhibit a more asymmetric structure than tropical cyclones and have a broader geographical impact. They typically originate over the western side of major oceanic basins, strengthening as they traverse the ocean. Winds can be very strong but only in rare cases reach tropical-cyclone-level speeds up to category 2 on the Saffir–Simpson scale. Many of the windstorms that hit northwest Europe are still in the most damaging phase of their life cycle as they pass over the major population centres 300–1,000 km inland (Hillier, 2017, p. 203). The 10 most damaging windstorms in the period 1979–2012 are listed in Table 1.4 (Roberts et al., 2014). If European storms drive the losses observed worldwide for extratropical cyclones, losses are also significant in North America (Hillier, 2017, fig. 3.7). For example, the 1993 'Storm of the Century', which had a death toll of 270 (exceeding the combined death toll of Hurricanes Andrew and Hugo by three times) and led to USD 6 billion in insured losses, was a snowstorm that generated record snowfalls over several US states and severe winds (i.e., a **blizzard**; Cardone et al., 1996). There are other types of severe windstorms involving ice or dust/sand. Infamous examples include the Great 2008 Chinese Ice Storm (Xie et al., 2014) and the 1935 Black Sunday Dust Storm (Lee and Tchakerian, 1995; see Section 1.3.4.3). Extratropical cyclones can also cause severe coastal flooding (Fig. 1.12a; Munich Re, 2002) and thunderstorms.

1.3.3.3 Severe Convective Storms (CS)

Severe convective storms are extreme thunderstorms with winds stronger than 25 m/s (90 km/h), hail larger than 2.5 cm in diameter, or tornadoes (e.g., Hillier, 2017, p. 209).

Other sub-perils include intense rainfall, which can trigger flash floods, as well as lightning strikes which can provoke fires (in the case of dry thunderstorms). In general, **thunderstorms** are observed during spring and summer in mid-latitudes (Taszarek et al., 2020) and all year long near the equator, as they develop from latent heat released by condensation of rising warm and moist air. Individual thunderstorm cells have local footprints (< 10 km) but can organize in large clusters or become a **supercell** (by the emergence of a mesocyclone, i.e., a cloud with rotation) leading to a wider spatial footprint. Supercells are the most dangerous, since they produce the largest hailstones and the most violent tornadoes.

Lightning, a basic aspect of thunderstorms, is due to the electrification of the cloud (Doswell, 2001). Although catastrophic lightnings are very rare, their high frequency makes them deadlier than tornadoes or hurricanes, on average, in the US (Ashley and Gilson, 2009). The deadliest lightning strike (after 1873) killed 21 people in a hut in Zimbabwe in 1975 (at the time eastern Rhodesia). Indirect deaths can be greater, for example when a lightning bolt triggers a plane crash or ignites some explosive material, with the most extreme event since the late nineteenth century being 469 people killed in a lightning-caused oil tank fire in Egypt in 1994 (Cerveny et al., 2017).

Strong **straight-line winds** during a severe convective storm comprise downbursts (including microbursts and macrobursts), gust fronts, and bow echoes (including *derechos*). A **downburst** is an area of strong wind caused by a downdraft hitting the surface and spreading out horizontally. A **gust front** forms when winds propagate away from the storm and create an outflow boundary. A **bow echo** is a bow-shaped line of storm cells associated with strong winds along its leading edge, with the ***derecho*** a long-lived bow echo (Przybylinski, 1995).

Hail is precipitation in the form of balls of irregular lumps of ice, always produced by convective storms. By convention, hail has a diameter of 5 mm or greater. Hail events have a footprint (i.e., **hail streak**) commonly a few tens of kilometres in width and length (Changnon, 1977). This is their high kinetic energy, which causes damage at impact. For instance, hailstones with diameters in excess of 5 cm may reach terminal fall velocities of 100 km/h or more (Knight and Heymsfield, 1983). Hail and lightning cause most of the insured losses related to severe convective storms (e.g., Hillier, 2017, p. 213). Hail causes significant losses to crops (Katz and Garcia, 1981) and property, such as mobile assets like cars (Hohl et al., 2002). The economic loss of a major Central European hailstorm can exceed EUR 1 billion (after adjusting for inflation), with infamous examples including the 1984 Munich hailstorm, hailstorm Wolfgang in 2009, and hailstorm Andreas in 2013 (e.g., Punge et al., 2014). The great 1984 Munich hailstorm, which resulted in some very serious injuries, reached H8 on the Tornado and Storm Research Organisation (TORRO) hailstorm intensity scale, which means hailstones up to 90 mm in diameter (Webb et al., 2009) (see Table 3.3). The deadliest hailstorm killed 246 people in India in 1888 (Cerveny et al., 2017).

Tornadoes occur in mid-latitude regions of both hemispheres, but violent tornadoes mostly strike the continental US (along the so-called 'Tornado Alley'). As a **tornado** converts the potential energy stored within the parent storm into kinetic energy close to the surface, concentrated via the tornado vortex, winds can reach up to 140 m/s (500 km/h) (see the Enhanced Fujita (EF) scale in Table 3.5). This make tornadoes the most destructive meteorological phenomenon on Earth. Yet, because of their very localized nature, they still account for much less in losses than hail. Tornadoes typically have short lifetimes of a few minutes, narrow widths between tens and hundreds of metres, and track lengths from less than a kilometre up to tens of kilometres. In very rare cases, tornadoes can last up to one hour while remaining on the ground for hundreds of kilometres (Fujita, 1973). During the 1950–2011 period, between USD 293 billion and 449 billion of total losses would have occurred in the US once adjusted to 2011 prices and population (Simmons et al., 2013). The 1925 F-5 Tri-State Tornado, which remained on the ground for 350 km, killed 689 people, injured almost 2,000, and flattened several towns (Fujita, 1973), making it the deadliest tornado in US history. The 2011 EF-5 Joplin tornado is the costliest tornado to date (USD 3 billion in insured losses) and the deadliest single tornado to hit the US (162 fatalities) since modern recordings began in 1950 (Paul and Stimers, 2012). Worldwide, the deadliest tornado occurred in Bangladesh in 1989 and killed approximately 1,300 individuals (Cerveny et al., 2017).

1.3.4 Climatological Perils

Climatological perils represent dangerous events caused by long-lived, regional to continental processes that occur in the atmosphere. They are distinguishable from meteorological (wind-related) perils, which involve shorter-lived processes. The perils here are temperature-related and include extreme temperatures (heatwaves and cold waves), droughts, and wildfires. According to a study by Zhao et al. (2021), more than 5 million people die each year globally because of extreme temperatures (90% from cold exposure and 10% from heat). Most occurrences correspond to high-probability–low-consequences accidents (i.e, many isolated events; Fig. 1.3). Only catastrophic events (as per the definition of catastrophe; Section 1.1.1) are considered below, with events lasting from a few days (e.g., a heatwave) to weeks (e.g., a wildfire) to months or years (e.g., a drought).

1.3.4.1 Heatwaves (HW)

Heatwaves are prolonged periods of excessively hot weather relative to seasonal average. They generally need to last for at least 3 days with minimum temperature – or maximum temperature – at the 90th percentile to be considered heatwaves (Perkins and Alexander, 2013). They occur by the arrival of very hot dry air and by anticyclonic conditions holding back rain-bearing depressions. They are therefore usually associated with droughts (Mueller and Seneviratne, 2013). Their main impact is on human health, with mortality observed to potentially more than double during such an event, compared to natural circumstances (Whitman et al., 1997). Heat stress may be measured in terms of an apparent temperature, which is a function of the relative humidity. In the most extreme case, any exceedance of a wet-bulb temperature of 35°C for extended periods would induce hyperthermia and be lethal to any individual (see definition of wet-bulb temperature in Section 2.4.2.2). This has not happened, but could in the future in some regions of the world due to climate change (Sherwood and Huber, 2010). Due to comorbidities, people, especially elderly people, can die from heat stress at much lower wet-bulb temperatures. The effects of heat can be exacerbated in cities, due to the urban heat island effect (Tomlinson et al., 2011). Heatwaves are major natural hazards, which are less visible than other disasters. This is due to their very limited impact on the built environment and the fact that heat-related deaths must first be distinguished from other causes of death and defined as excess deaths (Whitman et al., 1997). The extreme heatwave that hit Europe in summer 2003 caused more than 30,000 fatalities and more than EUR 13 billion in agriculture and forestry losses. France was the most impacted with almost 15,000 excess deaths reported (De Bono et al., 2004). Heatwaves can exacerbate health and emergency services. They can also overload the power grid, potentially leading to blackouts (due to overuse of air conditioning and refrigerators, and/or lack of river cooling water for nuclear reactors). They also exacerbate droughts and wildfires by the combined effects of high air temperature and solar radiation. High temperatures can also increase the likelihood of landslides at high altitude due to the melting of permafrost (De Bono et al., 2004).

1.3.4.2 Cold Waves (CW)

A **cold wave** is a period of excessively low temperatures relative to seasonal average. It is often associated with winter storms, including blizzards and ice storms. Its main subperil is frost. A common threshold to declare a cold wave is negative temperatures below, for example, −5°C over a period of several days (Huynen et al., 2001). The temperature threshold can also be defined as a low percentile of the background temperature distribution. Low temperature anomalies are associated with cold polar air masses, which can range up to several thousand kilometres in scale (Walsh et al., 2001). A cold wave leads to hypothermia and other sources of lethal factors, such as a reduction of immunity and cardiovascular stress (Huynen et al., 2001). It can also cause widespread damage to crops and livestock (Bolinger et al., 2022). Transportation systems can freeze, triggering a failure of supply chains, such as in the case of the Great 2008 Chinese Ice Storm (Xie et al., 2014). The cold, snow, and ice from the 2021 Valentine storm in the central US caused hundreds of deaths and USD 24 billion in damage, making it the costliest winter weather event in the US. The most notable aspects of this event were the extreme cold temperatures and the duration of the freeze, which led to forced power outages and near-record highs in regional natural gas prices (e.g., Bolinger et al., 2022). Most all-time records of extreme cold in the US, however, remain attributed to the Great Arctic Outbreak of 1899 (Kocin et al., 1988).

1.3.4.3 Droughts (Dr)

A **drought** is a long-lasting episode of precipitation deficit, characterized by water supply deficiency of the soil. In contrast to aridity, which is permanent, a drought is a transitory anomaly often defined as a tail event on the statistical distribution of precipitation in a given region. Although droughts and pluvial floods are generally treated separately due to being on the opposite sides of the hydrological spectrum, they are inherently interconnected (He and Sheffield, 2020). Various indices exist to assess the intensity and spatial

Fig. 1.15 Dust storm during the Dust Bowl caused by a severe drought, here approaching Stratford, Texas. Photographs taken on 18 April 1935 (NOAA, public domain).

extent of droughts (Palmer, 1965; Mishra and Singh, 2010). As previously indicated, there is a strong relationship between droughts and heatwaves, but droughts last much longer, that is, months to years instead of days. Droughts occur in virtually all climatic zones. Four main types of droughts exist: they can be meteorological (i.e., lack of precipitation), hydrological (i.e., lack of river irrigation), agricultural (i.e., water overuse by crops) or socio-economic (i.e., water overuse by populations) (Mishra and Singh, 2010). Of all the twentieth-century natural hazards, droughts have had the greatest detrimental impact (e.g., Mishra and Singh, 2010) and account for the maximum number of total deaths in that century in EM-DAT (e.g., Nadim et al., 2006), if excluding epidemics (see the caption of Fig. 1.4). The 1988 US drought and associated heatwave is the ninth costliest weather disaster in the US, just after hurricanes (Table 1.3; NOAA-NCEI, 2023). A drought can cause substantial losses in crops and livestocks, and can indirectly trigger a famine. It is also known to cause dust storms, with the most infamous example being the 1935 Black Sunday Dust Storm (Lee and Tchakerian, 1995), caused by the legendary 1930s Dust Bowl drought (Cook et al., 2014) (Fig. 1.15). This event served as inspiration for Christopher Nolan's science fiction movie *Interstellar*, in which he illustrates the devastating impact of an ecological disaster, both on farming and on health (Nolan, 2014; Wright, 2022). Water stress due to droughts can be a cause for conflict (i.e., water wars). It has been suggested that the Syrian conflict was in part due to a multi-year drought which triggered food insecurity with effects on political stability in the region (Gleick, 2014).

1.3.4.4 Wildfires (WF)

A **wildfire** is the uncontrolled burning of vegetation in a natural environment, such as a forest, grassland, bush land, or tundra (urban fires will be treated in Section 1.3.7). Ignition of a fire can be due to lightning or to human activity (Syphard and Keeley, 2015). Once ignited, a fire will continue to spread depending on fuel availability and flammability. The complex process corresponds to the release of energy in the form of heat owing to the thermal decomposition of the vegetation (i.e., combustion) and the transport of that energy to

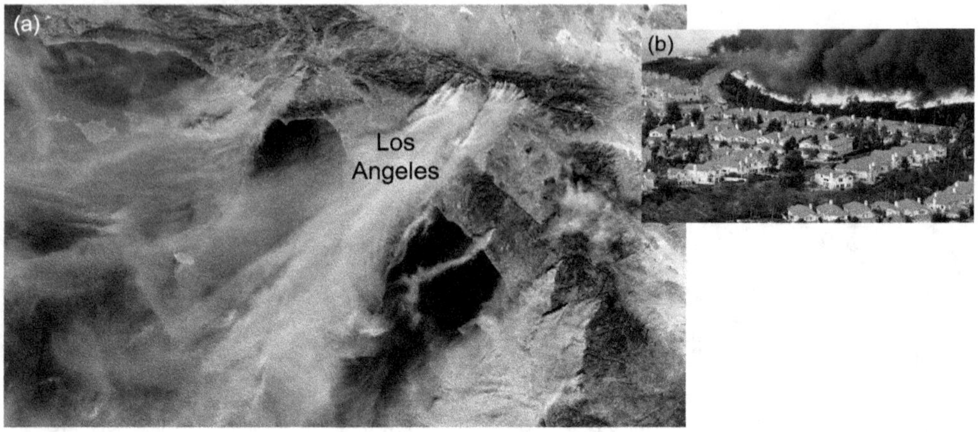

Fig. 1.16 The 2003 Cedar wildfire, Southern California. (a) A Moderate Resolution Imaging Spectroradiometer (MODIS) satellite image showing the extent of the wildfires (USGS, public domain). (b) Wildfire threatening Ramona, San Diego County (San Diego Fire-Rescue Department, public domain). A total of 3,361 homes and 26 lives were lost in this event (Keeley et al., 2004).

surrounding unburnt fuel for further ignition. This is a function of fuel moisture content and temperature, as well as environmental conditions, such as wind speed and topography (e.g., Sullivan, 2009). The availability of fuels and characteristics of fires are further altered by other perils, such as droughts (Littell et al., 2016) and pest infestations (Hicke et al., 2012), with both hazards enhanced by climate change. Combined with the effects of inappropriate land management, the incidence of large wildfires has risen in the last few decades. Fire suppression and logging activities (Lindenmayer et al., 2020) increase the fuel load and affect forest patterns, disturbing the role of wildfires in the regulation of complex ecosystems. Overall, and with an increase of the housing density and size of the wildland–urban interface, large insured losses from wildfires have recently become a real concern for insurers (Murnane, 2006). The 2003 Southern California Cedar wildfire, in which 26 lives were lost, is depicted in Figure 1.16. Wildfires in Portugal killed more than 120 people in 2017 while 100 people were lost to Greek wildfires in 2018, the deadliest in Greece's history (e.g., Turco et al., 2019). The 2017 North Bay fires were the deadliest and most destructive wildfires in California's history at the time, with 44 fatalities and nearly 9,000 structures lost (Nauslar et al., 2018). The sad record was however broken the next year with the Camp Fire, which resulted in 82 fatalities and the destruction of the city of Paradise, with 18,000 structures lost (Ager et al., 2021). The deadliest event however remains the Peshtigo fire of 1871, in Wisconsin, which killed more than 1,000 people (Pernin, 1971). Wildfires can trigger other perils. Extreme wildfires can create their own storms called **pyrocumulonimbus**, which can further amplify the fire. Positive feedback effects are winds and **pyrogenic lightning** (Dowdy et al., 2019). Wildfires also cause smoke pollution and increase the global carbon budget, further participating in climate change. The 2019–20 Australian mega-fires led to an unprecedented AUD 1.95 billion cost and almost 500 premature deaths due to smoke-related health consequences alone

(Johnston et al., 2020). Wildfires can also be followed by run-off-generated debris flows because of the changes in soil conditions due to the fire (Rengers et al., 2020). Following the large Thomas Fire, post-fire debris flows caused 23 fatalities in Montecito, California, in 2018 (Kean et al., 2019).

1.3.5 Biological Perils

Biological perils represent all the dangerous events caused by living organisms, which include epidemics and pest infestations. The former are mainly bacteria and viruses which have the potential to cause human deaths in large numbers. The latter are mainly insects which have the potential to destroy entire crops.

1.3.5.1 Epidemics (Ep)

Infectious diseases are illnesses caused by bacteria, viruses, fungi, or parasites that often spread via contaminated water or food. Once person-to-person transmission becomes possible, a chain reaction of propagation via exponential growth can occur, potentially leading to an epidemic (Fig. 1.17a). An **outbreak** is the sudden increase in incidence of a disease (possibly a new one) within a relatively small area. An **epidemic** is the increase, often sudden, in incidence of a disease above background level, in a larger area than an outbreak. An example would be the seasonal flu. Finally, a **pandemic** is an epidemic that spreads in many countries or continents (Smith, 2017). Epidemics are a common consequence of disasters and conflicts, due primarily to population displacements and reduced access to safe water and healthcare services for the survivors (Watson et al., 2007). If direct consequences on mortality and health-related costs are obvious, indirect economic losses of epidemics are more complex and relate to the dysfunctioning of the society due to behavioural changes and labour shortages (Helbing and Kuehnert, 2003). Many pandemics happened through history via pathogens such as influenza, smallpox, and measles, from the early plagues of Antiquity (Morgan, 1994) to the Black Death (or Bubonic Plague) responsible for the death of 30–50% of Europe's population in the Middle Ages (Smith, 2017). The deadliest epidemic since 1600 is the 1918 Spanish Flu, which caused 32 million deaths, according to the Global Epidemics Dataset (1500–2020) (Table 1.5; Marani et al., 2021a,b). The frequency of outbreaks has been observed to increase due to increasing interactions between humans and the natural environment that potentially lead to more **zoonotic diseases** (i.e., that spread between animals and people) (Jones et al., 2013). In the years prior to the COVID-19 pandemic, several events were a source of concern: the 2003 SARS 'mini pandemic' (Cherry, 2004); the 2014 West Africa Ebola outbreak, which caused a gross domestic product (GDP) loss of up to USD 32 billion (Huber et al., 2018); and the 2016 Zika outbreak in Brazil, with close to 5,000 suspected cases of microcephaly (i.e., infants born with an abnormally small head; Heukelbach et al., 2016). A 2013 survey indicated that global pandemic ranked as the biggest extreme risk facing insurers (e.g., Smith, 2017). By the end of 2023, the COVID-19 pandemic (Fig. 1.17b) caused almost 6.9 million fatalities worldwide (Dong et al., 2020) and illustrated the complexity of the

Table 1.5 The 10 deadliest epidemics in 1500–2023

Name/location	Period	Disease	Fatalities[a] [million]
Spanish Flu	1918–1920	Influenza	32.0
European diseases in the Americas	1520–1635	Smallpox, etc.	10.3
New Spain 'Great Pestilence'	1545–1548	Cocoliztli	10.0
COVID-19	2019–2023	Coronavirus	6.9
Mexico	1519–1520	Smallpox	6.5
Thirty Years' War typhus epidemic	1618–1649	Typhus	6.2
India	1903–1907	Plague	5.5
Sixth cholera pandemic	1899–1926	Cholera	5.0
British India	1888–1937	Smallpox	4.2
Third cholera pandemic	1846–1863	Cholera	4.2

[a] 1500–2020 numbers from the Global Epidemics Dataset (Marani et al., 2021a) (CC-BY 4.0, https://creativecommons.org/licenses/by/4.0/). Some names modified from the original data set and COVID-19 added to the ranking.

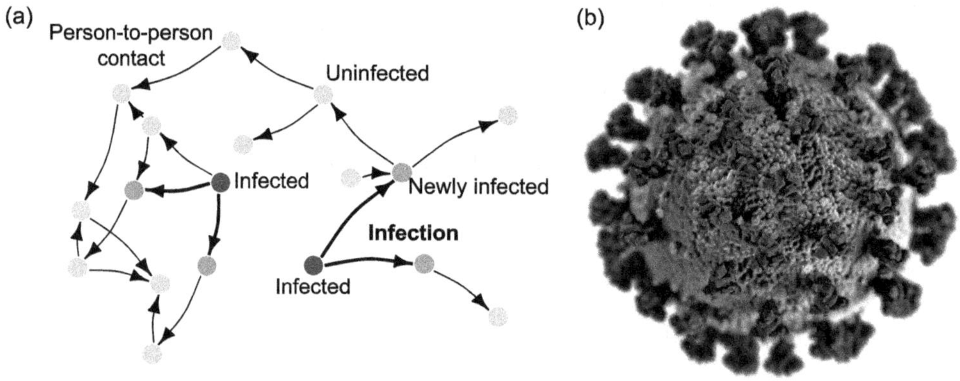

Fig. 1.17 Illustrating the epidemic process and the structure of a virus. (a) The infection process: The infection propagates from person to person, with the spread increasing geometrically. (b) Morphology exhibited by the severe acute respiratory syndrome coronavirus 2 (SARS-CoV-2), identified as the cause of the COVID-19 pandemic (Centers for Disease Control and Prevention, public domain).

many indirect consequences of a pandemic: consequences on populations included anxiety and depression (Hou et al., 2021). The consequences of lockdowns on the economy included supply chain cascade failures and business interruptions, followed by a significant reduction of the GDP of many countries related to the slowdown of most economic sectors (Guan et al., 2020).

1.3.5.2 Pest Infestations (PI)

A **pest** is a destructive insect (or other animal) that can attack agricultural crops, woodland crops, or livestock (one could include blight due to fungi and bacteria in this peril). An infestation corresponds to the case where the pest impact is severe enough to significantly affect the production yield in a vast area. It should also be sudden enough to be considered an individual catastrophe. Invasive insects cost a minimum of USD 70 billion per year globally (Bradshaw et al., 2016). This number however corresponds to the continuous (background) impact of insect crop consumption. A pest outbreak is an example of an emerging event in an ecological system occurring from the complex interactions between different trophic levels, between prey and predators. During an outbreak, the number of insects significantly increases against background numbers. The event can last for years but is distinguishable from normal periods. A well-known example is the dynamics of spruce budworm outbreaks in relation to avian predation and forest growth (Ludwig et al., 1978). The intensity of a pest infestation depends on the number of insects and their feeding rate. Outbreaks can coincide with warm weather and unusually heavy rainfall (Roussi, 2020). The frequency and size of such events is expected to increase with climate change (Bentz et al., 2010). In Canada, tree harvest can be more than halved during severe spruce budworm outbreaks (MacLean, 1984); other forest pests include forest tent caterpillar, mountain pine beetle, jack pine budworm, and gypsy moth (Chang et al., 2012). Locust swarms have devastating effects in many regions of the world. The Locust Plague of 1874 encompassed several US states and caused USD 200 million in crop damage at the time. Other great locust outbreaks have occurred in the past century on most continents (e.g., Vijayalakshmi and Meena, 2020). As of 2020, Kenya has been experiencing its worst locust swarms for 70 years (Roussi, 2020). An average swarm, which contains up to 40 million insects, can travel 150 km in a single day and can devour enough food to feed 34 million people (e.g., Devi, 2020). It can thus lead to famines by significantly reducing the food supply. A desert locust infestation, for example, contributed to the Great Famine of Mount Lebanon during World War I (Foster, 2015). A pest infestation can also increase the likelihood of wildfires by increasing the fuel load (i.e., dead trees; Hicke et al., 2012).

1.3.6 Extraterrestrial Perils

An **extraterrestrial peril** is any peril having its source from outside the Earth and its atmosphere. This includes asteroid and comet impacts, as well as geomagnetic storms. There are other extraterrestrial dangers which are too rare to be considered here, an example being a supernovae, which would have the potential to flood the Earth's atmosphere with X-ray (Smil, 2008, p. 13). In contrast to all the previously mentioned perils, whose maximum sizes are constrained by the finite limits of the Earth, extraterrestrial perils do not have such constraints. Extreme events, such as multi-kilometre-size asteroid impacts, would have the potential to kill hundreds of millions and destabilize global civilization, making them the 'ultimate environmental catastrophe' (Morrison, 2006). Fortunately, they do not enter in the typical catastrophe category (Fig. 1.3).

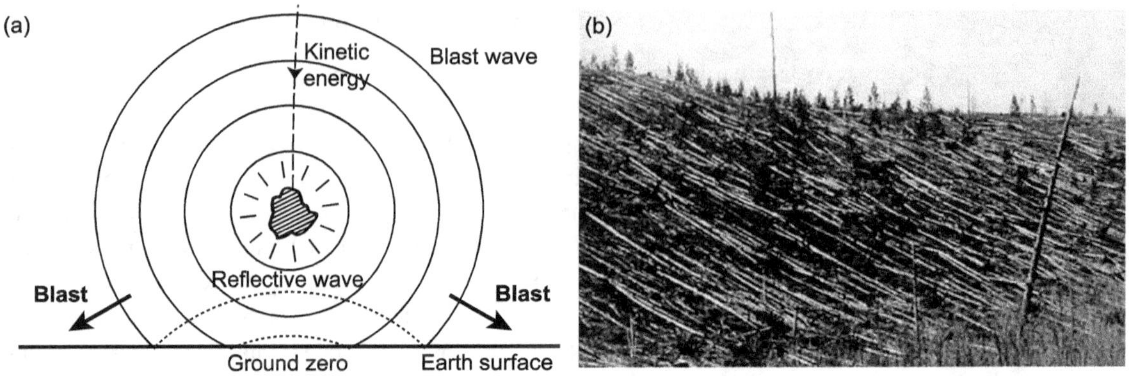

Fig. 1.18 Illustrating the airburst process and potential observed damage. (a) The airburst process. When the explosion occurs, a blast wave is generated that propagates outward. This overpressure front is accompanied with strong winds, and, potentially, fires due to the high temperatures. Modified from Glasstone and Dolan (1977, fig. 3.21), originally for a nuclear explosion. (b) The Tunguska event of 1908, a giant blast which occurred over the basin of the Podkamennaya Tunguska River in Central Siberia. It levelled a large fraction of trees over an area spanning over 2,000 km^2 (Soviet Academy of Science, 1927 Kulik expedition, public domain). If the original object had been composed of iron, it would have created a crater instead, similar in size to the famous Meteor Crater in Arizona (US).

1.3.6.1 Asteroid Impacts (AI)

An **impact event** is a collision at high velocity between an extraterrestrial object (i.e., a stony or iron asteroid or an icy comet) and the Earth (Shoemaker, 1983). The kinetic energy, which depends on the object's mass and speed, transforms into explosive energy at impact and can be comparable to an atomic bomb exploding. Objects with a diameter of less than 1 m are called meteoroids and seldom make it to the ground to become meteorites. Objects with a diameter of several metres to tens of metres ordinarily explode in the upper atmosphere, and are called **airbursts** (Fig. 1.18a). Greater objects impact the ground and create a crater. Humanity certainly experienced major impact events in the remote past (Section 1.2.1) but not in historical times. The 1908 Tunguska airburst, likely due to an object 65 m in diameter, and whose explosive yield was equivalent to 10–15 megatons (Mt) of TNT (about 1,000 times the energy released by the nuclear weapon launched over Hiroshima in 1945), fortunately occurred over a remote part of Central Siberia (Fig. 1.18b). A repeat of the Tunguska airburst over a major city, such as New York City, could kill millions and cause more than USD 1 trillion in property losses (Mignan et al., 2011). Impacts of this size are believed to recur every 1,000 years (Brown et al., 2002) but the likelihood of an impact over a populated area remains small. On average, it would cause a few thousand deaths (Chapman and Morrison, 1994). In 2013, a 500-kt airburst caused by a 20-m diameter asteroid occurred over Chelyabinsk, Russia, and led to minor non-structural damage and indirect injuries. It was the biggest impact over land since the Tunguska event (Popova et al., 2013). Local 'freak' injuries or deaths by stones falling from the sky (i.e., meteorites) remain extremely rare (La Paz, 1951). Impacts over water masses can generate a tsunami but are expected to remain a rare instance. For example, a 250-m-diameter impactor is likely to yield a 7-m wave at a distance of 1,000 km (Ward and Asphaug, 2000; Morrison, 2006).

1.3.6.2 Geomagnetic Storms (GS)

Geomagnetic storms are disturbances of the Earth's magnetosphere due to interactions with the strong solar winds typically occurring after coronal mass ejections from the Sun. They are dangerous since they can disrupt the operation of conducting networks, such as power systems, satellites and other communication systems, rail networks, and pipelines. In our hyperconnected society, this may lead to significant damage and indirect fatalities. Other so-called space weather storms include radio blackouts due to solar X-ray flux and solar radiation storms due to energetic particle flux (Eastwood et al., 2017). The 1989 Québec, Canada, geomagnetic storm led to the failure of the Quebec power grid (Boteler, 2003). In 2003, one of the largest observed solar flares on record caused a 90-minute blackout affecting 50,000 customers in Sweden. The occurrence today of an event similar in size to the powerful 1859 geomagnetic storm (the Carrington Event) could cause economic losses in the range of trillions of dollars in the US alone since it could take years to recover from the inflicted damage to the power grid. A Carrington-class event actually occurred in 2012 but was fortunately not directed toward the Earth (Eastwood et al., 2017).

1.3.7 Technological Perils

Technological perils represent all the dangerous events triggered by the failure of a critical infrastructure which can have a direct physical impact on humans and/or structures (other critical infrastructures with only a social or economic impact will be covered in Section 1.3.8). Perils include fires (Fi), explosions (Ex), and toxic spills (TS), as well as energy, material, and information flow cut-offs. Human-made catastrophes are often a complex combination of several perils with patterns differing between different types of critical infrastructures. Critical infrastructures can first be divided into two main categories: they can be localized sites, such as nuclear power plants, hydro-dams, or refineries, or they can be geographically distributed sites, such as power grids, transportation networks, or the internet (e.g., Argyroudis et al., 2020). Fires, explosions, and toxic spills are the most common events at oil and gas energy sites, according to ENSAD (Mignan et al., 2022a) and other data sets (Khan and Abbasi, 1999). **Toxic spills** can be chemical in nature at refineries and radioactive at nuclear power plants. The water mass released during dam overtopping or dam collapse can be identified as a **dam flood** (DF). **Energy supply failures** (Ef) include power blackouts (BO) and other ruptures at the energy production, storage, transport, or distribution level. Other types of flow cut-offs include transport system failure (Tf; e.g., supply-chain failure), water system failure (Wf), and information system failure (If, which can be caused by a cyberattack (CA)). There are many other types of site-specific perils, for example earthquakes being triggered at enhanced geothermal system (EGS) plants (Mignan et al., 2015a) (part of the case study in Part III).

The failure of a critical infrastructure, which may be coined a severe operational loss event in the context of operational risk (Beroggi and Wallace, 1994), can be human-made

Table 1.6 Complexity of technological perils simplified

Process	Site type examples	Main perils[a]
Single localized site		
Chemical engineering	Refinery, fertilizer warehouse	Explosion (Ex), fire (Fi), chemical toxic spill (TS)
Nuclear engineering	Nuclear power plant	Radioactive toxic spill (TS), explosion (Ex), fire (Fi)
Water engineering	Hydro-dam, levee	Dam flood (DF) by breach or overtopping
Geothermal engineering	Enhanced geothermal system	Earthquake (EQ)
Geographically distributed site		
Chemical engineering	Oil and gas networks/pipelines	Explosion (Ex), fire (Fi), chemical toxic spill (TS)
Electrical engineering	Power grid	Blackout (BO)
Transport engineering	Road, rail, maritime, air transport	Transport system failure (Tf) (e.g. supply-chain failure)
Water engineering	Water supply and wastewater networks	Water system failure (Wf)
Information engineering	Internet(-of-things), satellite system	Information system failure (If) (e.g., cyber-attack (CA))

[a] All sites that are part of the energy production, storage, transport, and distribution system can also lead to energy supply failure (Ef); other perils such as business interruption (BI) will be considered in Section 1.3.8.

or caused by a natural disaster. Human-made causes can be accidental (the most common cause; Mignan et al., 2022a) or malicious (e.g., a cyberattack; for terrorist attacks, see Section 1.3.8.5). Despite the safeguards built to protect critical infrastructures from disasters, the complexity of engineered structures makes a rare but severe accident an inherent and 'normal' phenomenon (Perrow, 2001). Here, the complexity of technological systems is simplified by associating individual perils with specific engineering processes (Table 1.6 and below).

When the critical infrastructure failure is triggered by a natural disaster, the disaster may be called a **Natech event**. Such an event corresponds to very serious secondary effects that take place when natural events trigger additional disasters at technological installations that house or process hazardous materials, such as nuclear power plants, chemical facilities, or pipelines (Krausmann et al., 2011; Petrova and Krausmann, 2011). Up to 5% of chemical accidents are caused by natural events (e.g., Krausmann et al., 2011). Infamous examples include the levee breach and subsequent flooding of New Orleans with toxic chemicals from industrial and residential areas, provoked by Hurricane Katrina's surge (Picou, 2009) and the Fukushima nuclear disaster due to the tsunami generated by the Tohoku earthquake (Norio et al., 2011). The latter is the most destructive Natech disaster on record, with total economic damages exceeding USD 210 billion. The tsunami caused severe accidents at the Fukushima-1 and Onagawa nuclear power plants, as well as explosions and fires at a refinery in Chiba and at a petrochemical plant in Sendai. It also destroyed many critical networks (power grid, water supply, communication system), thereby resulting in supply disruptions, in addition to severely affecting transport of people and goods (Petrova and Krausmann, 2011). Other examples will be given in the next paragraphs.

1.3.7.1 Chemical Site Failures

Chemical sites include refineries, other chemical processing facilities, and chemical material storage sites, such as fertilizer warehouses and weapon depots. Explosions and fires are interrelated (Mignan et al., 2022a), with the event severity a function of the characteristics and quantity of the hazardous material. An **explosion** is a sudden expansion in volume in which energy is transmitted outward as a shockwave. There are four main types of explosions: pressure vessel explosions, boiling liquid expanding vapour explosions, vapour cloud explosions, and explosive detonations. A **pressure vessel explosion** (PVE) is caused by the failure of a pressurized vessel. If the pressurized liquid had reached a temperature well above its boiling point, the PVE becomes a **boiling liquid expanding vapour explosion** (BLEVE) (Abbasi and Abbasi, 2007). The loss of pressure and the extremely rapid volume expansion of the gas creates the explosion. If the substance is combustible (e.g., methane, propane), a **vapour cloud explosion** (VCE) occurs, resulting from the ignition of a flammable vapour cloud (Tang and Baker, 1999). The 2005 BP Texas City refinery explosion, one of the most catastrophic incidents in the history of the process industries, killed 15 workers, injured 180, and caused economic losses of up to USD 1.5 billion (Isimite and Rubini, 2016). The detonation of high explosives leads to a larger explosion due to an even more violent chemical reaction. The greatest accidental explosion remains to this day the 1917 2.9-kt Halifax explosion of an ammunition ship, which was strong enough to generate a tsunami. The explosion killed more than 1,600 people and injured thousands more (Greenberg et al., 1993). A much more recent example is the 2020 0.5–1.1 kt Beirut explosion of ammonium nitrate, which caused at least 181 deaths and over 6,000 injuries (Fig. 1.19b; Rigby et al., 2020). Explosions typically lead to the release of toxic chemicals (toxic spill), an example being chlorine (Abbasi and Abbasi, 2007). The 1984 Bhopal catastrophe, which is often referred to as the world's worst industrial disaster, was a leak which exposed more than half a million people to toxic gases from the pesticide plant. At least 3,800 people were killed immediately and up to 20,000 more premature deaths were observed in the following decades (Broughton, 2005). The 2010 Deepwater Horizon BP oil spill in the Gulf of Mexico (Fig. 1.19a), the largest-volume marine oil spill in history, led to severe ecological damage, the disruption of industries and tourism in the Gulf Coast, and various health effects. The oil-drilling platform was destroyed by explosions and fires, killing 11 workers (Shultz et al., 2014). The second largest marine spill in US waters dates back to 1989 and was caused by the Exxon Valdez oil tanker, in Alaska (Atlas and Hazen, 2011).

1.3.7.2 Nuclear Site Failures

A **nuclear power plant** (NPP) generates electricity from the heat produced by fission in the nuclear reactor. It cannot explode like a nuclear weapon (Reed, 2018), but a core meltdown

Fig. 1.19 Examples of human-made explosions. (a) The blazing remnants of the Deepwater Horizon offshore oil rig on 21 April 2010, after the explosion (US Coast Guard, public domain). (b) Aftermath of the 2020 Beirut explosion (Mahdi Shojaeian, Mehr News Agency, CC-BY 4.0, https://creativecommons.org/licenses/by/4.0/, via Wikimedia Commons – crater ellipse added).

can release dangerous levels of radiation. Radionuclide contamination can affect the atmosphere, hydrosphere, biosphere, and pedosphere (the soil layer). The three main historical events have been the 1979 Three Mile Island accident (US), the 1986 Chernobyl disaster (Ukraine), and the 2011 Fukushima Daiichi nuclear disaster (Japan). The 1979 Three Mile Island accident released small quantities of xenon and iodine radioisotopes into the environment due to a reactor's partial meltdown. However, it is still uncertain if there has been an increase in cancer mortality due to radiation-related exposure (Talbott et al., 2003). The Chernobyl and Fukushima disasters have been rated on the International Atomic Energy Agency (IAEA) International Nuclear and Radiological Event Scale (INES) as Major Accidents (INES 7). The Chernobyl accident was, however, much more severe. The 1986 Chernobyl nuclear disaster released a total of 5,300 PBq (petabecquerels), exposing the local population and first responders to dangerous levels of radiation. It led to an epidemic of thyroid cancer and also increased the incidence of leukaemia and birth defects. There is no clear estimate of the deaths due to the accident but projections range from thousands to tens of thousands of cases of cancer mortalities. Direct deaths from acute radiation syndrome would have killed about 30 workers. The accident at the Fukushima Daiichi NPP was one of the biggest environmental disasters in recent years but released approximately 10% of the amount of radiation that was released at Chernobyl. Moreover, the majority of the radionuclides were transported offshore and deposited in the Pacific Ocean (Steinhauser et al., 2014). Projections suggest that there were 1,000 extra cancer deaths due to the Fukushima disaster (von Hippel, 2011).

1.3.7.3 Dam Floods (DF)

A **dam** is a barrier that retains water in a reservoir. The resource can be used as a water reserve, for water flow regulation, or for hydropower generation. **Levees** are specifically used to prevent flooding by a river or a coastline. Dam failure or dam overtopping can cause catastrophic floods due to the release of a large amount of accumulated potential energy.

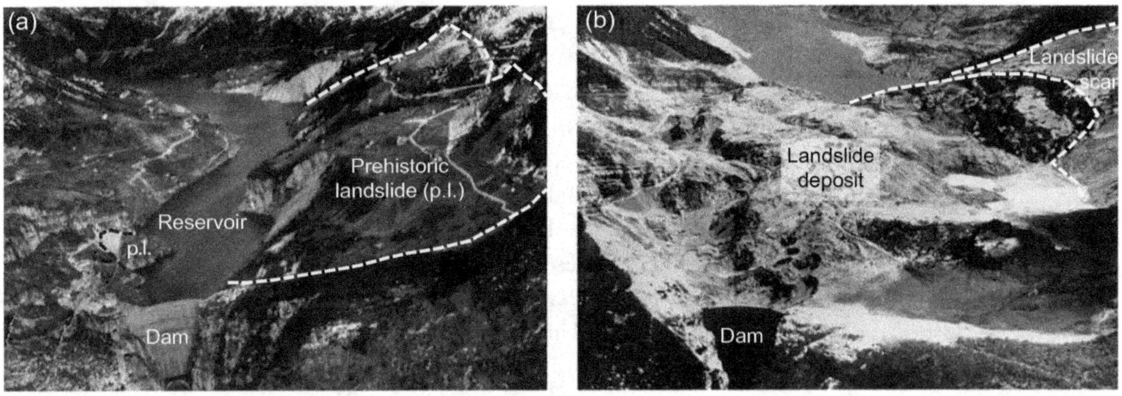

Fig. 1.20 The Vajont Dam, before and after the 1963 landslide. (a) Pre-landslide view of the Vajont Dam and reservoir in 1960. (b) Post-landslide view in 1963 (public domain via Wikimedia Commons, annotations following Ward and Day, 2011).

This type of event can be due to a natural disaster such as an earthquake damaging the dam, extreme rainfall increasing the water level, or a landslide provoking a tsunami in the artificial reservoir. Human error or malfunction of some of the elements of the dam (e.g., hydropower system, spillway, bottom outlet) can contribute to such a disaster (Matos et al., 2015; Argyroudis et al., 2020). The worst historical dam floods have been the 1975 Banqiao Dam failure (China), the 1979 Machchhu-II dam failure (India), the 1889 South Fork Dam failure (known as the Johnstown Flood; US), and the 1963 Vajont Dam overtopping (Italy). The Johnstown Flood of 1889 is considered the deadliest disaster to take place in the US prior to 1900. It was due to the catastrophic failure of the South Fork Dam with the generated torrent taking the life of over 2,200 people in Johnstown (Kaktins et al., 2013). In 1963, the mountainside of the Vajont Dam collapsed, sending a landslide into the reservoir which contained more than 100 million m^3 of water. A tsunami was generated and water overtopped the dam. Within minutes, about 2,000 people were killed in the villages down the valley while the dam remained intact (Kilburn and Petley, 2003) (Fig. 1.20). In 1975, extreme rainfall produced by Typhoon Nina led to a cascade of dam collapses. Two large dams (Banqiao Dam and Shimandan Dam), two medium dams, and 58 small dams failed from overtopping. The death toll reached over 26,000 (Xu et al., 2008). Heavy rainfall was also the cause, in 1979, of the collapse of the Machchhu-II dam and the subsequent death of more than 2,000 people by a destructive wave (Dhiman and Patra, 2019). An example of catastrophic levee failure is the flooding of New Orleans in 2005, where over 50 levees were breached by overtopping and erosion due to the record-level surge associated with Hurricane Katrina (Comfort, 2006; Sills et al., 2008).

1.3.7.4 Blackouts (BO)

A **power grid** is composed of generator nodes (i.e., power plants), transmission nodes, and distribution nodes connected via transmission lines. A **power outage** is a loss of electrical power affecting end users. A **blackout** can mean the failure of the electricity supply in large parts of a city or even of a country. Parts of a network still functioning are called island operation areas. A blackout is caused by outages cascading via transmission lines and nodes due to some power overload. This may be triggered by one local incident such as a tree fall, a lightning strike, a malfunction due to ageing, vandalism. Some externalities, such as an earthquake or a storm, can impact many parts of the power grid at once. High power demand (for heating or air conditioning) and reduced network service due to maintenance can exacerbate the problem. The 2003 North American blackout, which affected 50 million people in eight US states and two Canadian provinces, was initiated by an untrimmed tree too close to high-voltage transmission lines in the Midwest (e.g., Dueñas-Osorio and Vemuru, 2009). It caused hundreds of transmission lines and generating units to trip (Andersson et al., 2005). The disruption of power supply can result in losses in productivity across a number of electrified industries (Li et al., 2018) (see business interruptions in Section 1.3.8.1). Most communication systems would also fail in the event of a generalized blackout. Closed businesses and grounded public transportation during the 2003 Northeast Blackout led to USD 6.5 billion in reduced earnings (Anderson et al., 2007). The widest blackout occurred in India and left more than 600 million people without electricity for a day in July 2012 (Romero, 2012). The Texas freeze of February 2021, caused by Winter Storm Uri, produced an estimated economic loss of USD 130 billion in Texas alone, and USD 155 billion in total. It can be explained by a spike in energy demand due to the cold and by the closure of gas power plants and wind turbines due to frozen equipment. Some consequences of the blackout in freezing conditions included carbon monoxide poisoning and fires caused by people burning furniture to keep warm (Busby et al., 2021).

1.3.7.5 Transport System Failures (Tf)

The **transport system** is another critical infrastructure vital to a society. It includes road networks, rail networks, maritime and waterway networks, and air transport networks, as well as their many interrelations. Parts of a transport system can fail when a critical node or link is lost, or when a combination of various nodes and/or links are lost, which can also critically affect the network topology (Faturechi and Miller-Hooks, 2015). Failure here means lack of network serviceability. Due to its importance, many redundancies exist and an externality such as a natural disaster or malicious act is usually required to make it fail. If direct damage can lead to many casualties, the main economic losses due to transport system failure are often related to business interruptions due to people and goods not being able to go from point A to point B (Section 1.3.8.1). Supply-chain failure can lead to complex contingent business interruptions (see BI paragraph). Such events have intensified due to increasing urbanization and interdependencies (e.g., Faturechi and Miller-Hooks, 2015). Human-made events such as car accidents or labour interventions have a relatively short-term effect on transport network serviceability in contrast to physical damage inflicted by targeted attacks or technical failure. An example of catastrophic failure due to a design flaw is the 2007 I-35W Mississippi River bridge collapse which involved 111 vehicles, killed 13 people, and injured 145 more (Hao, 2010). Another design flaw provoked the 1986 space shuttle Challenger disaster (Feynman, 1988). Considering natural disasters, the 1989 Loma Prieta and 1994 Northridge earthquakes in California and the 1995 Kobe earthquake in Japan provoked spectacular highway bridge failures (Fig. 1.9; Chang and Nojima, 2001). The Kobe earthquake also caused the container port (ranked sixth worldwide at the time) to shut down, which permanently changed international container traffic in the region (Chang, 2000). The 2010 eruptions of Eyjafjallajökull, Iceland, caused enormous disruption to air travel across western and northern Europe, leading to a loss of more than EUR 3 billion over 1 month (Mazzocchi et al., 2010). The 1991 Mount Pinatubo eruption also led to more than USD 100 million in costs to aviation at the time (Casadevall et al., 1995). Damage to the road network caused by the 2010 Haiti earthquake-triggered landslides and rockfalls made rescue operations very difficult (DesRoches et al., 2011), potentially increasing the number of victims. Damage to the New York City transportation system caused by the flooding associated with Hurricane Sandy (2012) amounted to USD 7.5 billion (e.g., Faturechi and Miller-Hooks, 2015). For additional examples, see Argyroudis et al. (2019).

1.3.7.6 Water System Failures (Wf)

Water systems comprise water supply and wastewater handling networks, both critical for the normal functioning of a society. Water access is vital to humans and needed for food production, for power generation, and in every industrial process. Water shortages, linked, for instance, to droughts (Section 1.3.4.3), can have many devastating consequences from dramatically reduced crop yields to water wars (Starr, 1991). Considering the normal case where water is available, it is the failure of the water supply system which then leads to direct water flow interruption or to water contamination. Impacts include lack of drinking water, interruption of medical services, and business interruption. Water systems are prone to interruption during a disaster (Balaei et al., 2020), which can amplify some perils (e.g, fire, blackout) and/or lead to other ones (e.g., epidemic). Loss of water services is common after earthquakes as they can cause widespread pipe breaks. Loss of electrical power can further reduce the water system performance (Adachi and Ellingwood, 2008). Most of the losses following the 1906 San Francisco earthquake were due to fires, which could burn for days because of the lack of water to control them (Scawthorn et al., 2019). Contaminated water caused by water mains damaged during the 2010 Haiti earthquake triggered a cholera outbreak months later that killed approximately 8,500 people (e.g., Balaei et al., 2020). During a flood, wastewater treatment plants can overload and spill raw sewage into fresh water, which for example happened when Hurricane Rita struck Louisiana in 2005 (Matthews, 2016).

1.3.7.7 Cyberattacks (CA)

Cyberattacks can include theft of data or currency, ransoms, business interruption, or some other forms of system destabilization. Other cyber risks exist where the cause is unintentional (e.g., error, failed internal process; Cebula et al., 2014). The twentieth century for example ended with the fear of the Y2K Millennium bug and the risk of a global collapse of the computer system, which eventually did not materialize (Swiss Re, 2013, p. 51). Many of the perils previously mentioned can also trigger some failure of the information and communication systems. Attacks occur, by definition, via electronic communication networks and the virtual world with consequences on information systems and the flow of information. One particularity of cyberattacks is that they are not geographically bound and can cascade into greater events (Coburn et al., 2019). Cyber risks have only recently been rationalized in the context of human-made catastrophes, possibly because of their multi-type nature (Wheatley et al., 2020). Different processes are indeed involved depending on the type of attack. For example, a cyber-heist on a banking system is different from a distributed denial of service (DDoS) attack, itself different from a cyberattack on a power grid or other connected critical infrastructure. Interestingly, propagation of malware over information networks may be treated as an epidemic (Liu and Zhong, 2017). These events can also be weaponized in the context of cyber warfare. Attempts are being made to conceptualize cyber risk (e.g., data breaches) from criminality theories (Sen and Borle, 2015). The global cost of data breaches is estimated to be in the hundreds of billions of US dollars per year (e.g., Eling and Schnell, 2016) and are on the rise (Wheatley et al., 2020). In 1999, the mass-mailing computer virus *Melissa* made headlines for causing hundreds of millions of dollars in economic losses (Turrini and Ghosh, 2010). In 2007, DDoS attacks were orchestrated against Estonian organizations, showing the potential threats posed by cyberattacks (Herzog, 2011). The *Stuxnet* cyber worm, discovered in 2010 and considered a military-grade cyber missile, sabotaged an Iranian nuclear facility by targeting the frequency-converter drives of uranium enrichment centrifuges. It apparently infected over 60,000 computers in its spreading phase (Farwell and Rohozinski, 2011). The 2013 Yahoo! data breach was an extreme data breach event with over a billion records being lost (Trautman and Ormerod, 2017).

1.3.8 Socio-Economic Perils

Socio-economic perils are defined as all the dangerous events potentially causing human or economic losses from social and/or economic processes, which are all driven by human behaviour. **Economic disruptions** include business interruptions (BI) in different economic sectors, financial crises (FC), and banking system failures (Bf). These events can have the characteristics of catastrophes: they can be sudden and costly events. **Social degradation** can be defined as moving downward along **Maslow's hierarchy of needs** (Maslow, 1943), and is characterized by unemployment, lack of healthcare, lack of personal security, lack of shelter, and food insecurity (Fig. 1.21). This in turn increases mortality and tendency toward economic collapse. Combined with economic turmoil, social degradation can be amplified by public service failure (Sf), such as a drastic reduction of healthcare or emergency services. In such a negative environment prone to dissatisfaction, social unrest (SU, e.g., riots) can ensue. Terrorist attacks (TA) and armed conflicts (AC) represent the ultimate states of social and/or political instability, with famine (Fa) one of the worst possible outcomes. Since political aspects play a role in many of the aforementioned perils, socio-political perils are implicitly included in this section. It is to be noted that the banking system and public services are critical infrastructures whereby failure does not directly cause physical damage per se (in contrast to the critical infrastructures described in Section 1.3.7), hence the present classification. Most of the perils described in the previous sections, as stressors of the socio-economic system, can increase

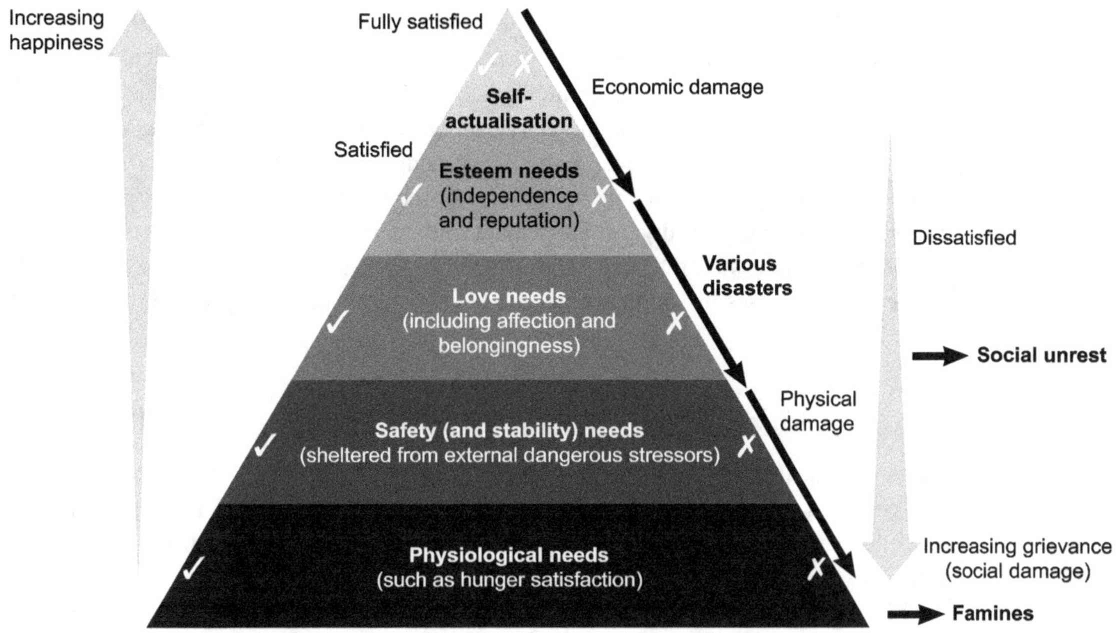

Fig. 1.21 Maslow's hierarchy of needs in the context of catastrophe risks, with pyramid text from Maslow (1943). **Economic disasters** (widespread BI, financial crises, banking system failures) lead to job suppressions and loss of independence/reputation. Loss of loved ones, belongingness, and shelter (no safety from further harm, public service failures) due to a catastrophe can destabilize the cohesion of a society. Social unrest can unfold due to increased grievance. The worst-case scenario is the occurrence of a famine when basic physiological needs cannot be fulfilled on a large scale.

the likelihood of socio-economic perils (Mignan and Wang, 2020). This will be illustrated by several examples in the next paragraphs.

1.3.8.1 Business Interruptions (BI)

Business interruption refers to the loss of revenue from the reduction of the economic output or profit of a firm. It can be due to the destruction of the capital stock via property damage, or to the closing of the business due to lack of energy and/or labour. **Contingent business interruption** corresponds to off-site disruptions linked to the firm, such as **supply-chain failure** (i.e., lack of input material). Business interruption is a common consequence of disasters and can produce economic losses of the same order as direct property losses (Rose and Huyck, 2016). Great catastrophes have a tendency to amplify losses associated with business interruption due to the interconnections between the different economic sectors, such as in the case of Hurricane Katrina (Hallegatte, 2008). Most of the historical cases of critical infrastructure failures previously described caused some level of business interruption. Infamous cases of contingent business interruption with a global impact include the 2010 Tohoku earthquake, the 2011 Thailand floods, and the

COVID-19 pandemic (Abe and Ye, 2013; Guan et al., 2020). Overall, the costliest disasters have macroeconomic consequences by leading to pronounced slowdowns in production (Noy, 2009). Widespread business interruption can cause losses of the order of a few per cents of a country's GDP when involved in a global supply-chain failure. Taking the example of Thailand, the 2011 floods pushed thousands of factories to suspend production, subsequently disrupting supply chains with major repercussions on electronic and car manufacturing companies worldwide. The Japanese Sony corporation alone predicted a USD 1.2 billion annual loss for the year 2011–2012 from that event (e.g., Courbage et al., 2012), while economic losses in Thailand amounted to almost 9% of the country's GDP (SwissRe sigma, 2012).

1.3.8.2 Financial Crises (FC) and Banking System Failures (Bf)

Economic catastrophes include financial crises and banking crises. A **financial crisis** is when financial assets decrease significantly in nominal value, its most extreme manifestation being a market crash (i.e., $\geq$ 10% devaluation within one quarter). A **banking system failure** is when many banks become insolvent and go bankrupt. Both types of events, of possible endogenous or exogenous origins, are intertwined and can lead to extreme economic losses (i.e., severe recessions; Boissay et al., 2016; Mahalingam et al., 2018) and social hardship. Several types of risks are involved: **market risk** (linked to pricing, asset value), **credit risk** (linked to non-payment of loan), and **operational risk** (linked to failed internal process or external event), among others (e.g., Basel Committee on Banking Supervision, 2021). **Derivative risk**, in particular, may threaten financial market stability when the growth of derivative contracts (structured debts, options, swaps, etc.) becomes excessive (e.g., Reichert and Shyu, 2003). An example of an endogenous process, a contagion of sell orders, is the financial crash of October 1987 on Black Monday, which led to the quasi-instantaneous evaporation of more than a trillion dollars worldwide (Sornette et al., 2004). The worst financial crises, the 1929 Wall Street Crash and the 2008 Subprime Mortgage Crisis, removed more than USD 10 trillion in anticipated GDP worldwide (Mahalingam et al., 2018). In the banking system, a contagion process can occur by a cascading of liquidity shocks or of debt defaults through the inter-bank market (Cetorelli and Goldberg, 2011; Walker et al., 2022). Exogenous events (natural and human-made catastrophes) are at the origin of less than 10% of financial crises. Hurricane Katrina, despite its USD 190 billion loss, moved the New York Stock Exchange by less than a single percentage point. The 2001 9/11 World Trade Center terrorist attack made the Standard and Poor's 500 (S&P 500) lose 11.6% in four trading days but most market indices quickly recovered (Mahalingam et al., 2018). The arrival of COVID-19 provoked a collapse in equity valuations in March 2020 and only strong fiscal and monetary interventions led to the recovery of the financial market (Bolton et al., 2021). Commodity markets can also react, up or down, due to speculation following a disaster (e.g., Sheu and Kuo, 2020). A liquidity shock in the banking system can happen following a disaster, the damage reducing the financial health of a bank and in turn limiting its lending capacity (e.g., 1995 Kobe earthquake; Hosono et al., 2016). When its footprint is wide, a disaster can cause a large number of non-performing bank loans and loan defaults, affecting bank solvency and increasing the likelihood of bank default (Klomp, 2014). However, empirical evidence of the impact of natural disasters on the banking sector remains scarce (Brei et al., 2019).

1.3.8.3 Public Service Failures (Sf)

Public services involved in the safety and health of people include emergency services (fire brigades, police, rescue teams) and the healthcare system. All can fail in the case of a disaster, which can potentially increase the event's severity and contribute to social unrest. This generally occurs in extreme circumstances, due to disruptions to the communication and decision-making capabilities of the complex inter-organizational network that the public sector's entities constitute (Kapucu and Van Wart, 2006). System overload is another possible cause of public service failure. Particularly during an epidemic, overload of the healthcare system can reduce the number of lives being saved (e.g., COVID-19 pandemic; Phua et al., 2020). The 2014 Ebola outbreak, the most widespread of this disease in history, was due to under-resourced healthcare capacity, which pushed the system to collapse (Philips and Markham, 2014). The flooding of New Orleans, caused by Hurricane Katrina (2005), paralysed the city and overwhelmed the public services. People became angry at the inadequate preparations and general disorganization, leading to some social unrest (Congleton, 2006).

1.3.8.4 Social Unrest (SU) and Stampedes (St)

Social unrest, also known as civil disorder, is a situation where collective acts of disobedience emerge in a society. **Riots**, in particular, are aggregate acts of violence against individuals and property, usually characterized by looting and arson (McPhail and Wohlstein, 1983). Individuals become violent and take the risk of being jailed when their level of grievance against the system is high (Epstein, 2002). When crowd violence cannot be contained by outnumbered law enforcement officers, 'crime pays' (or seems to in the eye of the offender), with theft and other violent acts rising (Becker, 1968; Epstein, 2002). This is distinguishable from background criminality, as these acts are sudden and concentrated. The Rodney King riots, which took place in Los Angeles in 1992, are the worst civil disturbances in modern US history; they resulted in 53 deaths, 2,300 injuries, more than 1,000 buildings destroyed by fire, thousands of jobs lost, and an estimated cost of USD 1 billion in damage (Matheson and Baade, 2004).

Impulsive movements of crowds, accidental in nature, can cause deadly **stampedes** (St) (Ngai et al., 2009). They occur at large crowd gatherings, which include riots or other events that may trigger panics (e.g., terrorist attacks, disaster-area evacuations). They are not social unrest events per se, although the term 'unrest' may also mean a state of agitation or disorganization. The worst stampedes have led to thousands of fatalities. Between 1990 and 2022, the deadliest stampedes occurred during religious mass gatherings, with 1,426 deaths at the Hajj in Mecca, Saudi Arabia, in 1990. The most recent in that period, the 2022 South Korea Halloween stampede, the worst stampede in the history of the country, caused 156 deaths and 170 crush injuries (Sharma et al., 2023).

1.3.8.5 Terrorist Attacks (TA)

Terrorism is defined as 'the threatened or actual use of illegal force and violence to attain a political, economic, religious or social goal through fear, coercion or intimidation' (e.g., LaFree and Dugan, 2007). Terrorist attacks are therefore intentional catastrophes (Ericson and Doyle, 2004). Attacks can be of multiple natures including the use of fighters (with firearms or non-firearms), conventional weapons, or non-conventional weapons (e.g., chemical, biological). They can occur anywhere, with private citizens and properties the most common targets. Mass fatality events (> 100 fatalities) are most commonly perpetrated with bombs and guns, and approximately 150 such events occurred between 1970 and 2019. Two-thirds of these terrorist attacks took place in the Middle East and Africa (North and sub-Saharan Africa) (Tin et al., 2021). The 9/11 attack on the World Trade Center (2001) was the costliest event in the history of insurance at the time, with USD 80 billion in losses, and resulted in the death of nearly 3,000 people (Kunreuther et al., 2005). It was also the first insured loss from terrorist activity to exceed USD 1 billion, dwarfing all previous events. Some of the most significant insured losses in the late twentieth century were due to three IRA bombings in 1992–1996 in London's financial district and Manchester (totalling USD 2.3 billion) and to the 1993 bombing of the World Trade Center (USD 0.7 billion; e.g., Ericson and Doyle, 2004). The 9/11 attack remains the deadliest in recent history, while the 2017 truck bombing in Mogadishu remains the deadliest in Africa, with almost 600 fatalities (Tin et al., 2021). The November 2015 coordinated attacks in Paris (including the Bataclan concert hall attack) were the most severe in Western Europe since the mid-2000s with 130 people killed (e.g., Nussio et al., 2019). While finalizing this textbook, on 7 October 2023 Hamas killed 1,400 people in Israel (Elyoseph et al., 2023). It triggered continued Israeli bombing in return, killing many thousands more in Gaza (Horton, 2023). Use of chemical and biological weapons remains extremely rare (Bogen and Jones, 2006). The 1995 Aum Shinrikyo terrorist attack in Tokyo employed sarin, a chemical weapon, while the 2001 anthrax letters affair is a case of bioterrorism (Szinicz, 2005). Terrorist attacks also cause indirect losses, such as business interruption. Post-9/11, fear led to a reduction in air travel and related tourism, amounting to USD 50 billion of direct BI and USD 60 billion of indirect BI two years after the event (Rose and Blomberg, 2010).

Mass school shootings may be considered a type of domestic terrorism. The majority of events have occurred in the US, with their number increasing every decade since the 1980s. Fifty-five deaths occurred from the 1940s to the 1990s. Sixty-five more deaths occurred between 2000 and 2018 (Katsiyannis et al., 2018). Almost 40 individuals were killed between 2019 and 2023, including 21 people (19 children) during the 2022 Uvalde tragedy in Texas (Jowarder, 2022). The 1999 Columbine High School shooting in Colorado, which left 12 students, a teacher, and the two shooters dead, garnered significant mass media attention, with 'Columbine' having become synonymous with school violence (King and Bracy, 2019). At the time, the Columbine school shooting was the deadliest in US history. It was surpassed in 2012 with the Sandy Hook Elementary School, Newton, Connecticut, shooting which led to 26 deaths (Katsiyannis et al., 2018). The deadliest shootings at college campuses have been the 1966 University of Texas tower shooting, with

17 deaths, and the 2007 Virginia Tech shooting, with 33 deaths (including perpetrators) (Stearns, 2008).

1.3.8.6 Armed Conflicts (AC)

Armed conflicts are sustained and severe fights between armed troops. A minimum threshold of 1,000 battle-deaths per year was used in the Correlates of War (COW) project (e.g., Sarkees and Schafer, 2000) to declare a war, and a threshold of 25 battle-deaths per year was used in the Uppsala data set to define an armed conflict (Gleditsch et al., 2002). A **war** can therefore be understood as an intense armed conflict. International armed conflicts refer to wars between two or more states while non-international armed conflicts refer to civil wars. Armed conflicts have complex origins involving many actors in a wide range of conditions. Their dynamics are also complex, depending on different military strategies (e.g., manoeuvre/attrition/punishment typology; Reiter, 1999) and on different uses of weapons (firearms, bombs, etc., conventional/non-conventional). Sabotage of critical infrastructures is also a relatively common practice (including, recently, cyberattacks; Section 1.3.7.7). By definition, armed conflicts are destructive. For example, the Kuwait oil spill orchestrated by Iraqi forces during the 1991 Gulf War led to an environmental catastrophe and to the largest oil spill in history (Sauer et al., 1993). The Uppsala data set lists a total of 259 armed conflicts for the period 1946–2014, counting the Ukraine conflict which erupted in 2014 (Pettersson and Wallensteen, 2015). World War II (WWII, 1939–1945) was the greatest disaster in history since the Black Death of the mid-fourteenth century in terms of absolute number of fatalities, with at least 40 million people dead worldwide, including civilian battle deaths, the use of atomic weapons against Japan, ethnic genocide, and starvation (e.g., Lacina et al., 2006). The Ukraine conflict started in 2014, but it was in 2022, with the invasion by the Russian military, that it took on greater proportions, making it the most prominent conflict in Europe since WWII. This ongoing event exemplifies the many consequences of war, beyond direct casualties, including critical infrastructure failures, toxic spills, supply-chain failures, and socio-economic degradation (Rawtani et al., 2022).

1.3.8.7 Famines (Fa)

Access to food is a fundamental condition for the survival of any living organism and, as such, is the most basic need in Maslow's hierarchy of needs (Fig. 1.21). Continued undernutrition leads to all fat reserves being exhausted, which further leads to **starvation** and, ultimately, death when proteins remain the only energy source for the body to function. Young children and the elderly are the most prone to die in this context. A **famine** is defined as the relatively sudden collapse in the level of food consumption of large numbers of people, causing high mortality (Scrimshaw, 1987). Famines were frequent in ancient history and an important factor for the fall of the Roman empire. They were also common in Europe during the Middle Ages. For example, two-thirds of the Italian population died in 1347 while one-third of the inhabitants of Paris died in 1438. Citing only a few well-documented famines that occurred in recent centuries, the 1846–1847 Great Famine of Ireland (potato blight) killed 1.5 million people, while four famines in China in the first half of the nineteenth century claimed 45 million lives (e.g., Scrimshaw, 1987, and references therein). Millions also died in the first half of the twentieth century in the Soviet Union and Asia (Devereux, 2009). Causes are various. For example, the Great Bengal Famine of 1943–1944 was due to inflationary prices and the Great Ethiopian Famine of 1984–1985 was due to droughts. Famines still persist in Africa in the early decades of the twenty-first century (Devereux, 2009). Lack of food due to floods, droughts, or pest infestations is not necessarily a sufficient cause for a famine. Socio-economic and socio-political perils, such as armed conflicts and failures in flows of goods (accidental or planned), play an important role. The overall impact is conditional on the population's level of poverty. Starvation pushes people to riot or even revolt against a system if still physically capable of such acts of rebellion (Scrimshaw, 1987). Examples include the French Revolution of 1789 and the Arab Spring in 2010–2012, both caused by soaring food prices (Soffiantini, 2020). Epidemics are often associated with famines due to many people becoming immunosuppressed in the same region at the same time. This explains why the Four Horsemen of the Apocalypse are war, famine, pestilence (i.e., epidemics), and death (Hardy, 1988).

1.4 Concluding Remarks

This opening chapter provided the basic concepts of catastrophe risk, which included the language of this scientific domain and its history, as well as a *tour d'horizon* of past catastrophes (Ex. 1.5). Encyclopaedic in nature, the last section described 42 perils in total (Table 1.7). This exploration consisted of a collation of historical facts combined with qualitative descriptions of the vast variety of perils that exist. Grim statistics of death counts were given to prove the gravity of the problem that the world is facing. The next chapters will provide numbers easier to deal with, such as event magnitudes, hazard intensities, and

Table 1.7 All 42 perils ordered alphabetically by their identifiers

ID[a]	Peril	Category	Notes
AC	Armed conflict	Socio-economic	War
AI	Asteroid impact	Extraterrestrial	Including airbursts, comet impacts
Bf	Banking system failure	Socio-economic	
BI	Business interruption	Socio-economic	Including contingent BI
BO	Blackout	Technological	Power outage, a type of Ef
CA	Cyberattack	Technological	A type of If (malicious origin)
CS	Convective storm	Meteorological	A type of WS, for example, with RS, HS, To
CW	Cold wave	Climatological	Including frost, ice
DF	Dam flood	Technological	Including breach, overtopping; including levee flood
Dr	Drought	Climatological	
Ef	Energy supply failure	Technological	Generic, for example, at production sites

Table 1.7 *(cont.)*

ID[a]	Peril	Category	Notes
EQ	Earthquake	Geophysical	
Ep	Epidemic	Biological	Including outbreak, pandemic
Ex	Explosion	Technological	
FC	Financial crisis	Socio-economic	Including market crash, recession
Fa	Famine	Socio-economic	
FF	Fluvial flood	Hydrological	River flood
Fi	Fire	Technological	
GS	Geomagnetic storm	Extraterrestrial	
HS	Hailstorm	Meteorological	
HW	Heatwave	Climatological	
If	Information system failure	Technological	Computer and communication systems, for example, internet
Li	Lightning	Meteorological	Thunderstorm
LS	Landslide	Geophysical	Including debris flow, lahar, avalanche
PI	Pest infestation	Biological	Including insect outbreak
PF	Pluvial flood	Hydrological	
RS	Rainstorm	Meteorological	
Sf	Public service failure	Socio-economic	Including emergency services, healthcare
SS	Storm surge	Hydrological	Coastal flood (meteorological origin)
St	Stampede	Socio-economic	
SU	Social unrest	Socio-economic	Riot
TA	Terrorist attack	Socio-economic	
TC	Tropical cyclone	Meteorological	A type of WS, for example, with RS, CS
Tf	Transport system failure	Technological	Including supply-chain failure
To	Tornado	Meteorological	
Ts	Tsunami	Hydrological	Coastal flood (non-meteorological origin)
TS	Toxic spill	Technological	Pollution, including chemical, radioactive
VE	Volcanic eruption	Geophysical	
Wf	Water system failure	Technological	Including water supply, wastewater system
WF	Wildfire	Climatological	Forest fire, bushfire
WS	Windstorm	Meteorological	Generic
xC	Extratropical cyclone	Meteorological	A type of WS, for example, with RS, CW

[a] The present classification supersedes the ones previously proposed by the author (Mignan et al., 2016; Mignan and Wang, 2020; Mignan, 2022a). It also differs from the Catastrophe Risk Evaluation and Standardizing Target Accumulations (CRESTA) peril identification, which uses a higher aggregation level: EM for earth movement (landslide, sink hole, asteroid, etc.), EW for extreme weather (drought, heatwave, freeze, etc.), FL for all floods, while tsunamis are part of EQ, etc. (CRESTA, 2023).

damage ratios. The topic is vast, as proven in the previous pages, and could potentially overwhelm readers. However, all aspects will be investigated in a clear and comprehensive manner in the next chapters, with perils described in the hazard domain in Chapter 2 and in the risk domain in Chapter 3. All the models will be kept relatively simple in Part I, requiring only basic knowledge of algebra and statistics. More sophisticated models, demanding some familiarity with calculus and graph theory, will be covered in Part II.

Further recommended reading

- *The Mathematics of Natural Catastrophes* (Woo, 1999)
- *Global Catastrophes and Trends* (Smil, 2008)
- *Natural Catastrophe Risk Management and Modelling* (Mitchell-Wallace et al., 2017, chapter 3)

1.5 Exercises

Exercise 1.1

A magician is planning to play Russian roulette at his next show.

(a) Explain to him that it is a priori a bad idea by defining the concepts of risk, of aleatory uncertainty, and of epistemic uncertainty for the specific case of a six-shot revolver.
(b) List three notable differences between this case and the risk associated with catastrophes in general.

Exercise 1.2

Answer the following question. Do we necessarily reduce risk by:

(a) reducing aleatory uncertainty,
(b) reducing epistemic uncertainty, or
(c) none of the above? Justify your choice.

Exercise 1.3

Figure 1.22 depicts the exposure and vulnerability footprints of an ad hoc region, as well as the hazard intensity footprint of an ad hoc event. Values can be 0 (none), 1 (low), 2 (medium), or 3 (high).

(a) Define loss in a simple way and draw the loss footprint for the present case (you can do it manually or with the code template provided in the file `exercise_notebook.ipynb`).
(b) Justify your proposed definition of loss by explaining where the lowest and highest losses are expected to occur.
(c) What information is missing to calculate risk?

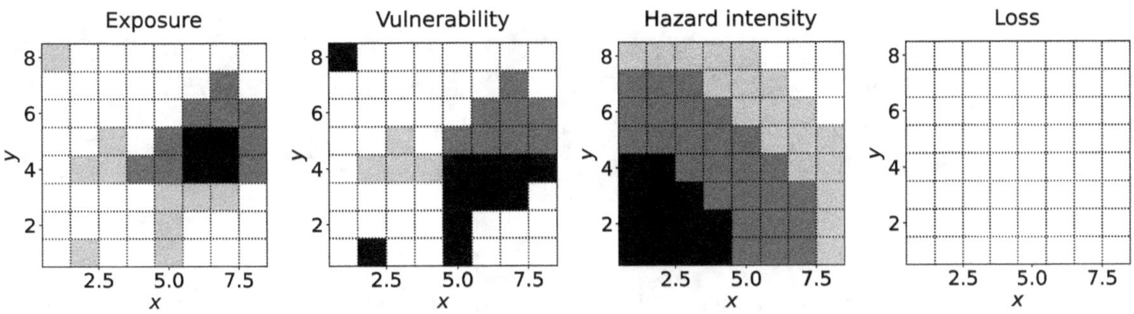

Fig. 1.22 Exercise 1.3: Ad hoc example illustrating the catastrophe risk process (loss map to be filled).

Exercise 1.4

The United Nations Office for Disaster Risk Reduction (UNDRR) and the International Science Council, in the context of the Sendai Framework for Disaster Risk Reduction 2015–2030, defined an *environmental hazard* class consisting of perils causing the 'degradation of the natural systems and ecosystem services upon which humanity depends [...], here defined as loss of utility' (UNDRR and ISC, 2020). Should this category, presently missing in Section 1.3, be added to the textbook taxonomy? Write a one-page essay arguing on the pros or cons. The official report, titled *Hazard Definition and Classification Review (Technical Report)* can be found at https://www.undrr.org/ to help develop the argument.

Exercise 1.5

Find one example of peril not mentioned in Chapter 1 (yet known to exist or to be physically plausible). In what category does it fall (according to Table 1.7)?

2 Probabilistic Hazard Assessment

This chapter will provide all the tools necessary to implement virtually any type of peril in a CAT model's hazard module. These tools comprise the creation of a set of simulated events per considered peril and of a matching catalogue of hazard intensity footprints for each one of these events. The combination of both products will allow us to generate hazard curves and hazard maps, the main metrics employed in probabilistic hazard assessment. Different perils represent varied physical processes which necessitate some domain expertise. The richness of the peril phenomenology was demonstrated in the previous chapter while describing geophysical, hydrological, meteorological, climatological, biological, extraterrestrial, technological, and socio-economic perils. Despite the general purpose of the standard CAT modelling framework, peril-specific CAT models are commonly developed in silos by dedicated experts, for example by geophysicists for an earthquake model or by meteorologists or climate scientists for a storm model. In view of the dozens of perils to be quantified in this textbook (42 in total, listed in Table 1.7), a more generalist approach must be employed throughout this textbook. The concept of catastrophe object will be introduced and a computational ontology will be developed, following the work of Mignan (2022a), to formally model 'the relevant entities and relations that emerge from [that object], and which are useful to our purpose' (Guarino et al., 2009). All the concepts necessary for the task will be introduced in Section 2.1. The subsequent sections will follow the structure of the proposed ontology, going from event source (Section 2.2) to event size distribution (Section 2.3) to, finally, event intensity footprint (Section 2.4). Lastly, Section 2.5 will expose the nuts and bolts of the hazard module with an application to probabilistic seismic hazard assessment.

The guiding principle of this chapter, indeed, of the entire book, is in alignment with a message from Nobel Prize laureate Richard Feynman: 'What I cannot create, I do not understand.' A second note found on his blackboard at the time of his passing also resonates with our mission: 'Know how to solve every problem that has been solved.' Together, these two statements epitomize the core objectives of this chapter in the context of catastrophe modelling.

2.1 Basic Concepts

The hazard module characterizes the sources of events expected in a specified spatiotemporal window for a given peril, as well as the frequency and intensity of these events. This is achieved through the creation of three key components: a stochastic event set (Section

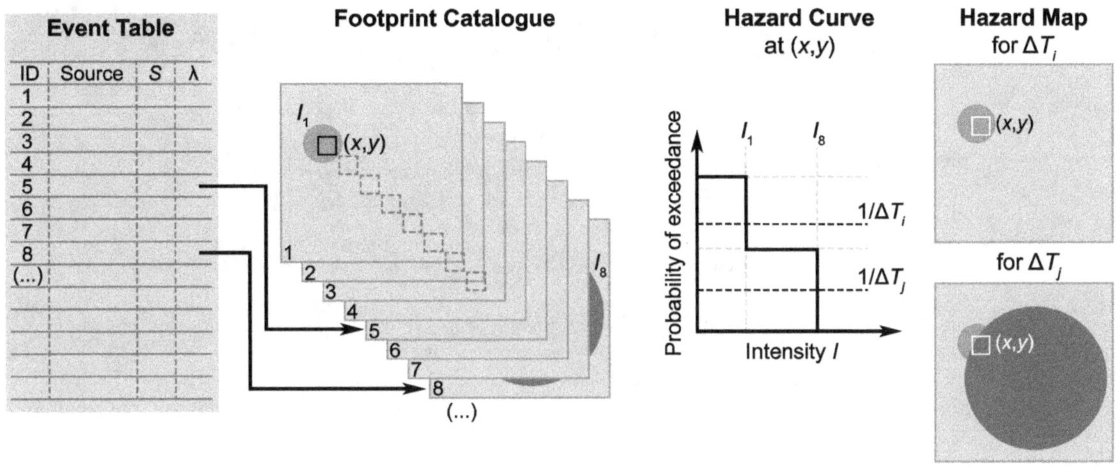

Fig. 2.1 Illustrating the main steps taken in the hazard module. For each simulated event listed in an event table, there is a matching intensity footprint. For illustration purposes, only two circular binary footprints (i.e., intensity I constant within the footprint, 0 outside) are here considered: a small and frequent event (ID $= 1$) and a large and rare event (ID $= 8$), with intensities $I_1 \ll I_8$ and rates of occurrence $\lambda_1 \gg \lambda_8$. A two-step hazard curve can be generated at the mentioned location (x, y) that shows the probability of exceeding a range of intensities I. On the final hazard map, the hazard value at (x, y) represents the intensity at which the probability $1/\Delta T$ is crossed on the matching hazard curve, with ΔT the selected return period. The hazard module will be detailed step by step in Section 2.1.

2.1.1), a matching catalogue of individual event intensity footprints (Section 2.1.2), and a suite of hazard metrics (hazard curves and hazard maps; Section 2.1.3). While specific models tailored to each peril are essential, the same overarching framework can be applied to most perils, as illustrated in Figure 2.1.

2.1.1 Stochastic Event Set

A **stochastic event set** (or probabilistic event set) is an ensemble of synthetic events, which are representative of the hazard within some specified region and return period. The term **stochastic** refers to a process which has a random component (for added clarity, often characterized by a specific probability distribution function; e.g., Eq. (2.1)). *Although a stochastic event set may include historical events, the majority of the events are stochastic projections that describe the statistical distribution of the hazard.* There are two approaches to generating such a set, leading to an event table or a simulation table.

An **event table** is a table of N unique events, which are described by their size (S_{ID}), source location, and annual rate of occurrence (λ_{ID}). A unique identifier (ID) is attributed to each event, which represents an idealized version of a potential event. Typically, at a given source (e.g., a fault for an earthquake), event sizes are defined incrementally on the range $[S_{min}, S_{max}]$. The minimum size S_{min} is defined by the user while the maximum size S_{max} is physically constrained (Section 2.2). For each specified size, the rate of occurrence is estimated from a size distribution model (Section 2.3). An ad hoc event table is represented

Table 2.1 An ad hoc event table

Event ID[a]	Source	Size S	Rate λ
EQ0001	Fault 1	Magnitude $M_{EQ0001} = M_{min}$	λ_{EQ0001}
EQ0002	Fault 1	Magnitude $M_{EQ0002} = M_{min} + 0.1$	λ_{EQ0002}
EQ0003	Fault 1	Magnitude $M_{EQ0003} = M_{min} + 0.2$	λ_{EQ0003}
...	...	...	...
EQ0038	Fault 1	Magnitude $M_{EQ0038} = M_{min} + 3.7 = M_{max}$(Fault 1)	λ_{EQ0038}
...	...	...	...
EQ2024	Fault N	Magnitude $M_{EQ2024} = M_{max}$(Fault N)	λ_{EQ2024}

[a] See Table 2.7 for an example of a more detailed earthquake event table.

by Table 2.1 for illustration purposes. The method to generate, for instance, an earthquake event table will be described in Section 2.5. When including the event loss, the table is referred to as the event loss table (ELT) (Chapter 3).

A **simulation table**, in contrast, takes the form of an event catalogue, which lists individual event occurrences for a total of N_{sim} simulations, with simulations commonly lasting one year (i.e., year-simulations). It can then also be called **Year Event Table** (YET; Varghese and Rau-Chaplin, 2013). For each simulated occurrence, both simulation ID and event ID are given. An ad hoc simulation table is represented in Table 2.2. When including the event loss, it is referred to as the year loss table (YLT) (Chapter 3). The number N_{sim} must be large enough to ensure that the rarest events are sampled in sufficient numbers. The rarest events to be considered in a CAT model depend on its purpose. Event **return periods** ($\Delta T_{\text{ID}} = 1/\lambda_{\text{ID}}$) of 1,000 years may be considered for common buildings, 10,000 years for important buildings (e.g., hospital, bridges), 100,000 years for critical buildings (e.g., large bridges, dams), and one million years for highly critical infrastructures (e.g., large dams, nuclear power plants) (e.g., Mignan et al., 2018). The method to generate a simulation table (YLT case) from an event table (ELT case) will be described in Chapter 3. However, it is important to mention that the number of event occurrences $X = k$ in a year-simulation (Table 2.2) is, in general, drawn from the **Poisson distribution**. The distribution, which assumes event independence, takes the form:

$$\mathbb{P}(X = k) = \frac{\lambda_{tot}^k \exp(-\lambda_{tot})}{k!}, \tag{2.1}$$

with X a random variable taking the value k (Grimmett and Stirzaker, 2001, pp. 46–48) and λ_{tot} the total event frequency. An advantage of the simulation table over the event table is its flexibility to encode aleatory uncertainty (Section 2.5) and other processes, such as event clustering (with the use of the negative binomial distribution instead of the Poisson distribution), time dependency, and cascading effects (Part II, Chapter 5).

Synthetic tropical cyclone catalogues were developed early on (Russell, 1971), while use of synthetic earthquake catalogues remained uncommon prior to the 2000s (Ebel and Kafka, 1999). In addition to their occurrence, stochastic events are also characterized by

Table 2.2 An ad hoc simulation table

Simulation ID	Number of events per year k^a	Event ID
42	1	EQ0039
72	1	EQ0002
...	...	...
4420	1	EQ0123
4501	2	EQ0001, EQ1023
...	...	...
999956	1	EQ0413

[a] Typically, for k events in a simulated year, k rows with the same simulation ID will be generated so that only one event is defined per row. Simulations with zero occurrences are omitted.

the random sampling of specific attributes from historical events, such as location and size. Examples will be given for tornadoes in Section 2.2.2 and for earthquakes in Section 2.5. At the end of this chapter, readers will have the skills to create stochastic events via guided exercises.

2.1.2 Hazard Intensity Footprint Catalogue

Once a stochastic event set has been created, the next step is assessing the impact of each event on the environment. This is quantified by the **intensity** $I(x, y)$ of the event at a given location (x, y), which is the physical quantity that potentially leads to damage. For an earthquake, this quantity can be a measure of ground-shaking, such as peak ground acceleration; for floods, it is commonly the water height; for explosions, it is the overpressure. By definition, the intensity I must be the input variable of a damage function (Chapter 3). The **intensity footprint** represents the area impacted by the event with $I(x, y) > 0^+$. Since one intensity footprint $I_{\mathrm{ID}}(x, y)$ is defined per event (assuming one model and that aleatory uncertainties will be added at a later stage), a catalogue of N footprints must be developed for an event table of length N (Fig. 2.1).

Most of the (mean) intensity footprints modelled in this chapter will be described by analytical expressions, representative of simple diffusion processes, and taking the general form:

$$I(x, y) = f_I\left(S, r(x, y)\right), \tag{2.2}$$

where S is the event size and r the distance from the event source at the (x, y) location (Section 2.4). The generic function f_I, as well as future functions f_S and f_D, are not to be confused with the probability density function, denoted simply f, which will be introduced soon. Further hazard processes will later be considered where footprints are defined from a threshold value being reached in a given environmental spatial layer (Section 2.4).

Advanced computational (numerical) methods may be needed to realistically model the intensity footprints of some other perils (Part II, Chapter 4).

2.1.3 Hazard Metrics

Probabilistic hazard assessment (PHA) aims to estimate the hazard intensity of a given peril for a range of likelihoods, from common to rare events. **Deterministic hazard assessment** is concerned with specific event scenarios. PHA consists of computing all the possible intensity footprints and their related frequencies of occurrence, which are then combined in hazard curves by location. Another common approach is the aggregation of all events in hazard maps for a specified return period ΔT. Such a map is obtained by reducing the information contained in a hazard curve to one point per location (Fig. 2.1).

2.1.3.1 Hazard Curve

A **hazard curve** represents the probability that a hazard intensity threshold will be exceeded over a specific period of time at a specific location, taking into account the contributions from all possible events. There are two main approaches to compute hazard curves, based on different implementations of the aleatory uncertainty that is linked to the hazard intensity I (i.e., hazard secondary uncertainty).

In the first, more traditional, approach, the hazard curve is directly computed from the combination of the event table and footprint catalogue. The probability of exceedance of the intensity value y in δt years at a given location is:

$$\mathbb{P}(Y > y|\delta t) = 1 - \exp\left[-\sum_{i=1}^{N} \lambda_i \delta t \int_y^{\infty} f(I_i) dI_i\right], \tag{2.3}$$

by assuming that the intensity and the number of events are independent, with Y the intensity random variable, λ_i the annual rate of event i, N the length of the event table, and δt often taken to be one year (and therefore omittable from the equation). f is the **probability density function (PDF)** of the mean event intensity I_i, which provides a way to describe the likelihood of deviating from I_i in different instances (i.e., the aleatory uncertainty). The concept of PDF will be formally introduced later in this chapter and an example of application of Eq. (2.3) will be given in Section 2.5. In the simplified process represented in Figure 2.1, no aleatory uncertainty is included, leading to a step function. The reader shall appreciate that the functional form of Eq. (2.3) infers event independence, as it derives from the Poisson distribution (Eq. (2.1)), with

$$\mathbb{P}(X > 0|\delta t) = 1 - \mathbb{P}(X = 0|\delta t) = 1 - \exp(-\lambda_{tot}\delta t) \tag{2.4}$$

the probability that at least one event with rate $\lambda_{tot} = \sum_{i=1}^{N} \lambda_i$ occurs in δt. Note finally that Eq. (2.3) is a **complementary cumulative distribution function** (CCDF), defined as $\bar{F}(x) = \mathbb{P}(X > x) = 1 - \mathbb{P}(X \leq x) = 1 - F(x)$ (or, CCDF = 1− cumulative distribution function (CDF); see Box 2.1 in Section 2.2.2 for the relationship between PDF and CDF, which can be ignored for the time being).

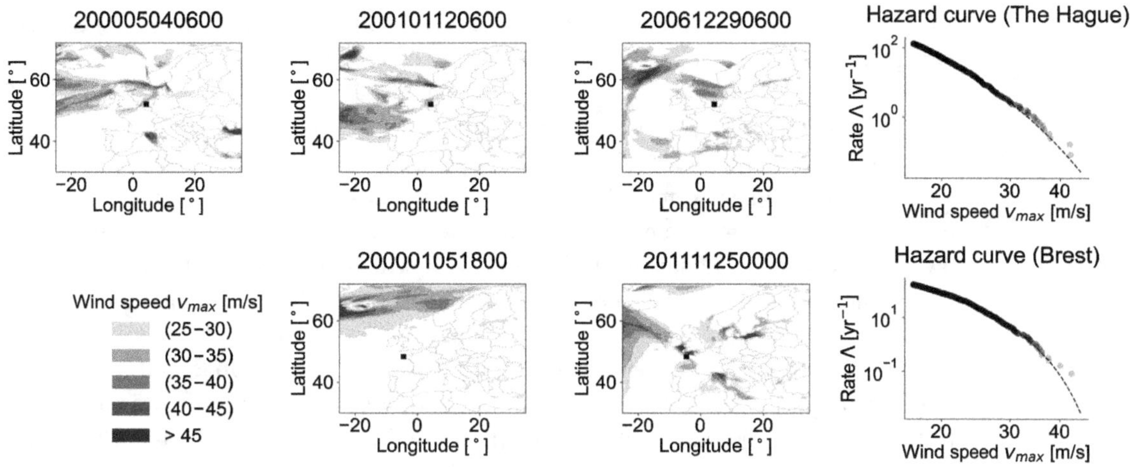

Fig. 2.2 Examples of hazard curves derived from simulated winter storms at two locations (The Hague, the Netherlands, and Brest, France – black squares on maps), and derived from 12 years of stochastic windstorm events developed by the Copernicus Climate Change Service (2022). Five examples of hazard intensity footprints $I(x, y)$ are also shown with maximum 3-s wind gust $I = v_{max}$ [m/s] as intensity measure. Note that hazard curves are defined here in terms of annual exceedance rate $\Lambda(v_{max})$ with dashed curves representing the Generalized Pareto distribution (Section 2.3.2).

The second approach to generate a hazard curve is from a simulation table (and matching footprint catalogue). Here, the annual probability of exceedance is:

$$\mathbb{P}(Y > y) = 1 - \exp(-\Lambda(y)) = 1 - \exp\left(-\frac{\sum_k \mathcal{H}(I_k - y)}{N_{sim}}\right), \tag{2.5}$$

where $\Lambda(y)$ is the annual rate of exceeding the intensity value y; $\mathcal{H}(x)$ is the **Heaviside step function**, which acts as a count function (with $\mathcal{H}(x \geq 0) = 1$, 0 otherwise); I_k is the simulated hazard intensity value at the considered location that is associated with the kth event hazard footprint; and N_{sim} is the number of year-simulations (Ebel and Kafka, 1999). In this approach, the aleatory uncertainty is described in each event intensity footprint instance, with I_k directly sampled from the intensity PDF. As N_{sim} increases, Eq. (2.5) tends toward Eq. (2.3). Such a comparison will be done in Section 2.5. Figure 2.2 shows some examples of hazard curves generated from simulations of European winter storms. Stochastic events were developed by the Copernicus Climate Change Service (2022) and are considered here to estimate the hazard at two coastal cities in a restricted 12-year period (2000–2011). In practical applications, a more extended time frame should be taken into account to enhance the precision of the distribution's tail, where potential damage may occur.

2.1.3.2 Hazard Map

A **hazard map** provides the intensity value y at each point of a geographic grid for a specific return period ΔT, with y derived from the hazard curve according to the equality

$\Delta T = 1/\mathbb{P}(Y > y)$. Such a map can be generated in a two-step procedure, first by computing a hazard curve per grid cell, and second by finding the intensity value matching the selected return period ΔT, as demonstrated in Figure 2.1. Common values for ΔT are 25, 50, 100, 250, 475, and 1,000 years (note that the 475-year return period is equivalent to a 10% probability of exceedance in 50 years; see Exercise section).

To illustrate the hazard module process, from stochastic event set definition to hazard map generation, a complete probabilistic hazard model will be developed for earthquakes toward the end of this chapter in Section 2.5. It will follow the main standards of the industry and use some of the same input data, here part of the fault source model of the 2013 European Seismic Hazard Model (ESHM13) (Basili et al., 2013; Woessner et al., 2015; Danciu et al., 2021), considered by the Global Earthquake Model (GEM) initiative (Pagani et al., 2020) for that part of the world. The simple models to be described in the next sections (for the earthquake case) will be used to generate the earthquake set and ground-shaking footprints. The approach will be much simpler than the one used in the standard OpenQuake engine (OQ-engine) (Pagani et al., 2014a) but will keep the same underlying philosophy.

2.1.4 Hazard as an Energy-Transfer Process

Hazard assessment is peril-specific and requires expertise from many different scientific disciplines. Yet, all perils follow the same general process consisting of some energy being transferred from a source to the environment via the emergence of events that follow a specific size distribution. The energy released by each event then diffuses in space, which ultimately leads to potential damage (Fig. 2.3). This level of abstraction, applicable to any

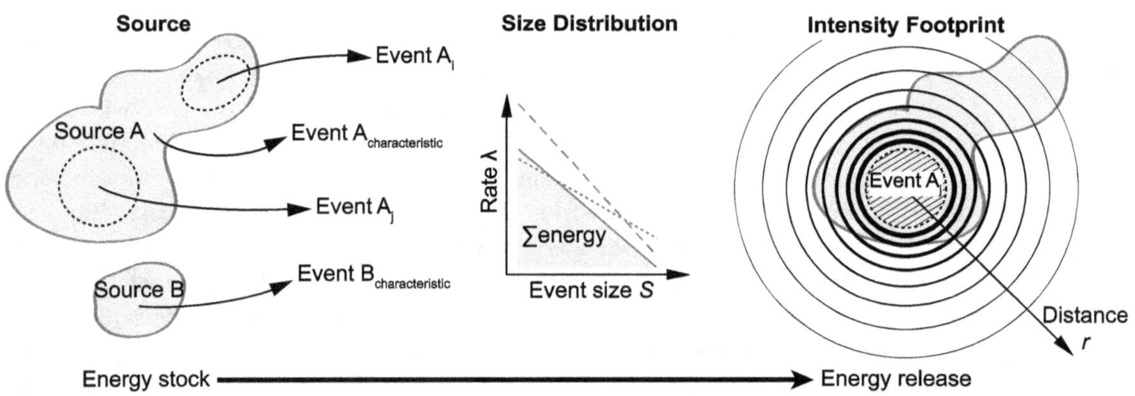

Fig. 2.3 Understanding the hazard process in terms of energy transfer. The energy stored in a source is released via events originating from that source. A **characteristic or maximum-size event** releases all the energy stored in the source while smaller events may also occur that release only part of the stored energy. The size distribution is often of the power-law family. The energy is finally redistributed in space and represents the potential cause of damage via the event intensity footprint. Modified from Mignan (2022a, fig. 7), CC-BY 4.0, https://creativecommons.org/licenses/by/4.0/.

peril, leads to the generic concept of **catastrophe object**. Note that non-diffusive processes will be investigated in Part II. The next sections will follow the same steps as shown in Figure 2.3, from the definition of the event source and the modelling of the event size S (Section 2.2), to the peril size distribution relating event rate λ to event size S (Section 2.3), to, finally, the modelling of the intensity footprint $I(x, y)$ for an event of size S in a spatial grid of coordinates (x, y) (Section 2.4). These steps represent the computational ontology associated with the catastrophe object. All the processes will be described analytically via empirical laws and physical equations. The following perils will be covered individually in the next three sections: earthquakes, volcanic eruptions, tropical cyclones, and asteroid impacts. Others will be described in one or two sections: landslides, fluvial floods, storm surges, tsunamis, tornadoes, hailstorms, rainstorms, lightning, heatwaves, wildfires, epidemics, explosions, blackouts, cyberattacks, fires, business interruptions, armed conflicts, and terrorist attacks. Many of the perils for which the intensity footprint will not be modelled in this chapter will be in Chapter 4 on numerical computational methods.

2.2 Event Source

The size S of an event is constrained by the source from which it originates. A **source** is here defined as the energy stock that drives the event. Each source represents a unique object with peril-specific characteristics. For instance, a tropical cyclone manifests itself from an atmospheric low-pressure system that moves along a track, while an earthquake corresponds to the rupture of a fault plane under tectonic loading – two very different objects. *All types of perils can however be described by five source categories* (Mignan, 2022a): *point sources* (Section 2.2.1), *line sources* (Section 2.2.2), *track sources* (Section 2.2.3), *and area and diffuse sources* (Section 2.2.4). Examples of localized sources, that is, point, line, and track sources, are shown in Figure 2.4. In contrast to line sources, **track sources** explicitly account for changes of event size over time. **Area sources** will be shown to represent, in many cases, a simplified version of diffuse sources, which will be addressed in Part II, as they require numerical methods for event size assessment and intensity footprint modelling. Source types are idealized representations of actual sources; therefore, a given peril may be represented by more than one source category. For each considered peril, only the most common size metrics will be presented, and of those, preferentially the ones most closely related to energy. Generally, only one model per peril will be provided to estimate the event size from the source parameters, based on first principles and other simple empirical relationships. References to more sophisticated models will be given for completeness, and the role of alternative models will be discussed in Part III when dealing with epistemic uncertainties.

Perils will be listed below in alphabetical order, independently of their physical origin.

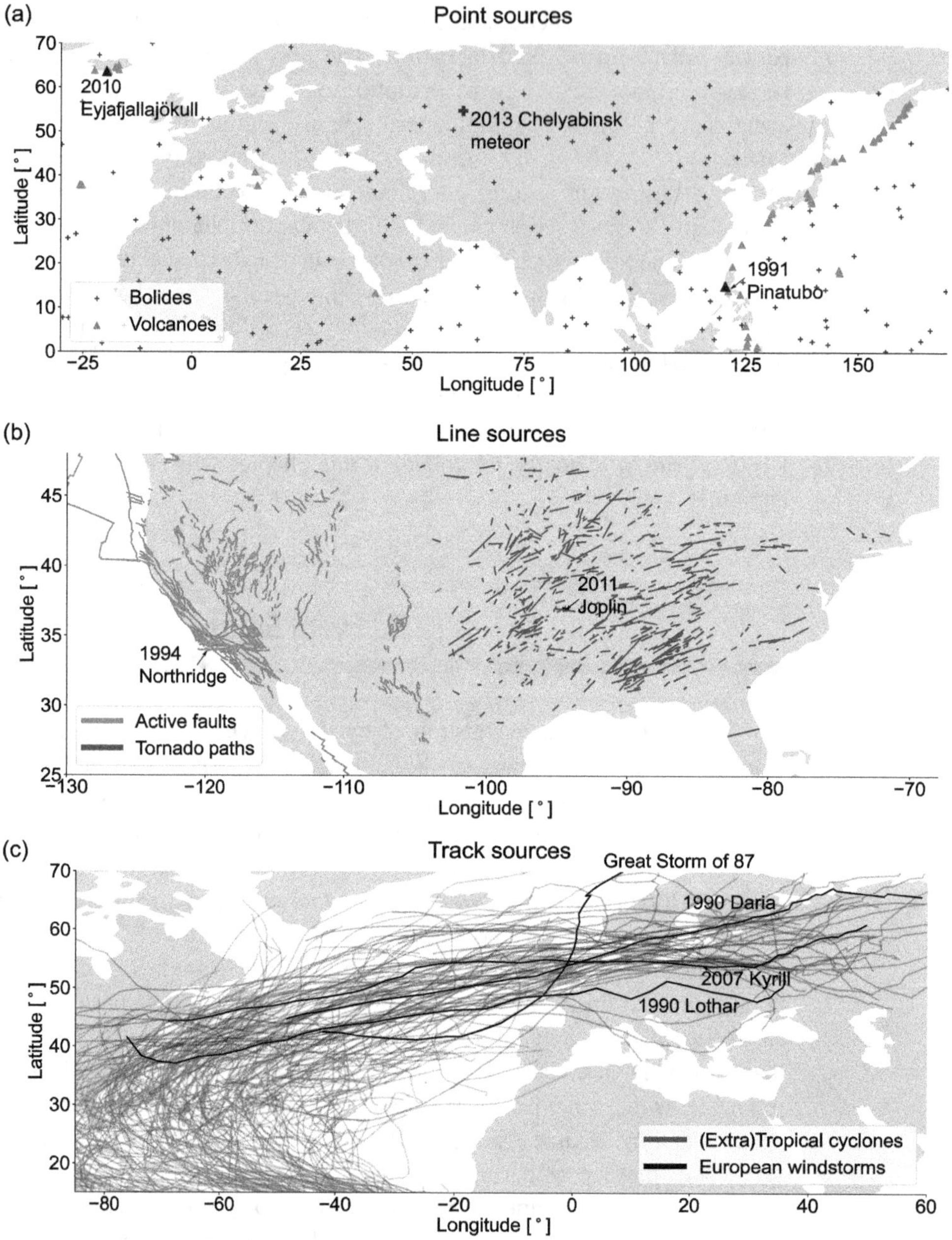

Fig. 2.4 Examples of localized sources: (a) Point sources: bolides in the 1988–2022 period (CNEOS, 2023) and volcanoes (LaMEVE; Crosweller et al., 2012). (b) Line sources: active faults (GEM Global Active Faults Database; Styron and Pagani, 2020 – CC-BY-SA, published with permission) and F4–F5 tornado paths in the period 1950–2021 (NOAA's Storm Prediction Center, 2023). (c) Track sources: Atlantic hurricanes and extratropical cyclones (IBTrACS; Knapp et al., 2010, 2018) and European windstorms (XWS; Met Office, University of Reading and University of Exeter, 2023; Roberts et al., 2014; CC-BY 4.0, https://creativecommons.org/licenses/by/4.0/) for the same period 1979–2012. A black and white version of this figure will appear in some formats. For the colour version, please refer to the plate section.

2.2.1 Point Sources

2.2.1.1 Asteroid Impacts (AI)

Asteroids and comets are celestial bodies that can impact Earth. They both have different compositions, hence different densities ρ, and different trajectory types, hence different velocities v. The typical density of stony objects is $\rho = 3$ g/cm^3 and of iron objects is $\rho = 7$ g/cm^3. Cometary material has a density close to that of ice, with $\rho = 1$ g/cm^3. A typical asteroid impact velocity is $v = 20$ km/s; it is higher for comets with $v = 50$ km/s (Hills and Goda, 1993). It is commonly assumed that the flux of objects hitting the Earth is uniformly distributed on the globe (Fig. 2.4a), although a slight latitudinal dependency is expected from the dynamics of the process (Le Feuvre and Wieczorek, 2008). The size of an impact event is defined in terms of its **kinetic energy** E [J],

$$E = \frac{1}{2}mv^2, \tag{2.6}$$

with m [kg] the mass of the body and v [m/s] the velocity. The event size is commonly expressed in tons of **TNT equivalent** (E_{TNT}) via the conversion factor 1 t $= 4.184 \times 10^9$ J (Ex. 2.1a). Equation (2.6) represents the pre-entry kinetic energy, called total impact energy in the data set of fireballs reported by US government sensors (CNEOS, 2023). Using an event size $S = E$ for intensity footprint modelling in Section 2.4 will, however, overestimate the hazard since the energy going into the blast (in an airburst or a ground impact) is only some fraction of the original, deposited energy. Improved calculations would explicitly examine atmospheric deceleration, disruption, and ablation processes, as well as ground penetration, usually requiring numerical simulation methods (Bland and Artemieva, 2006; Boslough and Crawford, 2008; Mathias et al., 2017). Those would require far more details about the source, such as flight path angle, cross-sectional area of the object, **drag** coefficient (the object's resistance to air), and ablation coefficient (Mathias et al., 2017). There is virtually no maximum limit to the size of the impacting body. The largest impacts led to the creation of the Moon 4.5 billion years ago (Canup and Asphaug, 2001) and to a number of major extinctions in geological times (Alvarez et al., 1980). For the smallest near-Earth objects that remain undetected, their size can be estimated as they cross the atmosphere as bolides. In this case, the kinetic energy is $E_{TNT} = E_r/\eta$ with E_r the observed radiated energy and η the integral luminous efficiency (Nemtchinov et al., 1997; Ex. 2.1b).

2.2.1.2 Explosions (Ex)

Human-made explosions have chemical, mechanical, or nuclear causes, which find their source in containers of flammable, pressurized, or nuclear material. Regarding accidents, the source is generally located at an industrial site or in a cargo within a transport network. However, malicious attacks using bombs (e.g., terrorist attacks, armed conflicts) may occur anywhere. For all types of explosions, the event size is primarily defined by the **energy of explosion**, which includes the **blast energy** that contributes to the shockwave, but also, depending on the scenario, different levels of thermal energy, kinetic energy in projectiles, and strain energy in vessel rupture.

First considering a chemical explosion involving flammable material (vapour cloud explosion (VCE)), the event size depends on the mass and thermal properties of the fuel. The blast energy E [J] is defined as

$$E = \eta m \Delta H_c, \tag{2.7}$$

where η is the explosion yield factor representing the fraction of available combustion energy participating in blast wave generation, m [kg] is the mass of fuel in the vapour cloud (i.e., in the fuel–air mixture) and ΔH_c [J/kg] is the heat of combustion of the flammable material. A range of $45 \leq \Delta H_c \leq 55$ kJ/g is observed for the common fuels, such as gasoline, diesel, methane, propane, butane, natural gas (for hydrogen, $\Delta H_c = 142$ kJ/g). A commonly accepted value for η is 0.03 (ranging from 0.01 to 0.1), indicative of hydrocarbons being fairly poor explosives (Casal, 2008, p. 125). The 2005 22-ton Buncefield explosion for example, which occurred at an oil storage facility in England and injured 43 people, involved 300 tons of gasoline. Using Eq. (2.7) with m calculated to be 6.5% of the total overflow mass $m_{tot} = 300$ ton, $\Delta H_c = 46.4$ kJ/g for gasoline, and $\eta = 0.1$, one obtains $E_{TNT} = 21.6$ t (Ottemöller and Evers, 2008). In the case of **high explosives**, both fuel and oxygen are contained in the same molecule. The chemical reaction is faster, meaning a high η. Examples include TNT (trinitrotoluene) and nitroglycerine. Other types of chemical explosions are possible; for example, confined ammonium nitrate decomposing on its own under high temperatures releases gases violently. A simple formula is $E_{TNT} = \eta_0 m$ with $0.3 \leq \eta_0 \leq 0.55$ (e.g., Lee et al., 2022). The 2020 1.1-kt Beirut port explosion was the worst such case, involving 2,750 tons of ammonium nitrate (Rigby et al., 2020).

Mechanical explosions are due to the sudden loss of containment of compressed gases (i.e., pressure vessel explosions (PVE) – or boiling liquid expanding vapour explosions (BLEVE) in the case of superheated liquids). The source can be any pressure storage tank. Several methods exist for estimating the energy of explosion (Crowl, 2003, appendix B), with the simplest one being:

$$E = \frac{(P - P_0)V}{c_\gamma - 1}, \tag{2.8}$$

where E [J] is the energy of explosion, P_0 [Pa] the ambient pressure, P [Pa] the burst pressure of the vessel, V [m^3] the volume of gas in the vessel, and c_γ the heat capacity ratio for the gas (Brode, 1959). For example, considering a 1–m^3 vessel containing nitrogen at 500 bar, an ambient pressure of 1 bar, and $c_\gamma = 1.4$, we get $E = 1.25 \times 10^8$ J (Crowl, 2003, appendix B) (recalling that 1 bar = 10^5 Pa). The blast energy represents between 40% and 50% (ductile vessel break with projectiles) and 80% (fragile vessel failure) of the explosion energy (Casal, 2008, p. 166). Given a fixed volume V, an increase in temperature (e.g., in a fire due to a VCE) causes P to increase also, which can lead, combined with other causes of loss of containment, to a chain of explosions (Part II, Chapter 5).

For completeness, considering the case of a nuclear explosion, a simplified equation of the energy yield E [J] is

$$E \approx \frac{m_{tot}}{8} \left(\frac{\Delta r \cdot \log n}{\tau} \right)^2, \tag{2.9}$$

where m_{tot} [kg] is the total mass of the spherical bomb (core + tamper), Δr [m] is the difference between the expanded radius and the initial radius of the sphere, τ [s] is the time required for a neutron born in a fission to subsequently strike and fission another nucleus, and n is the number of neutrons liberated per fission ($n = 2.637$ for uranium-235 and $n = 3.172$ for plutonium-239) (Reed, 2018). The study by Reed (2018) provides a spreadsheet to calculate the different parameters of Eq. (2.9) and includes, as a possible scenario, a terrorist attack. In numbers, the largest nuclear detonation in history was the Tsar Bomba test conducted in 1961 by the Soviet Union, with a 50-Mt yield (Adamsky and Smirnov, 1994); the atomic bombings of Hiroshima and Nagasaki by the US in 1945 were 16 kt and 21 kt in size, respectively (Kerr et al., 2005).

2.2.1.3 Fluvial Floods (FF)

A fluvial flood has a **catchment basin** as source, conditional on the occurrence of a rainstorm (or ice melting). It is represented by a point source, which is the upstream river point of convergence of all the surface water collected in the land area of the catchment (Fig. 2.5). The size S of a flood event is described by its peak flow rate, or peak discharge Q_p at that point (another type of point source is a manhole in urban flooding where the sewage system is overloaded; water system failure (Wf) peril). The **Rational formula**,

$$Q_p = c_r I_{RS} A, \tag{2.10}$$

conceptualizes the relationship between Q_p [m^3/s], rainfall intensity I_{RS} [m/s], and catchment physical characteristics, that is, its area A [m^2] and the **run-off coefficient** c_r (i.e., the fraction of rainfall that remains at the surface after reaching the ground and that contributes to the streamflow). This relationship assumes a steady and spatially uniform rainfall up to the **concentration time** t_c, which is the time needed for run-off to travel throughout the catchment (from the most remote point to the outlet) and thus for Q_p to be reached (Fig. 2.5a). t_c depends on additional characteristics of the catchment, such as slope, length of the main channel, and Manning's friction parameter (e.g., Gericke and Smithers, 2014). Both run-off and Manning's coefficients are functions of the land cover with the former going from 0.2 in grassland to 0.8 in urban areas (e.g., Dhakal et al., 2012), and the latter from 0.01 for smooth bare soil to, for example, 0.4 for grassed watersheds (Engman, 1986). Basically, Eq. (2.10) states that the outflow equals the inflow $I_{RS}A$ reduced by the factor c_r, with $c_r I_{RS}$ the effective precipitation or run-off rate. Globally, the fraction of precipitation that contributes to run-off over all land areas is approximately 0.36 (Pagano and Sorooshian, 2002). The Rational formula works reasonably well for small catchments but non-linear behaviours are expected for larger ones due to the complexity of landscape processes, and possible changes of c_r over time (e.g., changes in soil saturation). Empirical observations over large catchments globally reflect this non-linearity with formulas of the form $\mathbb{E}[Q_p] \propto A^\alpha$ with $\alpha \approx 0.6$ (Meigh et al., 1997, table 1). However, numerical modelling of overland flow is needed for a realistic assessment of Q_p. Although outside the scope of this introductory textbook, the process is well illustrated in the OverlandFlow component of the Landlab library where a two-dimensional hydrodynamic algorithm is used to simulate

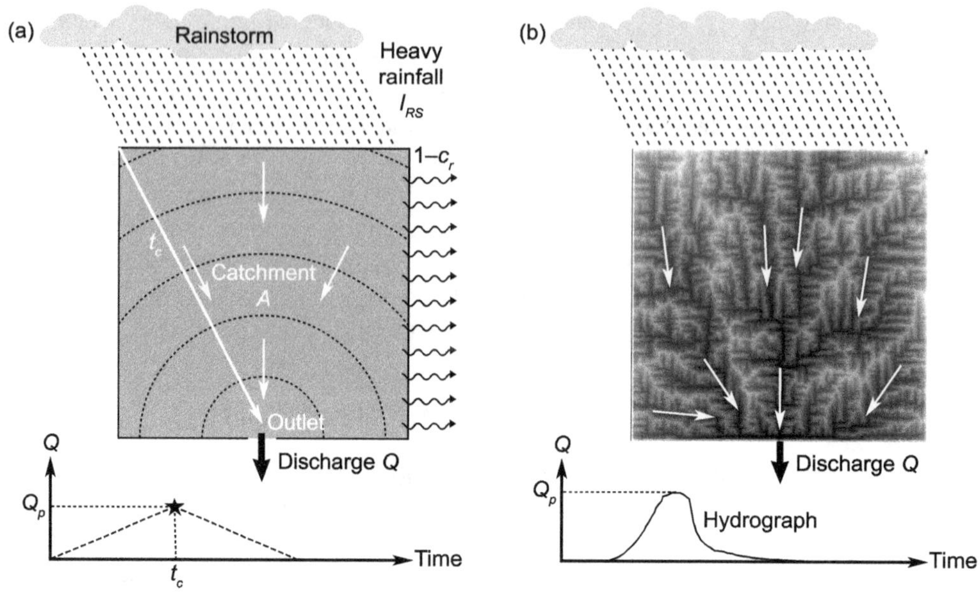

Fig. 2.5 Relationship between catchment characteristics and water discharge. (a) Concept behind the Rational formula. (b) Example of catchment simulated in Landlab (Hobley et al., 2017) (CC-BY 3.0, annotations added) and resulting (more realistic) hydrograph (redrawn from Adams et al., 2017).

water flow in a digital elevation model (Adams et al., 2017; Fig. 2.5b). The model can estimate the evolution of the discharge Q over time and not only Q_p, in contrast to the basic Rational formula which provides only the point (t_c, Q_p) on the **hydrograph** (i.e., graph of the rate of flow versus time). Moreover, Eq. (2.10) is known to underestimate Q_p. The importance of the hydrograph will become clear when numerically modelling flood footprints (Part II, Chapter 4). Finally, **flood energy** can be estimated from the hydrograph, with

$$E = \int_t (c_\gamma Q(t) c_s / W) \mathrm{d}t \tag{2.11}$$

in [J/m^2] and where $c_\gamma = 9{,}800$ N/m^3 is the specific weight of clear water, Q [m^3/s] the discharge at time t, c_s is the energy slope (i.e., channel bed slope corrected for roughness) and W [m] is the water-surface width (Costa and O'Connor, 1995).

2.2.1.4 Volcanic Eruptions (VE)

A **volcano** is a point source from which pyroclastic material may be discharged from pressurized magma into the atmosphere. Most volcanoes are located along the Pacific Ring of Fire (Figs. 1.8 and 2.4a). In practice, the size S of a volcanic eruption (explosive case) is directly represented by the volume (V) of ejected matter. It is the main parameter in the semi-quantitative **Volcanic Explosivity Index (VEI)**, which includes other variables

Table 2.3 Volcanic Explosivity Index (VEI)

VEI[a]	Ejected volume V	Column height h	Description	Historical cases[b] (V [km^3], h [km])
0	$< 10^4$ m^3	< 0.1 km	Gentle	
1	10^4–10^6 m^3	0.1–1 km	Effusive	
2	10^6–10^7 m^3	1–5 km	Explosive	
3	10^7–10^8 m^3	3–15 km	Severe	
4	0.1–1 km^3	10–25 km	Violent	1902 Pelée ($V = 0.1$)
				2010 Eyjafjallajökull ($V = 0.1, h = 6.4$)
5	1–10 km^3	> 25 km	Terrific	1980 St Helens ($V = 1.3$)
6	10–100 km^3	> 25 km	Cataclysmic	1883 Krakatau ($V = 20, h = 50$)
				1991 Pinatubo ($V = 11, h = 40$)
7	100–1,000 km^3	> 25 km	Paroxysmal	1815 Tambora ($V = 110, h = 43$)
8	> 1,000 km^3	> 25 km	Colossal	

[a] VEI table simplified from Newhall and Self (1982, table 1).
[b] Values from LaMEVE (Crosweller et al., 2012).

such as eruption cloud height and some qualitative description (from 'gentle' to 'colossal') (Newhall and Self, 1982; Table 2.3). For general comparison with other perils, it is worthwhile to also define S in energy terms. During a volcanic eruption, energy is released in several forms: thermal, kinetic, potential, seismic (via volcanic earthquakes), and blast energies, with the **thermal energy** definitively the most important one, potentially from 10 times to more than 1,000 times greater than all the others combined (e.g., Hédervári, 1963). The energy output of an explosive eruption can thus be defined as

$$E \approx E_{th} = \rho V c_h (T - T_0), \tag{2.12}$$

where E_{th} [J] is the thermal energy, ρ [kg/m^3] the material density, V [m^3] the material volume, T [°C] the temperature of the ejecta (or magmatic material) and T_0 [°C] that of the atmosphere, and c_h [J/(kg °C)] the specific heat capacity (Pyle, 1995). The material can be the magma participating in the explosion or the ejecta. For the 1815 Tambora eruption, the largest eruption recorded in history, Eq. (2.12) gives $E = 1.25 \times 10^{20}$ J, or $E_{TNT} = 30$ Gt, when using $\rho_{magma} = 2.7$ g/cm^3, $V_{magma} = 33.2$ km^3 and a specific energy $c_h(T - T_0)$ = 1,400 J/g (Foden, 1986) ($1{,}030 \leq \int c_h dT \leq 1{,}310$ J/g given in Pyle (1995, table 3) for various magma melts). Note that such calculations are subject to high uncertainties and that the event size will be fixed to $S = V$ for hazard assessment.

2.2.2 Line Sources

2.2.2.1 Earthquakes (EQ)

A **fault** is a planar rock fracture which shows evidence of relative movement. This cumulative displacement is due to successive earthquake occurrences on that source. A fault is characterized by its geometry, which is described by the plane's length, width, strike

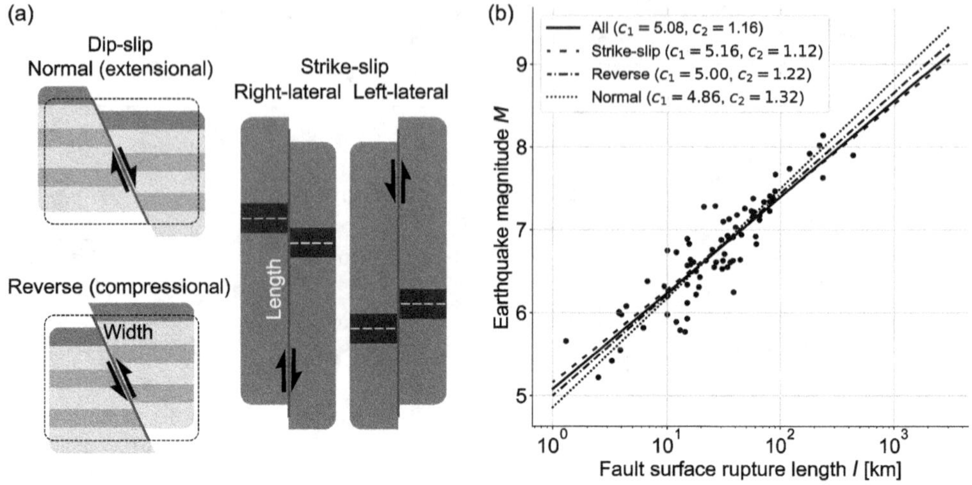

Fig. 2.6 Fault rupture characteristics and relationship to earthquake magnitude. (a) Fault mechanisms, in cross-section on the left and in map view on the right. (b) Empirical relationships between surface rupture length l and magnitude M with the parameters of Eq. (2.14), from Wells and Coppersmith (1994, table 2A) and data points from Wells and Coppersmith (1994, table 1).

(azimuth orientation with respect to the north), dip (angle between fault plane and horizontal plane), and rake (angle between slip vector and the horizontal). Different dips and rakes yield different fault mechanisms (Fig. 2.6a): strike-slip when the fault is vertical (90° dip) and dip-slip or oblique slip otherwise. A strike-slip mechanism can be left-lateral or right-lateral; a dip-slip mechanism can be normal (in extensional setting) or reverse (in compressional setting). Oblique slip is a combination of both. Although a fault segment (a plane) is an area source, it is common and simpler to use its trace projection at the surface instead (Fig. 2.4b). In seismic hazard assessment, the term 'area source' actually refers to a zone where potential fault locations are uncertain (see Box 2.5). The earthquake size S is defined in terms of magnitude M, which is a function of fault geometry and kinematics. Different magnitude scales exist, with the one directly related to the source characteristics the **moment magnitude** (M_w, here noted M). It is defined as

$$M = \frac{2}{3} \log_{10} M_0 - 6.0, \tag{2.13}$$

which is a function of the seismic moment $M_0 = c_\mu A d$ [Nm] with c_μ [Pa] the rock shear modulus, A [m^2] the rupture surface area, and d [m] the average displacement on the fault (Hanks and Kanamori, 1979). The moment magnitude can additionally be estimated empirically with

$$M = c_1 + c_2 \log_{10} l, \tag{2.14}$$

where l is the surface rupture length and c_1 and c_2 fitting parameters (Wells and Coppersmith, 1994; Fig. 2.6b). For additional relationships, see the review by Stirling et al. (2013).

The maximum possible magnitude M_{max} is bound by the maximum size of tectonic plate boundaries (McCaffrey, 2008; Mignan et al., 2015b). The largest earthquakes ever recorded by instruments, as already mentioned in Chapter 1, are the 1960 $M = 9.5$ Chile, the 1964 $M = 9.2$ Alaska, the 2004 $M = 9.1$ Sumatra, and the 2011 $M = 9.1$ Tohoku, Japan, all megathrust earthquakes in subduction zones. Events with significantly higher magnitudes would be possible only if triggered by an asteroid impact with $M_{EQ} = 3.9 + 0.7 \log_{10} E_{AI}$ [kton] and E_{AI} the impact energy (e.g., Hills and Goda, 1993). Considering a giant 10-km asteroid impact with an estimated yield of 75 Tt (Eq. (2.6)) would provoke an earthquake-equivalent magnitude greater than 11! The 1906 Great San Francisco earthquake ruptured 430 km of the Northern San Andreas Fault (NSAF) (Wells and Coppersmith, 1994). Assuming a shear modulus $c_\mu = 3.3 \times 10^{10}$ Pa, a rupture width of 10 km, and an average motion $d = 4.3$ m (Song et al., 2008), one gets a seismic moment $M_0 = 6.1 \times 10^{20}$ Nm, leading to $M = 7.9$ (Eq. (2.13)). For a strike-slip mechanism, Eq. (2.14) yields $M = 8.1$. The entire San Andreas fault is approximately 1,200 km long, which would give $M = 8.6$ with the same empirical relationship (close to the consensus estimate of 8.5; Parsons et al., 2018). This value is much lower than the impossible 9.6 earthquake depicted in the 2015 disaster film *San Andreas* featuring Dwayne Johnson (Flynn and Peyton, 2015). Finally, the **seismic energy** E [J] released by an earthquake can be derived from the magnitude M with

$$\log_{10} E = 5.24 + 1.44M \tag{2.15}$$

(Båth, 1965). According to this equation, the 1960 $M = 9.5$ Great Chilean (Valdivia) earthquake would have released 8.3×10^{18} J, or almost $E_{TNT} \approx 2$ Gt.

2.2.2.2 Storm Surges (SS)

The source of a storm surge may be represented as the projection of the impact of strong winds along a shoreline (Fig. 2.7a). For the case of steady onshore wind acting uniformly on a water body, the surge height h is proportional to the wind stress, which is equivalent to $h \propto \frac{v^2}{\theta_0}$ with v the wind speed and θ_0 the bottom slope of the continental shelf (Irish et al., 2008). A very simple relationship is

$$h_{max} = c v_{max}^2, \tag{2.16}$$

with h_{max} [m] the peak surge, v_{max} [m/s] the maximum wind speed, and here $c = 0.0011$ a generic coefficient fitted to the Saffir–Simpson hurricane scale (see Table 3.4; Simpson, 1974). These values are only illustrative since site-specific conditions of the land–water interface significantly alter the resulting surge. This is shown in Figure 2.7b, where Eq. (2.16) is compared to another relationship derived from a numerical model for Manhattan Island, New York City (Lin et al., 2010). One other shortcoming of Eq. (2.16) is that the v_{max} indicator is often not sufficient to estimate the actual storm surge. Other contributing source parameters include the storm's spatial extent (Irish et al., 2008) and the offshore storm-track characteristics. Note also that h_{max} does not reflect the overall size S of the storm surge, but the peak surge at a point along the line source (e.g., SURGEDAT; Needham and Keim, 2012), instead of the full profile along the shoreline (Fig. 2.7a). As it will be

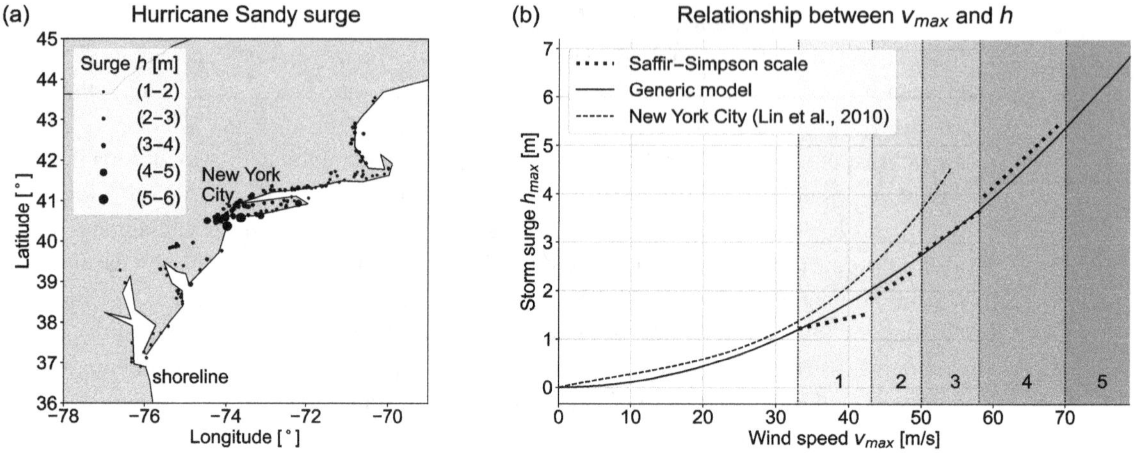

Fig. 2.7 Storm surge as a line source and relationship to wind speed. (a) Surge height h [m] observed along the Atlantic coast near the New York City shoreline during Hurricane Sandy (2012); data from the USGS (McCallum et al., 2013, table 3). (b) First-order relationship between wind speed v_{max} and storm surge height h_{max}. Generic model according to Eq. (2.16) and calibrated to the Saffir–Simpson scale with hurricane categories from 1 to 5 indicated (Simpson, 1974); fit for New York City (at the southern tip of Manhattan Island) of the form $h_{max} = c_1 v_{max} + c_2 v_{max}^2 + c_3 v_{max}^3$ with $c_1 = 0.031641, c_2 = -0.00075537$ and $c_3 = 3.1941 \times 10^{-5}$ (Lin et al., 2010).

shown in Section 2.2.3, the overall size of a storm can be assessed by integrating the kinetic energy per unit volume over the storm volume, which scales with the surge size (Powell and Reinhold, 2007). S could therefore be defined in terms of volume of surge propagation along the shoreline (Bass et al., 2017). For storm surge inundation modelling, information about $h(x, y)$ along the (shore)line source will be sufficient at first order. However, source definition (as per the proposed chapter classification) is often not pertinent to storm surge modelling in practice, as the coastal flood hazard is commonly directly assessed from the dynamical model of a storm (Jelesnianski et al., 1992).

2.2.2.3 Tornadoes (To)

Tornadoes originate from severe convective storms but, in hazard assessment, their source is directly defined from the path of their vortex. Although the actual source of a tornado is a track (Section 2.2.3), the simplest way to represent a tornado source is by a straight line (Fig. 2.4) or a limited number of connected lines, such as in the case of the 2011 Joplin tornado (Masoomi and van de Lindt, 2018). Because of their meteorological nature, tornadoes do not have fixed sources in space but instead **stochastic paths** representative of prevailing weather conditions. This distinction sets them apart from, for example, earthquakes, volcanic eruptions, or industrial explosions, which have well-localized sources. However, they are similar to asteroid impacts in that impact locations are random. Recall from Section 2.1.1 that the term 'stochastic' refers to a process which has a random component. *At the core of stochastic modelling is the Monte Carlo simulation method*, which is a crucial technique to master in CAT modelling. It is described in Box 2.1 and applied to

Table 2.4 Enhanced Fujita path-portion weight per declared EF tornado

Portion	v_{max} midpoint [m/s]	EF0	EF1	EF2	EF3	EF4	EF5[a]
EF0	33.5 (29–38)	1	0.772	0.616	0.529	0.543	0.538
EF1	44.0 (39–49)		0.228	0.268	0.271	0.238	0.223
EF2	55.0 (50–60)			0.115	0.133	0.131	0.119
EF3	67.5 (61–74)				0.067	0.056	0.070
EF4	82.0 (75–89)					0.032	0.033
EF5	97.0 ($\geq$ 90)						0.017

[a] Weights from Ramsdell and Rishel (2007, table 3-1), as given in Fricker and Elsner (2015, table 1) (CC-BY 4.0, https://creativecommons.org/licenses/by/4.0/).

the catastrophe context in Exercise 2.2. Due to the stochastic nature of the tornado process, source parameters are commonly sampled from historical data (Romanic et al., 2016). The event size S, defined as the maximum wind speed v_{max}, is a direct characteristic of the line source, alongside its starting point, end point, width (representative of the radius of maximum winds r_{max}), and forward speed. The event size can also be described by the Enhanced Fujita (EF) scale, as a function of v_{max} (e.g., Edwards et al., 2013; Table 2.4; see also Table 3.5). Figure 2.8 illustrates how stochastic tornado line sources can be generated from the 1950–2021 catalogue of NOAA's Storm Prediction Center (2023). Assuming a line source instead of a track source, according to the minimum assumption model of Schaefer et al. (1986), significantly overestimates the size of a tornado, especially the largest ones. In practice, the line source can be subdivided to account for different EF levels along the total source length. For example, an event declared an EF-5 is only of that category intensity along 15% of the total length of its track, while an EF-1 tornado is effectively EF-1 along more than 40% of its track (e.g., Standohar-Alfano and van de Lindt, 2015). In the energy domain, the size of a tornado is expressed in terms of kinetic energy where the mass m [kg] is defined by the air density $\rho = 1$ kg/m^3 times the total tornado volume [m^3]. To take into account the variation of v_{max} along the line source, the following form is used,

$$E = \frac{1}{2} m \sum_{i=0}^{k} w_i v_i^2, \tag{2.17}$$

with k the event's EF rating, v_i [m/s] the midpoint wind speed for EF rating i, and w_i the corresponding fraction of path area (Table 2.4; Fricker and Elsner, 2015). For the most energetic tornado observed in the period 2007–2013, the EF-4 Tallulah-Yazoo City-Durant tornado of 24 April 2010, one obtains a kinetic energy $E = 655$ TJ, or $E_{TNT} = 157$ kt. This estimate considers a 240-km length and a 2.82-km width and assumes a characteristic vortex height of 1 km to estimate the total volume (Fricker and Elsner, 2015).

2.2.2.4 Tsunamis (Ts)

With most tsunamis being caused by earthquakes (Fig. 1.13), tsunami source definition is provided in the line source section. However, the source can also take the form of a

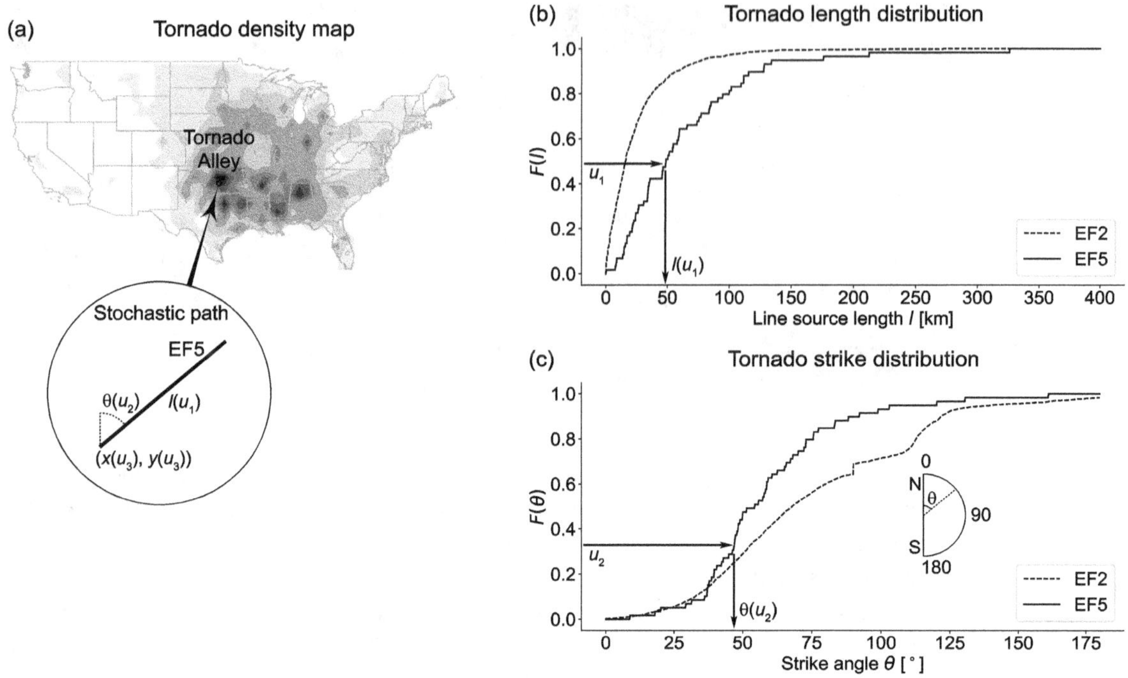

Fig. 2.8 Illustrating how a stochastic tornado line source can be generated, here from the 1950–2021 catalogue from NOAA's Storm Prediction Center (2023). (a) Density map of all EF2+ tornadoes in the US with the Tornado Alley highlighted. (b) Tornado path length distribution $F(l)$. (c) Tornado path strike angle distribution $F(\theta)$ showing that most tornadoes are principally oriented SW–NE / SWW–NEE. An example of sampling in space, length, and strike is done here for an EF5 tornado (see Box 2.1 for the general method).

point, such as an asteroid impact over water or an underwater volcanic eruption, or of a more complex source, such as a submarine landslide (Section 2.2.4). Figure 2.9 represents in cross-section the initial water mass movement. Geophysical events form a protrusion while asteroid impacts form a cavity. To be able to generate a vertical displacement of the surface of the water mass, the velocity of the triggering event must be fast enough in comparison to the tsunami wave velocity

$$v_{Ts} = \sqrt{g z_w}, \tag{2.18}$$

where $g = 9.81$ m/s^2 is the **gravitational acceleration** and z_w [m] the water depth (with $\max z_w = 10{,}000$ m in the deep oceanic trenches, $\max v_{Ts} = 313$ m/s, or 1,127 km/h). It follows that shallow events have the largest potential to generate a tsunami. Then, the tsunami size S is the **potential energy** E [J] of the displaced volume of water,

$$E = \rho_w V_w g h_w, \tag{2.19}$$

where $\rho_w = 1{,}025$ kg/m^3 is the density of sea water, V_w [m^3] is the displaced water volume, and h_w [m] is the height of the vertical water displacement.

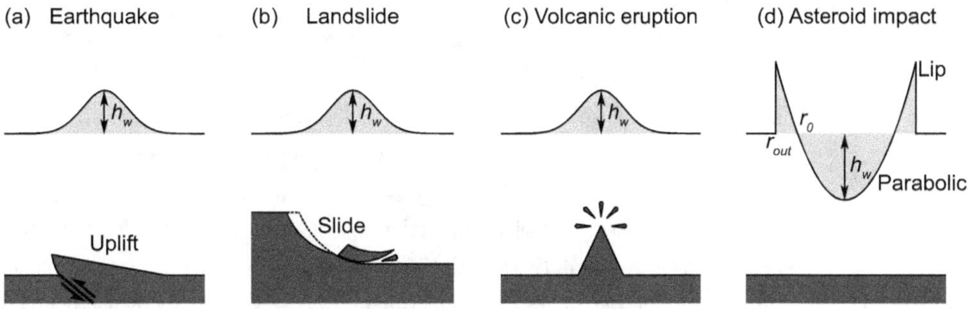

Fig. 2.9 Schematic representation of tsunami source types, with localized water volume change due to (a) an earthquake, (b) a landslide, (c) a volcanic eruption, or (d) an asteroid impact (not at scale). The light-grey areas represent changes in water levels in cross-section, the top black curves represent the water surface, and the bottom dark-grey areas represent the underwater terrain.

In the case of an underwater earthquake producing a vertical displacement d_{EQ}, we have $h_w \approx d_{EQ}$ by assuming the impressibility of water. The tsunami energy is then

$$E = \frac{1}{2}\rho_w g A_{EQ} d_{EQ}^2, \tag{2.20}$$

with A_{EQ} [m^2] the earthquake rupture area and d_{EQ} [m] the earthquake average vertical displacement according to the 'Waterberg' model (Lautrup, 2005). This represents an upper limit for when $v_{EQ}/v_{Ts} > 20$–50 (Trifunac and Todorovska, 2002). Considering the great tsunami caused by the 2004 Sumatra–Andaman earthquake, assuming $A_{EQ} = 1{,}200 \times 150 = 180{,}000$ km^2 and $d_{EQ} = 5$ m, we get $E = 2.26 \times 10^{16}$ J or $E_{TNT} = 5.4$ Mt (Lautrup, 2005).

Submarine landslides, due to their much smaller area than large earthquakes, create much more localized tsunamis and require the involvement of a large quantity of material (Trifunac and Todorovska, 2002). The maximum initial wave height h_{max} [m] of a landslide-induced tsunami is

$$\max h_w = \frac{\pi h_{LS}^2}{4 z_w} \text{ if } l_{LS} \gg z_w, \tag{2.21}$$

where h_{LS} [m] is the thickness of the submarine landslide, l_{LS} [m] is its length, and z_w [m] is the initial water depth (Pelinovsky and Poplavsky, 1996). Considering one of the largest known submarine landslides, the prehistoric Storegga slide at the edge of Norway's continental shelf, with average thickness of 114 m and water depth of 500 m, Eq. (2.21) yields max $h_w = 20.4$ m (Röbke and Vött, 2017).

The dynamics of an underwater volcanic eruption are poorly known, making the description of the tsunami generation process difficult outside numerical modelling (Ulvrová et al., 2014). An underwater eruption can combine the ejection of material (of volume V_{VE}), the creation of a crater, and the rise and collapse of a spherical vapour cavity. It means that, depending on the event size and depth, the initial water mass movement may be a protrusion (Fig. 2.9c; e.g., Heidarzadeh et al., 2022, fig. 12) or a parabolic cavity, like in the asteroid impact case (Fig. 2.9d; e.g., Ulvrová et al., 2014, fig. 6).

Box 2.1 The Monte Carlo Method (with Inverse Transform Sampling)

The Monte Carlo method is a straightforward approach to simulate stochastic events via the use of random numbers (Metropolis, 1987). A pseudo-random uniform variable (or **random number**) $0 \leq u \leq 1$ is first generated. This is denoted $u \sim \mathcal{U}_{[0,1]}$ and indicates that u follows the **uniform distribution** $\mathcal{U}$ with outcomes equally probable on the interval [0, 1] (with '$\sim$' meaning 'is distributed as'). We then wish to **sample** (i.e., to randomly select) a possible value of the random variable X, given u, from the **distribution function** of X (also known as **cumulative distribution function (CDF)**). When derived from observations, it is called **empirical cumulative distribution function (eCDF)** (see Box 2.4 for a pseudocode). This function, which describes the underlying stochastic process, is defined as the probability that X does not exceed the numerical value x, expressed as $F(x) = \mathbb{P}(X \leq x)$ (Grimmett and Stirzaker, 2001, p. 27). In the case where X is continuous, $F(x) = \int_{-\infty}^{x} f(x')dx'$ with f the probability density function (PDF) (Grimmett and Stirzaker, 2001, p. 89). Note that for a **probability mass function** (PMF – case where X is discrete; e.g., Eq. (2.1)), $F(x) = \sum_{i:x_i \leq x} f(x_i)$ (Grimmett and Stirzaker, 2001, p. 46). A numeric value x is finally sampled by defining $F(x) = u$ from which we obtain $x = F^{-1}(u)$ where F^{-1} indicates the functional inverse of F (e.g., Clauset et al., 2009), hence the name **inverse transform sampling**. Let us now illustrate this process with the **logistic function** $F(x) = 1/(1 + \exp[-w(x - c)])$ with the weight $w = 25$ high enough to have most of the data concentrated around $c = 2$. This yields $x = -\log(u^{-1} - 1)/w + c$ with 10 x samples depicted below:

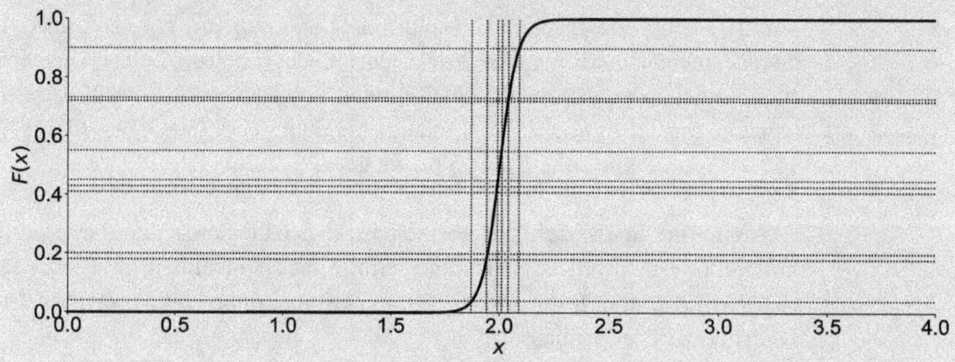

Finally, an impactor excavates a radially symmetric parabolic cavity, which yields the tsunami energy E [J]

$$E = \begin{cases} \frac{1}{6}\pi\rho_w g(h_w r_0)^2 \text{ for } r_{out} = r_0 \text{ (all water ejected)} \\ \frac{1}{3}\pi\rho_w g(h_w r_0)^2 \text{ for } r_{out} = \sqrt{2}r_0 \text{ (all water deposited into lip)} \end{cases}, \tag{2.22}$$

where h_w [m] is the depth of the cavity and r_0 [m] and r_{out} [m] are the inner and outer radii, respectively (Fig. 2.9d). With the tsunami energy a fraction of the kinetic energy of the impactor, that is, $E_{Ts} = \eta E_{AI}$ with $\eta = 0.155$ and E_{AI} calculated from Eq. (2.6), one can estimate h_w from the impactor source characteristics by assuming a ratio of cavity diameter ($2r_0$) to cavity depth of 3 : 1 (Ward and Asphaug, 2000).

2.2.3 Track Sources

2.2.3.1 Tropical Cyclones (TC)

The source of a tropical cyclone is an area of low pressure in motion. First considering the case of a stationary storm, the source can be considered a point represented by the cyclone eye. The two main characteristics of that source are the central pressure P_c and the radius of maximum winds r_{max} that delimitate the eye. At first order, P_c is a function of sea surface temperature, while r_{max} depends on pressure difference and latitude (Vickery et al., 2000). The difference between P_c and the ambient pressure P_n is the principal parameter for gauging the size of a tropical cyclone. In the absence of storm forward motion, the surface pressure P at a radial distance r from the centre of the eye can be approximated by the formula

$$P = P_c + (P_n - P_c) \exp\left(-\frac{r_{max}}{r}\right)^B, \tag{2.23}$$

where $1 \leq B \leq 2.5$ is **Holland's dimensionless parameter** that dictates the radial pressure profile shape (Holland, 1980; Fig. 2.10a). This horizontal difference in pressure across the cyclone drives the cyclonic winds, which leads to the following wind profile,

$$v(r) = \left(\frac{B}{\rho_a}(P_n - P_c)\left(\frac{r_{max}}{r}\right)^B \exp\left[-\left(\frac{r_{max}}{r}\right)^B\right] + \frac{r^2 c_f^2}{4}\right)^{1/2} - \frac{r c_f}{2}, \tag{2.24}$$

where v [m/s] is the wind speed at radius r, P_c [Pa] is the central pressure, P_n [Pa] is the ambient pressure, c_f [rad/s] is the Coriolis parameter, and $\rho_a = 1.15$ kg/m^3 is the air density (note that $c_f = 2\Omega \sin\phi$ with $\Omega = 7.292 \times 10^{-5}$ rad/s the rotation rate of the Earth and ϕ [rad] the latitude). The spatial extent of a tropical cyclone can then be delimited as the radius of winds in excess of the maximum tropical storm speed threshold of 33 m/s (Fig. 2.10b). However, the size S of an event is, in practice, the maximum wind speed

$$v_{max} = v(r = r_{max}) = \sqrt{\frac{B(P_n - P_c)}{\rho_a e}}, \tag{2.25}$$

with e the base of natural logarithms (Holland, 1980). The category of a tropical cyclone can then be assessed from v_{max} on the Saffir–Simpson hurricane scale (see Table 3.4). In energy terms, Emanuel (2005) proposed to define the size of a tropical cyclone as the total energy E [J] derived from the total power dissipated over the storm lifetime τ,

$$E = 2\pi \int_0^\tau \int_0^{r_{out}} \rho_a c_D v^3 r dr dt, \tag{2.26}$$

where r_{out} [m] is the outer storm limit and $c_D = 2 \times 10^{-3}$ (Emanuel, 1999) is the surface drag coefficient, which represents the resistance by the surface to the wind.

Since the source is moving with a forward speed v_f, it must be defined as a track source with v_{max} a function of the time t. The evolution of P_c and v_{max} along the storm track are shown in Figure 2.10c for Typhoon Yvette (1992), which exemplifies the anti-correlation between the two variables (Eq. (2.25)). Along the track, the wind speed progressively

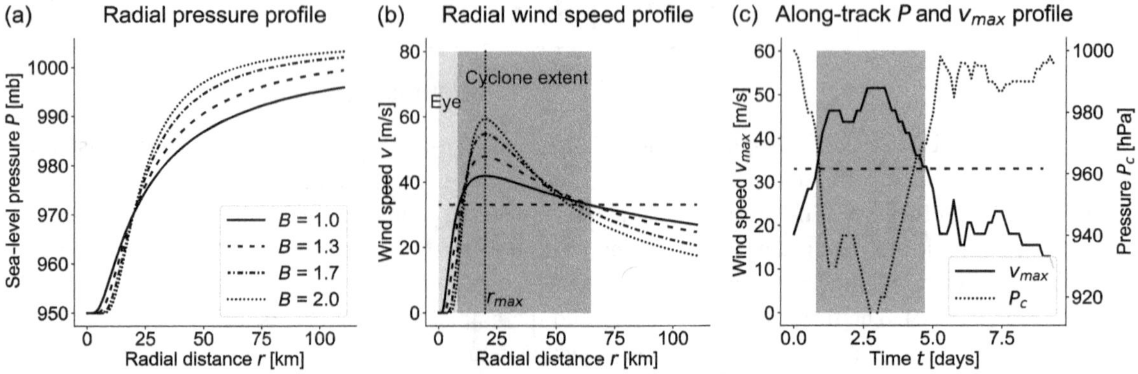

Fig. 2.10 Pressure and wind speed profiles of a tropical cyclone. (a) Radial pressure profile of a stationary storm calculated from Eq. (2.23) with $P_n = 1{,}005$ mb, $P_c = 950$ mb, $r_{max} = 20$ km, and $\phi = 30°$. (b) Matching gradient-level azimuthal wind speed profile calculated from Eq. (2.24). (c) Along-track profile of P_c and v_{max} for Typhoon Yvette (1992), with data from IBTrACS (Knapp et al., 2010, 2018).

increases as the tropical cyclone grows, and then decreases, as the storm makes landfall, weakens, and dies off. The wind speed along-track profile takes the simple form

$$v(t) = v_{max} e^{-c_d t}, \tag{2.27}$$

where v_{max} is the maximum wind speed prior to landfall, $t = 0$ is the time of landfall, and c_d is a site-specific decay constant (Kaplan and DeMaria, 1995). This **exponential decay** model can also be expressed in terms of pressure difference (Vickery and Twisdale, 1995). As with any other peril caused by a weather system, there is no fixed source in space, only stochastic sources representative of the weather conditions. Modelling of tropical cyclone tracks from the statistics of historical records (Vickery et al., 2000) is similar in concept to the one previously described for tornado sources (Fig. 2.8). This is done by sampling from the statistical distribution of the storm genesis points, directions, translation speeds, and wind speeds (Rumpf et al., 2007). More details will be given on v_f and the evolution of r_{max} over time when the intensity footprint of a tropical cyclone will be fully modelled in Section 2.4.

2.2.3.2 Other Meteorological Perils, Especially Hailstorms (HS)

Extratropical cyclones are part of a very complex coupled weather system, making their modelling a challenge (Walker et al., 2020). Indeed, they do not have the nearly circular vortex of tropical cyclones. The detailed description of their source remains outside the scope of this introductory textbook, as relatively complex algorithms, including filtering techniques, are needed to track these storms (Hodges, 1995). Examples of European winter storms simulated by Copernicus Climate Change Service (2022) were shown in Figure 2.2 for completeness.

Severe convective storm source modelling is also not covered here since both track and straight-line wind profiles are generally not explicitly defined when events are derived from historical meteorological data for specific regions (Dotzek et al., 2009). However, the other sub-perils of convective storms (i.e., lightning, hail, tornado) are considered. The tornado source was defined in Section 2.2.2.3. Lightning will be described as a natural trigger of wildfires in Section 2.2.4.2. It can just be noted here, for comparison with other perils, that the energy expended in lightning flashes in a single thunderstorm cell is in the order of 10^9-10^{10} J (Williams and Lhermitte, 1983).

Considering hail as an individual peril (i.e., hailstorm), the source may be defined as a track, in this case an area source in motion with the hail streak in the shape of an ellipse. Track trajectory and ellipse characteristics (axis lengths and orientation) are modelled as stochastic realizations, similarly to other meteorological perils. Due to the limited size of hail streaks, stochastic events can be defined as aggregated hail streaks in larger weather systems (Punge et al., 2014). The size S of an event is determined in terms of its kinetic energy with the velocity v [m/s] the terminal fall speed function of the average physical characteristics of hailstones, that is,

$$v = \sqrt{\frac{8g\rho_h r}{3\rho_a c_D}}, \tag{2.28}$$

where ρ_a is the density of air, $\rho_h = 900$ kg/m^3 is the density of hail, r [m] is the hailstone radius, $c_D = 0.6$ is the drag coefficient, and $g = 9.81$ m/s^2. The total energy is obtained by summing over all the spherical hailstones in the elliptic hail streak area according to their diameter distribution (Katz and Garcia, 1981).

2.2.4 Area Sources and Diffuse Sources

The source types previously defined (points, lines, and tracks) are all localized in space, and their characteristics are sufficient to fully describe the event size S. Source and size information will be the primary information necessary for hazard intensity footprint modelling in Section 2.4. However, some perils cannot be simplified and implemented into the idealized hazard process shown in Figure 2.3. This is the case when the source, coined a **diffuse source**, is distributed throughout the environment, and acts as an extended energy stock for the event to grow over time (Fig. 2.11). Examples of perils include landslides, wildfires, and epidemics. Their respective diffuse sources are the terrain (providing more soil), the vegetation (providing more fuel), and the population (providing more susceptible individuals). All events initiate as so-called **seed events**, which can be described as area sources but do not constrain the size of the final event, which requires dynamic modelling. Area sources for seed events are defined below and will be extended to diffuse sources in Part II (Chapter 4).

2.2.4.1 Landslides

The source of a landslide is the set of terrain patches experiencing mass movement initiated by an unstable slope. The size S_0 of a seed event can then be defined as the area A or volume V of each unstable patch or of all patches combined. The area is a simpler metric assessable from a map of the **factor of safety** F_S, which is the ratio of resisting forces to

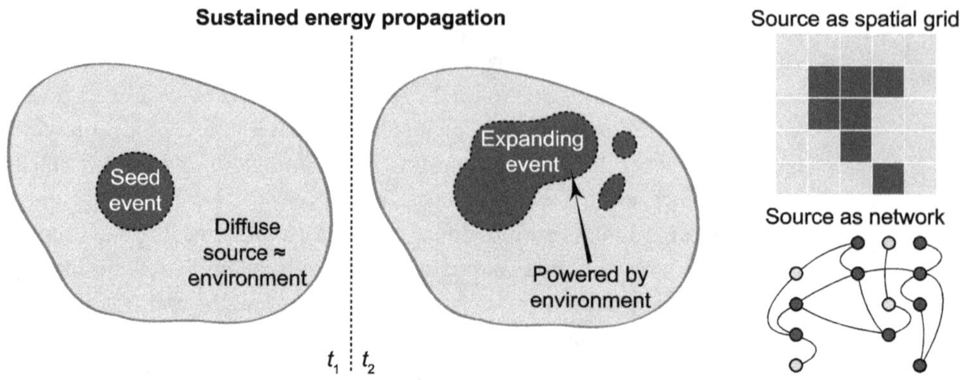

Fig. 2.11 Specific case of the environment as a diffuse source. Some event sources cannot be reduced to a localized source. This is the case when the surrounding environment acts as a source, providing continual energy to the event to grow from a seed event. In comparison to Figure 2.3, here, the intensity footprint $I(x, y)$ has the same spatial extent as the event of size, or area, S. Diffuse sources will be discretized in Part II (Chapter 4) as spatial grids (e.g., landslides, wildfires) or networks (e.g., blackouts, transport system failures).

driving forces. If $F_S = 2$, the slope has resisting forces twice as large as the driving forces, making it stable. If $F_S = 1.5$, the slope's strength is only 50% greater than the driving forces, making the situation critical to failure but not failing yet. An unstable condition is met for $F_S < 1$. The value is controlled by water saturation, soil properties, and terrain morphology. A simple formula (for an infinite slope) is

$$F_S = \frac{\frac{c_s}{\rho_s g h} + \cos\theta[1 - c_w \frac{\rho_w}{\rho_s}]\tan\phi_f}{\sin\theta}, \tag{2.29}$$

where h [m] is the depth perpendicular to surface (or thickness) of the sliding body, c_s [N/m^2] (or [Pa]) is the soil cohesion, ρ_s [kg/m^3] is the soil density, $\rho_w = 1{,}000$ kg/m^3 is the unit weight of water, θ [°] is the slope angle, ϕ_f [°] is the soil's internal friction angle, $c_w = h_w/h$ is the relative wetness representing the ratio between water column height h_w and h, and $g = 9.81$ m/s^2 (Pack et al., 1998). Figure 2.12a shows the factor of safety as a function of slope angle. The main driver of a landslide is heavy rainfall, which increases the relative wetness, here illustrated by raising c_w from 0 (dry condition) to 1 (water saturated). In this process, the slope angle for landslide triggering goes down from 35° to 25° for the given soil conditions. In general, a range $25° \leq \theta \leq 40°$ is expected for rainfall-triggered events (Crosta, 1998). Higher effective cohesion is obtained in the presence of a vegetation root system, which hints on how a wildfire, by destroying part of the root system, can increase the likelihood of a landslide. Figure 2.12b shows the F_S map, or landslide susceptibility map, for a simulated fractal topography (Fournier et al., 1982) (to be made explicit in Part II, Chapter 4). Figure 2.12c shows the resulting seed events, delimited by the contour line $F_S = 1^-$. In energy terms, a landslide releases potential energy which is transformed into kinetic energy. The maximum velocity of the mass movement would be $v = \sqrt{2gh}$ with h the elevation change in the idealized case where all the potential energy is converted into kinetic energy (i.e., free-fall case). In reality, only a fraction is converted

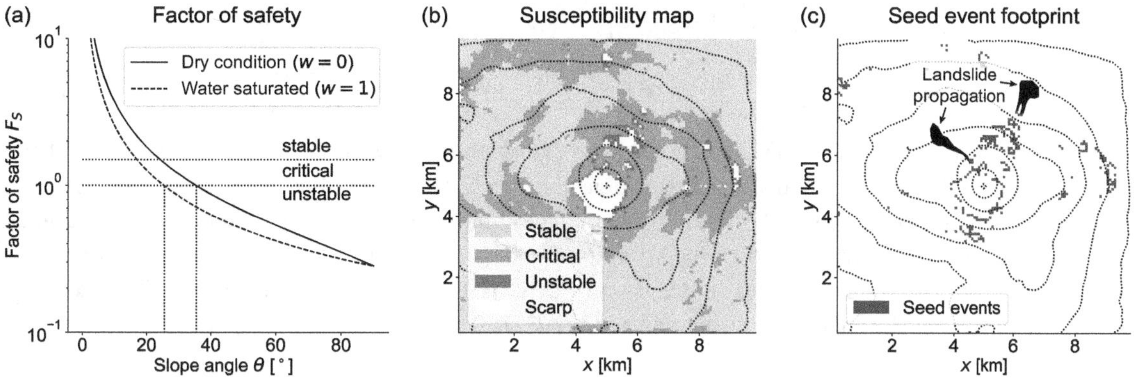

Fig. 2.12 Landslide susceptibility defined from the factor of safety (F_S). (a) Relationship between F_S and slope angle θ (Eq. (2.29)) with parameters $h = 2$ m, $c_s = 10$ kN/m^2, $\phi_f = 20°$, and $\rho_s = 1{,}800$ kg/m^3. (b) F_S map for a simulated topography with fractal dimension $D_f = 2.3$ (isolines represented by dotted curves) and a soil with relative wetness $c_w = 0.2$. (c) Landslide seed event map, with size $S_0 = A$ and A the area delimited by the contour line $F_S = 1^-$. Two examples of landslides are sketched, for which size $S > S_0$ is due to the phase of propagation to be modelled in Part II.

($\approx$ 24%; Ekström and Stark, 2013), with the rest dissipated in the form of heat by friction and fracture energy.

2.2.4.2 Urban Fires (Fi) and Wildfires (WF)

The source of a fire is some combustible material combined to some spark that ignites the fire. Once a fire is ignited, it will propagate throughout the environment as long as the heat of combustion is sufficient to ignite neighbouring areas and the fuel is not exhausted. This process depends on the characteristics of the fuel, topography, and weather, with parameters such as terrain slope, wind speed, moisture content, and temperature (Rothermel, 1972). Fires in both wildland and urban areas may be modelled as elliptical area sources (Anderson et al., 1982), with the fire spread rate R_v (e.g., [m/min]) defined as

$$R_v = R_{v,0}(1 + |\phi_s + \phi_w|)\frac{1 - c_\epsilon}{1 - c_\epsilon \cos\phi}, \tag{2.30}$$

with $R_{v,0}$ the spread rate on flat terrain without wind, ϕ_s the direction of maximum slope, ϕ_w the wind direction, c_ϵ the eccentricity function of wind speed and terrain slope, and ϕ an arbitrary direction. $R_{v,0}$ is a function of the combustible characteristics (e.g., Trunfio, 2004). Over time, as the environmental conditions change (e.g., wind direction), the fire front grows elliptically by following **Huygens' wavelet principle**, which states that all points on the fire front may be regarded as new (elliptical) sources (Anderson et al., 1982). Since the orientation of the ellipses can change, the shape of the fire becomes more complex, which is modelled numerically. The trigger of a fire, if human-made, can be a flame, an explosion, or an electric spark. If natural, the main source is a lightning strike. In this case, the occurrence of a seed event depends on the ignition rate λ_0 [min^{-1}]

$$\lambda_0 = \eta\lambda_{Li} = \eta c_1(1 - c_2)\left(3.44 \times 10^{-5} h^{4.9}\right), \tag{2.31}$$

which is a function of the convective cloud top height h [km], a scaling factor $c_1 = 0.107$ for the global climatological average, the intra-cloud to cloud-to-ground ratio c_2 function of the cloud depth and $\eta = 0.001$ the ratio of lightning strikes λ_{Li} that ignite a fire (conditional on the convective storm being a dry thunderstorm) (e.g., Krause et al., 2014). Energy is released during a fire in the form of heat by the process of combustion. The size of an event S is then the **combustion energy** E [J],

$$E = mc_v, \tag{2.32}$$

where m [kg] is the burnt mass and $c_v = 7.2 \times 10^6$ J/kg is the caloric value of wood. m can be estimated from the wood density $\rho_{wood} = 400$–800 kg/m^3 depending on the tree species, the density of wood per hectare, and the event burnt area A [ha].

2.3 Event Size Distribution

The probabilistic nature of a hazard is described by the occurrence rate $\lambda(S)$, with S the event size (quantified as a function of the source characteristics in the previous section). Estimating λ represents the second step in the general hazard process described in Figure 2.3. The size distribution of a given peril is commonly defined in terms of the exceedance rate, $\Lambda(S) = \lambda(> S)$, and is described by one of two main categories of functions $f_S(S)$: the power law when events are representative of a self-similar (fractal) process (Section 2.3.1), and the functions from extreme value theory when events are the extrema of a process. This group consists of the generalized extreme value (GEV) and generalized Pareto distribution (GPD) families (Section 2.3.2). Although model selection often depends on a given discipline's most accepted approach for event definition and event count (Harris, 2005; Malamud and Turcotte, 2006), there is nonetheless a tendency to use the power law for discrete events (e.g., asteroid impacts, earthquakes, volcanic eruptions, wildfires) and extreme value theory for events defined from a continuous quantity (e.g., storms from maximum wind speed, floods from maximum discharges). A common property of these statistical distributions, under specific conditions for GEV and GPD, is a tail heavier than the exponential distribution, making catastrophic events the dominant factor of the size–frequency distribution (so-called **heavy-tailed distributions**). This is illustrated by the giant-catastrophe metaphor in Box 2.2.

Perils will be described below in alphabetical order with a statistical model parameterized to each empirical size distribution, so that a stochastic event set $\{\text{ID}, S_{\text{ID}}, \lambda_{\text{ID}} \text{ for } N_{\text{ID}} \text{ events}\}$ can later be generated for each peril considered. When available, global data will be used for model calibration. Although the focus will be on power law and extreme value statistics, additional heavy-tailed distributions may be used in practice to provide more flexibility in the description of natural processes. The Gamma

Box 2.2 The Giant-Catastrophe Metaphor

The **normal (or Gaussian) distribution** describes the 'normal' processes of everyday life, such as people's height. Its CCDF is defined as $\bar{F}(S) = \left\{1 - \text{erf}\left[(S - \mu)/(\sigma\sqrt{2})\right]\right\}/2$, where μ is the mean, σ the standard deviation and erf the **error function** with $\text{erf}(x) = (2/\sqrt{\pi})\int_0^x \exp(-z^2)dz$. Due to its exponential tail, the probability of an event having a size $S \gg \mu$ quickly tends to zero, as proven by the **68–95–99.7 rule** (i.e., percentage of values lying within $\pm 1\sigma$, 2σ, or 3σ, respectively). Considering the normal height distribution of men being $\mu = 178 \pm 7.6$ cm (Roser et al., 2013) and assuming the annual rate of males reaching their maximum height being equivalent to the number of male births per year (i.e., 70 million), one gets about $\Lambda(h) = 5$ individuals/year exceeding a height $h = 218$ cm (which is the size of Kevin Peter Hall, who impersonated the Predator). The chance to see someone 10 cm taller falls to $\Lambda(h) = 1.6 \times 10^{-3}$ or once every 600 years. Seeing a healthy person reach the size of Robert Wadlow (1918–1940), the tallest person in recorded history ($h = 272$ cm), becomes virtually impossible with $\Lambda(h) \approx 10^{-27}$. But what if the height followed a heavy-tailed distribution? How often would we see, say, a 15-m giant? Would it even be possible to see a 100-m tall giant? This is an important thought experiment since such metaphoric giant monsters actually come visit us from time to time in the form of catastrophes. Considering a power exponent $\alpha = 2.5$ (Eq. (2.33)) and the Predator $\Lambda(h)$ point as anchor, the return period of King-Kong (assuming $h = 15$ m) would be 4 years and the return period of Godzilla (assuming $h = 100$ m), only 63 years! And one knows how much damage they can do!

distribution, the stretched exponential, etc. allow one to cover the wide spectrum of possibilities between the exponential and the power law (e.g, Sornette, 2006). They will, however, not be covered here.

2.3.1 Power-Law Distribution

A **power law** is a relationship that takes the functional form

$$f_S(x) \propto x^{-\alpha}, \tag{2.33}$$

with $\alpha > 1$ the power exponent (or scaling parameter). This parameter typically lies in the range $1.5 < \alpha < 3$. In practice, empirical phenomena obey a power law only for values $x \geq x_{min}$ (Clauset et al., 2009). It will soon become apparent that *most perils follow a power-law distribution*. This has been explained by a fundamental process of cascade of energy dissipation on all length scales (i.e., process of fractal energy dissipation) from some uniform input energy (Mandelbrot, 1974). Note that a **fractal**, previously defined as a **self-similar** (mathematical) object, has the property to show similar or identical patterns at different scales. For the power law, we have indeed $f_S(cx) \propto c^{-\alpha}x^{-\alpha} \propto f_S(x)$ when scaling x by a constant factor c. This may also be related to the concept of self-organized criticality (SOC) (Bak et al., 1987, 1990), to be described in Part II (Chapter 4). SOC has been proposed as one explanatory hypothesis, among others (King, 1983), for the emergence of power laws in both natural (Malamud and Turcotte, 2000) and human-made hazards (e.g. Roberts and Turcotte, 1998). Although the power law has been considered by many as a

Box 2.3 Basic Power-Law Statistics

It was implicit in the giant-catastrophe metaphor (Box 2.2) that most data samples remain close to the mean in the Gaussian distribution while the mean strongly depends on extreme samples in a heavy-tailed distribution. This can be formalized by looking in more detail at the power-law function $f_S(x)$ (Eq. (2.33)), for which the PDF (Eq. (2.34)) satisfies the condition

$$\int_{x_{min}}^{+\infty} f(x)dx = \int_{x_{min}}^{+\infty} c \cdot f_S(x)dx = 1$$

by definition (with c a normalization constant). Then, the first two **statistical moments** μ_k, the mean ($\mu = \mu_1$) and the variance ($\mathrm{Var} = \mu_2 = \sigma^2$) are calculated from

$$\mu_k = \langle x^k \rangle = \int_{x_{min}}^{+\infty} x^k f(x)dx \propto +\infty^{k+1-\alpha},$$

which shows that the mean is finite only for $\alpha > 2$ and the variance only for $\alpha > 3$. With $1.5 < \alpha < 3$ often observed for hazards, it means the potential for 'surprise' extreme events emerging from the tail. This is exactly the basis for the **Noah effect** which 'designate[s] the observation that extreme precipitation can be very extreme indeed' (Mandelbrot and Wallis, 1968). Similarly, in the **black swan** metaphor of Nassim Taleb, 'if you are dealing with quantities from Extremistan [i.e., Black Swans], you will have trouble figuring out the average from any sample since it can depend so much on one single observation. The idea is not more difficult than that' (Taleb, 2007, p. 34). This also relates to the **catastrophe principle** which is defined as $\mathbb{P}(\max(X_1, X_2, \cdots, X_n) > y) \sim \mathbb{P}(X_1 + X_2 + \cdots + X_n > y)$ as $y \to \infty$, meaning that 'an unexpectedly large sum of random variables is likely the result of a catastrophe, that is, a result of one unexpectedly large event' (Nair et al., 2022, p. 58).

universal property of Nature due to its link to fractality, other distributions may often also apply (Cui and Perera, 2008; Clauset et al., 2009; Corral and González, 2019).

The PDF of the power law (Eq. (2.33)) is

$$f(x) = \frac{\alpha - 1}{x_{min}} \left(\frac{x}{x_{min}} \right)^{-\alpha} \tag{2.34}$$

(see demonstration in Box 2.3). However, the power-law behaviour is best visualized with its CCDF:

$$\bar{F}(x) = \mathbb{P}(X > x) = \int_x^{\infty} f(x')dx' = \left(\frac{x}{x_{min}} \right)^{-(\alpha-1)} \tag{2.35}$$

(i.e., the **Pareto distribution** with $\alpha - 1$ the Pareto shape parameter, also known as the tail index). Finally, the exponent parameter can be estimated from

$$\hat{\alpha}_{MLE} = 1 + n \left[\sum_{i=1}^{n} \log \frac{x_i}{x_{min}} \right]^{-1}, \tag{2.36}$$

where x_i, $i = 1, ..., n$, are the observed values of x such that $x_i \geq x_{min}$ (e.g., Clauset et al., 2009, eq. (3.1)) (Ex. 2.3a and b). MLE here denotes the **maximum likelihood estimation**

Table 2.5 Power-law size distribution parameters for different perils

ID	Peril	Size S	a	b^a	Description
AI	Asteroid impacts	Energy E_{TNT} [kton]	0.565	0.88	Global (1994–2022)
BO	Blackouts	Number of customers affected	6.466	1.12	US (1984–2002)
EQ	Earthquakes	Seismic energy [J]	12.029	0.72	Global (1970–2018)
TA	Terrorist attacks	Fatalities	2.703	1.45	Global (1968–2006)
TC	Tropical cyclones	Power dissipation index [m^3/s^2]	32.323	2.93	Global (1990–2022)
Ts	Tsunamis	Run-up height [m]	0.852	1.01	Global (1900–2022)
VE	Volcanic eruptions	Bulk deposit volume V [km^3]	−1.155	0.66	Global (1000–2014)
WF	Wildfires	Burnt area [ha]	8.553	1.23	Global (2000–2017)

[a] Only the a and b values re-estimated for the data sets shown in Figure 2.13 are listed in this table. Beware that estimates are highly dependent on the choice of S_{min} and are provided here only for illustration purposes. A proper data completeness analysis should be done to estimate S_{min} and the power law tested against other models (Clauset et al., 2009; Corral and González, 2019). See the text for other values found in the literature, including for additional perils. Note again that $\alpha = b + 1$ and that other statistical models may provide better fits on a case-to-case basis.

method that seeks to find the parameter values that maximize the product of individual likelihoods ($\mathcal{L}$) of the observed data given the model (here with $\mathcal{L}(\alpha) = \prod_{i=1}^{n} f(x_i|\alpha)$). Note that $\hat{\alpha}_{MLE}$ can be biased due to finite sample size, which becomes more pronounced toward the tail, where extremes are; the Hill estimator is often used to address the issue (Hill, 1975). Box 2.3 goes deeper into the statistics of the power law, explaining why catastrophes are surprising events, and how power laws relate to the Noah effect (Mandelbrot and Wallis, 1968), black swans (Taleb, 2007), and catastrophe principle (Nair et al., 2022).

The size distribution of hazards is rarely described probabilistically but instead in terms of an annual exceedance rate Λ for an event size value S. It can be derived from the CCDF, such that

$$\Lambda(S) = \Lambda(S_{min})\bar{F}(S) = \Lambda(S_{min})\left(\frac{S}{S_{min}}\right)^{-(\alpha-1)}. \tag{2.37}$$

In practice, the empirical power-law distribution takes the form

$$\log_{10}\Lambda(S) = a - b\log_{10}S \tag{2.38}$$

with $a = \log_{10}(\Lambda(S_{min})S_{min}^{\alpha-1})$ the **event productivity** and $b = \alpha - 1$ the **event size ratio** (or tail index), that is, the slope of the size distribution in a log–log plot. Table 2.5 provides a summary of the parameters that Eq. (2.38) can adopt for various perils, while Figure 2.13a shows the matching empirical power-law distributions. Note that data points are plotted by first calculating the empirical CCDF $1 - \hat{F}(S)$ (or CCDF $= 1 -$ eCDF; Ex. 2.3c; see Box 2.4 for a pseudocode). The next paragraphs provide an overview of the power distributions that different perils follow (with power exponent b systematically given).

2.3.1.1 Armed Conflicts (AC) and Terrorist Attacks (TA)

Lewis Fry Richardson, pioneer in weather forecasting (Section 1.2.3), was the first to introduce the concept of power-law scaling to the study of conflicts (Richardson, 1948). He

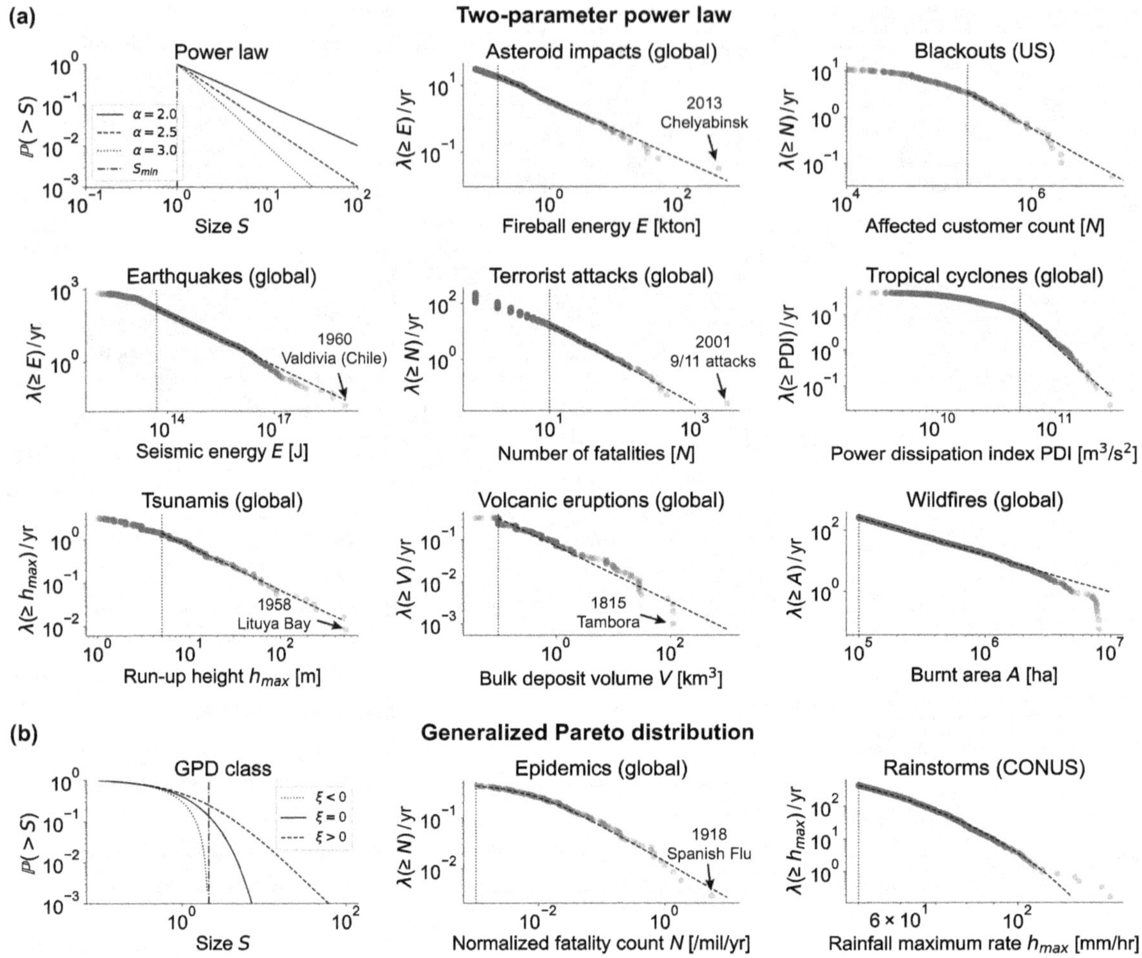

Fig. 2.13 Empirical size distribution of various perils. (a) Modelled by a two-parameter power-law distribution (Eq. (2.38); Table 2.5) with power exponent estimated using Eq. (2.36): Asteroid impacts (CNEOS, 2023), blackouts (Clauset et al., 2009), earthquakes (ISC-GEM; ISC, 2023; Storchak et al., 2013, 2015; Di Giacomo et al., 2018; CC-BY-SA, with permission), terrorist attacks (Clauset et al., 2007), tropical cyclones (IBTrACS; Knapp et al., 2010, 2018), tsunamis (NGDC/WDS, 2022), volcanic eruptions (LaMEVE; Crosweller et al., 2012), and wildfires (FRY; Laurent et al., 2018a,b; CC-BY 3.0). (b) Modelled by the generalized Pareto distribution (Eq. (2.50)) with parameters estimated using Python's `scipy.stats.genpareto.fit()` (Virtanen et al., 2020): epidemics (Marani et al., 2021a,b; CC-BY 4.0, https://creativecommons.org/licenses/by/4.0/) and rainstorms (Zhou et al., 2019; CC-BY 4.0, https://creativecommons.org/licenses/by/4.0/). Note that the y-axis is here defined by the cumulative rate $\lambda(\geq S)$ instead of $\lambda(> S) = \Lambda(S)$ in order to display the most extreme event.

found a power exponent $b \approx 1$ for various types of conflicts (war, banditry, gang warfare) with size S the number of fatalities. Cederman (2003) found $b \approx 0.5$ for interstate wars taking place between 1820 and 1997 and the 1465–1965 European great power wars. Clauset et al. (2009) calculated $b = 0.7$ for 1816–1980 wars. In the case of terrorism worldwide

Box 2.4 Pseudocode to Plot a CCDF from Empirical Data

```
Si = sampling(N) #e.g., from power law distribution
# or Si from real observations

# empirical cumulative distribution function (eCDF)
eCDF = sequence_natural_num(N) / N #i.e., (1, 2, ..., N)/N

# plot complementary cumulative distribution function (CCDF)
CCDF = 1 - eCDF
Si_sorted = sort(Si) #in ascending order
plot_scatter_points(Si_sorted, CCDF)
```

from 1968 to 2006, the values $a = 2.703$ and $b = 1.45$ are obtained (Fig. 2.13a), close to $b = 1.38$ found by Clauset et al. (2007). Overall, the power exponent b is observed to be in the range 0.5–1.5 when considering all types of conflicts.

2.3.1.2 Asteroid Impacts (AI)

The flux of small near-Earth objects colliding with the planet follows a power law with $S = E_{TNT}$ [kton], the kinetic energy of bolides (Eq. (2.6)). Only objects large enough lead to an impact; the smaller ones disintegrate in the upper atmosphere. Brown et al. (2002) found $a = 0.5677$ and $b = 0.90$ from 1994–2002 satellite data. Corral and González (2019) more recently obtained $b = 1.02$ for updated data from CNEOS (2023). Considering the same data set but updated to 2022 yields $a = 0.565$ and $b = 0.88$ (Fig. 2.13a). Applying this parameterization to the 1908 Tunguska event, which had an estimated yield of 10 Mt, yields a return period of Tunguska-type events of about 1,000 years. It also indicates that the Earth is, on average, struck by an object of energy $E_{TNT} \approx 4$ kt every year (Ex. 2.1c).

2.3.1.3 Blackouts (BO)

The distribution of blackouts also follows Eq. (2.38), with various measures possible for size S. In their review, Dobson et al. (2007) found $0.6 \leq b \leq 1.0$, with S defined in terms of unserved power or energy (based on data from China, New Zealand, North America, Norway, and Sweden). Using the number of customers affected as the measure for S, Clauset et al. (2009) obtained $b = 1.3$ for the US in the period 1984–2002. For the same data set shown in Figure 2.13a, best parameter fits are $a = 6.466$ and $b = 1.12$.

2.3.1.4 Cyberattacks (CA)

The size distribution can be described as a power law with S the number of personal identity losses or the data breach volume. In Maillart and Sornette (2010), a value of $b = 0.7$ was found for data from the Open Security Foundation for the 2000–2008 period. Extreme

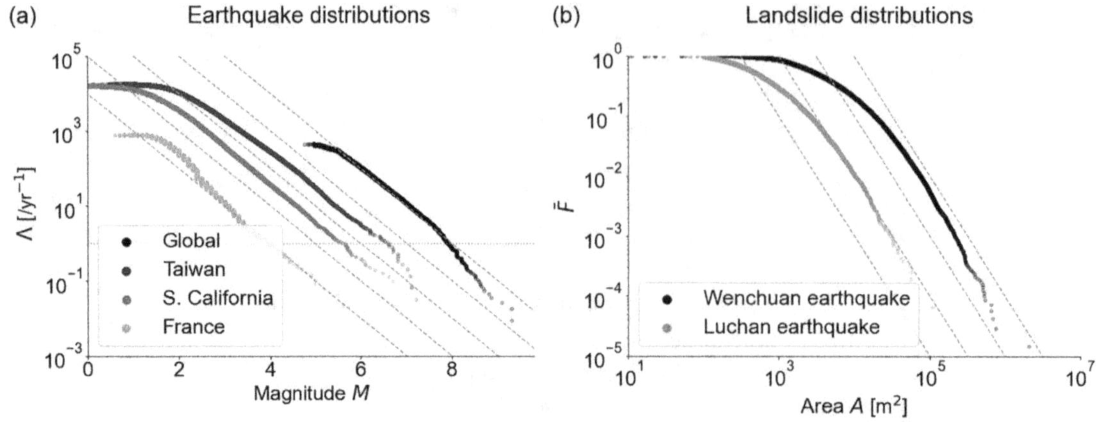

Fig. 2.14 Size distributions of earthquakes and landslides. (a) Earthquake case: global (ISC-GEM; ISC, 2023; Storchak et al., 2013, 2015; Di Giacomo et al., 2018, CC-BY-SA, with permission), Taiwan (Wu et al., 2008), Southern California (Hauksson et al., 2012), and France (SI-HEX catalogue from Cara et al., 2015; BCSF-Rénass, 2022; CC-BY 4.0, https://creativecommons.org/licenses/by/4.0/). The dotted lines represent the Gutenberg–Richter relationship (Eq. (2.39)) with $b = 1$ as a visual guide. (b) Landslide case: triggered by the 2008 $M = 7.9$ Wenchuan earthquake (Li et al., 2014) (data courtesy of the authors) and the 2013 $M = 6.6$ Lushan earthquake (Xu et al., 2015). The dotted lines represent the power law (Eq. (2.38)) with $b = 2$ as a visual guide. Xu et al. (2008), for example, found $b = 2.07$ for the Wenchuan landslides.

value distributions have also been proposed to fit the distribution of cyber losses, with various studies reviewed by Jung (2021). Comprehensive public databases spanning many years have yet to be developed to improve statistical inference (Neto et al., 2021).

2.3.1.5 Earthquakes (EQ)

The size distribution of earthquakes follows the **Gutenberg–Richter law**

$$\log_{10} \Lambda(M) = a - bM \tag{2.39}$$

with M the earthquake magnitude (Gutenberg and Richter, 1944). The b-value is close to 1 in tectonic regimes (Fig. 2.14a). It means that the number of events decreases roughly by a factor of 10 for each unit increase in magnitude and that one earthquake of magnitude $M = a$ is expected per year. About one earthquake with a magnitude ≥ 8 can be expected globally every year. For two seismically active regions such as Southern California and Taiwan, one can expect magnitudes close to 6 and to 7 annually, respectively. In a less seismically active region such as France, one can roughly expect one $M = 4$ event per year (Fig. 2.14a). Equation (2.39) is an exponential function but is equivalent to a power law in the seismic energy domain E [J], with $\log_{10} E \propto 1.44M$ (Eq. (2.15)). For earthquakes occurring globally in the period 1970–2018 (Storchak et al., 2013), one obtains $a = 12.029$ and $b = 0.72$ for Eq. (2.38) with $S = E$ (Fig. 2.13a; Ex. 2.3d).

2.3.1.6 Hydrological Perils (including FF and Ts)

Although the size distribution of hydrological perils is commonly described within the context of extreme value theory (Section 2.3.2), use of a power law has been advocated by some. Malamud and Turcotte (2006) found $1.1 \leq b \leq 3.7$ with size S the maximum discharge Q_p [m^3/s] within a given time interval at different rivers in the US. Regarding tsunamis, the size S may be taken as the run-up height h_{max} [m]. Burroughs and Tebbens (2005) found $0.8 \leq b \leq 1.3$ for different locations in Japan. Considering the 1900–2022 NCEI/WDS Global Historical Tsunami Database (NGDC/WDS, 2022), one gets $a = 0.852$ and $b = 1.01$ (Fig. 2.13a).

2.3.1.7 Landslides (LS)

The size distribution of landslides is described as a power law, with the size S the area A or the volume V. Conversion between area and volume can be performed with the empirical relationship $V = cA^{1.5}$ (e.g., Malamud et al., 2004). Considering $S = A$, the power exponent b is observed to be in the range 1.1–1.4, independently of the landslide trigger, in different regions of the world (Van Den Eeckhaut et al., 2007a; Jafarimanesh et al., 2018). Although global inventories of landslides exist, they rarely contain quantitative information on the landslide size, but instead semi-quantitative estimates (e.g., small to very large) or fatality counts (Petley et al., 2005; Kirschbaum et al., 2015; Guha-Sapir et al., 2022). Since mass movement is a secondary hazard and thousands to tens of thousands of events can be triggered by a unique earthquake or rainstorm, landslide inventories exist per trigger event (Fig. 2.14b).

2.3.1.8 Tropical Cyclones (TC)

Similarly to other meteorological perils, the size distribution of tropical cyclones is typically described within the context of extreme value theory (Section 2.3.2). However, defining storm size in terms of a power dissipation index (i.e., $\text{PDI} = \int_0^\tau v_{max}^3 dt$ simplified from Eq. (2.26); Emanuel, 2005), a power-law distribution may apply. Corral and González (2019) found a relatively high power exponent with $2.28 \leq b \leq 4.15$ (for 1986–2016 hurricanes from the Hurricane Database HURDAT2), which was shown to be uncertain and unstable, questioning the power-law nature of the empirical distribution. Using the same proxy on IBTrACS (Knapp et al., 2010, 2018) yields $a = 32.323$ and $b = 2.93$. This behaviour is rather marginal, as it extends for less than one order of magnitude (Fig. 2.13a), as observed by Corral and González (2019).

2.3.1.9 Volcanic Eruptions (VE)

Explosive volcanic eruptions follow a power law of the form of Eq. (2.38) with the size S the erupted volume V [km^3]. Considering all volcanic eruptions which occurred since year 1000 in LaMEVE (Crosweller et al., 2012) and the size in terms of bulk deposit volume, the values $a = -1.155$ and $b = 0.66$ are obtained (Fig. 2.13a). A recent review of large

VEI eruptions by Newhall et al. (2018) indicates that VEI-7 events recur every 500–1,000 years. With the parameterization obtained from LaMEVE for the period 1000–2014, a return period of 300 to 1,300 years is found (see V–VEI relationship in Table 2.3).

2.3.1.10 Wildfires (WF)

Wildfires also follow a power-law distribution, with the size S defined by the burned area A. The first quantified empirical evidence that wildfires follow Eq. (2.38) was given by Malamud et al. (1998) and this behaviour has been verified since then by many others, as reviewed in Cui and Perera (2008), who found $1.1 \leq b \leq 1.8$ for China and the US. Clauset et al. (2009) obtained $b = 1.2$ for US federal land, and Corral and González (2019) found $b = 0.82$ for Angola and $b = 1.28$ for Canada. Using the FRY catalogue (Laurent et al., 2018a,b), one gets $a = 8.553$ and $b = 1.23$ (Fig. 2.13a).

2.3.2 Extreme Value Distributions

Modelling the size distribution of perils with the power-law function (Eq. (2.33)) is a commonly accepted approach, due to its strong links to the physics of complex natural systems (Sornette, 2006). Of most interest in CAT risk modelling is the behaviour of the tail of the distribution where extremes occur. Statistically, unobserved levels of extremes can be extrapolated from observed maxima values. This is done via **extreme value theory**, which provides two main classes of models to describe extremes: the generalized extreme value distribution family and the generalized Pareto distribution family (Coles, 2001). The former is used when considering annual maxima (i.e., **block maxima method**). This is particularly useful in hydrological and meteorological analyses where floods and storms represent the maxima of the water flow discharge and of the wind velocity field, respectively. The latter instead applies when extreme events are defined as all the values above a fixed threshold (**peak-over-threshold method**). Note that a power-law tail may be retrieved in both cases.

Considering a quantity X and its sequence of i.i.d. values $X_1, X_2, \dots, X_n$ in a given time period (e.g., 1 year),

$$X_{max} = \max\{X_1, X_2, \dots, X_i, \dots, X_n\} \tag{2.40}$$

is the maximum value (e.g., annual maximum) over the n-observation period (called a block). The quantity of interest could be the daily water discharge Q in flood hazard assessment, for example. In practice, the behaviour of X_i is unknown, making exact calculation of X_{max} impossible. The extreme value paradigm considers the case $n \to \infty$, which leads to the **generalized extreme value** (GEV) class of distributions:

$$\mathbb{P}(X_{max} \leq S) = \exp\left[-\left\{1 + \xi\left(\frac{S - \mu}{\sigma}\right)\right\}^{-\frac{1}{\xi}}\right] \tag{2.41}$$

with the ‘size’ value S usually termed the **return level** z_p, μ the location parameter, $\sigma > 0$ the scale parameter, and ξ the shape parameter (Coles, 2001, eq. (3.2)). $\xi < 0$ corresponds to the **Weibull distribution** and $\xi > 0$ to the **Fréchet distribution**. It should be noted that the extreme value distribution of a quantity behaving as a power law is the Fréchet distribution with $\xi = 1/b$ (or $\xi = 1/(\alpha - 1)$ – see Section 2.3.1 for the definition of b and α). The subset of the GEV family with $\xi = 0$ is interpreted as the limit of Eq. (2.41) as $\xi \to 0$, leading to the **Gumbel distribution**:

$$\mathbb{P}(X_{max} \leq S) = \exp\left[-\exp\left\{-\left(\frac{S-\mu}{\sigma}\right)\right\}\right] \tag{2.42}$$

(Coles, 2001, eq. (3.3)).

Considering the probability of exceedance instead,

$$\mathbb{P}(X_{max} > S) = 1 - \mathbb{P}(X_{max} \leq S) = \frac{1}{\Delta T}, \tag{2.43}$$

with ΔT the return period, and then inverting Eqs (2.41) and (2.42), gives the so-called **return level plot**:

$$S = \begin{cases} \mu - \frac{\sigma}{\xi}\left[1 - \{-\log(1 - \frac{1}{\Delta T})\}^{-\xi}\right] & \text{for } \xi \neq 0 \\ \mu - \sigma \log\left\{-\log(1 - \frac{1}{\Delta T})\right\} & \text{for } \xi = 0 \end{cases} \tag{2.44}$$

(Coles, 2001, eq. (3.4)). Introducing $y_p = -1/\log(1 - 1/\Delta T)$ in Eq. (2.44) finally yields:

$$S = \begin{cases} \mu + \frac{\sigma}{\xi}\left[y_p^{\xi} - 1\right] & \text{for } \xi \neq 0 \\ \mu + \sigma \log y_p & \text{for } \xi = 0 \end{cases}, \tag{2.45}$$

an example of which is shown in Figure 2.15.

Although widely advocated in hydrology and meteorology, direct use of the GEV distribution for parameter estimation has been discouraged since it does not use the full data set (e.g., Sornette, 2006, p. 28). An alternative representation of extremes that uses all the available data consists of analysing excesses over a high threshold (so-called peak-over-threshold method, or threshold excess approach). In that case, the **generalized Pareto distribution** (GPD) applies, for which CCDF is defined as

$$\mathbb{P}(X > S|X > \mu) = \left[1 + \xi\left(\frac{S-\mu}{\sigma}\right)\right]^{-\frac{1}{\xi}}, \tag{2.46}$$

with μ the location parameter (i.e., the chosen high threshold), σ the scale parameter, and ξ the shape parameter. For $\xi > 0$, the GPD is equivalent to the Pareto distribution with power exponent (or tail index) $b = \alpha - 1 = 1/\xi$ (Section 2.3.1). For $\xi < 0$, the distribution has an upper bound $S_{max} = \mu + \sigma/|\xi|$. If $\xi = 0$, the GPD is equivalent to the **exponential distribution**:

$$\mathbb{P}(X > S|X > \mu) = \exp\left(-\frac{S-\mu}{\sigma}\right). \tag{2.47}$$

Since $\mathbb{P}(X > S) = \mathbb{P}(X > \mu)\mathbb{P}(X > S|X > \mu)$ (following **Bayes' theorem** $\mathbb{P}(A|B) = \mathbb{P}(B|A)\mathbb{P}(A)/\mathbb{P}(B)$ where A and B are events, and here with $\mathbb{P}(X > \mu|X > S) = 1$), Eqs (2.46) and (2.47) become:

$$\mathbb{P}(X > S) = \frac{1}{n} = \begin{cases} \zeta_\mu\left[1 + \xi\left(\frac{S-\mu}{\sigma}\right)\right]^{-\frac{1}{\xi}} & \text{for } \xi \neq 0 \\ \zeta_\mu \exp\left(-\frac{S-\mu}{\sigma}\right) & \text{for } \xi = 0 \end{cases}, \tag{2.48}$$

where size S is exceeded on average once every n observations (i.e., a return period), and with $\zeta_\mu = \mathbb{P}(X > \mu)$ (Coles, 2001, eqs. (4.11) and (4.12)). The matching return level plot is, finally,

$$S = \begin{cases} \mu + \frac{\sigma}{\xi}\left[(\zeta_\mu n)^\xi - 1\right] & \text{for } \xi \neq 0 \\ \mu + \sigma \log(\zeta_\mu n) & \text{for } \xi = 0 \end{cases} \tag{2.49}$$

(Coles, 2001, eqs. (4.13) and (4.14)). For direct comparison with Eq. (2.37) in Figure 2.13, one may define the annual exceedance rate Λ from the GPD's CCDF with

$$\Lambda(S) = \begin{cases} \Lambda(S_{min})\left[1 + \xi\left(\frac{S - S_{min}}{\sigma}\right)\right]^{-\frac{1}{\xi}} & \text{for } \xi \neq 0 \\ \Lambda(S_{min}) \exp\left(-\frac{S - S_{min}}{\sigma}\right) & \text{for } \xi = 0 \end{cases} \tag{2.50}$$

and $S_{min} = \mu$.

2.3.2.1 Epidemics (Ep)

Marani et al. (2021a) developed a global data set of epidemics and showed that epidemic size S [‰/year] follows a GPD, with $\sigma = 0.0113$ ‰/year and $\xi = 1.40$ for a fixed $\mu = 10^{-3}$ ‰/year. ζ_μ is naturally estimated as the sample proportion of observations (or events) exceeding μ, which is $\zeta_\mu = 0.38$ in this case. Their results are shown in Figure 2.13b using Eq. (2.50). This size distribution has a power-law tail with exponent $b = 1/\xi = 0.71$.

2.3.2.2 Fluvial Floods (FF)

Flood events are commonly defined from the annual discharge maxima $Q_p = \max(Q)$ derived from records of daily discharge measurements Q. It means that those data are obtained using the block maxima approach, inferring a GEV distribution. In this approach, an event is defined as a 50-year flood or a 1,000-year flood, for instance. In contrast to the perils considered in Section 2.3.1, the size distribution is commonly not determined from a catalogue of historical events but by site-specific data, in this case, discharges Q at a gauging station. Figure 2.15 shows an application of the GEV to the well-studied Potomac River flow (Smith, 1987; Katz et al., 2002; Gilleland and Katz, 2016).

2.3.2.3 Meteorological Perils (including TC, xC, and RS)

Similarly to fluvial floods, windstorms can be defined from the annual maxima of a continuous quantity measured locally. Recorded daily maximum gust speeds can be used to assess the maximum wind speed v_{max} observed during 1 year at a given weather station (Cheng and Yeung, 2002; Hofherr and Kunz, 2010). This again involves the GEV distribution. Another approach consists of analysing the size distribution of storms from an event catalogue. The peak-over-threshold method is used then, with the empirical wind speed v_{max} distribution at a given location estimated from the storm tracks crossing a specified radius. The size distribution of wind speeds is commonly accepted to be of the Weibull type (Simiu and Heckert, 1996; i.e., finite upper tail), although the Gumbel distribution has

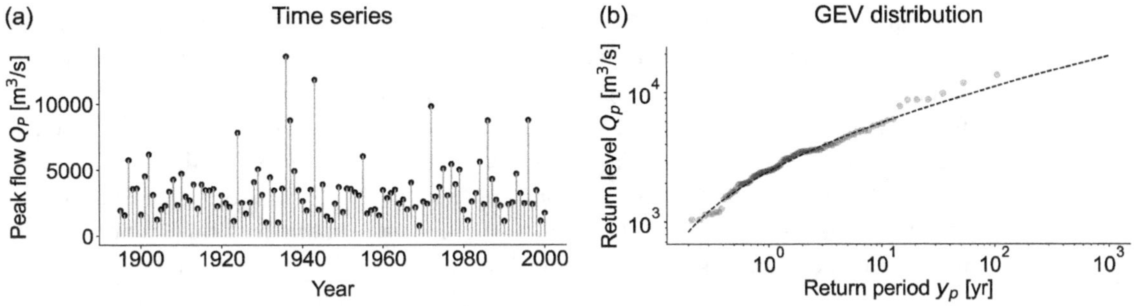

Fig. 2.15 GEV application to the Potomac River peak discharge. (a) Time series of the period 1895–2000 with data from the USGS, as used by Katz et al. (2002) and provided by Gilleland and Katz (2016). (b) Empirical distribution in the form of a return level plot with parameters $\mu = 2,462$ m^3/s, $\sigma = 1,171$ m^3/s, and $\xi = 0.19$ (Fréchet distribution; Eq. (2.45)), estimated using the R package `extRemes` (Gilleland and Katz, 2016).

also been found to provide reasonable fits (Cheng and Yeung, 2002; Hofherr and Kunz, 2010). A Fréchet distribution, however, is virtually never observed (Harris, 2005; i.e., no power-law tail behaviour). For coastal cities in Florida, Malmstadt et al. (2010) found $-0.42 < \xi < -0.70$ for $\mu = 25, 30$, or 35 m/s. The upper wind speed found at Miami, for example, is 70.3 m/s (using $\sigma_{35} = 24.7$ m/s, $\xi = -0.699$, and $\mu = 35$ m/s). A negative-shape parameter has also be found for European windstorms with, for example, $-0.23 \leq \xi \leq -0.10$ (Della-Marta et al., 2009). Since the measurement v_{max} can represent both the size S and the hazard intensity I for storms (Section 2.4), the event size distribution at a given location is equivalent to a hazard curve (Hofherr and Kunz, 2010), as shown previously in Figure 2.2. For that example, $-0.11 \leq \xi \leq -0.00$ was found for two European coastal towns (fit limited to a 12-year observational period). Considering the case of rainstorms using the Extreme Precipitation Database for the continental US (2013–2017) (Zhou et al., 2019), one finds for the GPD $\sigma = 11.82$ mm/h and $\xi = -0.05$ for $\mu = 50$ mm/h with the size S the rainfall maximum rate h_{max} (Fig. 2.13b).

2.4 Event Intensity Footprint

The third step in the hazard process, as depicted in Figure 2.1, involves the modelling of the impact of each stochastic event on the environment. This modelling is accomplished by assessing the hazard intensity $I(x, y)$ across a geographical grid of coordinates (x, y), considering both the event's size S_{ID} and its distinctive source characteristics identified by its ID. The rate $\lambda(S_{\mathrm{ID}})$, as previously defined, is a crucial attribute of the footprint $I_{\mathrm{ID}}(x, y)$, necessary for conducting probabilistic hazard assessments. At this stage, note that I still represents a mean intensity value, prior to the inclusion of any secondary aleatory uncertainty. All of the steps (including aleatory uncertainty encoding) will be wrapped up and applied to the context of probabilistic seismic hazard assessment in the concluding section of this chapter (Section 2.5).

Despite the diverse range of perils and the various disparities in their respective hazard processes, the intensity footprint $I(x, y)$ for any peril can effectively be modelled using one of three primary approaches (Mignan, 2022a): (1) an analytical expression of spatial diffusion, wherein the intensity value I diminishes as one moves away from the event source (Section 2.4.1); (2) a threshold model, where I serves as a passive function influenced by the variability in the underlying environment (Section 2.4.2); and (3) a numerical model of dynamic propagation, where I can potentially increase within the diffuse source (briefly introduced in Section 2.4.3 and further elaborated upon in Part II, Chapter 4).

2.4.1 Analytical Expressions of Static Spatial Diffusion

Although every hazard process is dynamical in nature, the process can be simplified to a static footprint in many cases. The energy originally stocked in a source and suddenly released in the form of an event will naturally diffuse in space (Fig. 2.3), which can be approximated by an analytical expression. For both point and line sources, the (mean) intensity footprint is

$$I(x, y) = f_I(S, r), \tag{2.51}$$

with (x, y) the spatial coordinates, S the event size, and r the distance to the event source. For track sources, the (mean) intensity footprint is defined from the maximum intensity observed at each location over the duration of the event, with

$$I(x, y) = \max_t f_I(S, r, t) \tag{2.52}$$

and t the time lapses. For a point source, $r = \sqrt{(x - x_0)^2 + (y - y_0)^2}$ with (x_0, y_0) the source coordinates; for a line source, $r = \min_i \sqrt{(x - x_{0,i})^2 + (y - y_{0,i})^2}$ with $(x_{0,i}, y_{0,i})$ the source coordinates (i.e., r as **Euclidean distance**). For an area source, $(x_{0,i}, y_{0,i})$ would correspond to the surface's outer contour, although area sources are usually representative of a propagating event, not a diffusive one. A track source is commonly a point source in motion.

2.4.1.1 Earthquakes (EQ)

Earthquake ground motion depends mainly on two parameters, the earthquake magnitude M and the distance r from the fault rupture. The general formulation of a **ground motion prediction equation** (GMPE) is

$$\log I = c_1 + c_2 M + c_3 r + c_4 \log r, \tag{2.53}$$

with I the seismic shaking intensity and c_1, c_2, c_3, and c_4 empirical parameters (Ambraseys et al., 1996; Youngs et al., 1997). A compilation of published GMPEs can for instance be found in the `openquake.hazardlib.gsim` package of the OpenQuake-engine (GEM, 2022). The intensity is often taken as the peak ground acceleration (PGA) ([g] or [m/s^2]), but can also be peak ground velocity, peak ground displacement, spectral acceleration, or simply felt intensity. Figure 2.16a shows examples of earthquake PGA footprints for three magnitude levels for one of the simplest possible GMPEs,

$$\log_{10} I = -1.34 + 0.23M - \log_{10} r, \tag{2.54}$$

with I [g] the mean PGA and $r = \sqrt{d^2 + z^2}$ [km], d being the distance to the surface projection of the fault rupture and z the fault depth (Ambraseys and Bommer, 1991). Note that when the rupture plane geometry is directly modelled (area source type), the distance to rupture is used instead with different parameterizations (e.g., Chiou and Youngs, 2014). Significant deviations are often observed, which can be due to source-related or site-specific effects, such as rupture directivity (Somerville et al., 1997) or **basin amplification** (a phenomenon where seismic waves are amplified because they get trapped and resonate in a geological basin; Fig. 2.16b; Bard et al., 1988). Those effects can be considered in GMPEs via correcting factors, based for example on **seismic microzonation** for site amplification (i.e., maps that consider local characteristics to assess ground-shaking variations). Physics-based waveform simulations, such as the CyberShake initiative (Graves et al., 2011), can be used to complement traditional GMPEs.

2.4.1.2 Overpressure Blasts (including AI and Ex)

Both human-made explosions and asteroid impact overpressure fields are derived from similar empirical laws. A common approach in industrial risk analysis is first calculating the dimensional scaled distance r_Z according to the **Hopkinson–Cranz scaling law**

$$r_Z = \frac{r}{\sqrt[3]{m_{TNT}}}, \tag{2.55}$$

which is a function of the explosive mass m_{TNT} [kg] and the distance r from the explosion source. Various empirical relationships have been proposed to model the overpressure field, one example being

$$P = \frac{1772}{r_Z{}^3} - \frac{114}{r_Z{}^2} + \frac{108}{r_Z}, \tag{2.56}$$

with P [kPa] the peak overpressure and r_Z [m] according to Eq. (2.55) (Mills, 1987; Fig. 2.17a). Considering the 2005 Buncefield VCE yield of 22 tons, Eq. (2.56) yields $P \geq 4$ psi (total damage) for $r \leq 151$ m (conversion between **pound per square inch (psi)** and kilopascal is 1 psi = 6.89 kPa).

For the specific case of asteroid impacts, the kinetic energy of the celestial body transforms into destructive explosive energy at impact. As the blast can be spherical or hemispherical (Fig. 1.18), burst altitude is a necessary parameter in addition to the impact energy to estimate the overpressure field at ground level. A simple binary severity footprint is given by

$$r_{4psi} = 2.09h - 0.449h^2 E_{TNT}^{-1/3} + 5.08 E_{TNT}^{1/3}, \tag{2.57}$$

with r_{4psi} [km] the 4-psi overpressure radius (total damage threshold; Chapter 3), E_{TNT} [Mt] the impact energy, and h [km] the burst altitude (Mathias et al., 2017). In addition to blast overpressure, large events can also release damaging levels of thermal radiation

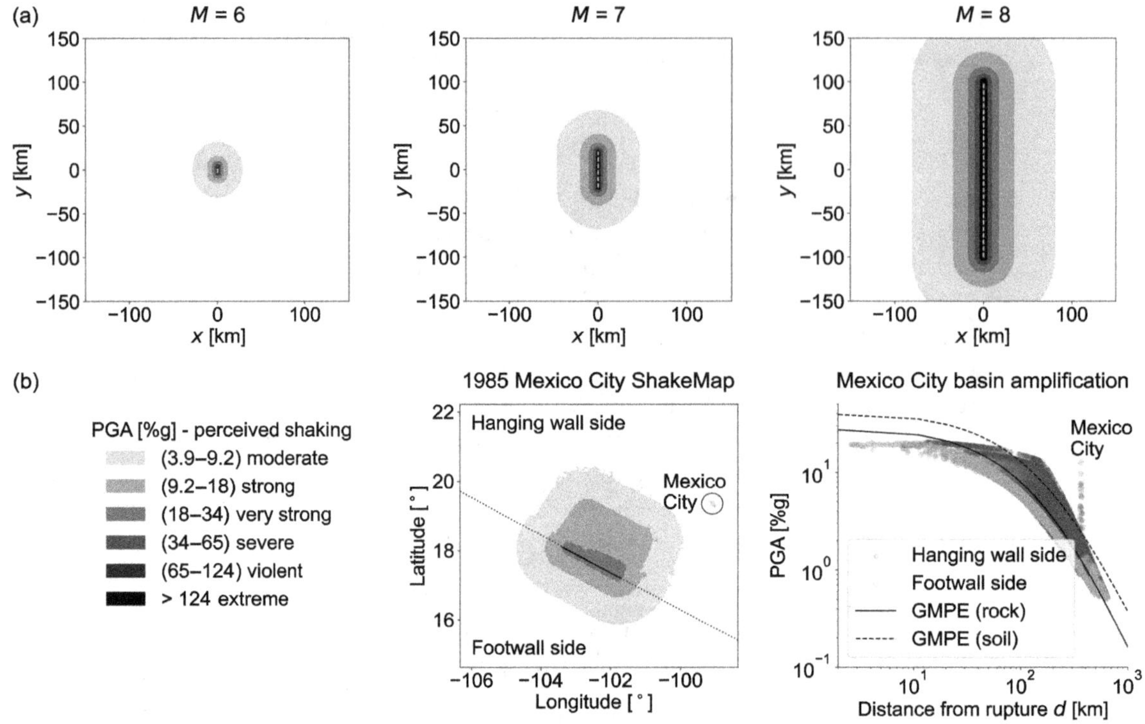

Fig. 2.16 Examples of earthquake ground-motion footprints. (a) PGA footprints according to Eq. (2.54) for magnitudes $M = 6$, 7, and 8, and a fault depth $z = 1$ km (rupture lengths $l = 8, 41$, and 200 km from $l = 10^{-3.22+0.69M}$; Wells and Coppersmith, 1994, table 2A). (b) Ground-shaking associated with the 1985 Mexico earthquake, exemplifying the phenomenon of basin amplification. Data from USGS ShakeMap (Allen et al., 2008) and GMPE curves from Youngs et al. (1997). The PGA categorization follows Allen et al. (2008, fig. 1), with potential damage going from very light to very heavy (Chapter 3).

(not considered here). Applying Eq. (2.57) to the 1908 Tunguska event with $E_{TNT} = 10$ Mt and $h = 5$ km yields $r_{4psi} = 16$ km, consistent with the 1908 tree damage footprint (Mignan et al., 2011). Footprints of three different yields are shown in Figure 2.17b with a good match shown between Eqs (2.56) and (2.57) when fixing $h = 0$ (Fig. 2.17a). For example, for the 2005 Buncefield event, Eq. (2.57) leads to $r_{4psi} = 142$ m, proving that both empirical relationships are in very close agreement. Empirical relationships are often calibrated to data from nuclear tests (Glasstone and Dolan, 1977).

2.4.1.3 Tornadoes (To)

The mean wind field for a stationary tornado is calculated as the sum of the tangential and radial velocities v_{tan} and v_{rad} [m/s], with both wind velocity components v based on the **Rankine vortex model**,

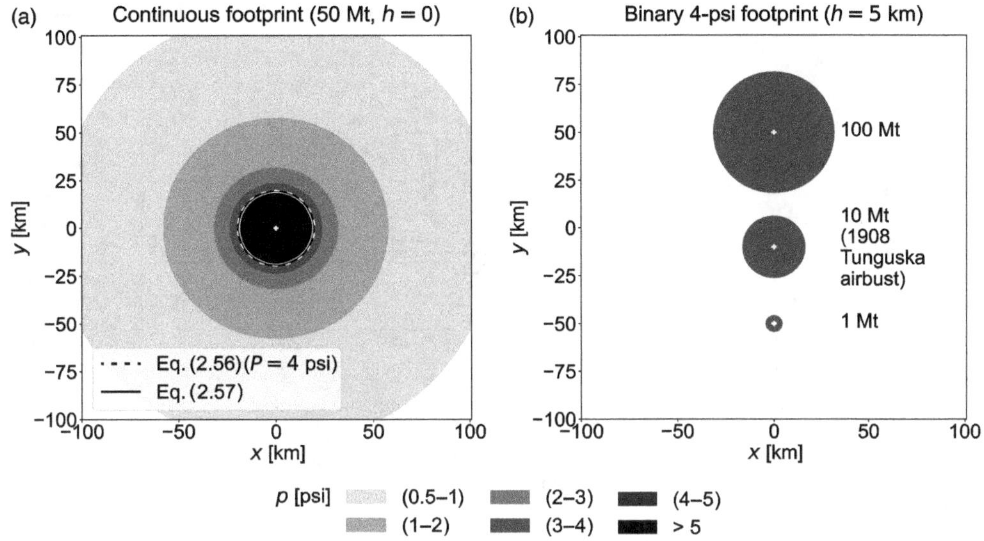

Fig. 2.17 Comparing overpressure footprints of explosions and asteroid impacts. (a) Footprint for a 50 Mt (nuclear) explosion (Eq. (2.56)). The 4-psi contours show the good match between Eqs (2.56) and (2.57) (with $h = 0$). (b) Spatial scale of yields $E_{TNT} = 1, 10$, and 100 Mt and an explosion height of $h = 5$ km (applicable to asteroid airbursts and nuclear explosions).

$$v = \begin{cases} v_{max} r r_{max}^{-1}, & \text{if } r \leq r_{max} \\ v_{max} r^{-1} r_{max}, & \text{if } r > r_{max}, \end{cases} \tag{2.58}$$

with r [m] the distance from the tornado origin, v_{max} [m/s] the maximum (tangential or radial) velocity, and r_{max} [m] the radius of the maximum (tangential or radial) velocity. The wind field for the tornado in motion finally consists of adding the forward motion velocity component v_f to the wind vector (Holland et al., 2006). An example of wind field is shown in Figure 2.18a–c. When the tornado footprint is defined from the Rankine vortex model, the tornado source must be defined as a track, with the vortex a point source in motion (Eq. (2.52)). However, the following expression can be used instead to approximate the maximum wind speed v_{max} perpendicular to the tornado line source:

$$v_{max} = \begin{cases} \left[\left(\sin\left(\frac{\pi x}{2r_{max}}\right) v_{tan} + \cos\left(\frac{\pi x}{2r_{max}}\right) v_{rad} + v_f\right)^2 + \right. \\ \left.\left(\cos\left(\frac{\pi x}{2r_{max}}\right) v_{tan} - \sin\left(\frac{\pi x}{2r_{max}}\right) v_{rad}\right)^2\right]^{\frac{1}{2}}, & \text{if } r \leq r_{max} \\ \left[(v_{tan} r_{max} x^{-1} + v_f)^2 + (v_{rad} r_{max} x^{-1})^2\right]^{\frac{1}{2}}, & \text{if } r > r_{max} \end{cases} \tag{2.59}$$

with x the distance perpendicular to the tornado's translational vector (Burow et al., 2020). The good match between Eq. (2.59) and $\max_x(v)$ is shown in Figure 2.18d, and the final hazard intensity footprint in Figure 2.18e.

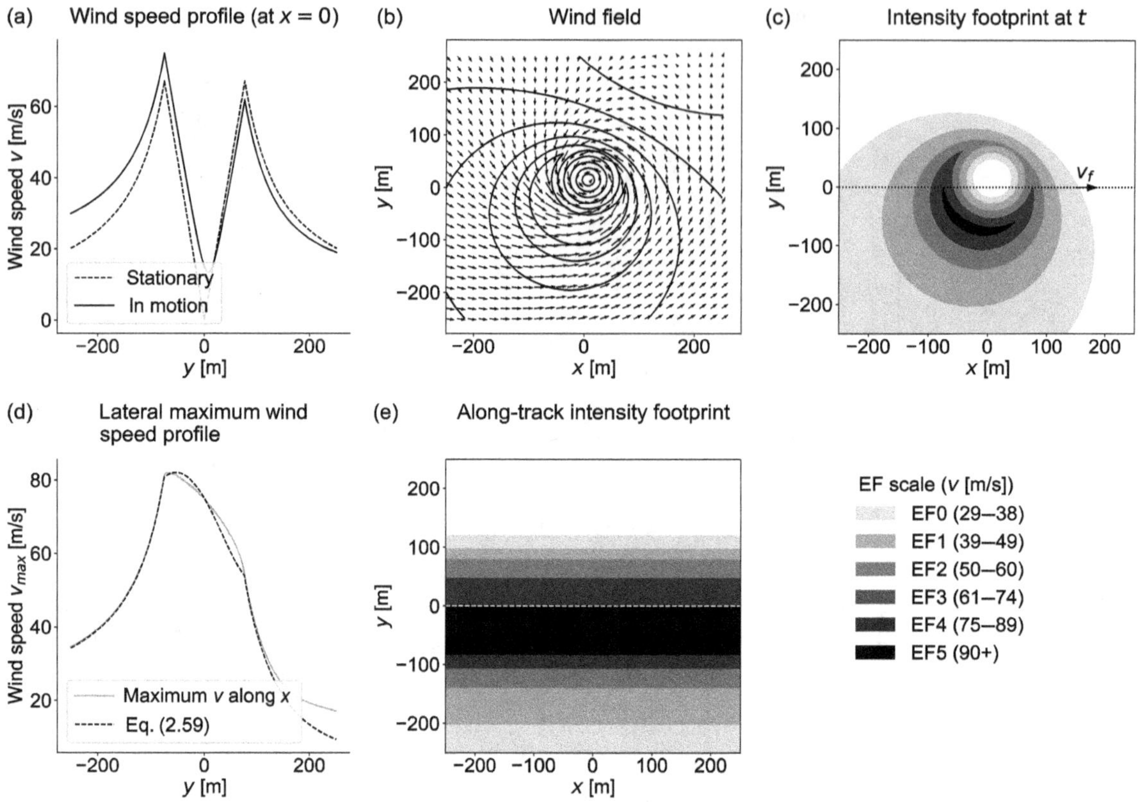

Fig. 2.18 Example of tornado wind field with matching intensity footprint, with the following parameterization: $v_{tan,max} = 60$ m/s, $v_{rad,max} = 30$ m/s, $v_f = 15$ m/s, and $r_{max} = 75$ m (Holland et al., 2006), and vortex centred on $(x_0 = 0, y_0 = 0)$. (a) Wind speed profile with the velocity v the sum of the tangential velocity v_{tan}, radial velocity v_{rad}, and forward velocity v_f vectors (Eq. (2.58)). (b) Wind field with matching wind vectors and lines of equal wind speed (**isotachs**). (c) Hazard intensity footprint at instant t defined as $v(x, y, t)$. (d) Lateral maximum wind speed v_{max} profile and approximation for line source. (e) Final hazard intensity footprint $v_{max}(x, y)$ for a line source (dashed white line).

2.4.1.4 Tropical Cyclones (TC)

The Holland (1980) parametric wind profile for a stationary tropical cyclone, previously described in Section 2.2.3.1 (Eq. (2.24); Fig. 2.10), could be used to calculate the intensity footprint along the event track after adding the forward motion component (v_f) to the wind field. Although the Holland (1980) model remains historically important, it has been shown to not fully reflect the variability observed in tropical cyclones due to the strong constraints of parameter B (Willoughby and Rahn, 2004). A model was then proposed by Willoughby et al. (2006) to remedy this problem, at the cost of increased complexity (including a transition region between inner and outer regions, as well as a dual exponential to capture two different decay rates outside the eye wall of the storm). Since this model requires numerical integration, it is not further mentioned here. Instead, the revised Holland model (Holland,

2008) is considered, which consists of allowing the B parameter (now B_s) to vary according to the following empirical relationship:

$$B_s = -4.4 \times 10^{-5}(P_n - P_c)^2 + 0.01(P_n - P_c) + 0.03\frac{\partial P_c}{\partial t} - 0.014\phi + 0.15v_f^x + 1,$$
$$\text{with } x = 0.6\left(1 - \frac{P_n - P_c}{215}\right), \tag{2.60}$$

where $P_n - P_c$ [hPa] is the pressure drop to the cyclone centre, $\partial P_c/\partial t$ [hPa/h] is the pressure change along the track, ϕ [°] is the latitude, and v_f [m/s] is the storm forward speed. The revised Holland model is, for example, used in the CLIMADA risk platform (Aznar-Siguan and Bresch, 2019). Figure 2.19 shows the different steps leading to the modelling of a tropical cyclone intensity footprint, here taking the track of Hurricane Katrina as source. Second-order variations of the velocity field v, not included here, are due to the frictional

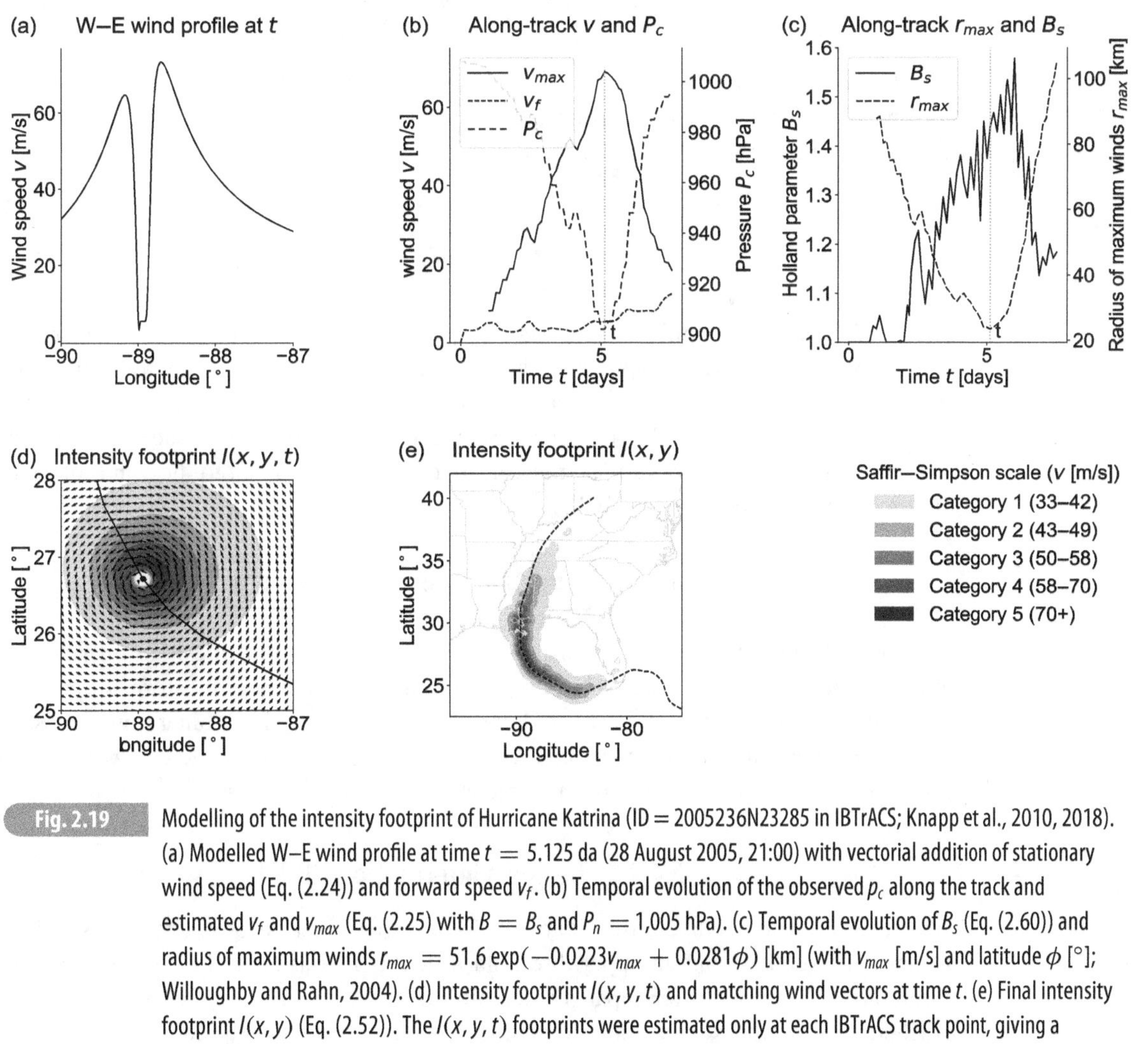

Fig. 2.19 Modelling of the intensity footprint of Hurricane Katrina (ID = 2005236N23285 in IBTrACS; Knapp et al., 2010, 2018). (a) Modelled W–E wind profile at time $t = 5.125$ da (28 August 2005, 21:00) with vectorial addition of stationary wind speed (Eq. (2.24)) and forward speed v_f. (b) Temporal evolution of the observed p_c along the track and estimated v_f and v_{max} (Eq. (2.25) with $B = B_s$ and $P_n = 1{,}005$ hPa). (c) Temporal evolution of B_s (Eq. (2.60)) and radius of maximum winds $r_{max} = 51.6\exp(-0.0223v_{max} + 0.0281\phi)$ [km] (with v_{max} [m/s] and latitude ϕ [°]; Willoughby and Rahn, 2004). (d) Intensity footprint $I(x, y, t)$ and matching wind vectors at time t. (e) Final intensity footprint $I(x, y)$ (Eq. (2.52)). The $I(x, y, t)$ footprints were estimated only at each IBTrACS track point, giving a discontinuous aspect to the footprint, here illustrating the process of footprint generation for a track.

drag of wind over the terrain, which depends on the roughness length. It is almost zero on exposed open terrain (e.g., water surface) and increases as the density and height of obstacles increases (e.g., large city centre; Mendis et al., 2007).

2.4.1.5 Volcanic Eruptions (VE)

Apart from local pyroclastic flows, the principal direct hazard associated with explosive volcanic eruptions is the fall of tephra. The tephra ash load at a given location is calculated by the pressure $P = \rho_{ash} g h$ [Pa] where ρ_{ash} is the density of ash ($\rho_{ash} = 900$ kg/m^3 when dry), $g = 9.81$ m/s^2, and h [m] is the tephra layer thickness. h decreases exponentially with distance r, as

$$h = h_0 \exp\left(-\log(2)\frac{r}{r_{1/2}}\right), \tag{2.61}$$

with h_0 the maximum thickness and $r_{1/2}$ the half-distance (Pyle, 1989). For a circular footprint, $r_{1/2}$ can be estimated from the event size S, that is, the volume V of tephra according to the following geometric formula:

$$r_{1/2} = \log(2)\sqrt{\frac{V}{2\pi h_0}}. \tag{2.62}$$

Note that h_0 may vary significantly between eruptions in the range of metres. In the case of the 1980 Mount St. Helens eruption, h was observed to be 5 cm $r = 80$ km away from the volcano. Considering $V = 2.9$ km^3 and the observed maximum thickness $h_0 = 1$ m, one obtains a distance of $h = 64$ km to reach the same ash depth, which is at the same order of magnitude. Second-order variations, not considered here, are linked to the wind intensity and direction (Bonadonna et al., 2015).

2.4.2 Threshold Models of Passive Event Emergence

Some hazard intensity footprints can be generated by applying a threshold on an **environmental layer**, such as a water level relative to the topography (e.g., storm surge) or a temperature threshold (e.g., heatwave). A threshold model may have binary outcomes, in which case a value above the threshold means potential harm and a value below means no harm. The threshold defines the spatial contour of the hazard footprint.

2.4.2.1 Business Interruptions (BI)

There exists a lower damage threshold of 5–10% that must be breached to result in a business interruption, and an upper threshold, often as low as 50%, to cause the facility to completely shut down for repair or demolition (Rose and Huyck, 2016). The spatial distribution of business interruption then depends on the relationship between the hazard intensity footprint of the trigger event and the businesses' locations.

2.4.2.2 Heatwaves (HW)

A heatwave footprint is defined as all the regions where the temperature exceeds a threshold T_{th} for a minimum period of time Δt_{min} ($\Delta t_{min} = 3$ consecutive days in Perkins and Alexander, 2013). Above T_{th}, the temperature potentially leads to harm to the population cohorts susceptible to heat. The threshold can be defined as a percentile (e.g., 90th percentile or above; Perkins and Alexander, 2013) or as a fixed value (e.g., 35°C for daytime and 20°C for night-time, so-called tropical nights; Collins et al., 2000). Figure 2.20, for example, depicts the early days of the 2003 European heatwave by declaring an event as at least 3 consecutive days with air temperatures exceeding 35°C. Jiang et al. (2015) defined a heatwave as land surface temperature (LST) greater than 325 K ($T_{LST} = 51.85$°C), which is equivalent to an air temperature $T_{th} \approx 38$°C for the empirical relationship $T = 0.74 T_{LST}$ found for the Nevada Ranges (Mutiibwa et al., 2015). This value is however only illustrative in nature since it also depends on the altitude (Mutiibwa et al., 2015), land-cover type, season, and latitude (Good et al., 2017). An extreme threshold that would lead to potential harm to any human being, if observed for extended periods, is $T_w = 35$°C with T_w the **wet-bulb temperature** (Sherwood and Huber, 2010). It is measured by covering a standard thermometer bulb with a wet cloth and fully ventilating it. It can be calculated using the following formula:

$$T_w = T \operatorname{atan}\left(0.151977\sqrt{\eta_w + 8.313659}\right) + \operatorname{atan}(T + \eta_w) - \operatorname{atan}(\eta_w - 1.676331) + 0.00391838\eta_w^{\frac{3}{2}} \operatorname{atan}(0.023101\eta_w) - 4.686035, \tag{2.63}$$

where T [°C] is the dry-bulb temperature and η_w [%] the relative humidity (Stull, 2011). T_w has been observed to never exceed 31°C although the 35°C threshold would be reached in extreme global warming scenarios (Sherwood and Huber, 2010). For a humidity $\eta_w =$

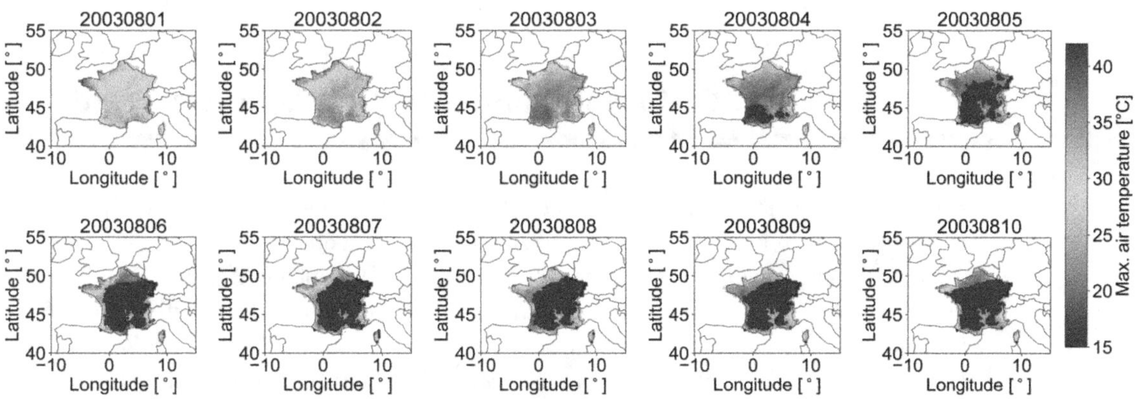

Fig. 2.20 The 2003 European heatwave footprint defined with a threshold method. The darker points delineate the heatwave and represent the locations (here in France) where the maximum daily air temperature exceeds the threshold $T_{th} = 35$°C for at least 3 consecutive days. Data from the Gridded Agro-Meteorological Data in Europe (Toreti, 2014), from the Agri4Cast Resources DATAPORTAL (CC-BY 4.0, https://creativecommons.org/licenses/by/4.0/). A black and white version of this figure will appear in some formats. For the colour version, please refer to the plate section.

50 %, the air temperature should be $T_{th} = 45°C$ to reach hyperthermia levels (52°C for $\eta_w = 30\%$, etc.).

2.4.2.3 Storm Surges (SS)

Coastal flooding by storm surge can be assessed with a static approach. The so-called **bathtub model** defines a flooded area as all the locations below a certain elevation that are hydrologically connected to the coast, with the elevation threshold based on the size S, or surge height h along the shoreline. In other words, it is a projection of a horizontal flood surface onto the topography. This model tends to overestimate flood extents compared to more realistic hydrodynamical models (Ramirez et al., 2016) but is mentioned here as it remains a valuable tool, for example, for global analyses (Dasgupta et al., 2011). In the case of a levee breach, the water volume is dispersed into the flood plain, with the footprint geometry also constrained by a threshold (Muir Wood et al., 2005).

2.4.3 Numerical Models of Dynamic Event Propagation

This section serves only as a placeholder for all the perils that require numerical approaches for the modelling of their hazard intensity footprints. All the following perils will be considered in Part II (Chapter 4), which is dedicated to these methods (in alphabetical order): armed conflicts, blackouts, epidemics, fluvial floods, landslides, pest infestations, stampedes, social unrest, and wildfires. All of these perils strongly interact with the complex environment, hampering any simple analytical representation. This includes perils with a diffuse source (i.e., environment as source; Section 2.2.4) and perils in which event intensity interacts with the environment (e.g., water flow over terrain).

2.5 Application to Probabilistic Seismic Hazard Assessment

All the steps previously defined will now be wrapped up with an illustrative application to probabilistic seismic hazard assessment (PSHA; Cornell, 1968; McGuire, 2004). A stochastic event set will first be generated in Section 2.5.1, which will require a fault source model from which stochastic earthquake ruptures will be defined. Data will be taken from a small subset of the crustal fault source model of the European Database of Seismogenic Faults (EDSF; Basili et al., 2013) with the geographical information converted to a generic (x, y) grid. Considering a generic region will allow us to make simple model choices without the need to be representative of any specific area in Europe. For example, it will be assumed that an earthquake of magnitude $M \geq 5$ will occur on average every year to assess event occurrence rates λ. A catalogue of ground-motion footprints $I(x, y)$ will then be developed in Section 2.5.2 for the stochastic set previously created. The final results will be presented in the form of seismic hazard curves and seismic hazard maps in Section 2.5.3.

2.5.1 Stochastic Rupture Set (Event Table)

The concept of a stochastic event set was defined in Section 2.1.1 as a table of unique events, described by their identifier (ID), size S_{ID}, source location, and occurrence rate λ_{ID} (event table case). Generating stochastic events requires some practical know-how, which – in the case of earthquakes – involves the modelling of so-called **floating ruptures** (i.e., ruptures occurring on a part of a fault line source). The proposed approach will be much simpler than the one implemented in the OpenQuake-engine (OQ-engine), the open-source hazard and risk calculation engine developed by the GEM initiative (Pagani et al., 2014a). The EDSF source model was used in the European Seismic Hazard Model (ESHM13– Woessner et al., 2015; ESHM20– Danciu et al., 2021), which is part of the regional sub-models collated by GEM. Although primarily aimed at illustrating how to use the models described in the previous sections of this chapter, what follows shares the same philosophy as public platforms such as OQ-engine and other proprietary softwares developed by CAT model vendors.

2.5.1.1 Source Model

A **source model** is a set of sources distributed in space. In GEM, the earthquake source model is a collection of seismic sources describing the seismic activity in a region of interest. It specifies the number of sources (i.e., fault segments), their characteristics (geometry, mechanism), and the average annual occurrence rate of earthquakes over a magnitude range (since different faults are characterized by different levels of seismic activity). Using the same notation as in GEM (Pagani et al., 2014b), a seismic source model (SSM) can be defined as a set of N seismic sources

$$\mathrm{SSM} = \{\mathrm{Source}_1, \mathrm{Source}_2, ..., \mathrm{Source}_N\}, \tag{2.64}$$

with $N = 762$ in the EDSF for crustal faults. For the present exercise, only $N = 5$ sources will be considered (see rectangular box in Figure 2.21). Selected attributes from the EDSF source model are given in Table 2.6. The effective length (EFFECTIVEL) was originally used to calculate the range of possible magnitudes (MINMW-MAXMW) under different source–magnitude relationships. Note that GEM uses a logic tree approach for epistemic uncertainty quantification, which will be covered in Part III. For the sake of simplicity, only one model is used in this chapter per modelling step. The total length (TOTALL or l_{tot}) is used instead of the effective length, assuming that the generic faults can rupture along their entire length, with the maximum possible magnitude M_{max} calculated from Eq. (2.14) (Section 2.2.2.1). Since all the faults have a reverse mechanism (FAULTTYPE = RR), the considered empirical relationship is $M_{max} = 5.00 + 1.22 \log_{10} l_{tot}$ (Fig. 2.6; Wells and Coppersmith, 1994).

2.5.1.2 Generation of Stochastic Earthquake Ruptures

Typically, a source can host events smaller than the characteristic event of size S_{max} (Fig. 2.3). *To represent the range of potential damage in a given region, events must be defined for a range of sizes*, increasing from S_{min} to S_{max} by a fixed increment ΔS. S_{min} should

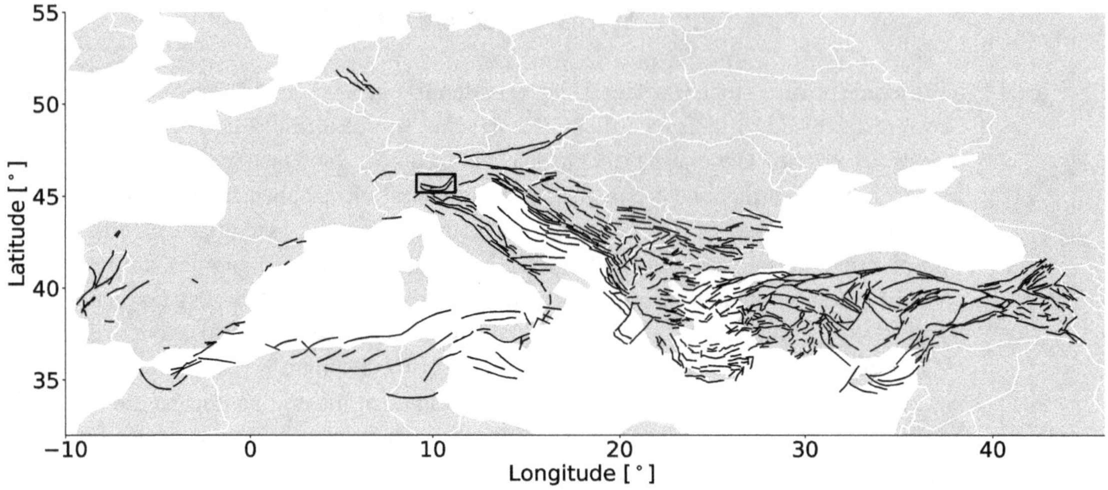

Fig. 2.21 Map of crustal sources in the European Database of Seismogenic Faults (EDSF; Basili et al., 2013, CC-BY 4.0, https://creativecommons.org/licenses/by/4.0/). The black rectangle (in northern Italy) includes the five sources described in Table 2.6 and used as input in our generic PSHA analysis.

Table 2.6 Excerpt from the EDSF crustal fault source model

IDSOURCE[a]	FAULTTYPE	TOTALL	EFFECTIVEL	MINMW	MAXMW
ITCS002	RR	104.2	41.1	6.59	6.93
ITCS010	RR	78.1	28.5	6.28	6.66
ITCS048	RR	93.1	28.5	6.28	6.66
ITCS072	RR	40.6	30.6	6.34	6.71
ITCS073	RR	65.1	33.9	6.42	6.79

[a] The five sources located in the box shown on the map of Figure 2.21. Source: Basili et al. (2013) (CC-BY 4.0, https://creativecommons.org/licenses/by/4.0/).

represent an event that could lead to minor damage (Chapter 3). Including smaller events would increase the computational cost while not adding any value to the CAT model. For earthquakes, a smaller M suggests a shorter rupture. In this particular case, each fault source undergoes a discretization process to generate a number of distinct earthquake ruptures of smaller lengths $l(M)$, which are 'floated' along the meshed source (with the rupture moved incrementally across the mesh until the full set of possible rupture locations is obtained; Pagani et al., 2014b). A much simpler approach is used here. First, to limit the number of events considered in this illustrative example, a relatively high minimum magnitude $M_{min} = 6.5$ is used, which means that the hazard space is not fully covered. The range of possible magnitudes per source j is $[6.5, M_{max}(l_{tot,j})]$ with $\Delta M = 0.1$. Second, our floating rupture method consists of randomly distributing $n(M_i) = l_{tot,j}/l(M_i)$ events per source j and per magnitude increment M_i (e.g., five 20-km long ruptures and two 50-km long ruptures generated along a fault 100 km long). This leads here to a stochastic set of $n_{\mathrm{ID}} = 84$ ruptures (Fig. 2.22). This is a much lighter approach than using a mesh, but done

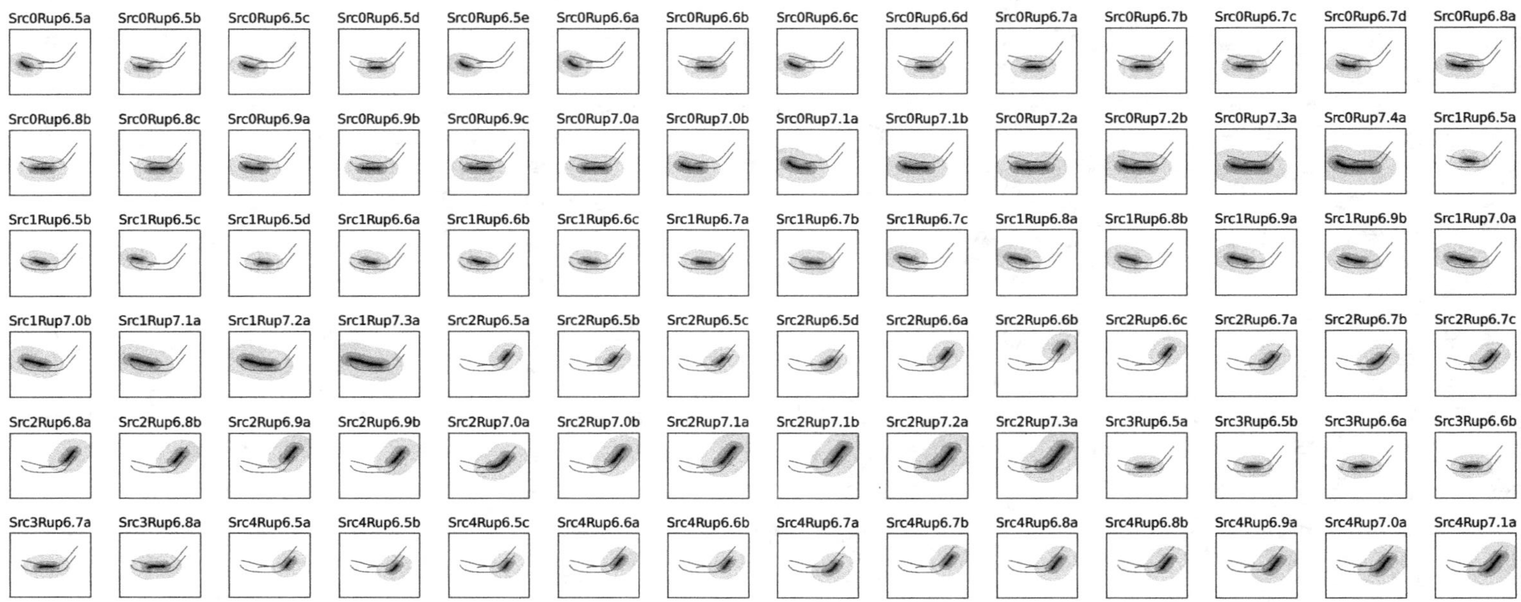

Fig. 2.22 Earthquake ground-motion footprint catalogue, with footprints defined using Eq. (2.54) for the $n_{ID} = 84$ ruptures randomly floated on the five sources (Table 2.6). The number n_{ID} comes from several considerations: the number of sources, of magnitude bins, and of ruptures of a given size per source. Same PGA scale as in Figure 2.16.

Box 2.5 Stochastic Ruptures for Unknown Fault Locations

It is still possible to model earthquake ruptures when fault sources remain unknown or are poorly constrained. 'An **[earthquake] area source** is defined by the boundary of a geographical region and is modelled by uniformly distributing the original occurrence rates among a collection of point sources that are equally distributed on a regular grid within the polygon area' (below delimited by dashed contours) (Pagani et al., 2014a, p. 697). At each point, a stochastic plane is then generated given a magnitude and the probable fault geometry (dip, strike, etc.). An example is given below for Kyrgyzstan by using data from Ullah et al. (2015) (CC-BY 4.0, https://creativecommons.org/licenses/by/4.0/). Observe the rupture planes (surface projections) increasing in size for different magnitudes.

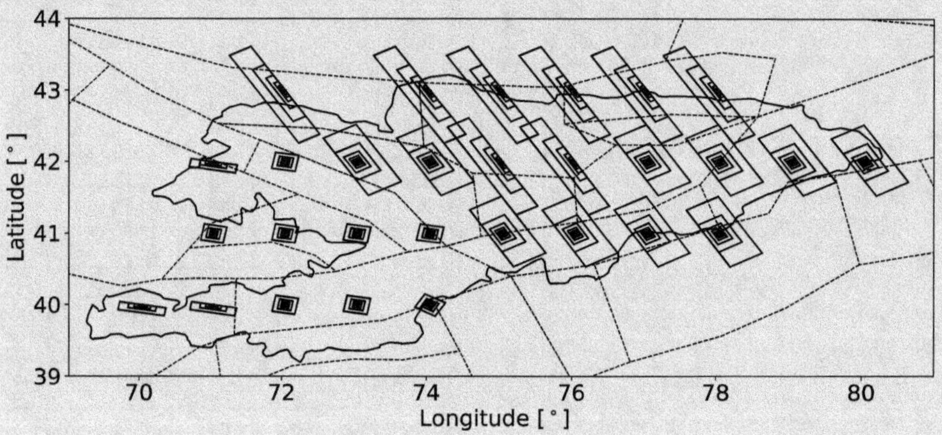

at the cost of not systematically testing all possibilities. *A property of a stochastic event set is that events can be defined in many different ways. It should just be representative of the space of possibilities for robust probabilistic hazard assessment.* The richness of options due to epistemic uncertainties is depicted in Boxes 2.5 and 2.6.

2.5.1.3 Event Rate Model

For the present example, magnitudes vary in the range $6.5 \leq M_i < 7.5$. For each stochastic event, the occurrence rate λ_{ID} is derived from a $\Lambda(S)$ empirical relationship (Section 2.3). The occurrence rates defined in the stochastic set are non-cumulative and normalized by the number of events of a given size S_i. In PSHA, the **double-truncated Gutenberg–Richter magnitude–frequency distribution** is used instead of Eq. (2.39) (Pagani et al., 2014b) and is defined as follows:

$$\lambda(M_i) = 10^{a-b\left(M_i-\frac{\Delta M}{2}\right)} - 10^{a-b\left(M_i+\frac{\Delta M}{2}\right)}, \tag{2.65}$$

where λ is the annual occurrence rate for a specific magnitude band centred on the magnitude bin M_i of width ΔM. A similar approach is used for other perils but with event sizes potentially binned in the $\log_{10}$ scale. The annual occurrence rate $\lambda(M_i)$ here is uniformly

Box 2.6 **Runaway Ruptures: the 2023 *M*7.8 Türkiye–Syria Earthquake**

Earthquake ruptures are known to potentially propagate over multiple fault segments, with the *M*7.8 earthquake of 6 February 2023 in the Republic of Türkiye – the strongest in the country in more than 80 years – and Syria one such example (Dal Zilio and Ampuero, 2023). Complex ruptures, once observed, are commonly included in PSHA models alongside simpler stochastic events. Inclusion of stochastic runaway ruptures in PSHA models remains rare (Field et al., 2014). However, a lightweight algorithm was proposed by Mignan et al. (2015b), which, years before the 2023 Türkiye–Syria earthquake, anticipated this type of rare rupture sequence in the Anatolian peninsula. A follow-up **counterfactual** analysis by Woo and Mignan (2018), which aimed to provide valuable insight into rare catastrophes with little or no historical precedent, proved its value a couple of years later. The map below represents the observed 2023 rupture traces (Reitman et al., 2023) (USGS) next to the EDSF map (Basili et al., 2013) (CC-BY 4.0, https://creativecommons.org/licenses/by/4.0/) with two valid stochastic runaway ruptures, according to Mignan et al. (2015b), highlighted in bold (note that it was a doublet of earthquakes; see Part II; Chapter 5). A description of the algorithm, which assesses rupture bending, branching, and jumping, remains outside the scope of this textbook.

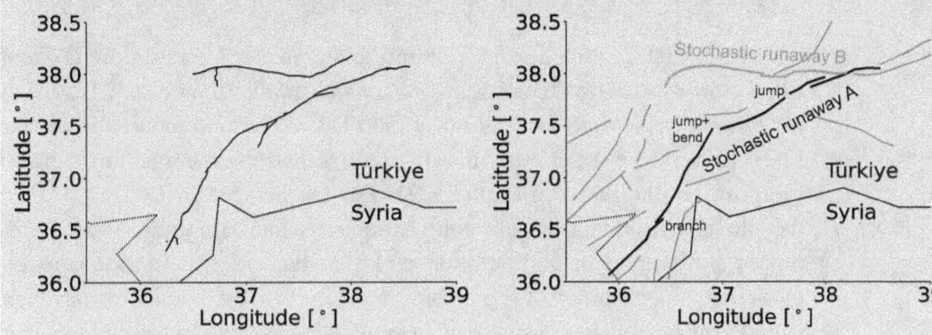

distributed over the N_{rup} ruptures of matching length l_i by assuming a similar seismic activity on all faults, that is, $\lambda_{\text{ID}}(M_i) = \lambda(M_i)/N_{rup}(M_i)$. A weighting strategy must be used if faults have different levels of activity, which is most often the case and described by the slip rate in seismic source models (e.g., Mignan et al., 2018). An excerpt of the final event table is shown in Table 2.7 with ad hoc Gutenberg–Richter parameters $a = 5$ and $b = 1$ (i.e., one earthquake of magnitude 5 expected annually).

Table 2.7 Excerpt from the generated event table

ID	Source[a]	M	Rate λ
EQ01	Src0Rup6.5a (on ITCS002)	6.5	0.000405
EQ02	Src0Rup6.5b (on ITCS002)	6.5	0.000405
...	...	...	...
EQ83	Src4Rup7.0a (on ITCS073)	7.0	0.000330
EQ84	Src4Rup7.1a (on ITCS073)	7.1	0.000306

[a] See full list of rupture sources in Figure 2.22.

2.5.2 Earthquake Ground-Shaking Footprint Catalogue

Once the stochastic event set is defined, a hazard intensity footprint $I(x, y)$ must be modelled for each one of the $n_{\text{ID}} = 84$ events. The simple GMPE of Section 2.4.1.1 is used with an ad hoc depth fixed to $z = 1$ km (Eq. (2.54); Ambraseys and Bommer, 1991). The entire catalogue of 84 (median) intensity footprints is displayed in Figure 2.22 (before implementation of secondary aleatory uncertainties).

2.5.2.1 Random Fields

A **random field** corresponds to an (x, y) map (or surface) of a quantity that includes the natural spatial variations $\epsilon(x, y)$ of that quantity. In the case of earthquakes, the empirical GMPE includes a standard deviation (σ) that quantifies this variability. The relationship is defined by

$$\log I = f_I(M, r) + \epsilon, \tag{2.66}$$

with $\epsilon \sim \mathcal{N}(0, \sigma^2)$ the noise representing the aleatory uncertainty component, and intensity I now stochastic (recall that '$\sim$' means 'is distributed as'; Box 2.1). Owing to the logarithmic scale used in Eq. (2.66), the PDF to be implemented in the hazard curve (Eq. (2.3)) is therefore the log-normal distribution and the median (instead of mean) intensity footprints of the maps in Figure 2.22. For the present GMPE, $\sigma = 0.27$ (derived from the model's residuals distribution; Ambraseys and Bommer, 1991). Since event occurrences are explicit in a simulation table, in that case, their footprints can be altered to model the stochasticity of the hazard intensity process at each occurrence. They are then called **ground-motion fields** (Pagani et al., 2014a). A simple approach is again used in this illustrative example with

$$\log I_{\text{ID}}(x, y) \sim \mathcal{N}(\mu = f_I(M_{\text{ID}}, r(x, y)), \sigma^2), \tag{2.67}$$

by excluding the role of any spatial correlation between different locations (x, y). Defining spatially correlated ground-motion fields is, however, important in seismic risk assessment applications (Park et al., 2007). Note also that if hazard computations are performed here with no truncation of the log-normal distribution of ground motion, applying a truncation at $\pm n\sigma$ (with $n = 2$ or 3) is common practice in PSHA to avoid unphysical estimates of the expected level of shaking (e.g., Beauval and Scotti, 2004) (Ex. 2.4c). This is a subject of controversy since $n\sigma$ then becomes the controlling factor in PSHA. Indeed, the larger the value, the larger the exceedance probability is (Eq. (2.3); e.g., Wang, 2009). Figure 2.23 shows three examples of ground-motion fields for one rupture of the event table. Such instances of event footprints enter in the computation of the hazard curve when using Eq. (2.5). By comparison, the footprints shown in Figure 2.22 exclude aleatory uncertainties.

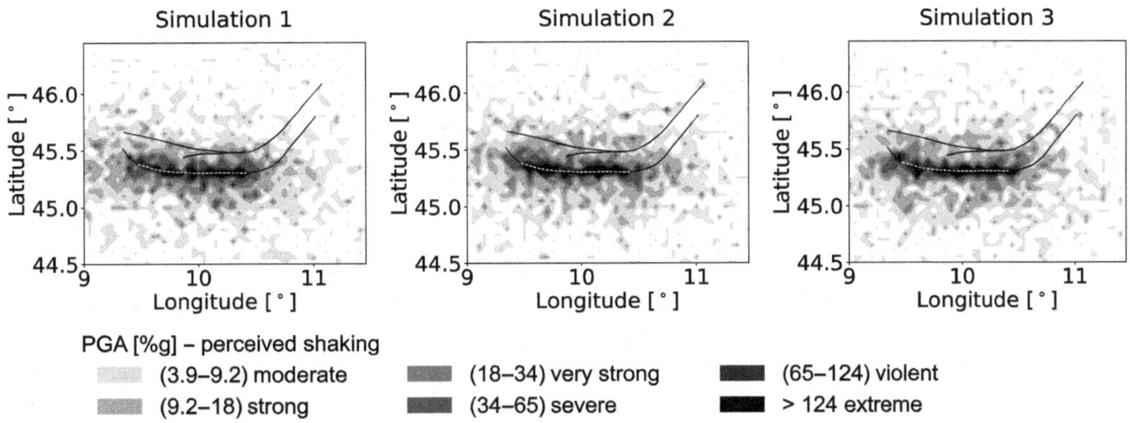

Fig. 2.23 Examples of ground-motion fields for a given rupture (Src0Rup7.3a on source ITCS002), using Eqs (2.54) and (2.67), and $\sigma = 0.27$. In contrast to the intensity footprints previously modelled in Section 2.4, the present footprints include secondary aleatory uncertainty.

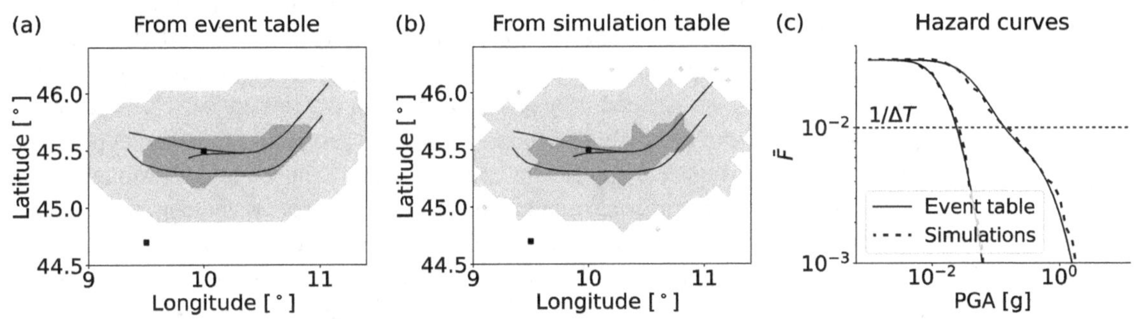

Fig. 2.24 Hazard maps and hazard curves for our ad hoc PSHA. (a) Hazard map derived from the event table (with $\Delta T = 100$ yr). (b) Hazard map derived from the simulation table (with $\Delta T = 100$ yr and $n_{sim} = 10^6$ year-simulations). Same PGA scale as in Figure 2.23. (c) Two hazard curves at the locations represented by the squares in (a) and (b).

2.5.3 Hazard Metrics

Seismic hazard curves and hazard maps can be developed by following the methods described in Section 2.1.3. A seismic hazard curve integrates the contributions from all possible earthquakes at location (x, y) to estimate the annual probability that a particular ground motion will be exceeded at that site (Ex. 2.4). Seismic hazard maps are shown in Figure 2.24 for a return period of $\Delta T = 100$ years. The hazard map obtained directly from the event table with log-normal PDF used in Eq. (2.3) is represented in Figure 2.24a with two examples of hazard curves shown in Figure 2.24c (solid curves). Results are finally compared to the ones obtained using a simulation strategy (via the generation of a YET –

not shown). The resulting hazard map is shown in Figure 2.24b with the hazard curves for the same two locations shown in Figure 2.24c (dotted curves; Eq. (2.5)). Hazard maps and hazard curves are shown to be comparable for both approaches, but with slight random fluctuations observed for simulation-based metrics.

2.6 Concluding Remarks

In conclusion, as we reflect on the journey through this chapter, we recall Richard Feynman's insightful words, 'Know how to solve every problem that has been solved.' While our exploration may not have provided solutions to every challenge in hazard assessment, it has certainly laid a solid foundation for an understanding of numerous perils, especially within the context of an introductory textbook. Despite the complexity and the wide variety of the processes involved, almost two dozen perils were described quantitatively in a harmonized manner. This achievement can be attributed to *the systematic categorization of the catastrophe object, characterized by its source* (point, line, track, area/diffuse; Section 2.2), *its size distribution* $\lambda(S)$ (power law and extreme value distributions; Section 2.3), *and its intensity footprint* $I(x, y)$ (static diffusion, passive threshold; Section 2.4 – sustained propagation to be covered in Part II). This chapter has equipped the reader with a valuable toolbox of equations rooted in fundamental principles of physics, with a particular emphasis on energy-transfer considerations. Alongside this theoretical foundation, the practical application of these concepts was illustrated through a PSHA example in Section 2.5. With this knowledge in hand, you now possess the essential tools and models required to construct a *hazard pipeline for a basic catastrophe model from scratch (i.e., Source* $\rightarrow S(Source) \rightarrow \lambda(S) \rightarrow I(x, y, S, Source) \rightarrow \bar{F}(I)$; Section 2.1). While this chapter was relatively long in comparison to the ones to follow due to the diverse range of perils covered, it was a necessary step for acquiring the skills to tackle each peril individually. As we look ahead, the next chapter will maintain a similar uniformity in its approach by describing various perils using standardized damage and loss measures. This will streamline the risk assessment process and further enrich your understanding of catastrophe modelling.

Further recommended reading

- *An Introduction to Statistical Modeling of Extreme Values* (Coles, 2001)
- *Probability and Random Processes* (3rd ed.) (Grimmett and Stirzaker, 2001)
- *Seismic Hazard and Risk Analysis* (McGuire, 2004, chapters 1–5)
- *Critical Phenomena in Natural Sciences* (Sornette, 2006, chapter 4)
- *Natural Catastrophe Risk Management and Modelling* (Mitchell-Wallace et al., 2017, chapter 4)

2.7 Exercises

Exercise 2.1

(a) Calculate the kinetic energy E_{TNT} released by a medium-size asteroid impact in Mt (megaton TNT equivalent), by assuming the following parameters: a spherical object with radius $r = 25$ m, a rock density $\rho = 3$ g/cm^3 and a velocity $v = 20$ km/s. What type of celestial body is it (a stony asteroid, an iron asteroid, or a comet)? Justify your answer.

(b) The 2013 Chelyabinsk airburst had the following recorded characteristics: a total irradiated energy $E_r = 90$ kt and a velocity $v = 19.16$ km/s. The density of the object was estimated to be $\rho = 3.3$ g/cm^3 based on X-ray computed tomography (Popova et al., 2013). Considering that $\eta = E_r/E_{TNT} = 0.021 \log_{10} E_r + 0.0055v + 0.022$ with E_r in kt, v in km/s, and E_{TNT} the kinetic energy (Nemtchinov et al., 1997), calculate the mass [kg] and diameter [m] of the celestial body.

(c) Determine the annual exceedance rate $\Lambda(E_{TNT})$ of impacts occurring on Earth with E_{TNT} the total impact energy estimated in (a). What is the probability that a similar or greater-size event will occur in the next 1,000 years?

Exercise 2.2

Monsters from an alternate dimension have wreaked havoc, terrorizing the city of Montvillard and its surrounding areas for over a century. Your objective is to construct a stochastic event set to assess the local monster hazard. We have documented 21 previous sightings during the twentieth century, which are catalogued in Table 2.8. Figure 2.25 presents the corresponding map and time series for your reference. To simplify the calculations, these creatures have been categorized into three groups based on their size S, which correlates with their severity (or their potential for destruction): Gremlin-type ($S = 50$ cm; small mischievous events responsible for unexplained accidents), troll-type ($S = 2$ m; mid-size events with extensive crushing power), and dragon-type ($S = 10$ m; large-size events capable of triggering fires, toxic breath, and other damage spells).

(a) Compute and visualize the empirical cumulative distribution function (eCDF) for both the spatial distribution and size distribution of these events.

Table 2.8 Exercise 2.2: Catalogue of past monster occurrences

t	03	04	04	08	08	28	33	37	51	54	55	64	67	68	73	83	86	88	89	93	97
x	3	2	5	3	4	2	3	4	4	4	3	3	4	1	3	4	3	2	3	5	2
y	2	1	3	2	3	3	2	2	3	1	3	3	4	4	3	2	2	2	2	5	4
S	0.5	2	0.5	0.5	0.5	0.5	2	0.5	2	10	2	0.5	10	0.5	0.5	0.5	0.5	0.5	2	0.5	2

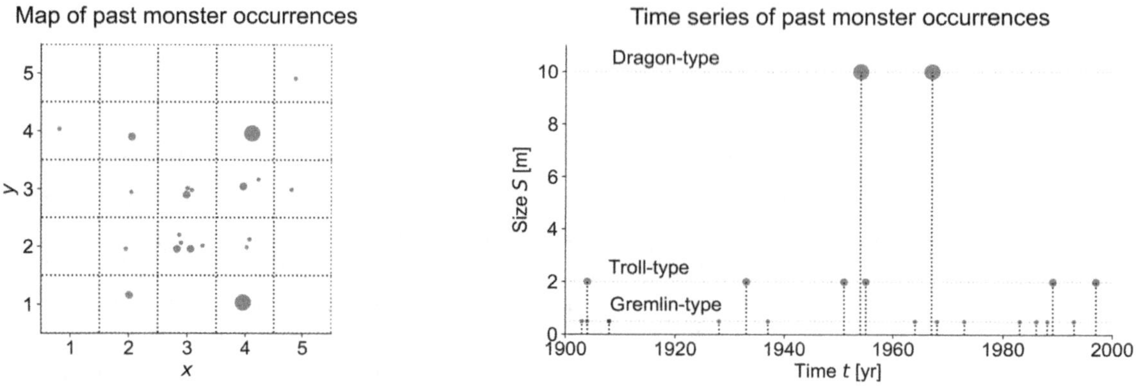

Fig. 2.25 Exercise 2.2: Map and time series of past monster occurrences.

Hint: Refer to the eCDF pseudocode in Box 2.4 for assistance. For the spatial distribution, consider defining $5 \times 5 = 25$ location identifiers with coordinates (x, y) as a look-up table.

(b) Generate a stochastic set of $n = 10$ monsters and depict them on a map similar to the one shown in Figure 2.25. Repeat the operation a second time for $n = 1{,}000$. Are your results in agreement with the observed distributions of monsters? Do you see any limitation with this simple approach?

Hints: The sampling method is described in Box 2.1. Use the eCDF plots previously generated to understand how to obtain a sample from a random number u (a sampling function for u is given in the file `exercise_notebook.ipynb`). The code to generate Figure 2.25 is also provided in that file.

Exercise 2.3

(a) Run the power-law sampling function provided in `exercise_notebook.ipynb` to generate $N = 10,000$ event sizes S with power exponent $\alpha_{true} = 2.5$ and minimum size $S_{min} = 1$. Let's assume, for the sake of illustration, that events are wildfires and event size is defined as the burnt area. Define a function to calculate $\hat{\alpha}$ by using the maximum likelihood estimator given in Eq. (2.36). Verify that $\hat{\alpha} \approx \alpha_{true}$.

(b) What is a suitable minimum sample size, denoted as N_{min}, required to achieve an estimate $\hat{\alpha}$ within a range of $\alpha_{true} \pm 0.1$? Provide a justification for your answer, supported by a graphical representation.

(c) Considering the parameterization used in (a) and assuming that these events occurred between years 2000 and 2023 (inclusive), plot the size distribution in terms of annual exceedance rate $\Lambda(S)$ (Eq. (2.37)).

Hint: see Box 2.4 for a pseudocode that plots a CCDF from empirical data. Add the power-law fit line and provide the estimates of a and b defined from $\log_{10} \Lambda(S) = a - b \log_{10} S$ (Eq. (2.38)).

(d) The global size distribution of earthquakes was found to be $a_E = 12.029$ and $b_E = 0.72$ in the seismic energy space, that is, $\log_{10} \Lambda(E) = a_E - b_E \log_{10} E$ (Eq. (2.38)). Considering the Gutenberg–Richter law $\log_{10} \Lambda(M) = a_M - b_M M$ (Eq. (2.39)) with M the earthquake magnitude, as well as the empirical relationship $\log_{10} E = 5.24 + 1.44M$ (Eq. (2.15)), estimate a_M and b_M. It was stated in Section 2.3.1 that 'about one earthquake with a magnitude equal or greater to 8 can be expected globally each year'. Does this seem correct? Justify your answer.

Exercise 2.4

(a) Develop a seismic hazard curve (Eq. (2.3)) for a site located $d = 10$ km away from the surface trace of an active fault on which earthquakes of magnitude $5 \leq M \leq 7.5$ recur following the Gutenberg–Richter law with $b = 1$ and $a = 3.5$ (Eq. (2.39)). Use the simple GMPE of Ambraseys and Bommer (1991) to calculate the intensity $I = f_I(M, r)$ with r the distance to the rupture, located at a depth $h = 2$ km (Eq. (2.54)). Then, plot the annual hazard curve. Use the code template provided in the file `exercise_notebook.ipynb`.

Hints: the steps are as follows: (1) discretize magnitudes in bins M_i of width $\Delta M = 0.1$, (2) calculate the event rates λ_i (in non-cumulative form; Eq. (2.65)), (3) calculate the mean hazard intensity I_i for each event, (4) generate the hazard curve using the log-normal distribution for the PDF $f(I_i)$ for a range of intensity increments y (recall that the standard deviation in log-space is $\sigma = 0.27$ for Eq. (2.54). Log-normal distribution provided in `exercise_notebook.ipynb`).

(b) It is common in PSHA to create hazard maps for a 475-year return period. This specific return period corresponds to a 10% probability of exceedance in 50 years, which is a widely accepted standard in seismic design. This probability represents the ground-motion intensity that a building should be designed to withstand. To demonstrate that the ground-motion intensity is the same for a 475-year return period and a 10% probability of exceedance in 50 years, perform the following steps:

- Calculate the exceedance probability in 50 years for a return period of 475 years and verify that it is indeed 10%.
- Turn your existing code for calculating hazard curves into a function with the time window δt as input.
- Use your function to calculate the two hazard curves and confirm that the ground motion for 475-year return period and for 10% probability of exceedance in 50 years are the same.

(c) It has been mentioned previously that truncating the log-normal distribution of ground motion is common practice in PSHA to avoid unphysical estimates of the expected level of shaking. However, this practice comes at the cost of potentially underestimating the hazard at the tail of the hazard curve. You have been provided with a truncated version of the log-normal distribution in the file `exercise_notebook.ipynb` to be implemented in your hazard curve function. Calculate the percentage change in the expected ground-motion level when a 2σ truncation is applied compared to the untruncated distribution for two return periods: 1,000 years and 10,000 years. Plot the two hazard curves for direct comparison.

3 Probabilistic Risk Assessment

Risk represents a compound measure of the likelihood and magnitude of adverse consequences affecting structures, individuals, and valuable assets (Section 1.1.2). To comprehensively assess risk, it is first imperative to establish an inventory of the assets exposed to potential hazards (Section 3.1). Estimating the expected damage resulting from a given hazard load is the second crucial step in the assessment process (Section 3.2). *The application of damage functions to exposure data forms the basis for calculating loss estimates*, which, in the case of economic losses, are further subjected to financial analysis (Section 3.3). These three facets collectively constitute the core elements of the three risk modules within a standard CAT model, each of which is juxtaposed with the hazard module discussed in the preceding chapter.

Figure 3.1 represents an enhanced rendition of Figure 1.6, offering a more detailed overview of a typical CAT model workflow. In this diagram, the hazard intensity footprints $I_{\mathrm{ID}}(x, y)$, previously modelled, will serve as input to generate damage footprints $D_{\mathrm{ID}}(x, y)$ per stochastic event with identifier ID. The peril-specific damage functions $f_D(I)$ – taking the form of fragility curves or vulnerability curves – depend on the characteristics of the asset inventory (Θ_ν), for example building type. *To ensure consistency, the mean damage ratio (MDR) will be used as the main measure for damage footprints*, with the final loss footprints simply expressed as $L_{\mathrm{ID}}(x, y) = D_{\mathrm{ID}}(x, y) \times \nu(x, y)$, where $\nu(x, y)$ represents the exposure footprint.

In contrast to many hazard data being publicly available (Chapter 2), exposure data are only partially available, while damage and loss data remain proprietary in most cases. As a consequence, a city will be simulated at the end of this chapter to represent the economic exposure of an ad hoc region (Section 3.3.3). This initiative forms an integral part of the **CAT Risk Modelling Sandbox**, which has been developed in conjunction with this textbook. The sandbox is designed to empower readers with the practical application of the quantitative methods developed in Chapters 2 and 3. It presents a relatively straightforward yet reasonably realistic template for hands-on learning. This resource takes the form of a tutorial, complete with well-documented code examples, accessible via the online resource website associated with the textbook. This will be complemented by the 'Build Your Own CAT Risk Model from Scratch' project, which will conclude Part I (Boxes 3.7 and 3.8).

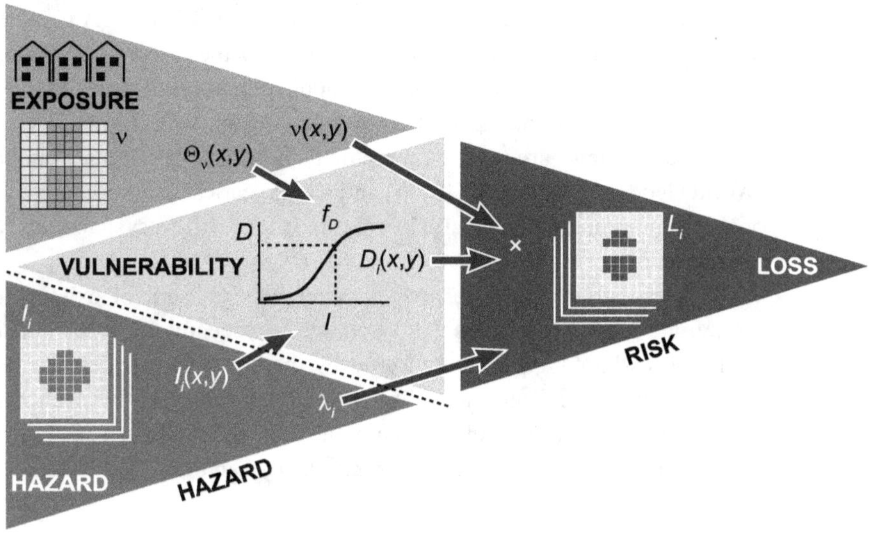

Fig. 3.1 Workflow of the CAT risk model framework, modified from Mignan (2022a) (CC-BY 4.0, https://creativecommons.org/licenses/by/4.0/). This chapter focuses on the following three risk modules: exposure, vulnerability, and loss.

3.1 Exposure Module

Exposure data consist of the assets considered at risk in a given region, which is representable by a spatial footprint $\nu(x, y)$. The asset value (ν) in CAT models is commonly the ground-up value, or economic value, which, for a building, is the financial value of the total structural rebuild (insured and uninsured values combined). Other possible quantities are content value (fixed inventory versus movable inventory e.g., in the case of floods) and business revenue (e.g., for business interruption modelling). A set of attributes $\Theta_\nu(x, y)$ provides additional information for use in the vulnerability module, including construction type, building height, year built. Insurance financial structures are also included to differentiate between economic and insured values (e.g., Jones et al., 2017) (to be defined in Part III). The asset ν can also represent the population for injury and fatality risk models. This is often approximated by the number of inhabitants per building, which is a function of daytime versus night-time. Considering the individual displacements of people in CAT models is undesirable since it is too computationally costly, although agent-based models can be useful in specific applications (Part II, Chapter 4). Some more specialized asset value types exist, such as crop yield in agricultural production risk assessment. Infrastructures, such as transportation networks and power grids, have their own exposure characteristics.

The location (x, y) of assets can be the centroid of polygons or the coordinates on a regular spatial grid. A polygon may represent the perimeter outline of a building at the highest level of resolution to the boundaries of a country in global low-resolution analyses. In between, various resolutions are possible, reflecting different administrative systems (e.g., postal code level, city level, or CRESTA zone level). **CRESTA** (Catastrophe Risk Evaluation and Standardizing Target Accumulations) is the standard established by the insurance industry for the harmonization of geographical data aggregation. If the location is provided as an address, coordinates (x, y) are retrieved using geocoding tools. In this textbook, two coordinate systems are used, geographical coordinates (longitude and latitude [°]) from the World Geodetic System (WGS-84) and simple cartesian coordinates [km] when geographical information is unnecessary. Economic values will be systematically given in USD (and assuming parity with EUR).

The descriptions of exposure characteristics Θ_ν and of existing exposure footprints $\nu(x, y)$ are given in Sections 3.1.1 and 3.1.2, respectively. To be consistent with the general physics-based approach followed by this textbook, models for the development of a synthetic exposure footprint will be presented throughout this section in successive boxes (Boxes 3.1–3.3). The resulting synthetic footprint will be described in detail in Section 3.3.3, with the exposure characteristics $\Theta_\nu(x.y)$ and exposure value $\nu(x, y)$ grids provided on the online resource website.

3.1.1 Exposure Characteristics

Exposure characteristics, here represented by the set Θ_ν, describe how an asset behaves under a given hazard intensity load (for additional insurance-related characteristics, see Part III). They are input parameters to the damage functions to be later defined in the vulnerability module (Section 3.2). These characteristics, which indicate how vulnerable an asset is, are termed in the CAT risk modelling jargon **primary modifiers** for common attributes and **secondary modifiers** when peril-specific (Jones et al., 2017, p. 20). Secondary modifiers are used to quantify the potential impact of structural features or of disaster risk reduction measures, which are generally peril-dependent. Since the asset value ν depends in part on Θ_ν, models $\nu = f_\nu(\Theta_\nu)$ can be used when direct exposure value estimates are lacking. Due to the potentially large number of attributes and their specific relationships to damage functions, a taxonomy must often be developed for the proper propagation of information from the exposure module to the vulnerability module (e.g., Acevedo et al., 2020, table 3). Examples can be found in HAZUS (FEMA, 2021) and GEM (Gamba, 2014). In general, retrieving information about each single object at risk in a given region is impossible; therefore, classes are defined to describe different groups of objects.

3.1.1.1 Occupancy Class

The main **occupancy classes** are residential, commercial, and industrial, which correspond to the main built-up land categories in land-use classifications (Anderson et al., 1976). Other classes include hospitals, schools, and agricultural lands, for instance. A simpler occupancy classification, used by the USGS' Prompt Assessment of Global Earthquakes for Response (PAGER) (Wald et al., 2010), combines urban/rural and residential/non-residential classes (Jaiswal and Wald, 2008). Occupancy classes follow specific spatial patterns, which are described in Box 3.1. With economic value data often lacking, the

Box 3.1 Modelling Urban Land-Use Patterns

Urban land-use patterns result from a complex process of interactions within the evolving built-up land and the natural land. However, the underlying rules are relatively simple, based on attraction (+) and repulsion (−) effects, distance-decay effects, and site-specific factors (White et al., 1997). For example, commerce is attracted to commercial zones while being far away has a repulsive effect due to competition. Housing in vicinity of a commercial area, but not next to it, is attractive while there is a strong repulsion to being close to an industrial zone. A river is attractive while a railway is repulsive for the residential class, whereas close proximity to railways and roads is attractive to industry, etc.

Class	Neighbour	Nearby	Mid-distance	Far away	Very far
		Commercial attraction to:			
Commercial	+++	+	−−−	−−−	−−−
Residential	++	+	+	+	+
Road	+++	0	0	0	0
		Industrial attraction to:			
Industrial	+++	++	+	0	0
Road	++	+	0	0	0
Railway	++	+	0	0	0
		Residential attraction to:			
Commercial	−−	++	++	+	0
Industrial	−−	−	0	+	+
Road	−	+	+	+	+

The probability that a site (x, y) will change its occupancy class from state h to state j can be defined as

$$p_{hj} = c_{1,j}\left(1 + [-\log u]^{\alpha}\right)\left(1 + \sum_{k,i,d} w_{kd}\mathbf{I}_{id}\right) + c_{2,j}$$

where $0 \leq c_{1,j} \leq 1$ is the suitability of the site (0 in water, if slope too steep), u is a random number corresponding to a stochastic perturbation adjustable via α, w_{kd} is the weight applied to neighbour sites i with state k at distance d, $\mathbf{I}_{id}$ is the identity matrix, and $c_{2,j}$ is an inertia parameter ($c_{2,j} > 0$ if $j = k$, $c_{2,j} = 0$ otherwise) (White et al., 1997). Based on the rules previously defined and encoded in w_{kd}, sites with the same occupancy class have a tendency to cluster and to be located in attractive areas for their specific purposes (Section 3.3.3).

average construction cost v_{cost} [USD/m^2] may be estimated globally from the GDP [USD] per capita with a power-law function of the form

$$v_{cost} = c_1 \mathrm{GDP}^{c_2} \tag{3.1}$$

with $c_1 = (24.1, 33.6, 30.8)$ and $c_2 = (0.385, 0.357, 0.325)$ for residential, commercial, and industrial classes, respectively (Huizinga et al., 2017). Although subject to very high epistemic uncertainties, these functions will be used later on to define the economic value of our synthetic exposure footprint. HAZUS determines the structure replacement cost from value tables for different types of occupancies (e.g., USD 1,292 per m^2 for a warehouse, USD 2,813 for a bank, etc. in the US) upon which adjustment factors are applied (from economy to luxury class, location, etc.) (FEMA, 2021).

3.1.1.2 Construction Type

The **construction type** includes the material and the method of construction. The main types are concrete (C), masonry (M), reinforced concrete (RC), steel (S), and wood (W). Further characteristics include the types of walls and frames and whether specific designs have been used (e.g., earthquake-resistant or flood-resistant designs). International conventions can be found in the HAZUS and European Macroseismic Scale 1998 (EMS-98) classifications (Grünthal, 1998; FEMA, 2021). For floods, clay and prefabricated are some of the main building types alongside masonry and reinforced concrete (Schwarz and Maiwald, 2008). If damage functions are often developed for a specific construction type, the RISK-UE classification proposed by Lagomarsino and Giovinazzi (2006) for seismic risk provides an elegant approach to describe the building type by one parameter $\Theta_\nu = \kappa_D$ (coined 'vulnerability index'). The matching damage function $D = f_D(I, \kappa_D)$ will be defined in Section 3.2. The correspondence between construction type and κ_D can be found in Lagomarsino and Giovinazzi (2006, table 3). For example for masonry, $\kappa_D = 0.87$ on average for rubble stone, 0.84 for adobe, and 0.74 for simple stone; for reinforced concrete, $\kappa_D \leq 0.64$ on average (with the value being anti-correlated to the resistance of the structure to shaking).

3.1.1.3 Building Height

Building height h_b (or the number of storeys from which h_b may be estimated) is another important parameter to take into account for damage assessment. A low-rise, mid-rise, high-rise classification is commonly used for simplicity. The building height plays different roles for different perils. In the case of earthquakes, it determines the response of the structure to ground-shaking at a particular period. The **fundamental period of vibration** (T_b [s]) of a building is directly linked to its **resonance** (Fig. 3.2a), as it represents the natural frequency at which the structure tends to vibrate most easily (hence amplifying the effects of external forces). It can be estimated empirically with

$$T_b = c_1 h_b^{c_2}, \tag{3.2}$$

where h_b [m] is the building height, and c_1 and c_2 are coefficients depending on the building typology (Lagomarsino and Giovinazzi, 2006; Crowley and Pinho, 2009). Damage is then assessed for the spectral acceleration I_{SA} of matching period $T_{SA} = T_b$ instead of the PGA ($T_{SA} = 0$). Tall buildings also behave differently from low- and medium-rise structures under wind load (Mendis et al., 2007). For floods, the damage ratio reflects the inundation

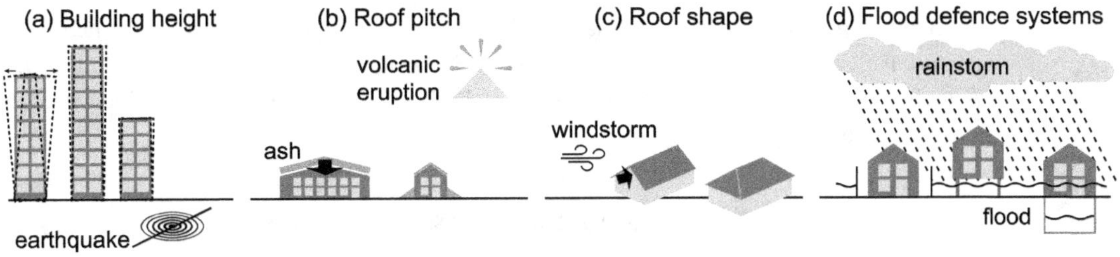

Fig. 3.2 Exposure characteristics and relationship to different hazard loads. (a) Resonance during ground-shaking, a function of the fundamental period of vibration of buildings, itself a function of height. (b) Ash load is greater on low-angle roof pitches where ash can accumulate during a volcanic eruption. (c) Gable roofs (left) are more sensible than hip roofs (right) to strong winds. (d) Various defence systems against inundation.

depth relative to the height of the building. Distinction must also be made between storeys above and below ground.

3.1.1.4 Other Building Characteristics

The previously described exposure characteristics (occupancy, construction type, and height) are the three main parameters used for damage function definition (Khanduri and Morrow, 2003; Brzev et al., 2013; FEMA, 2021). For example, the building type code MLRI corresponds to masonry / low-rise / industrial in HAZUS (FEMA, 2021). Some other useful parameters can be retrieved from building polygon data when available, such as area and shape. Damage potential can also depend on building age (or year built), with the parameter also used to estimate the economic value of the building by applying a depreciation model (FEMA, 2021). Secondary modifiers can include roof features: the roof pitch is a controlling parameter in volcanic ash load, a classification is low ($< 6°$), medium ($6–35°$), and high ($> 35°$) (e.g., Blong et al., 2017) (Fig. 3.2b). The roof shape (gable versus hip) is important in the case of windstorms (e.g., Stathopoulos, 2003) (Fig. 3.2c). Other secondary modifiers include roof anchoring against winds, seismic retrofitting measures, and secondary defence systems against inundation (Fig. 3.2d). As previously indicated, insurance policy terms – to go beyond ground-up losses – will be described in Part III.

3.1.1.5 Infrastructure Characteristics

Other objects at risk are utility networks, transportation infrastructures, and critical facilities (Pitilakis et al., 2014, p. 6). The main infrastructures are power grids, oil and gas systems, water supply and wastewater networks, road and railway networks, harbours, healthcare, and firefighting systems. Each component of an infrastructure must be described, since a local failure may lead to system failure. For a power grid, this includes the transmission lines, the generation plants, substations, etc.; for an oil system, it includes production facilities, refineries, pipelines, etc.; and for a road network, it includes bridges and tunnels (Pitilakis et al., 2014, table 1.1). Some parameters (of Θ_ν) are common to most elements (e.g., construction material) while others are component-specific. Some components have dedicated damage functions (e.g., SYNER-G database; Pitilakis et al., 2014). A technique for modelling a road network, which will be utilized to create our synthetic exposure footprint, is detailed in Box 3.2.

3.1.1.6 Agricultural Exposure Characteristics

Exposure in the agricultural sector includes crops, livestock, and forestry. Crops are particularly exposed to floods (Merz et al., 2010), hailstorms (Sánchez et al., 1996), climatological perils (Murthy et al., 2015), and pests (Tonnang et al., 2017). Land may be simply classified in terms of arable land (crops) versus grassland. Crops may also be differentiated by crop type (e.g., corn, grapes, oats, and tomatoes are among 40 classes defined in HAZUS; FEMA, 2021), with different species potentially reacting differently to a given stress (Sánchez et al., 1996). Soil data, including water content and the presence or not of irrigation support for droughts (Murthy et al., 2015), may be considered part of the exposure characteristics (Θ_ν). Exposure value (ν), in terms of production yield, directly depends on the crop type.

3.1.1.7 Population Exposure Characteristics

For hazard intensity loads which have a physical impact on the body, such as crushing, all individuals can be treated the same (Elsayed, 1997). For other perils such as epidemics or heatwaves, Θ_ν generally includes the age, sex, and health status to define population cohorts (Smith, 2017). Census statistics can be used to estimate the number of inhabitants as a function of building area and number of storeys. The vulnerability of people may then directly depend on the vulnerability of the building (see fatality functions in Section 3.2).

3.1.2 Exposure Footprint

An **exposure footprint** defines the spatial distribution of asset values $\nu(x, y)$ to be used as input for loss estimation in a CAT model. Asset attributes $\Theta_\nu(x, y)$ are also defined for each asset location (Section 3.1.1). High-quality exposure footprints are in most cases proprietary and limited to a subset of countries. Various initiatives exist to make exposure data publicly available, but so far suffer from incompleteness, heterogeneity, and spatial resolution issues. The synthetic exposure footprint to be described in Section 3.3.3 will be a convenient generic data set for the student of risk (see Box 3.3 for a description of the approach).

3.1.2.1 Aggregated Exposure Data

Population exposure data are more easily accessible than economic exposure data, with more than half a dozen global data sets publicly available (Gunasekera et al., 2015, table 1), including the Global Human Settlement Layer (GHSL) (Schiavina et al., 2022) and WorldPop databases (Bondarenko et al., 2020). With the development of open-source CAT risk

Box 3.2

Modelling Road Networks

A road network can be simulated by following a set of simple rules, as defined by Koenig and Bauriedel (2009). Their model (a cellular automaton; Part II, Chapter 4) was selected for both its simplicity and its reasonable representation of a transportation system. A regular spatial grid with coordinates (x, y) is considered, with each cell represented by a state $\mathcal{S} = 0$ (empty). The road network is initiated at a primary node, with $\mathcal{S}(x_0, y_0) = 1$ (occupied). For every occupied cell, all empty cells of the eight neighbour cells are set to $\mathcal{S} = 2$ (closed). All empty cells neighbouring a closed cell are set to $\mathcal{S} = 3 + u$ (candidate), with u a random number. A new node is defined where $\mathcal{S} = \max \mathcal{S}$ at each new iteration i. In the first edge creation, the first two nodes are linked. In later iterations, two nodes are linked if the following two conditions are met: (1) the distance from the new node to other nodes is less than R and (2) the shortest path between (x_0, y_0) and (x_i, y_i) is equal to $\min(\max l, X)$ with l the list of shortest paths verifying condition 1 and X a second controlling parameter. The process is repeated n times. In the context of our synthetic exposure (Box 3.3), land-use masks $u(x, y) = 0$ are defined for cells containing water or with terrain slope too steep. An edge crossing a water cell is then defined as a bridge.

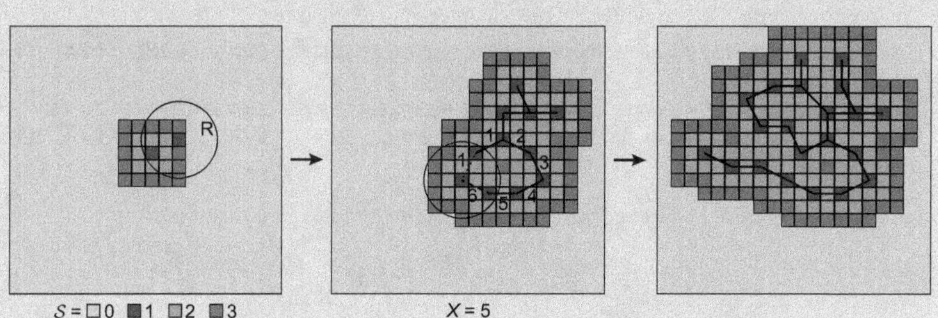

platforms, such as CLIMADA for weather and climate perils (Aznar-Siguan and Bresch, 2019) or GEM for earthquakes (Pagani et al., 2020), there is an increasing need for global economic exposure data to be publicly available for real-world applicability. LitPop has been developed for CLIMADA (Eberenz et al., 2020) and the Global Exposure Database GED4GEM has been developed for GEM (Gamba, 2014). Both are available globally at a 30-arcsec ($\approx$ 1 km) resolution. LitPop provides only economic value while GED4GEM provides structural value, population counts during daytime and night-time, occupancy class, and direct link to specific vulnerability curves (see, e.g., the related European Exposure Model; Crowley et al., 2021a). The taxonomy of GED4GEM (for Θ_ν) is being expanded to additional perils (floods, volcanic eruptions, windstorms, tsunamis, and droughts) via the GED4ALL project with data hosted by the Global Facility for Disaster Reduction and Recovery (GFDRR). At the time of writing of this textbook, the GED4ALL database remains at a very early stage of development (Silva et al., 2018). However, its exposure data model covers many perils and exposure types, already making it a valuable resource (GFDRR, 2023). Exposure footprints obtained from selected databases are shown for the island of Malta in Figure 3.3. Construction type is also generally available only in aggregated form. The distribution of construction types per country (and four occupancy

Box 3.3 Modelling Built-Up Land Footprints

A simplified version of the Slope, Land use, Exclusion, Urban, Transportation, and Hill-shade (SLEUTH) model of urban growth (Clarke et al., 1997; Candau, 2002; Clarke, 2014) (another cellular automaton) is used to generate a synthetic exposure footprint. A regular spatial grid is defined with state $\mathcal{S}(x,y) = 0$ (non-urban). Any time a cell is proposed to be urbanized ($\mathcal{S} \rightarrow 1$), a test of terrain slope suitability has to be passed. There are then four steps: (1) n random cells attempt to turn to $\mathcal{S} = 1$ with n function of a dispersion coefficient (spontaneous new growth step). (2) A newly urbanized cell becomes a new spreading centre if it passes the breed coefficient test, in which case two neighbour cells are selected for urbanization (new spreading centre step). (3) Any cell in a cluster of at least three urban cells urbanizes an empty neighbour cell if it passes the spread coefficient test (edge growth step). (4) k random urbanized cells search for a road, with k a function of the breeding coefficient. A road is searched within a radius r, with r a function of a road gravity coefficient. If found, a cell m steps away along the road is urbanized, with m a function of the dispersion coefficient (road-influenced growth step). The process is repeated t times. All constraints are defined as randomized tests which are passed or not. The set of rules is defined by five coefficients. In the context of a synthetic city, the road network grows at each iteration following the model described in Box 3.2 (at a lower grid resolution) and the occupancy class is reassessed following the model shown in Box 3.1.

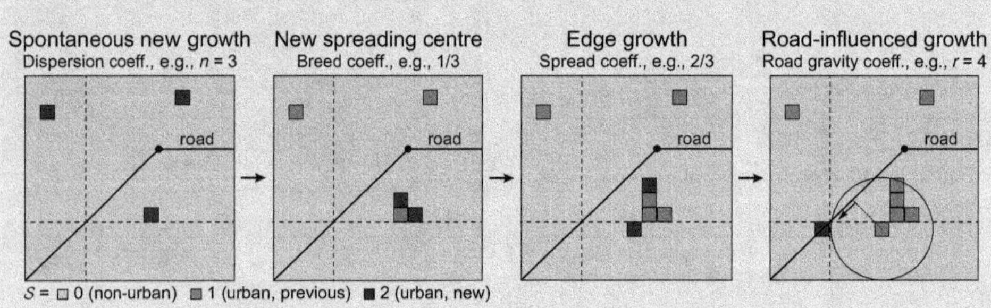

classes) is given in PAGER's global building inventory database (Jaiswal and Wald, 2008). A building type footprint is available at a 500-m resolution for several African countries in GED4ALL (Silva et al., 2018).

The aforementioned exposure databases were developed using a top-down (e.g., LitPop) or combined top-down/bottom-up (e.g., WorldPop, GED4GEM) strategy. The top-down approach, or downscaling (Box 3.4), involves the spatial disaggregation of readily available information – such as national GDP for economic value assessment – by using spatial constraints, which are also easily accessible. For example, the exposure footprint can be constrained by water masks and the topography. This approach, if systematically used without corrections, leads to a very approximate representation of the true exposure distribution. The bottom-up approach, which includes ground surveys, helps to refine the results.

3.1.2.2 Building-Level Exposure Data

Assessing the exposure at the building level has yet to be achieved at the global scale. One of the most extensive mapping initiatives viewed as an essential component of Digital Earth

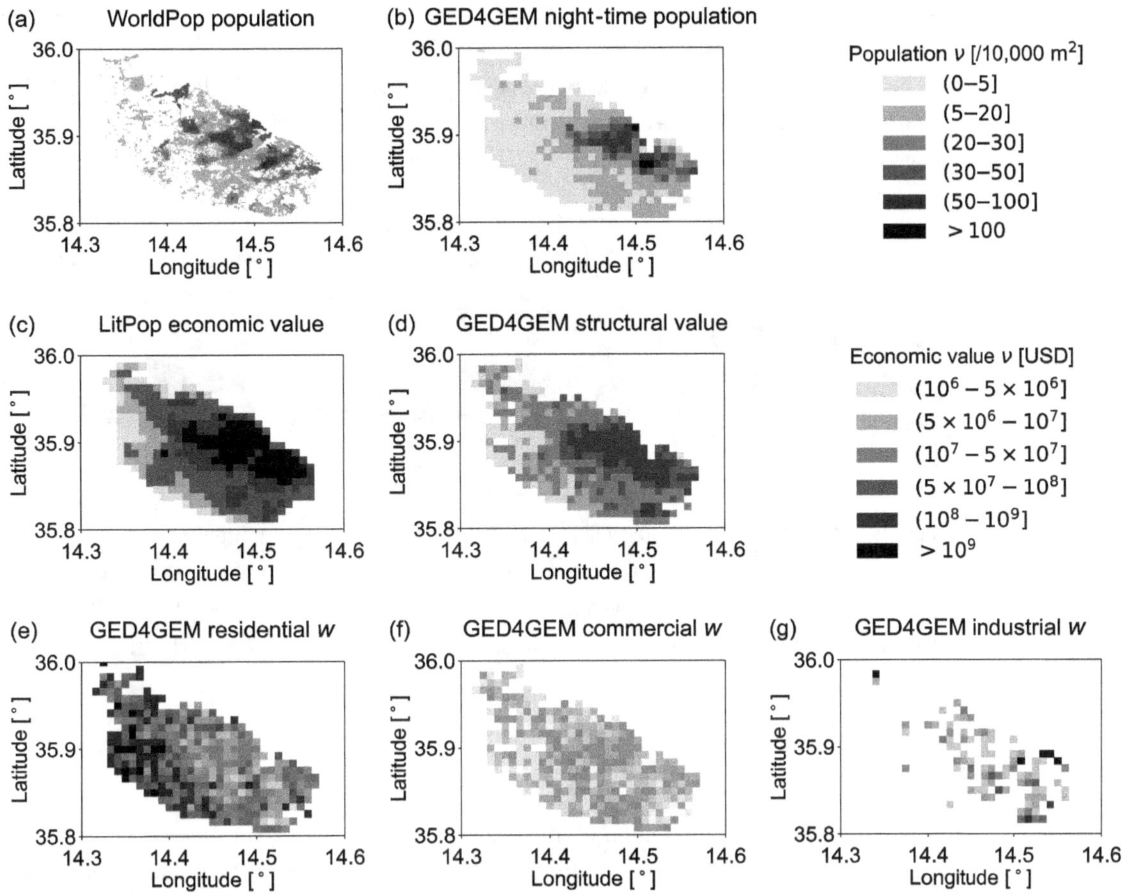

Fig. 3.3 Exposure footprints for Malta obtained from different databases. (a) Population count at 3-arcsec ($\approx$ 100-m) resolution from WorldPop database (Bondarenko et al., 2020) (CC-BY 4.0, https://creativecommons.org/licenses/by/4.0/). (b) Population count from GED4GEM. (c) Economic value at 30-arcsec ($\approx$ 1-km) resolution from the LitPop database (Eberenz et al., 2020) (CC-BY 4.0, https://creativecommons.org/licenses/by/4.0/). (d) GED4GEM structural value. (e–g) GED4GEM weights w for residential, commercial, and industrial occupancy classes (part of θ_ν). All GED4GEM data at 30-arcsec resolution accessed from the European Exposure Model Data Repository (Crowley et al., 2021a) (CC-BY 4.0, https://creativecommons.org/licenses/by/4.0/), as used in the European Seismic Risk Model (ESRM20) (Crowley et al., 2021b).

(Mooney and Corcoran, 2014) is the OpenStreetMap (OSM) crowdsourcing project, which provides geographical features of buildings, streets, and other infrastructures (Haklay and Weber, 2008). OSM has been shown to be suitable for disaster mapping and management (e.g., Goldblatt et al., 2020). Depending on the level of data completeness at a given location, some building attributes (Θ_ν) can be retrieved, such as building height, material type, construction year, or roof shape. Interestingly, the GED4ALL exposure attributes have been mapped to the existing OSM keys (GFDRR, 2023). OSM applications to CAT risk modelling, however, remain very limited since building data completeness is less than 20%

Box 3.4 Downscaling

The technique of **downscaling** (e.g., Murakami and Yamagata, 2019) consists of the following. Knowing, for example, the population N_{pop} in a given region as well as the land-use distribution, the population density is simply $\rho_{pop} = N_{pop}/A_{urban}$ in the urban area A_{urban} and zero elsewhere (where land-use classes are water, grassland, etc.).

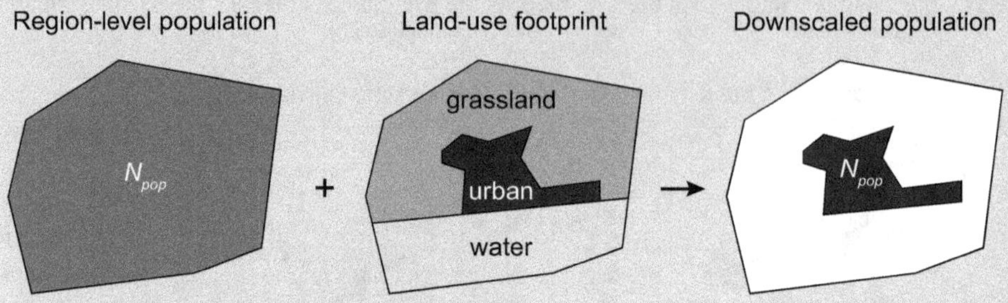

In practice, the exposure is disaggregated proportionally to a weight $w(x,y)$ function of some spatial variable(s). The LitPop downscaling technique, for example, uses GDP as a proxy for total asset value, with combined population and nightlight intensity footprints as a spatial constraint (Eberenz et al., 2020). The correlation between nightlight intensity and GDP has been verified by a number of studies (Ebener et al., 2005).

for 75% of cities globally (Zhou et al., 2022). OSM also does not give exposure values $\nu(x,y)$ and remains primarily a building-polygon stock resource. Machine learning, and especially deep learning, can nowadays automatize the modelling of building polygons (by segmentation) and of $\Theta_\nu(x,y)$ from satellite imagery. The Microsoft Global ML Building Footprints database is such an example, with almost a billion building footprints detected (Microsoft, 2023). It can be hoped that a global exposure $\{\nu(x,y), \Theta_\nu(x,y)\}$ database will one day become available at the building level in that context. Traditionally, local field surveys can be conducted to assess the construction material of each building in hazardous areas (Reese et al., 2011; Acevedo et al., 2020) but, in general, only the distribution of building types is available from the city level (e.g., from national institutes of statistics) to some regional level (e.g., GED4ALL, PAGER).

It is clear from this brief overview that many types of exposure databases are available to the student of risk. None are perfect but they answer different needs and some described here will be used for damage footprint modelling in Section 3.2. For loss assessment using generic vulnerability curves $D = f_D(I)$, simple databases of asset values $\nu(x,y)$, such as LitPop, may be used (Fig. 3.3c). When a damage function requires the occupancy class (part of Θ_ν) as input, GED4GEM appears to be a good option for damage footprint modelling. When building type is required as input, the country-level building type distribution offered by PAGER provides a way to weigh different damage outcomes.

3.2 Vulnerability Module

The vulnerability component of a CAT model is the interface between hazard, exposure, and loss (Fig. 3.1). Different definitions of the term vulnerability have been proposed, such as degree of loss, potential for loss, capacity against hazards, or susceptibility to hazards (Fuchs et al., 2007). Considering the CAT risk modelling pipeline, **vulnerability** provides a mapping from hazard intensity footprints $I(x, y)$ to damage footprints $D(x, y)$ via a damage function $f_D(I, \Theta_\nu)$, where Θ_ν describes the exposure characteristics that make the asset more or less vulnerable to the intensity load I. The different types of damage functions, which include fragility curves and vulnerability curves, are defined in Section 3.2.1. Special emphasis will be placed on damage to structures, but the development of fatality cost functions will also be discussed. Three main approaches exist to establish damage functions: expert judgement, empirical fits, and engineering-based models. All will be described with examples from different perils in Sections 3.2.2–3.2.4. A list of vulnerability curves will be provided for selected intensity loads (commonly representing different perils) in Section 3.2.5.

3.2.1 Basic Concepts

There are two main types of damage functions: fragility curves and vulnerability curves (Fig. 3.4). A **fragility curve** expresses the probability that a damage state (DS) will occur or be exceeded given an intensity load I. It therefore represents the physical damage and involves engineering methods to describe and assess it. A **vulnerability curve** provides an estimate of a mean damage measure given an intensity load. In CAT modelling, it is taken as the **mean damage ratio** (MDR), a value between 0 and 1, defined as the ratio of the repair cost of the object at risk to its replacement value. It therefore represents

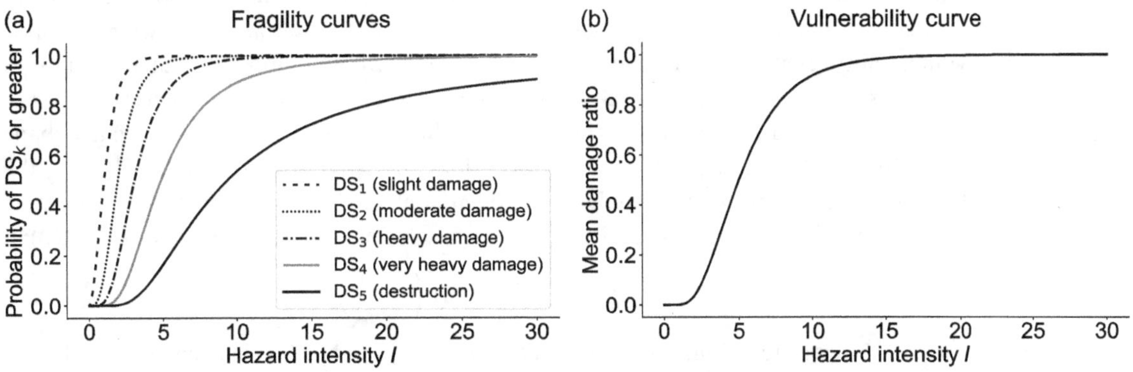

Fig. 3.4 Examples of fragility curves and of a vulnerability curve for an ad hoc hazard intensity measure I and with DS$_k$ a damage state.

the statistical average of the overall damage in economic terms, which can be assessed from insurance claim data. Depending on the scientific discipline, different terminologies may be used. For instance, vulnerability curves are called impact functions in weather and climate risk research. Damage footprints $D(x, y)$ are defined in terms of MDR, which requires **consequence functions** for converting damage state probabilities into an MDR value when the damage is assessed from fragility curves. With the scientific literature on earthquake damage modelling likely to be the most extensive across all natural hazards (Gentile et al., 2022), concepts below will often be defined from the seismic engineering field. However, they naturally apply to other perils (see a review for vulnerability to wind in, e.g., Walker (2011)).

3.2.1.1 Fragility Curve

A fragility curve is defined as a cumulative probability distribution of the form

$$\mathbb{P}(X \geq \mathrm{DS}_k) = f_D(I, \Theta_\nu), \tag{3.3}$$

where X is a random variable and DS_k a damage state. One may consider only one **damage state**, such as building collapse or the failure of a specific element. Multiple damage states are usually classified as follows: no damage (DS_0 or simply omitted), slight damage (DS_1), moderate damage (DS_2), heavy damage (DS_3), very heavy damage (DS_4), and destruction (DS_5) (e.g., EMS-98 classification for earthquake damage; Grünthal, 1998). Blong (2003a) proposed a similar five-level system (light / moderate / heavy / severe / collapse) for multiple perils (Section 3.2.2). For slight to moderate damage, the building can be occupied and repaired. For heavy (or extensive) damage, it is considered repairable but cannot be occupied. For very heavy damage, it cannot be occupied but is also not repairable. Therefore, DS_4 and DS_5 are often combined together (defining a complete damage state), the only difference being collapse or not of the structure (e.g., Memari et al., 2018). For volcanic ash fall, damage states of the roof are commonly considered (Spence et al., 2005). For wind damage, the following failure states may be considered: pane uplifted, root-to-wall connection failure, window glass failure, etc. (Li and Ellingwood, 2006). For human vulnerability, death and different levels of injuries delimitate different damage states. A damage state can be assessed post-disaster at the building level by a field survey, as well as by an aerial or a satellite survey (Section 3.2.3). The function f_D, typically taking a S-shape (i.e., **sigmoid function**), can be defined by the cumulative log-normal distribution, the logistic function, and other formulas (Lallemant et al., 2015). In practice, the **cumulative log-normal distribution** is widely used (e.g., Pitilakis et al., 2014; Gentile et al., 2022):

$$\mathbb{P}(X \geq \mathrm{DS}_k) = \frac{1}{2}\left[1 + \mathrm{erf}\left(\frac{\log I - \mu}{\sigma\sqrt{2}}\right)\right], \tag{3.4}$$

with erf the error function, I the hazard intensity, and μ and σ the mean and standard deviation of $\log I$, respectively. It follows that, for a given damage state DS_k and building type (described by its construction material, height, occupancy, and/or other characteristics), the set $\Theta_\nu = (\mu, \sigma)$ is sufficient to completely describe potential damage. $I_m = \exp(\mu)$ represents the median hazard intensity that induces a damage equal to or greater than DS_k and σ represents the spread due to aleatory uncertainty. The cumulative log-normal distribution has several convenient characteristics compared to other sigmoid functions, such as having a lower zero-bound on the hazard intensity axis I (Lallemant et al., 2015), but it may not necessarily fit the data better than other functions (Porter, 2021).

3.2.1.2 Vulnerability Curve

A vulnerability curve expresses the damage D expected on an asset of characteristics Θ_ν given a hazard intensity load I as

$$D = f_D(I, \Theta_\nu), \tag{3.5}$$

with D a continuous damage outcome. The terms damage index or central damage value are sometimes used (Blong, 2003a). If D is taken as the MDR, it provides a direct mapping between hazard and ground-up loss footprints with $L(x,y) = f_D(I(x,y), \Theta_\nu) \times \nu(x,y)$ and ν the exposure value. Similarly to fragility functions, vulnerability curves are commonly modelled by an S-shaped function, including the cumulative log-normal distribution, the hyperbolic tangent, the logistic regression, etc. For some perils, only the convex or concave side of an S-curve is needed, leading to the use, for example, of the exponential function (e.g., wind) or the square root (e.g., flood). In some cases, a step function (or vulnerability table) is used (see a list per peril in Section 3.2.5). It was previously indicated that damage states DS_4 and DS_5 can be combined as a complete damage state where the structure is not repairable (being collapsed or not). For a vulnerability curve, this is translated by defining an MDR threshold above which the property is not worth repairing and is a total loss, with the MDR fixed to 1 (a threshold of 0.8 is given in Walker (2011)).

3.2.1.3 Fragility-to-Vulnerability Conversion (Consequence Functions)

The relationship between fragility and vulnerability, or between damage state and MDR, can be illustrated by a simplified example with a binary outcome. Considering a set of buildings under the same intensity load having a probability of collapse $\mathbb{P}(X \geq DS_5) = 0.5$, half of the buildings will be totally destroyed (MDR = 1) while the other half will remain intact (MDR = 0), on average. Taking all buildings together, this is equivalent to an MDR of 0.5. In reality, buildings experience different levels of damage, which requires, first, to know the equivalence between damage state and MDR value and, second, to integrate over all possible damage states. This is done via so-called consequence functions f_c (also termed damage-to-loss functions or damage-to-impact functions), used in a weighted sum to obtain the average damage

$$D = \sum_{i=1}^{n} f_c(DS_i) \cdot \mathbb{P}(DS_i), \tag{3.6}$$

where D is the MDR, f_c is a consequence function, $\mathbb{P}(DS_i)$ is the probability of (or fraction of buildings) experiencing DS_i, and n is the number of damage states (see, e.g., Gentile et al. (2022) for a relatively similar formulation). In other words, Eq. (3.6) integrates the distribution of possible damage states for an intensity I into one metric, the MDR. Heuristics must often be used to convert a damage state into an estimate of the normalized

Table 3.1 Examples of consequence functions

DS_k	Damage	Kappos et al. (2006)	Blong (2003a)	Generic[a]
0	None	0	0	0
1	Slight/light	0.005	0.02	0.01
2	Moderate	0.05	0.10	0.10
3	Heavy	0.20	0.40	0.30
4	Very heavy	0.45	0.75	0.60
5	Collapse	0.80	1	1

[a] Defined roughly as the average of the previous two consequence functions MDR $= f_c(DS_k)$ (defined for earthquakes and for tropical cyclones, respectively). Due to the approximative nature of consequence functions and possible heterogeneities in damage state descriptions, the generic function is used only for illustration purposes in the context of the present textbook.

economic cost of repair. Table 3.1 shows examples of consequence functions developed for earthquakes (Kappos et al., 2006) and tropical cyclones (e.g., Blong, 2003a), as well as a generic consequence function derived from them. Consequence functions are also used to translate damage states into additional loss metrics, including casualties and business interruption (quantified in terms of downtime in days).

3.2.1.4 Vulnerability-to-Fragility Conversion

Fragility curves for different damage states may be derived from a continuous damage curve by using the **binomial distribution**:

$$\mathbb{P}(X = \mathrm{DS}_k) = \frac{n!}{k!(n-k)!}\left(\frac{D_\mu}{n}\right)^k \left(1 - \frac{D_\mu}{n}\right)^{n-k}, \tag{3.7}$$

with $\mathbb{P}(\mathrm{DS}_k)$ the probability of occurrence of DS_k, n the number of damage states, $0 \leq k \leq n$, and $0 < D_\mu < n$ a mean damage value (Braga et al., 1982; Lagomarsino and Giovinazzi, 2006; Lagomarsino et al., 2021). The shapes of the derived fragility curves $\mathbb{P}(X \geq \mathrm{DS}_k)$ have been shown to be similar to the cumulative log-normal distribution (Lagomarsino and Cattari, 2014). One drawback of this approach however is the lack of control on the fragility curve's spread (σ in Eq. (3.4)); it has been proposed to use the beta distribution instead to obtain some flexibility (Lallemant and Kiremidjian, 2015). The binomial formulation has been proposed for seismic risk assessment, with D_μ the continuous version of a damage state range (RISK-UE method; Eq. (3.8)). However, it could be assessed for any peril intensity in theory.

3.2.2 Expert Judgement Approach

An estimate of the probability of damage or of the mean damage value under a given hazard intensity load can be obtained by **expert judgement**. This relies on the opinion and experience of the expert, which is both subjective and approximative. Yet, this approach is simple, with possible biases reduced by increasing the number of experts and by expertise level weighting (Porter et al., 2007) (see Part III for a discussion on subjective choices). This heuristic technique is often sufficient for preliminary assessments and communication of risk measures to non-technical audiences, since the number of input variables is kept to a minimum. The scales listed below, based on expert judgement, provide a simple link between observed hazard intensities and expected damage levels for different perils.

3.2.2.1 Modified Mercalli Intensity Scale (EQ)

Based upon Giuseppe Mercalli's intensity scale of 1902, the **Modified Mercalli Intensity (MMI) scale** is a seismic intensity scale used for measuring the intensity of earthquake shaking and its relationship to expected damage levels (Wood and Neumann, 1931). The scale, defined in Table 3.2, shows that building damage occurs at MMI-VI and above. The European macroseismic scale (EMS) is similar to the MMI scale but calibrated to European buildings (Grünthal, 1998). The Japan Meteorological Agency (JMA) seismic intensity scale is similar in essence but on a 0–7 scale instead of the Roman numbering system I–XII (reduced to I–X+ by the USGS; Allen et al., 2008). **Ground-motion-to-intensity conversion equations** (GMICE) are used to estimate the macroseismic intensity as a function of instrumental measures of ground motion, such as peak ground acceleration (PGA) (Eq. (2.54)). They take the form $\text{MMI} = c_1 + c_2 \log_{10} I$, with I the PGA [m/s^2] and for example, $c_1 = 6.54$ and, $c_2 = 4.51$ in Faccioli and Cauzzi (2006). MMI is the input of the following **hyperbolic tangent function** in the RISK-UE method, for which parameterization is based on expert judgement:

$$D_\mu = \frac{n}{2}\left[1 + \tanh\left(\frac{I + 6.25\kappa_D - 13.1}{Q}\right)\right], \tag{3.8}$$

Table 3.2 Abbreviated Modified Mercalli Intensity scale

MMI	Shaking[a]	Damage levels and other observations[b]
I	Not felt	–
II	Weak	Felt indoors by few
III	Weak	Felt indoors by several, hanging objects may swing slightly
IV	Light	Felt indoors by many, outdoors by few, hanging objects swing
V	Moderate	Felt by nearly everyone, objects may move
VI	Strong	Felt by all, slight damage in poorly built buildings
VII	Very strong	Slight to moderate damage in ordinary buildings
VIII	Severe	Considerable damage in ordinary buildings
IX	Violent	Considerable damage, partial collapse
X	Extreme	Most structures destroyed with foundations, rails bent
XI	Extreme	Few, if any, structures standing, bridges destroyed, ground disturbances
XII	Extreme	Total damage. Waves seen on ground surfaces

[a] Classification according to Allen et al. (2008).
[b] Simplified from Wood and Neumann (1931).

with $n = 5$ the number of damage states, $0 < D_\mu < 5$ the mean damage value, I the MMI, and $\Theta_v = (\kappa_D, Q)$ the building characteristics (Ex. 3.1). κ_D is the vulnerability index representative of the construction type (Section 3.1.1.2) and $Q = 2.3$ is a ductility index (Lagomarsino and Giovinazzi, 2006; Lagomarsino et al., 2021). Expert judgement can be used to define new κ_D values or refine existing ones for specific regions (Mignan et al., 2015a, table A3).

3.2.2.2 TORRO International Hailstorm Intensity Scale (HS)

An 11-point **hailstorm intensity (H) scale**, developed and published in 1986 by the Tornado and Storm Research Organisation (TORRO), describes the expected damage for hailstones of different diameter ranges based on historical observations (Webb et al., 2009). Intensities H1 to H3 mainly affect crops while higher intensities are necessary to observe damage to structures, first to vehicles and glass windows, and then to buildings (roofs, chimneys, walls). Serious injuries can be expected at intensity H7. Fatalities become possible at H9 in so-called super hailstorms (Table 3.3).

3.2.2.3 Saffir–Simpson Hurricane Scale (TC) and Enhanced Fujita Scale (To)

Two important damage scales have been developed to communicate about the risk of strong winds caused by hurricanes and tornadoes. Originally developed by Herb Saffir and Bob Simpson, the **Saffir–Simpson hurricane wind scale** (Table 3.4) provides an estimate of the damages expected for increasing maximum wind speeds (v_{max}), with event severity going from Category 1 to Category 5 (Simpson, 1974) (recall from Section 2.2.2.2 that it also provides a scale for storm surge height h). Similarly, the Fujita–Pearson scale, introduced by Ted Fujita in collaboration with Allen Pearson, does the

Table 3.3 Abbreviated TORRO hailstorm intensity (H) scale

Scale	Intensity	Diameter [mm]	Damage levels[a]
H0	Hard hail	5	No damage
H1	Potentially damaging	5–15	Slight damage to crops
H2	Significant	10–20	Significant damage to crops
H3	Severe	20–30	Severe damage to crops, damage to glass structures
H4	Severe	25–40	Widespread glass damage, vehicle bodywork damage
H5	Destructive	30–50	Damage to tiled roofs, risk of injuries
H6	Destructive	40–60	Grounded aircrafts dented, brick walls pitted
H7	Destructive	50–75	Severe roof damage, risk of serious injuries
H8	Destructive	60–90	Severe damage to aircraft bodywork
H9	Super hailstorms	75–100	Structural damage. Risk of fatal injuries
H10	Super hailstorms	> 100	Structural damage. Risk of fatal injuries

[a] Source, courtesy of TORRO (torro.org.uk): www.torro.org.uk/research/hail/hscale (here with short damage descriptions).

Table 3.4 Abbreviated Saffir–Simpson scale

Category	v_{max} [m/s]	Surge h [m]	Damage levels[a]
1	33–42	1.2–1.5	Damage to trees and unanchored mobile homes
2	43–49	1.8–2.4	Damage to exposed mobile homes, signs, roofs, windows
3	50–58	3–4	Damage to small buildings, mobile homes destroyed
4	58–70	4–5.5	Extensive damage to roofs, some roof failures
5	≥ 70	> 5.5	Roof failures, some complete building failures

[a] Simplified from Simpson (1974).

same for tornadoes (Fujita, 1971, 1973). The **Enhanced Fujita (EF) scale** (Table 3.5), which became operational in 2007, reflects better examinations of tornado damage surveys, aligning wind speeds more closely with observed damage on structures and vegetation (e.g., Edwards et al., 2013).

Damage state descriptions differ between perils, with emphasis on building foundation for ground-shaking, on roofs and light structures for strong winds, and on crops and roofs for hailstones. Based on a review of existing damage scales (Blong, 2003b), Blong (2003a) provided peril-specific damage descriptions for tropical cyclones, tornadoes, hail, earthquakes, landslides, wildfires, floods, and tsunamis following a five-level damage scale. Damage scales are not described in this section if they do not directly provide a match between some hazard intensity I and expected damage (e.g., landslide damage scale; Alexander, 1989). Other scales have a resolution inadequate for standard CAT models, such as the **Torino Scale** of asteroid impacts: no consequence ($E_{TNT} \leq 1$), local damage ($1 < E_{TNT} \leq 100$), regional damage ($100 < E_{TNT} \leq 10^5$), and global damage ($E_{TNT} > 10^5$) levels given per kinetic energy range E_{TNT} [Mt] (Binzel, 1997).

Table 3.5 Abbreviated Enhanced Fujita (EF) scale

Level	v_{max} [m/s][a]	Damage levels and other observations[b]
EF0	29–38	Light damage (damage to chimneys, TV antennas)
EF1	38–49	Moderate damage (damage to roofs, windows, trailer houses)
EF2	50–60	Considerable damage (trailer houses and weak buildings destroyed)
EF3	61–74	Severe damage (roofs and walls torn off, trains overturned, cars lifted off the ground)
EF4	74–89	Devastating damage (frame houses levelled, steel structures badly damaged)
EF5	> 89	Incredible damage (reinforced concrete structures badly damaged, car-sized missiles)

[a] Value ranges from Edwards et al. (2013, table 2).
[b] Descriptions simplified from Fujita (1971).

3.2.3 Empirical Approach

An **empirical vulnerability function** is one derived by regression analysis of observations of pairs of hazard intensity and damage level (I, D) for samples from the same asset class. These data are obtained from field surveys of actual damage, where D values represent damage states DS_k, or from loss claims from which D, the MDR, is derived. The set of exposure characteristics Θ_υ represents the empirical parameters of the fit. This approach is particularly suited when damage data abound. Post-earthquake surveys can, for example, include tens of thousands of data records. Such building-level data are rarely publicly available, although the Da.D.O. (Observed Damage Database) now provides such information for some Italian earthquakes (Dolce et al., 2019). Recently, the automatization of damage detection and classification from satellite imagery by innovative machine learning techniques are able to provide a new source of damage data, such as the xBD large-scale data set that provides building-level damage data for a selection of earthquakes, tsunamis, floods, volcanic eruptions, wildfires, hurricanes, and tornadoes (Gupta et al., 2019). Although insurance claim data are proprietary, some can make it to the public domain, such as flood claim data from the National Flood Insurance Program (NFIP) (FEMA, 2023).

3.2.3.1 Vulnerability Curve Fitting from Damage Data

In a first example, a vulnerability curve is developed from the observed damage footprint of the 2011 Joplin, Missouri, tornado. The xBD data set provides a detailed damage map of this event with the damage state DS_k assessed for almost 15,000 buildings (Fig. 3.5a). Damage states DS_0 (no damage), DS_1 (minor damage), DS_2 (major damage), and DS_3 (destroyed) were labelled by the xBD team from DigitalGlobe's high-resolution satellite images (Gupta et al., 2019). To obtain pairs (I, D) for this catastrophe, an intensity footprint is modelled here using Eq. (2.59) (Burow et al., 2020). Input parameters for the Joplin tornado were obtained from Karstens et al. (2013) based on a calibration to observed tree-fall patterns. The correspondence between intensity footprint and damage footprint is shown in Figure 3.5b. The MDR was then calculated using Eq. (3.6) and consequence function f_c, here defined as $\text{MDR} = (0, 0.005, 0.8, 1)$ for (DS_0, DS_1, DS_2, DS_3), with the summation done for all buildings located within $0.001° \times 0.001°$ grid cells. A vulnerability curve of the form $D = (cv_{max})^3/(1 + (cv_{max})^3)$ was finally fitted to the data. This functional form was chosen to match the Emanuel (2011) model (Eq. (3.14); Section 3.2.5). The result is shown in Figure 3.5c.

3.2.3.2 Vulnerability Curve Fitting from Insurance Claim Data

In a second example, flood damage claim data are used to develop a vulnerability curve. The National Flood Insurance Program big data are not used in this section since they require a lot of data cleaning (Wing et al., 2020). Instead, a much simpler data set is used for illustration purposes, which describes the residential damages due to the Meuse flood of 1993 in the Netherlands, which caused a total of 254 million guilder (price level 1993; EUR 180 million in 2016) in direct damages (Wagenaar et al., 2017). This example aims

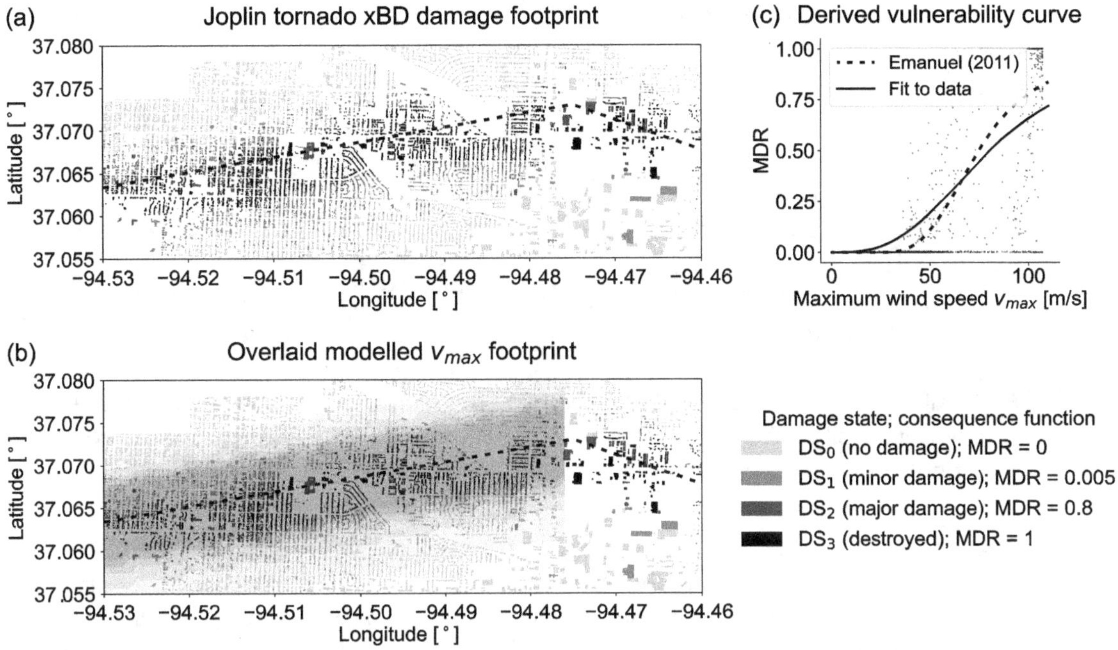

Fig. 3.5 Damage footprint of the 2011 Joplin tornado and vulnerability curve fit. (a) Damage footprint from the xBD data set (Gupta et al., 2019) (CC-BY-NC – published with permission). (b) Overlaid maximum wind speed v_{max} intensity footprint (Eq. (2.59)), with $r_{max} = 300$ m, $v_f = 11$ m/s, $v_{rad} = 86$ m/s and $v_{tan} = 43$ m/s from Karstens et al. (2013) and track coordinates from NOAA's Storm Prediction Center (2023) with mid-point fixed to $(-94.475°, 37.073°)$. Intensity modelled only on the track subsection where the tornado was at its maximum intensity (EF5). (c) Pairs of (v_{max}, D) data points with D estimated from Eq. (3.6), and fit $D = (cv_{max})^3/(1 + (cv_{max})^3)$ with $c = 0.012$ compared to the Emanuel (2011) model (Eq. (3.14)). A black and white version of this figure will appear in some formats. For the colour version, please refer to the plate section.

to show the high uncertainties often associated with insurance claims. Figure 3.6 shows a square root fit to predict content losses L as a function of flood depth h (Eq. (3.10)). The **coefficient of determination** (denoted as R^2 and pronounced 'R squared') is a statistical measure that represents the proportion of data explained by the model (defined as $R^2 = 1 - (y_{obs} - y_{pred})^2/(y_{obs} - \bar{y}_{obs})^2$ with y_{obs} observations, y_{pred} predictions, and $\bar{y}_{obs}$ the mean of the observations). When considering all content loss claims, only 3% of the data is explained by this simple model. When considering only residences without a basement, the model can explain 12% of the data. This shows the importance of considering additional exposure attributes Θ_υ for damage function calibration. Using a dozen parameters, Wagenaar et al. (2017) were able to increase R^2 to 17% with a machine learning model.

3.2.4 Engineering-Based Approach

The **engineering-based, or mechanical-based, approach** consists of estimating the damage experienced by a structure by simulating its physical response to a given hazard

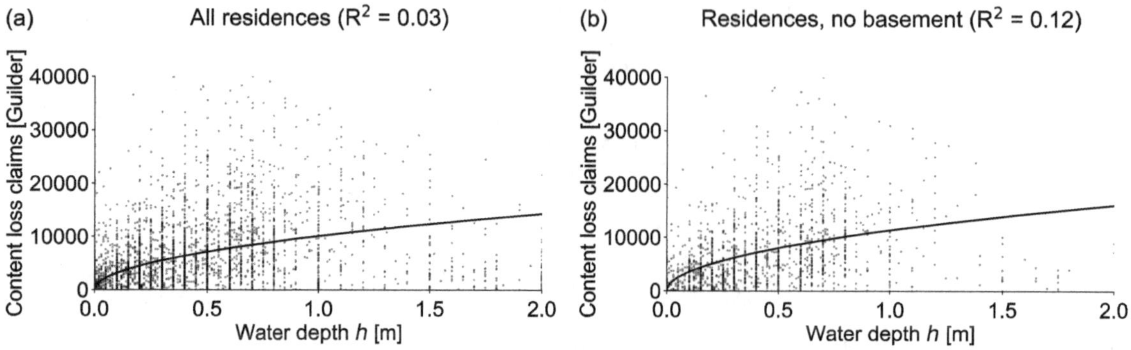

Fig. 3.6 Vulnerability curves fitted to the 1993 Meuse flood claims, the Netherlands. Claims defined as content losses L versus inundation depth h with data from Wagenaar et al. (2017) (CC-BY 3.0). A simple square root model $L = c\sqrt{h}$ is tested. (a) Considering all residences in the data set. (b) Considering only residences without a basement.

intensity load. It is especially useful when data are lacking or whenever resources are available for the development of such a model. Many models exist based on various numerical methods whose descriptions are outside the scope of this introductory textbook. To illustrate how to derive fragility curves from a physical perspective, a simple analytical mechanical model will be presented, the capacity spectrum method, which is popular in seismic risk assessment.

3.2.4.1 Capacity Spectrum Method

A mechanical model is formalized by defining the **capacity curve** of the structure under external load, with deformation as x-axis (relating to structural damage D) and applied force as y-axis (relating to hazard intensity load I). Figure 3.7a represents a bilinear capacity curve constrained by three parameters: the stiffness c_K, yield strength I_y, and ductility capacity d_μ, which are functions of the structure's characteristics. The **yield displacement capacity**, $d_y = I_y/c_K$, delimitates the elastic regime from the plastic one while the **ultimate displacement capacity**, $d_u = d_y + d_\mu$, represents the point of failure. Note that the capacity curve is defined by the yield capacity point and ultimate capacity point in HAZUS for different building types (FEMA, 2022, tables 5–7). Hazard intensity measures defined in Section 2.4 were often already defined in the units of a **force** [M L T^{-2}] or related, such as a **stress** (or pressure, i.e., force normalized by area [M L^{-1} T^{-2}], e.g., blast overpressure) or an acceleration (i.e., force normalized by mass [L T^{-2}], e.g., peak ground acceleration). Volcanic ash footprints were defined in terms of ash thickness but will soon be translated into a pressure (Section 3.2.5). Wind speed pressure can be assessed from the wind speed (e.g., Memari et al., 2018), etc. Intuitively, in the ideal case, the damage state can be related to the encountered deformation, or displacement d, with failure (DS_5) reached at $d = d_u$ while slight damage (DS_1) can be expected at d_y. In practice, in the earthquake case, DS_1, DS_2, DS_3, and DS_4 are identified on the capacity curve as $d_1 = 0.7d_y$, $d_2 = 1.5d_y$, $d_3 = 0.5(d_y + d_u)$, and $d_4 = d_u$, respectively (with no distinction between very

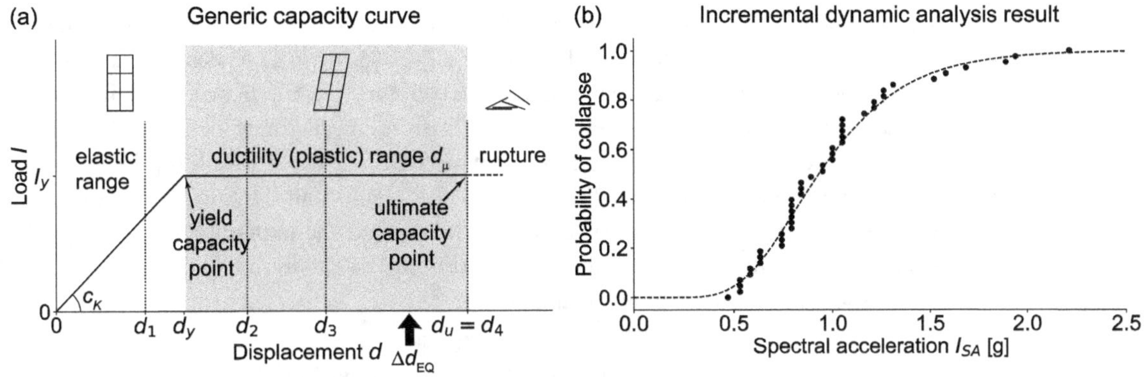

Fig. 3.7 Engineering-based approach to fragility curve modelling (earthquake case). (a) Capacity spectrum method: fragility curves can be derived from the analysis of the building capacity curve, which shows how the hazard intensity load leads to structural deformation, with displacement on the capacity curve equivalent to some damage state $DS_k = d_k$ – see text for details. (b) Incremental dynamic analysis: results from the collapse performance assessment of an eight-storey infill-frame building obtained by Lallemant et al. (2015) with cumulative log-normal distribution fit.

heavy damage and collapse). d is fixed to $0.7d_y$ instead of d_y for DS_1 to account for some non-linear behaviour (Lagomarsino and Giovinazzi, 2006). The building deformation due to an earthquake is equal to the building drift ratio $nh_s\Delta d_{EQ}$, where n is the number of storeys of height h_s and Δd_{EQ} is the **maximum interstorey drift ratio** (which is the relative translational displacement between two consecutive floors divided by h_s). Δd_{EQ} can be estimated from the earthquake ground motion via some empirical relationship of the form $\log \Delta d_{EQ} = c_1 + c_2 \log I_{SA}(T_{SA})$ (Baker and Cornell, 2006). The spectral acceleration I_{SA} per period T_{SA} is estimated alongside the PGA ($T_{SA} = 0$) in dedicated GMPEs (Akkar and Bommer, 2010). In the simplest case, $I_{SA}(T_b)$ can be considered instead of the full spectral form of ground-shaking (with the fundamental period of the structure T_b estimated from the building height h_b; Eq. (3.2)). Note also that the stiffness is equal to $c_K = 4\pi^2/T_b^2$ when the load axis of the capacity curve is defined in terms of spectral acceleration (Lagomarsino and Cattari, 2013). The cumulative log-normal distribution (Eq. (3.4)) can be used to define a fragility curve for DS_k with $\mu = \log d_k$ and σ a function of the ductility capacity and other parameters (Lagomarsino and Cattari, 2013; Pitilakis et al., 2014). One might consider using the binomial distribution (Eq. (3.7)) for illustration purposes with no control on σ (Mignan et al., 2018). Details about the engineering approach can be found elsewhere (Lagomarsino and Cattari, 2013). Wind load models used in HAZUS are briefly described in Vickery et al. (2006).

3.2.4.2 Fitting a Fragility Curve from Collapse Performance Assessment

Incremental dynamic analysis (IDA) is a numerical method, which offers a comprehensive evaluation of the seismic performance of structures (e.g., Pitilakis et al., 2014). Although the description of the technique is outside the scope of this textbook, some results obtained by IDA are here briefly described for completeness. The data, shown in Figure 3.7b, were provided by Lallemant et al. (2015, table I) to demonstrate how to fit a fragility curve using different statistical techniques. The data were obtained from the collapse performance assessment of an eight-storey infill-frame building. Forty-four earthquake histories were used to estimate at which spectral acceleration I_{SA} the modelled building would collapse. Figure 3.7b shows the IDA results and the fitted cumulative log-normal distribution with parameters $\hat{\mu} = \mathbb{E}[\log X]$ (with $\mathbb{E}$ the **expectation**, or simply the mean) and $\hat{\sigma} = \sqrt{\mathrm{Var}(\log X)}$ according to the **method of moments**, which is simple but has some shortcomings (Lallemant et al., 2015).

3.2.5 Vulnerability Functions per Peril

This section provides a list of vulnerability curves for selected perils with the MDR as sole damage measure, since it is the most straightforward metric in CAT risk modelling. Considering the characteristics (Θ_ν) of different building and structure element typologies would be far too vast an initiative when considering many different perils. For example, hundreds of fragility and vulnerability curves are available for earthquakes in HAZUS (FEMA, 2022), SYNER-G (Pitilakis et al., 2014), and the European Seismic Risk Model (ESRM20) (Romão et al., 2021) combined, with the parameter set $\Theta_\nu = (\mu, \sigma)$ of the cumulative log-normal distribution (Eq. (3.4)) commonly given per structure type. The number of damage functions available in the literature for other perils is, however, smaller – see parameterization reviews for wind in Memari et al. (2018) and for inundation in Tarbotton et al. (2015). Here, emphasis will be given on the definition of generic vulnerability curves to provide the student of risk a simple way to map the hazard intensity footprints of the previous chapter into some loss footprints in the present chapter (secondary uncertainties will not be defined but could be parameterized for a given probability distribution, usually a beta distribution; e.g., Jones et al., 2017, p. 42 – see Part III). Models will be listed in alphabetical order per peril. When derived from damage state descriptions, the generic consequence function of Table 3.1 will be used and a cumulative log-normal distribution (Eq. (3.4)) fitted to have a simple two-parameter model instead of a step function. All generic vulnerability curves, which are again only illustrative in nature, are shown in Figure 3.8. Note that some information on the impact of the different hazard loads on people is also provided for completeness.

3.2.5.1 Blasts (AI, Ex, and TA)

Damage from blast waves is assessed from the overpressure P. Most blast–damage relationships are available in tabular form and based on general observations made from explosions in the chemical process industry (Ramiro and Aísa, 1998, table 5.3) or, for greater explosions, from nuclear tests (Glasstone and Dolan, 1977, table 5.145). With virtually no data available for asteroid impacts, the same damage functions are used for this peril (Mignan et al., 2011). With large terrorist attacks often involving bombs, blast damage assessment is also necessary in that case (Stewart et al., 2006). Overlapping ranges of overpressure values P are usually given to describe different levels of damage for mixed

Table 3.6 Generic damage scale for blasts

DS	Damage level	I [psi]	I [kPa][a]	Descriptions given in the literature[b]
1	Slight/light	0.1–1.0	0.7–6.9	Shattered windows, minor structural damage
2	Moderate	1.0–2.0	6.9–13.8	Roof deflections, cracked and pitted walls
3	Heavy	2.0–3.0	13.8–20.7	Partial collapse of house roofs and walls
4	Very heavy	3.0–5.0	20.7–34.5	Collapse of roofs and walls
5	Collapse	5.0–10.0	34.5–68.9	Complete collapse of entire structures

[a] Recall that 1 psi $\approx$ 6.89 kPa.
[b] Combined descriptions, some simplified, with overpressure ranges averaged from Ramiro and Aísa (1998, table 5.3) and Huang et al. (2016, table 1).

construction types, making most blast damage tables very approximative. Some empirical fragility curves have been developed for local building stocks, for example for the 2013 West, Texas, fertilizer plant explosion, which killed 15 people and damaged 150 buildings (Huang et al., 2016) and for the 2020 Beirut explosion (Sadek et al., 2022). Table 3.6 lists intensity–damage pairs derived from the literature while using a five-level damage classification to allow direct comparison with the generic consequence function. A cumulative log-normal distribution fit gives approximately $\mu = \log(20)$ and $\sigma = 0.4$ (for P [kPa]; Fig. 3.8a). Mathias et al. (2017) employed a simpler Heaviside step function for probabilistic asteroid impact risk assessment, with destruction obtained at a 4-psi threshold. Regarding the impact of overpressure on people, 1% of eardrum rupture can be expected at $P = 2.4$ psi and 90% at $P = 12.2$ psi. Death from lung haemorrhage has a probability of 1% at $P = 14.5$ psi and of 90% at $P = 25.5$ psi (Ramiro and Aísa, 1998, table 5.3).

3.2.5.2 Earthquakes (EQ)

Earthquake damage is estimated from the full spectral form of ground-shaking or at a specific spectral acceleration period in the engineering-based methods (Section 3.2.4). For the sake of both simplicity and illustration, a generic damage function based on PGA is used instead. Specifically, the choice was made here to develop a vulnerability curve from the standard USGS ShakeMap scale, where damage levels have been estimated by expert judgement (Allen et al., 2008). The equivalency between a five-level damage scale and the ShakeMap description is given in Table 3.7. This is approximated by a cumulative log-normal distribution with parameters $\mu = \log(6)$ and $\sigma = 0.6$ (for PGA [m/s^2]; Fig. 3.8b). Injuries and fatalities depend on the expected building damage and are estimated via **casualty matrices**, which provide the rate of fatality and of different levels of injury per damage state. Considering fatalities only, the probability of dying from a collapsed building is about 10%, as apparent from the consequence function proposed by Galanis et al. (2018) and formulated as:

$$\mathbb{P}(\text{death}) = 0.00001\mathbb{P}(\text{DS}_2) + 0.00002\mathbb{P}(\text{DS}_3) + 0.0002\mathbb{P}(\text{DS}_4) + 0.1\mathbb{P}(\text{DS}_5), \quad (3.9)$$

Table 3.7 Generic damage scale for earthquakes

DS	Damage level	PGA [%g]	PGA [m/s^2]	ShakeMap equivalence[a]
1	Slight/light	3.9–18	0.38–1.77	Very light–Light
2	Moderate	18–34	1.77–3.33	Moderate
3	Heavy	34–65	3.33–6.37	Moderate/Heavy
4	Very heavy	65–124	6.37–12.16	Heavy
5	Collapse	> 124	> 12.16	Very heavy

[a] Based on ShakeMap (Allen et al., 2008, fig. 1).

with $\mathbb{P}(\text{DS}_k)$ denoting the probability of a given damage state. The parameterization was assigned from the maximum values of indoor fatality rates for all building types in HAZUS by Galanis et al. (2018).

3.2.5.3 Floods (FF, PF, SS, Ts)

The inundation depth is the main variable of fluvial and pluvial flood vulnerability functions. Although other metrics can be used, such as flood duration, inundation depth is readily recorded, leading to a lot of data for model calibration (Section 3.2.3). For storm surges and tsunamis, flow velocity can play an important role in expected damage (Koshimura et al., 2009). For static damage, the square root function is commonly applied with

$$D = c\sqrt{h} \text{ for } D < 1 \text{ and } D = 1 \text{ otherwise,} \tag{3.10}$$

where D is the MDR, h [m] is the inundation depth, and c is an empirical parameter. Intensity–damage relationships may also be provided as step functions (e.g., Apel et al., 2009). Generic damage functions for flood depth have been developed by the European Commission's Joint Research Centre (JRC) by averaging damage functions available in the literature for different continents (Huizinga et al., 2017). These vulnerability curves are here further averaged over different continents, and provided for residential, commercial, and industrial occupancies in Table 3.8. They are however almost undistinguishable from each other and can be fitted with Eq. (3.10) using $c = 0.45$ (Fig. 3.8c). Fatality curves are also defined as a function of h [m] with, for example, mortality $D = \exp((h - 4.58)/0.69)$ in zones of rapidly rising water and $D = \exp((h - 10.7)/1.59)$ otherwise (Jonkman and Kelman, 2005).

3.2.5.4 Hailstorms (HS)

Hail damage is a function of the hailstones' kinetic energy E and has been shown to be well described by a logistic function of the form:

$$D = \frac{\exp(c_0 + c_1 E_A)}{1 + \exp(c_0 + c_1 E_A)}, \tag{3.11}$$

Table 3.8 Generic vulnerability tables for floods

Depth h [m]	Residential MDR	Commercial MDR	Industrial MDR	Generic MDR[a]
0	0	0	0	0
0.5	39	37	33	36
1.0	56	56	52	55
1.5	67	68	65	67
2.0	78	79	73	77
3.0	89	89	84	87
4.0	94	94	94	94
5.0	98	97	98	98
6.0	100	100	100	100

[a] All MDR values [%] derived from Huizinga et al. (2017).

where D is the MDR and E_A [J/m^2] is the entire hail kinetic energy per surface unit. For crop damage, parameters are, for instance, $c_0 = -3.730$ and $c_1 = 7.483 \times 10^{-3}$ (for one stage, among others, of wheat growth) in Katz and Garcia (1981) and $c_0 = -2.348$ and $c_1 = 0.00541$ (various crop varieties during one hailstorm) in Schiesser (1990). A similar function has been used to assess car damage based on insurance claims (Hohl et al., 2002). The E_A variable may not be a sufficient measure since it averages over all hailstone sizes. Distinction can therefore be made between high seasons (mid-summer, higher rate of large hailstones) and low seasons (before and after) to reduce epistemic uncertainties when fitting Eq. (3.11) (Hohl et al., 2002).

3.2.5.5 Landslides (LS)

Landslide damage is generally assessed from the height of the deposited soil. A generic vulnerability curve is taken as the **Weibull distribution**

$$D = 1 - \exp\left[c_1\left(\frac{h+c_2}{c_2} - 1\right)^{c_3}\right], \tag{3.12}$$

with D the MDR and h [m] the landslide depth (i.e., deposited height), and the parameters $c_1 = -1.671$, $c_2 = 3.189$, and $c_3 = 1.746$ fitted by Papathoma-Köhle et al. (2015) for more than 250 observations in European mountainous areas (Fig. 3.8d). This model was selected since it was specifically developed for decision-making in a CAT model context. Other possible functional forms include second-order polynomial functions, the logistic function, or the exponential function (Totschnig et al., 2011). Since landslides are more likely to impact cellars and ground floors than any other storey, a normalized intensity h/h_b may be used where h_b is the building height to avoid overestimating the vulnerability of tall buildings (Totschnig et al., 2011). For rapid debris flows, flow velocity also contributes to the damage (Jakob et al., 2011).

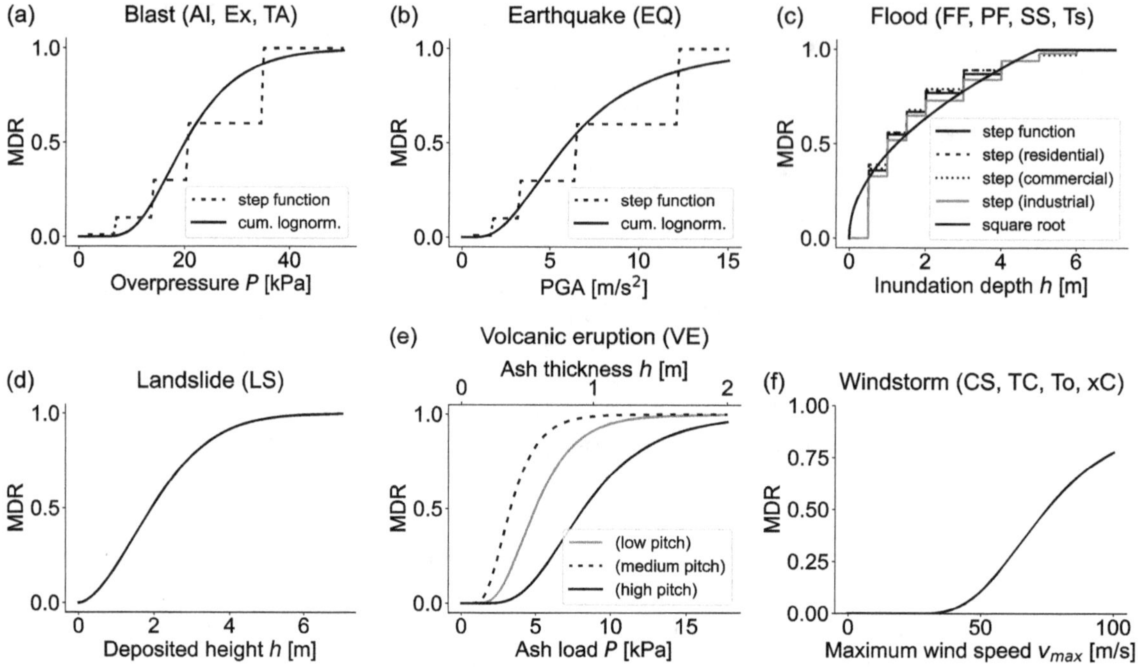

Fig. 3.8 Generic vulnerability curves for different hazard intensity loads I (for illustration only, subject to high epistemic uncertainty). (a) Blast case with step function (Table 3.6) and cumulative log-normal distribution fit ($\mu = \log(20)$, $\sigma = 0.4$). (b) Earthquake case with step function (Table 3.7) and cumulative log-normal distribution fit ($\mu = \log(6), \sigma = 0.6$). (c) Flood case with step functions (Table 3.8) and square root function fit (Eq. (3.10), $c = 0.45$). (d) Landslide case with Weibull distribution (Eq. (3.12); $c_1 = -1.671, c_2 = 3.189, c_3 = 1.746$). (e) Volcanic eruption case with cumulative log-normal distribution (Eq. (3.13); $\mu = (1.6, 1.2, 2.1)$, $\sigma = (0.44, 0.42, 0.42)$). (f) Windstorm case with Eq. (3.14) ($v_{thres} = 25.7$ m/s, $v_{half} = 74.7$ m/s).

3.2.5.6 Volcanic Eruptions (VE)

Damage due to volcanic ash load can be modelled with the cumulative log-normal distribution

$$D = \frac{1}{2}\left[1 + \text{erf}\left(\frac{\log P - \mu}{\sigma\sqrt{2}}\right)\right], \tag{3.13}$$

where D is the MDR and P [kPa] the volcanic ash load. **Ash load** P [Pa] is estimated from $P = \rho_{ash} g h$ where ρ is the density of ash ($\rho_{ash} = 900$ kg/m^3 in the dry context), $g = 9.81$ m/s^2, and h [m] is the ash thickness. Generic vulnerability curves are here defined for three roof pitch ranges: low ($< 6°$), medium ($6-35°$), and high ($> 35°$). Values are derived from Blong et al. (2017) with $\mu = (1.6, 1.2, 2.1)$ and $\sigma = (0.44, 0.42, 0.42)$ (for P [kPa]; Fig. 3.8e). These values were averaged over different building types and occupancies. Some implicit correlations include buildings of low roof pitch angle being engineered, increasing their overall resistance to ash load compared to the medium-pitch class. An MDR of 100% indicates roof collapse, and not necessarily building collapse.

3.2.5.7 Windstorms (WS, including CS, TC, To, and xC)

Damage surveys after tornadoes and hurricanes tend to show similar styles of building response to wind speed, irrespective of the wind phenomenon (Marshall, 1993). The generic vulnerability function is here taken from Emanuel (2011) with the function

$$D = \frac{v_n^3}{1 + v_n^3}, \tag{3.14}$$

where

$$v_n = \frac{\max[(v_{max} - v_{thres}), 0]}{v_{half} - v_{thres}}, \tag{3.15}$$

v_{max} [m/s] is the observed maximum wind speed, $v_{thres} = 25.7$ m/s (= 50 kts) is the wind speed at and below which no damage occurs, and v_{half} is a calibration parameter (Fig. 3.8f). Equation (3.14) reflects that damage increases as the cube of the maximum wind speed (e.g., Hillier, 2017, p. 210). This model was selected as it is already used in the CLIMADA platform for weather and climate CAT modelling (Aznar-Siguan and Bresch, 2019). Moreover, a $v_{half} = 74.7$ m/s calibration yields a damage function reasonably close to the HAZUS models for different building types (Sealy and Strobl, 2017). Note that extreme wind speed is measured from the **gust**, which is a sudden rush of wind occurring within a few seconds, taken at a 3-s gust wind speed by US wind engineers (e.g., Li and Ellingwood, 2006).

3.3 Loss Module

The final component of a CAT risk model is the loss module, which evaluates, first, the ground-up losses from the measures of damage combined with the exposure value and, second, the insured losses by applying insurance policy rules. For the insurance aspects of CAT risk management, the reader should refer to Part III (Chapter 6). In Parts I and II, only ground-up losses are considered, with the direct economic losses (L) to physical assets estimated from

$$L(x, y) = D(x, y) \times \nu(x, y), \tag{3.16}$$

where D is the MDR and ν is the replacement value of the exposed assets. Footprint coordinates (x, y) may represent individual building geolocations or aggregated data in a spatial mesh. Although not explored in this section, casualties, or human losses, can be modelled in a similar fashion by using consequence functions – similar to the ones given in Section 3.2.5 (e.g., Eq. (3.9)) – combined with the population exposure data sets mentioned in Section 3.1.2.

The outcomes of the CAT risk modelling pipeline are summarized in loss tables (Section 3.3.1). These *loss tables serve as the basis for calculating various risk metrics, including the average annual loss (AAL) and exceedance probability (EP) curves* (Section 3.3.2). For sake of illustration, fictitious loss tables will be included for the calculations. Furthermore, the CAT Risk Modelling Sandbox will be introduced in Section 3.3.3. This will assist students of risk in developing more realistic loss tables based on the actual realization of the entire CAT modelling pipeline. The 'build your own CAT risk model from scratch' project will also be presented with pseudocodes of a CAT model blueprint detailed in Boxes 3.7 and 3.8 (for codes in various programming languages, refer to the **CAT Risk Modelling Starter Kit**, available online). This will conclude Part I on the fundamentals of CAT modelling with the sandbox and starter kit providing the practical means of applying the lessons learned in Chapters 2 and 3.

3.3.1 Loss Tables

The **loss table** is a fundamental output that provides the loss distribution computed by the CAT model. Two types of tables are defined below: the event loss table (ELT) and the year loss table (YLT). They both consist of the tables generated in the hazard module, respectively the event table and the simulation table (or year event table (YET)) (Section 2.1.1), but with an event loss column added. The **event loss** represents the total ground-up loss $L_{\text{ID}} = \sum L(x, y)$ caused by the stochastic event (ID), taken as the sum of all losses in the loss footprint $L(x, y)$. Loss tables can provide loss statistics at different resolutions depending on the needs of the user, which is controlled by the selection of specific (x, y) locations, from one unique location (e.g., a critical infrastructure), to a portfolio of buildings, to the entire exposure footprint.

The **event loss table (ELT)** adds the mean losses L_{ID} to the event table, which already includes a unique identifier (ID) per stochastic event, the event occurrence rate λ_{ID}, and other event characteristics in the hazard domain (which can be omitted in the final ELT). In addition, the standard deviation of the loss (σ_{ID}) must be included to represent all secondary uncertainties occurring at the hazard and damage modelling levels (to be covered in Part III). When annual event rates are small with $\lambda_{\text{ID}} \ll 1$ (which is most often the case), the probability of occurrence $p_{\text{ID}} \approx \lambda_{\text{ID}}$ may be used instead in the ELT (e.g., Grossi et al., 2005). An example of a fictitious ELT is given in Table 3.9, which will be used as input to compute some risk metrics in the next section.

The **year loss table (YLT)** adds the total loss L_{ID} to the simulation table (or YET), which already lists individual event occurrences for a total of N_{sim} year-simulations. A YLT therefore consists of a list of simulation IDs i, event IDs j, and matching loss realizations $L_{i,j} = L_{\text{ID}=j} + \epsilon$, with ϵ the secondary uncertainties encoded in each simulated event. A YLT, which is derived from Monte Carlo simulations, is particularly useful to compute the combined effects of multiple event occurrences within a year. A pseudocode is provided in Box 3.5 to generate a YLT from an ELT (omitting secondary uncertainties for the sake of simplicity). The process consists of calculating the overall ELT frequency, sampling k events per year according to the Poisson distribution (Eq. (2.1)) and defining simulation identifiers accordingly (i.e., same identifier repeated k times given a year), and, finally, attributing the event's ID according to its relative frequency in the ELT (Ex. 3.3). This last step requires the calculation of the **normalized exceedance frequency** ($\text{EF}_{i,norm}$) of event i:

Table 3.9 A fictitious event loss table (ELT)

Event ID	Event rate λ_{ID}	Event loss L_{ID} [USD]
1	0.101	300,000
2	0.287	100,000
3	0.001	20,900,000
4	0.001	22,600,000
5	0.064	700,000
6	0.001	25,000,000
7	0.028	1,800,000
8	0.001	22,200,000
9	0.003	14,300,000
10	0.106	300,000
11	0.044	1,100,000
12	0.005	9,100,000
13	0.129	200,000
14	0.221	150,000
15	0.010	5,200,000

$$\text{EF}_{i,norm} = \frac{\sum_{j=1}^{i} \lambda_j}{\sum_{j=1}^{N} \lambda_j}, \tag{3.17}$$

where λ_j is the rate of event j and N is the number of events in the ELT. It takes a form similar to an eCDF, allowing sampling to be done, as already learned in Chapter 2 (Box 2.1).

3.3.2 Risk Metrics

Different risk indicators are available to assess the risk outcome of a CAT model from a loss table. The most common ones, described below, are the average annual loss (AAL), the exceedance probability (EP) curve, the value at risk (VaR), and the tail value at risk (TVaR) (Ex. 3.2). For a comprehensive list of risk metrics, see Jones et al. (2017, section 1.10). For illustration purposes, the following metrics will be estimated for the ELT shown in Table 3.9.

3.3.2.1 Average Annual Loss (AAL)

The annual expected loss for the set of N events listed in an ELT, denoted as the **average annual loss (AAL)**, is defined as the sum of the expected losses by each of the individual events for a given year, that is,

$$\text{AAL} = \sum_{i=1}^{N} \lambda_i L_i, \tag{3.18}$$

Box 3.5 Generating YLTs from an ELT

```
ELT = {'ID', 'rate', 'loss', 'EF'} # loss in descending
order

# calculate overall ELT frequency
rate_tot = sum(ELT['rate'])

# simulate number of events per year
k = Poisson_sample(rate_tot, Nsim)

# define simulation IDs
simIDs = [] # an empty list
simID = 1 # initialize
for event_count in k: {
simID_copies = repeat(simID, event_count times)
append simID_copies to simIDs
simID = simID + 1
}

# sample events according to ELT rates
n = sum(k)
u = random_number(n)
EF_norm = ELT['EF'] / rate_tot # normalised exceedance freq.
IDs = []
for i in 1 to n: {
ID = ELT['ID'] where first EF_norm > u[i]
append ID to IDs
}

# use ELT as look-up table to add losses
YLT = {'simID' = simIDs, 'ID' = IDs}
YLT['loss'] = ELT['loss'] where ELT['ID'] = YLT['ID']
```

with λ_i the annual occurrence rate and L_i the mean loss of event i. The AAL may also be termed annual mean loss (AML) or pure premium (Jones et al., 2017). For the example of Table 3.9, the AAL is USD524,450. For a YLT, it is

$$AAL = \frac{1}{N_{sim}} \sum_{k=1}^{N_{sim}} \sum_{j=1}^{n_k} L_{j,k} = \frac{L_{tot}}{N_{sim}}, \tag{3.19}$$

with N_{sim} the number of year-simulations, n_k the number of events per year-simulation, $L_{j,k}$ the loss realized for the jth event in year k, and L_{tot} the sum of all losses realized in the YLT (Jones et al., 2017, eq. (1.9)).

3.3.2.2 Exceedance Probability (EP) Curve

The **exceedance probability** (EP) curve represents the probability that a given loss level L_i is exceeded in a year:

$$\mathrm{EP}(L_i) = \mathbb{P}(X > L_i) = 1 - \mathbb{P}(X \leq L_i), \tag{3.20}$$

or

$$\mathrm{EP}(L_i) = 1 - \prod_{j=1}^{i}(1 - p_j), \tag{3.21}$$

with p_i the annual probability of occurrence of event i and losses L_i sorted in descending order in the ELT (i.e., $L_1 = \max L_i$) (Grossi et al., 2005). Equation (3.21) translates into one minus the probability that all the events above the L_i level have not occurred (Box 3.6). Another formulation defined from the rates λ_i is

$$\mathrm{EP}(L_i) = 1 - \exp(-\mathrm{EF}_i) = 1 - \exp\left(-\sum_{j=1}^{i}\lambda_j\right), \tag{3.22}$$

where EF is the **exceedance frequency**. Equations (3.21) and (3.22) are equivalent for $p_i = 1 - \exp(-\lambda_i)$ assuming event independence (i.e., a Poisson distribution). As clearly explained by Jones et al. (2017), 'it is always the case that if at least one event in a year exceeds the given level, the maximum event in that year must also have exceeded the same given level'. The exact term for EP is the **occurrence exceedance probability** (OEP), that is, the probability that the maximum event loss in a year exceeds a given level. The EP curve (i.e., OEP curve) is obtained when considering all possible levels $0 < L_i \leq \max L_i$. The EP curve derived from Table 3.10 is shown in Figure 3.9. A distinction must be made between OEP and the **aggregate exceedance probability** (AEP), which is the probability that the sum of event losses in a year exceeds a given level. Both OEP and AEP curves are easily derived from a YLT (Part II, Chapter 5). From an ELT, Panjer recursion or Fast Fourier Transform (FFT) algorithms must be used to convolve individual ELT entries and compute aggregate losses (Embrechts and Frei, 2009). In practice, results may be described in terms of **loss return period** instead, which is calculated as the reciprocal of the EP (return period = 1 / EP, with EP $\ll$ 1. To be exact, it should be return period = 1 / EF with EF the exceedance frequency; Jones et al., 2017, p. 27).

Box 3.6 — EP Curve: Going from Eq. (3.20) to Eq. (3.21)

The definition of an EP curve is $EP(L_i) = \mathbb{P}(X > L_i)$. With losses sorted in descending order and two events X_1 and X_2 above the L_3 threshold, it follows that $EP(L_3) = \mathbb{P}(X_1 \cup X_2) = \mathbb{P}(X_1) + \mathbb{P}(X_2) - \mathbb{P}(X_1 \cap X_2) = p_1 + p_2 - p_1p_2$, since events are independent. This approach becomes cumbersome when more events are involved (e.g., $\mathbb{P}(X_1 \cup X_2 \cup X_3) = \mathbb{P}(X_1) + \mathbb{P}(X_2) + \mathbb{P}(X_3) - \mathbb{P}(X_1 \cap X_2) - \mathbb{P}(X_1 \cap X_3) - \mathbb{P}(X_2 \cap X_3) + \mathbb{P}(X_1 \cap X_2 \cap X_3)$). Instead, one defines $EP(L_3) = 1 - \mathbb{P}(X \leq L_3) = 1 - \mathbb{P}(\neg X_1 \cap \neg X_2) = 1 - [(1 - p_1)(1 - p_2)] = p_1 + p_2 - p_1p_2$ for the two-event case (QED). Generalization to any number of events is now possible, leading to Eq. (3.21).

Table 3.10 Fictitious ELT with EP calculations

Event ID	Event rate λ_{ID}	Loss L_{ID} [USD][a]	EF[b]	EP
6	0.001	25,000,000	0.001	0.001000
4	0.001	22,600,000	0.002	0.001998
8	0.001	22,200,000	0.003	0.002996
3	0.001	20,900,000	0.004	0.003992
9	0.003	14,300,000	0.007	0.006976
12	0.005	9,100,000	0.012	0.011928
15	0.010	5,200,000	0.022	0.021760
7	0.028	1,800,000	0.050	0.048771
11	0.044	1,100,000	0.094	0.089717
5	0.064	700,000	0.158	0.146150
10	0.106	300,000	0.259	0.228177
1	0.101	300,000	0.365	0.305803
13	0.129	200,000	0.494	0.389819
14	0.221	150,000	0.715	0.510808
2	0.285	100,000	1.002	0.632856

[a] In descending order for EP calculations.
[b] Note that the EF can be greater than one.

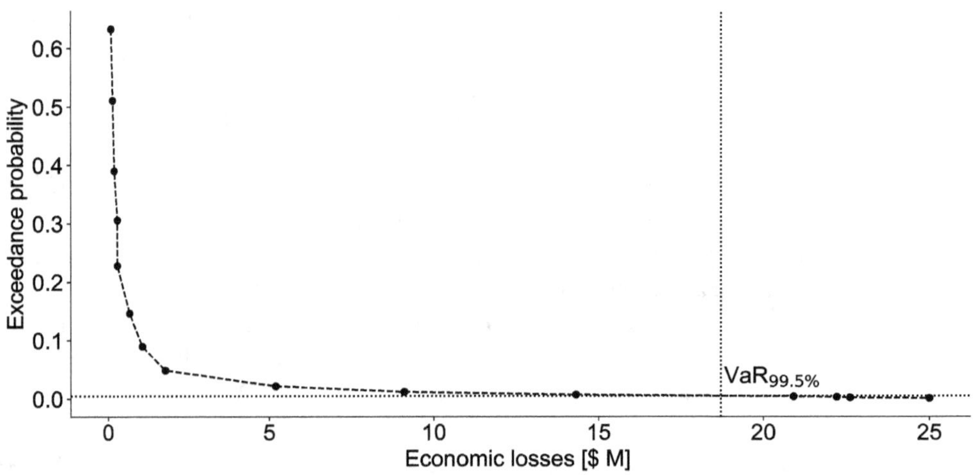

Fig. 3.9 Example of exceedance probability (EP) curve derived from an ELT (Tables 3.9 and 3.10).

3.3.2.3 Value at Risk (VaR)

The **value at risk (VaR)** is the loss value at a specific quantile q of the considered loss distribution (in terms of CDF) satisfying

$$\mathbb{P}(X > \mathrm{VaR}_q) = 1 - q \tag{3.23}$$

for a random variable X of loss. For example, supposing that an insurer specifies its acceptable risk level as the 99.5% VaR, it would correspond to the 0.5% level (1-in-200 year return period) on the EP curve. In other words, 99.5% of the risk can be absorbed while there is a small chance, 0.5%, of exceeding the matching loss value (Part III). The VaR measure is also commonly used in the banking industry to assess the risk (e.g., Reichert and Shyu, 2003). For our toy ELT, $\text{VaR}_{99.5\%} = \text{USD}18.7$ million (USD20.9 million at the resolution of the ELT). In a YLT, we have $\text{VaR}_q = y_k$ with $k = qn + 1$, n the number of events, and the YLT ordered by ascending losses $y_1 \leq y_2 \leq \cdots \leq y_n$ (Klugman et al., 2012, p. 454). Even if still widely used, VaR does not satisfy one of four criteria for coherence (subadditivity), which led to developing the next metric instead, the TVaR (Klugman et al., 2012, p. 41). Note that the **probable maximum loss** (PML), which is the value of the largest loss expected to incur from a specific event (i.e., deterministic worst case scenario), is often used informally to refer also to the VaR (or TVaR) in probabilistic CAT risk modelling (e.g., Woo, 2002; Grossi et al., 2005; Smolka, 2006). Confusingly, it may be defined as the EP loss for any given threshold (Mitchell-Wallace, 2017, p. 62).

3.3.2.4 Tail Value at Risk (TVaR)

Another commonly used tail risk measure is the **tail value at risk (TVaR)**, which is the expected value of loss above a specific level, given that the level has been exceeded. The value of TVaR is always greater than the VaR value at the same quantile (Jones et al., 2017, p. 29) with

$$\text{TVaR}_q(X) = \mathbb{E}(X|X > \text{VaR}_q) \tag{3.24}$$

(e.g., Klugman et al., 2012, eq. (3.12)). It is obtained from an ELT as the average of losses above the VaR. In the example of Table 3.9, $\text{TVaR}_{99.5\%} = \text{USD}23.3$ million (USD22.0 million at the resolution of the ELT). In a YLT, it is equal to

$$\text{TVaR}_q(X) = \frac{1}{n-k+1} \sum_{j=k}^{n} y_j, \tag{3.25}$$

with $k = qn + 1$, n the number of events, and the YLT ordered by ascending losses $y_1 \leq y_2 \leq \cdots \leq y_n$ (Klugman et al., 2012, p. 454). This risk measure is also called **tail conditional expectation** (TCE) (e.g., Woo, 2002).

3.3.3 The CAT Risk Modelling Sandbox

By this stage in the textbook, the reader possesses all the essential tools required to construct a CAT risk model. These tools encompass the creation of a stochastic event set along with its catalogue of hazard intensity footprints, the generation of loss footprints through damage functions, and the compilation of loss statistics within a loss table. Due to the extensive coverage in Part I (Chapters 1–3) and the abundance of information offered across numerous perils, navigating through these steps may initially appear a daunting task. To assist the students of risk in their journey into the realm of catastrophe risk assessment, two valuable applications are proposed:

- **CAT Risk Modelling Sandbox**: As part of the author's Generic Multi-Risk (GenMR) digital template initiative (Mignan, 2022b), this textbook introduces a virtual environment where various perils coexist within a generic region. This interactive platform serves as a guide, simplifying the complicated steps involved in CAT model development while maintaining a realistic context (Ex. 3.4). All the necessary materials are conveniently accessible in the online supplementary material accompanying this textbook. For the sake of completeness, a brief description of the sandbox (or GenMR digital template) is provided below.
- **'Build your own CAT risk model from scratch' project**: Following Richard Feynman's philosophy that 'what I cannot create, I do not understand', an opportunity is offered for in-depth comprehension. While the CAT Risk Modelling Sandbox provides a comprehensive suite of tools, including online codes and tutorials, some readers may desire a deeper understanding of the underlying models. To ensure that the fundamental modelling framework is graspable by all, pseudocodes are given in Boxes 3.7 and 3.8. These pseudocodes will guide the reader through the development of a rudimentary CAT model from the most elemental blueprint imaginable. This project serves as the culminating exercise of Part I, where readers are tasked with constructing a CAT model for a single peril of their choice (Ex. 3.5). An online CAT Risk Modelling Starter Kit complements the pseudocodes by providing templates in several programming languages.

3.3.3.1 Simulated Environment

The **GenMR digital template**, first described in Mignan (2022b), is a microcosm simulation of the complex Earth system for catastrophe dynamics research and development, multi-risk prototyping in the Generic Multi-Risk (GenMR) framework (Part II, Chapter 5; Mignan et al., 2014), and catastrophe risk eduction. Version 1.1, described here, is composed of environmental layers that consist of sets of variables $\Theta(x, y)$ defined in a spatial grid of coordinates (x, y). Each layer may be altered by environmental objects located within the layer. The parsimonious complex Earth system finally consists of a stack of interacting environmental layers defined in the natural, technological, and socio-economic systems. The natural environment layers comprise the topography with elevation $z(x, y)$, the soil with depth $h_s(x, y)$, and the land use with state $\mathcal{S}(x, y)$. The fractal topography is constrained by a west-dipping slope. Environmental objects, which are associated with peril sources, alter the topography via intrusion/extrusion rules. For example, a volcano is represented by a cone, a reverse fault by an ellipsoid tectonic hill, and a river valley by the intrusion of a damped sine river. More details are given in the supplementary material and in Mignan (2022b). Land use states $\mathcal{S}$ are water, grassland, forest, and urban. The urban area is developed using a hybrid cellular automaton, which was described in Boxes 3.1–3.3. The urban layer has three states: residential, commercial, and industrial, and includes a road network. The exposure footprint is defined from Eq. (3.1). Figure 3.10 shows an instance of the virtual region for a default parameterization (see online tutorials for details). *Despite the minimalism of the simulated environment, it models the important*

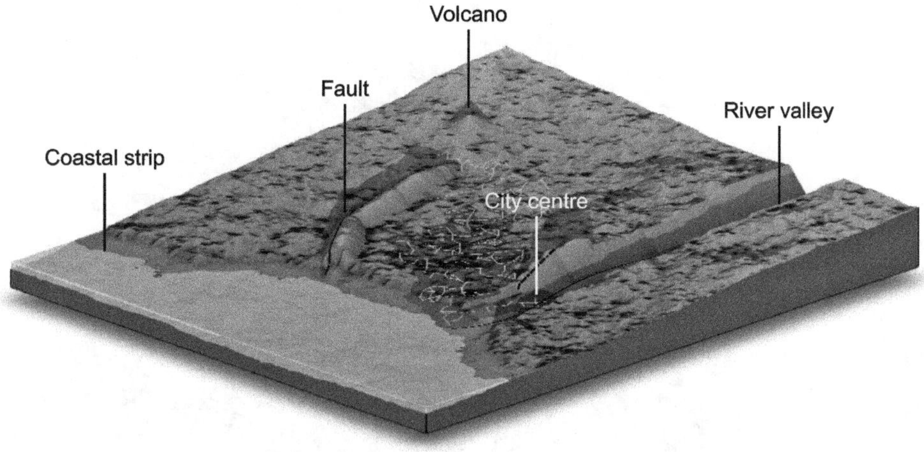

Fig. 3.10 One instance of the GenMR digital template (default parameterization) with topography layer and land-use layer (here consisting of a water mass, built area, and road network). Simulation rendered with Rayshader (Morgan-Wall, 2023). A black and white version of this figure will appear in some formats. For the colour version, please refer to the plate section.

spatial correlations that exist between different environmental layers, perils, and assets at risk. The topography, for example, is an important variable: urbanization occurs preferentially on flat areas, which are prone to flooding; landslides are more likely to occur in mountainous areas, where they can be triggered by earthquakes, etc.

3.3.3.2 Event Footprint Modelling

The sources of several perils are defined alongside the virtual environment when they interact with it. This is the case for earthquakes (EQ), volcanic eruptions (VE), fluvial floods (FF), and coastal floods (storm surges, SS). Additional perils considered in the CAT Risk Modelling Sandbox are, in alphabetical order: asteroid impacts (AI), explosions (Ex), landslides (LS), rainstorms (RS), wildfires (WF), and windstorms (WS) (more perils will be added in future releases). Source characteristics, event size distributions, hazard intensity footprints, and damage functions are defined based on the models described throughout Part I (and in Chapter 4 for dynamic models). The choices made for the creation of the stochastic event set(s) and generic model parameterizations are described in the online supplement. Figure 3.11 shows an example of intensity footprint $I(x, y)$ and loss footprint $L(x, y)$ for a synthetic earthquake. Repeating the process for n events leads to the creation of an event footprint catalogue and of loss tables. Both the ELT and the YLT are generated and the common risk metrics (AAL, EP curves, tail risk measures) are provided in the sandbox. The online interactive tutorials let the reader change the parameterizations, such as adding or removing perils, changing their rate of occurrence, etc. (Ex. 3.4). All the codes are provided with links to the textbook equations.

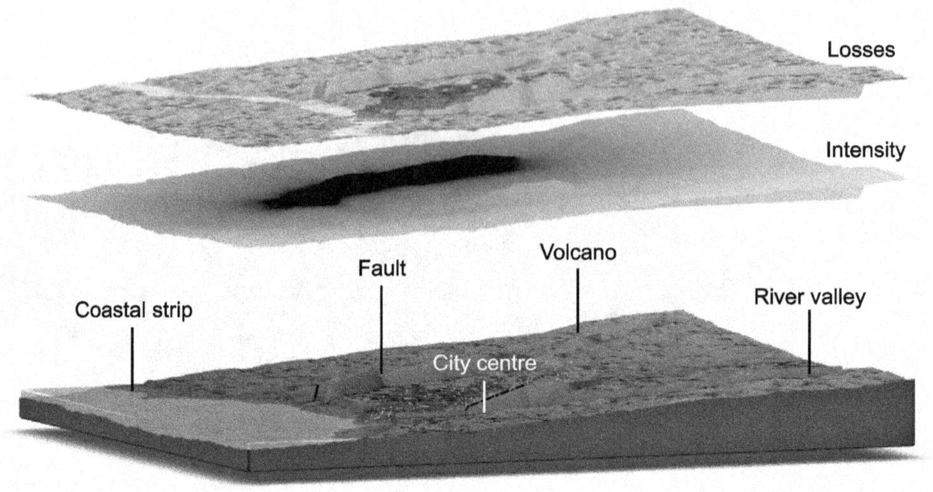

Fig. 3.11 Intensity footprint and loss footprint in the GenMR digital template (default parameterization), here for a large earthquake. Simulation rendered with Rayshader (Morgan-Wall, 2023). A black and white version of this figure will appear in some formats. For the colour version, please refer to the plate section.

Box 3.7 Build Your Own CAT Risk Model from Scratch (Hazard Part)

```
# define environment (i.e., grid for footprints)
x = sequence(xmin, xmax, by dx)
y = sequence(ymin, ymax, by dy)
grid = mesh(x,y)

## HAZARD ASSESSMENT ##
# define source model (e.g., 2 point-sources)
Src = {ID = (src1, src2), x0 = (src1_x0, src2_x0),
y0 = (src1_y0, src2_y0), S = (src1_S, src2_S)}

# define size distribution (with ** as power operator)
Smax = max(Smax1 = func(src1_S), Smax2=func(src2_S))
Si = 10**sequence(log10(Smin), log10(Smax), by dS_log10)
Si1 = 10**sequence(log10(Smin), log10(Smax1), by dS_log10)
Si2 = 10**sequence(log10(Smin), log10(Smax2), by dS_log10)
ratei = 10**(a - b * (log10(Si) - dS_log10 / 2)) -
10**(a - b * (log10(Si) + dS_log10 / 2))

# define event table
EventTable = {ID = 'PERIL' + sequence(1, N1+N2),
```

Box 3.7 (cont.)

```
Src = (repeat(src1, N1), repeat(src2, N2)), S = (Si1, Si2),
rate = (ratei[1:N1], ratei[1:N2])}
# correct rate, function of stochastic set definition, e.g.:
Nevent_perSi = (repeat(2, 2*N1), repeat(1, N2-N1)) # if N1<N2
EventTable[rate] = EventTable[rate] / Nevent_perSi

# define intensity I_grid footprint catalog
for i in 1:(N1+N2) {
ind = (Src[ID] == EventTable[Src][i])
r_grid = sqrt((grid[x] - Src[x0][ind])**2 + (grid[y] -
Src[y0][ind])**2)
I_grid[i,:,:] = func_intensity(EventTable[S][i], r_grid)
}
# -> calculate hazard metrics (see solution to exercise
2.4)
```

Box 3.8 Build Your Own CAT Risk Model from Scratch (Risk Part)

```
## RISK ASSESSMENT ##
# calculate damage D_grid and loss L_grid footprints
for i in 1:(N1+N2) {
D_grid[i,:,:] = func_vulnerability(I_grid[i,:,:], Theta_nu)
L_grid[i,:,:] = D_grid[i,:,:] * nu_grid
}

# update event table as loss table
for i in 1:(N1+N2) {EventTable[loss][i] = sum(L_grid[i,:,:])}
ELT = EventTable
# -> calculate risk metrics (see solution to exercise 3.2)
```

3.3.3.3 Build Your Own CAT Risk Model from Scratch

The final exercise of Part I, which is the student project of this textbook, consists of building a toy CAT risk model from scratch. For guidance, Boxes 3.7 and 3.8 provide the blueprint for a simple CAT modelling pipeline in the form of pseudocodes. The CAT Risk Modelling Starter Kit, provided in the supplementary material, contains this blueprint in several programming languages (including Python and R), as well as some other basic tools (e.g., for random sampling). The goals and requirements are described in Exercise 3.5.

3.4 Concluding Remarks

This chapter marks the conclusion of Part I, which explored the fundamental aspects of catastrophe risk modelling. It showed how data are defined and flow through the exposure (Section 3.1), vulnerability (Section 3.2), and loss (Section 3.3) modules within the framework of a standard CAT model. In conjunction with the comprehensive insights provided in Chapter 2 on the hazard module and hazard models for various perils, Chapter 3 equipped readers with the essential tools required for crafting their own CAT model. This was formalized in the form of a practical demonstration in the CAT Risk Modelling Sandbox and by the provision of a template to initiate the work in the 'build your own CAT risk model from scratch' project. Part II will describe more advanced modelling aspects which are not yet systematically integrated into CAT models, but represent important processes in understanding catastrophe risks. These encompass cellular automata, agent-based models, and system dynamics to describe some specific perils (Chapter 4); the modelling of cascading effects for super-catastrophes; and the consideration of long-term trends such as global warming and resource depletion, which can exacerbate risk factors (Chapter 5). For readers eager to learn more about risk management strategies, they may jump directly to Part III with minimal loss of comprehension.

Further recommended reading

- *Seismic Hazard and Risk Analysis* (McGuire, 2004, chapter 6)
- *Catastrophe Modeling: A New Approach to Managing Risk* (Grossi et al., 2005)
- *Loss Models: From Data to Decisions* (4th ed.) (Klugman et al., 2012)
- *Natural Catastrophe Risk Management and Modelling* (Mitchell-Wallace et al., 2017, chapters 1,4)

3.5 Exercises

Exercise 3.1

An earthquake of magnitude 6.5 just occurred 10 km away from where you live (and at a depth of 2 km).

(a) What happens when the seismic waves reach you? (what kind of shaking do you feel? Is there any damage?). What would have happened if it had only been a magnitude 4.5 event?

(b) The building where you live is made of adobe masonry. What are the probabilities of minor damage and of building collapse?

Hints: Sketching a modelling pipeline can be helpful as a starting point. All the equations and parameters necessary to answer these questions can be found in Chapters 2 and 3. Knowing how to navigate through the textbook will be helpful for the project described in Exercise 3.5.

Exercise 3.2

Consider the event loss table (ELT) shown in Table 3.11.

Table 3.11 Example of a fictitious ELT with two columns to fill

Event ID	Event rate λ_{ID}	Loss L_{ID} [USD][a]	EF	EP
9	0.002	15,000,000	..	..
1	0.003	10,800,000	..	..
8	0.004	10,100,000	..	..
4	0.006	6,900,000	..	..
6	0.011	4,600,000	..	..
2	0.012	4,300,000	..	..
3	0.020	2,700,000	..	..
5	0.176	200,000	..	..
7	0.274	100,000	..	..

[a] Already in descending order for ease of calculations.

(a) Calculate the exceedance frequency (EF) and exceedance probability (EP).
(b) Provide the following risk measures: the average annual loss (AAL), a plot of the EP curve, the value at risk $\text{VaR}_{99\%}$, and the tail value at risk $\text{TVaR}_{99\%}$.

Exercise 3.3

Generate a year loss table (YLT) from the ELT of the previous exercise for $N_{sim} = 10^5$ year-simulations:

(a) Follow the structure of the pseudocode provided in Box 3.5 to write your code (a function to sample from a Poisson distribution is given in `exercise_notebook.pynb`).
(b) Calculate the AAL, $\text{VaR}_{99\%}$, and $\text{TVaR}_{99\%}$ from the YLT and compare with the results previously obtained for the ELT.

Exercise 3.4

The loss tables so far considered were fictitious. The CAT Risk Modelling Sandbox provided in the online supplementary material shows how to generate loss tables from a stream of synthetic data going through the standard CAT modelling pipeline (Section 3.3.3).

(a) Run the online tutorial (for the default parameters) and observe how events populate the virtual environment for different perils and how they impact it, from hazard intensity footprints $I(x, y)$ to loss footprints $L(x, y)$.

(b) Change some of the parameters and inspect the impact on risk measures. Write a one-page report on your findings.

Hint: Consider reading the underlying codes to understand how the models described in the textbook can be formulated in a CAT model application. This forms a valuable resource for the next and final exercise of Part I.

Exercise 3.5 'Build Your Own CAT Risk Model from Scratch' Project

It is now time to develop your own CAT risk model from scratch!

(a) Choose a peril and a region. The region can be a generic (x, y) grid with an ad hoc exposure footprint, but it can also be a real region (which will then require some further research on the specificities of that region).
(b) Use the pseudocode template given in Boxes 3.7 and 3.8 or directly the code templates given in the online CAT Risk Modelling Starter Kit to build a toy CAT model in the programming language of your choice.
(c) Once the peril-specific CAT model is developed, generate the following results: a catalogue of hazard intensity footprints, a hazard map and hazard curves at three locations, a catalogue of damage footprints (in terms of MDR), an ELT, and, finally, the AAL and EP curves.
(d) Write a report on the method and the findings, and present your results to other students of risk.

Hints: Some perils are simpler than others to model. Point-source perils are some of the simplest ones to implement, for example. All the necessary models to define the entire CAT risk modelling pipeline are available for a limited number of perils in the textbook. It may be wise to consider one of these perils. However, any peril may be chosen – but may potentially require further research on your side. Note for instance that additional perils, requiring more sophisticated techniques to model, are considered in Part II. You can also copy and modify some of the codes of the CAT Risk Modelling Sandbox, but they must be integrated within the framework shown in Boxes 3.7 and 3.8.

PART II

ADVANCED CAT MODELLING

4 Dynamic Hazard Modelling

The hazard intensity models described in Chapter 2, for the most part, were simple analytical expressions of the form $I(x, y) = f_I(S, r)$, with I the hazard intensity, S the event size, and r the distance from the event source to any given location (x, y). Although different equations were defined for different perils, all represented a process of energy dissipation, with intensity I decreasing with distance r. The footprints $I(x, y)$ were said to be static because they were formulated as being instantaneous. In reality, energy dissipation is a **dynamic process**, that is, a process that evolves over time t. Although all the perils so far considered could be reanalysed from a dynamic perspective, it is not always justified in the context of risk modelling, since a static footprint often offers a reasonable approximation for the purpose of damage assessment. However, for some specific perils, the dynamics of the process must be considered for their proper characterization. This is the subject matter of the present chapter.

The main perils that require dynamic modelling involve events defined from the environment itself (i.e., diffuse sources) instead of from local sources. Examples include landslides, wildfires, and epidemics, which have for source the terrain, the vegetation, and the population, respectively. In such cases, *energy does not dissipate from a local energy stock (such as an active fault or an atmospheric depression) but propagates through an extended source from which the event can grow by being powered by the environment* (Fig. 2.11). Some other perils, which are defined from local sources, may also require dynamic modelling. This is the case for floods, where dissipation (by soil infiltration or evaporation) is secondary to the propagation of the water mass along the terrain's gradient.

A **dynamical system** (with spatial coordinates already discretized) is characterized by an **ordinary differential equation** (ODE) or a set of ODEs, typically expressed in the form $\mathrm{d}x/\mathrm{d}t = f(t, x)$, where x is the dependent variable and time t is the independent variable. In general, with one variable x, a dynamical system can exhibit behaviours such as growth, decay, or equilibrium; with two variables, oscillations; and finally, with three or more variables, a **chaotic** behaviour characterized by strong sensitivity to initial conditions (Strogatz, 2018). Examples of chaotic systems include the three-body problem of Poincaré (beautifully captured in Liu Cixin's science fiction masterpiece of the same name; Cixin, 2014) and the famous 'butterfly effect' in weather dynamics (Lorenz, 1963). While low-variable systems may exhibit chaotic characteristics, **complex systems** – which display periods of stability and chaos – involve many variables and interactions, leading to the **emergence** of systemic patterns that may not be predictable based solely on the properties of individual elements (such as the emergence of a power-law size distribution; Section 2.3.1). The changes observed in complex systems – sudden and apparently random – can be considered catastrophic, with the so-called catastrophe theory an early, yet aborted, attempt to classify such processes (Thom, 1975). Very rarely in a complex system, natural or human-made, are analytical solutions available. Therefore, *two computer simulation approaches will be used to model dynamic hazard: (1) the algorithmic approach, which includes cellular automata and agent-based models (ABM*; Section 4.1), *and (2) numerical integration in the field of system dynamics* (Section 4.2). Both approaches have pros and cons but aim at solving the same dynamic process (Rahmandad and Sterman, 2008).

For the sake of clarity, cellular automata and ABMs will generally be used when intensity footprints $I(x, y)$ are required, and system dynamics will be used when a hazard assessment may be carried out at a spatially aggregated level. The selection will also depend on model availability and simplicity. It follows that cellular automata and ABMs will be applied to landslides, fluvial floods, wildfires, blackouts, stampedes, and social unrest, while system dynamics will be applied to epidemics, pest infestations, and armed conflicts. A comparison of the two approaches will also be made for selected perils.

4.1 Cellular Automata and Agent-Based Models

4.1.1 Basic Concepts

A **cellular automaton** is a discrete model that consists of iteratively applying a set of local rules on the elements (or cells) of a regular grid. Despite the simplicity of the approach, it has been shown to mimic many complex phenomena observed in Nature, including catastrophes. This field of research, pioneered in the mid-twentieth century thanks to the advent of the electronic computer (Ulam, 1962; von Neumann, 1966), only became popular in the 1970s with John Conway's Game of Life, a two-dimensional model that crudely simulates life (Gardner, 1970). A systematic investigation of cellular automaton classes, started by Wolfram (1983) and culminating with Wolfram (2002), led to the iconic images that show how complex patterns self-organize from basic rules, such as Rule 30 (Box 4.1).

A cellular automaton is represented as a regular grid, or lattice, in which cells are in a finite number of states, and upon which a fixed set of local deterministic rules is applied at each discrete time step (e.g., Itami, 1994). The grid can be in any finite number of dimensions, although it is generally two-dimensional. Square grids are commonly used but other options include hexagonal and irregular triangular grids. At time t_0, an initial state is assigned to each cell. At each iteration t_i, a new state is determined for each cell based on the set of local rules of interactions between the cell and its neighbours. Typically, the rules for updating a cell's state are the same for all cells, remain the same over time, and are applied to the whole grid simultaneously. There are two main approaches to define cell neighbours on a two-dimensional lattice: the **von Neumann neighbourhood**, named after John von Neumann, is composed of a central cell and its four adjacent cells. The **Moore neighbourhood** is composed of a central cell and the eight cells that surround it. It is named after Edward F. Moore, who was another pioneer of cellular automaton theory. Many variants exist, depending on the type of application (Fig. 4.1).

Box 4.1 Game of Life and Rule 30

Two models which have popularized the cellular automaton are John Conway's Game of Life (Gardner, 1970) and Wolfram's one-dimensional elementary cellular automaton (Wolfram, 2002). In the former, a set of four rules leads to emerging lifelike structures with the properties to move, reproduce, and die. The rules are as follows: (1) any live cell with fewer than two live neighbours dies (death); (2) any live cell with two or three live neighbours remains alive (survival); (3) any live cell with more than three live neighbour cells dies (overcrowding); and (4) any dead cell with exactly three live neighbours becomes alive (birth). An iteration from the game is shown below on the left where patterns can be still, oscillate, or slide at the next iterations. Wolfram (1983) defined one-dimensional elementary cellular automata with binary outcomes depending on the three previous cells, giving $2^3 = 8$ possible combinations. Rule 30, which displays a chaotic behaviour, is defined as: $(111, 110, 101, 100, 011, 010, 001, 000) \rightarrow (0, 0, 0, 1, 1, 1, 1, 0)$. Wolfram's cellular automaton images, such as the one shown below on the right for 250 iterations of Rule 30, have become iconic in this field of research (Schiff, 2007, p. xii).

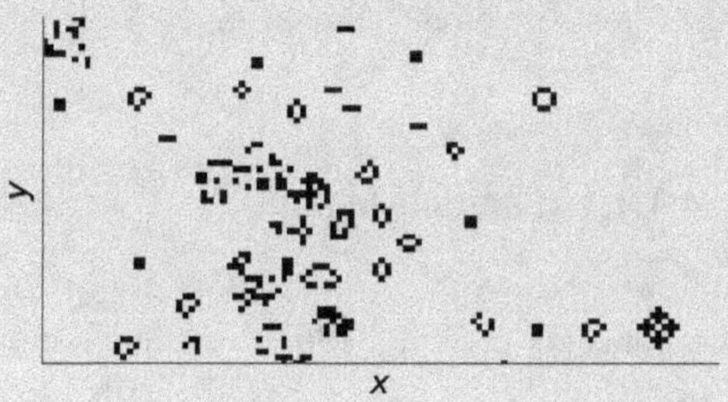

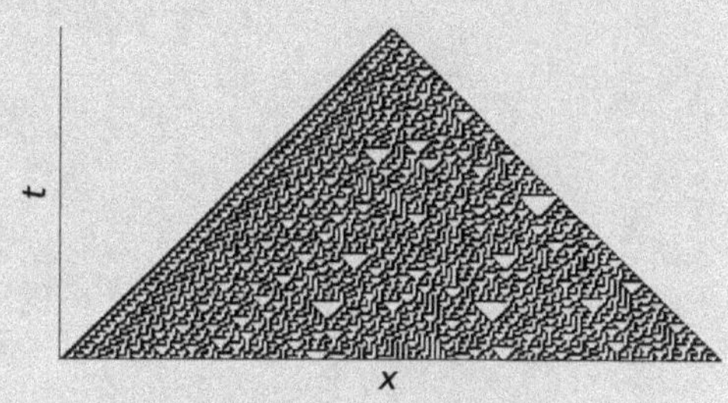

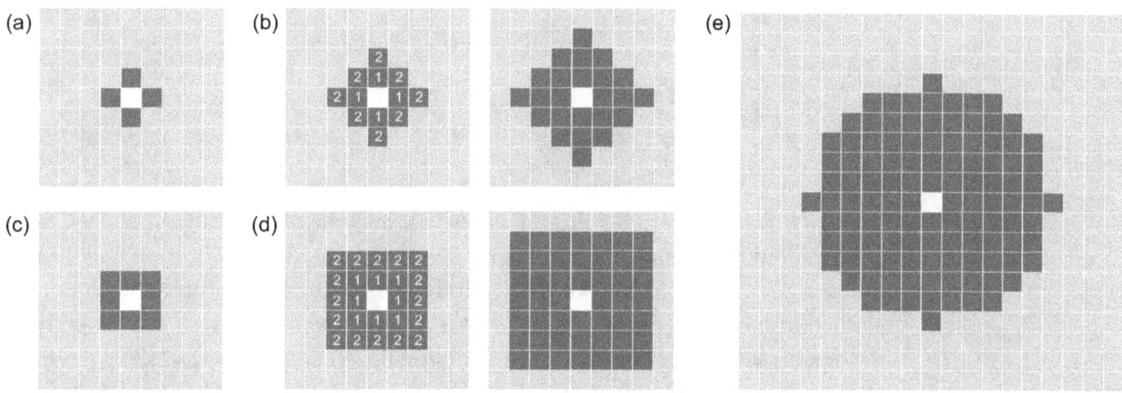

Fig. 4.1 Cellular automaton neighbourhoods. (a) von Neumann neighbourhood, or 4-neighbourhood, composed of a central cell (white) and its four adjacent cells (dark grey) on a two-dimensional square lattice. (b) Extension of von Neumann neighbourhood to ranges $r = 2$ and $r = 3$ (in terms of **Manhattan distance**). (c) Moore neighbourhood, or 8-neighbourhood. (d) Extension of Moore neighbourhood to ranges $r = 2$ and $r = 3$ (in terms of **Chebyshev, or chessboard, distance**). (e) Variant, with neighbourhood composed of 112 cells, used in the land-use cellular automaton model in Box 3.1 (White et al., 1997) (with 18 possible Euclidean distances).

It was discovered in the 1980s that a number of cellular automata display **self-organized criticality** (SOC), a property of dynamical systems with a critical point as attractor (Bak et al., 1987). A **critical point** can be thought of as a tipping point where the system undergoes a shift in its behaviour, such as a phase transition. In a complex system, the stable and chaotic regimes are separated by such critical points (Thurner et al., 2018). In SOC, therefore, the dynamic system always tends (is attracted) to that point, meaning that minor disturbances have the potential to trigger large chain reactions (such as catastrophes) at any time. In such a system, because events of any size can occur with the same patterns or structures observed regardless of the scale (i.e., **scale invariance**), a power-law size distribution naturally emerges. The concept was put forward by Bak et al. (1987) in their sandpile model, which is described below. The observation that many perils follow a power-law distribution (Section 2.3.1) promoted the use of SOC to describe both natural (Malamud et al., 1998; Malamud and Turcotte, 2000) and human-made perils (Roberts and Turcotte, 1998; Dobson et al., 2007). A virtually infinite number of variants exist, providing the flexibility needed for the modelling of multiple peril mechanisms (Sections 4.1.2–4.1.4). Another type of SOC model is the forest-fire model (Bak et al., 1990; Drossel and Schwabl, 1992; Section 4.1.5).

4.1.1.1 Sandpile Model

The **sandpile model**, which epitomizes the SOC process (Bak et al., 1987), is one of the simplest models to illustrate the concepts of both cellular automata and emergence in a complex system. It consists of updating the state $\mathcal{S} = z$ of the cells of coordinates (x, y) using the following synchronous rules:

$$\begin{cases} z(x,y) \rightarrow z(x,y) - 4 \\ z(x \pm 1, y) \rightarrow z(x \pm 1, y) + 1, \quad \text{if } z(x,y) \geq z_c, \\ z(x, y \pm 1) \rightarrow z(x, y \pm 1) + 1 \end{cases} \tag{4.1}$$

where z_c is the critical value (Fig. 4.2a). The system is set up with random initial conditions $z(x, y, t_0)$. The addition of a perturbation (i.e., $z \rightarrow z + 1$) at a random location may trigger the condition $z(x, y) \geq z_c$. What potentially follows is a chain reaction across multiple cells, so-called avalanches (Fig. 4.2b; Ex. 4.1a). As previously mentioned, a property of SOC is the emergence of a power-law size distribution $\bar{F}(S) \propto S^{-(\alpha-1)}$ (Eq. (2.35)), with S the avalanche area and α the power exponent (Fig. 4.2c). The Gutenberg–Richter law (Eq. (2.39), a power law in the energy domain) was the first peril size distribution to be related to a SOC process (Bak and Tang, 1989). This model is called a sandpile model because of its resemblance to an actual pile of sand. The randomly dropped particles in the model are analogous to the addition of particles to a sandpile, and the model avalanches are analogous to sand avalanches down the sandpile's sides. The following perils will be modelled as sandpile variants: blackouts (Section 4.1.2), floods (Section 4.1.3), and landslides (Section 4.1.4).

Fig. 4.2 The sandpile model. (a) Cellular automaton rules according to Eq. (4.1). The height represents the state value $\mathcal{S} = z$ for the cell of coordinates (x, y) on the square lattice. Modified from Mignan (2011, fig. 3). (b) Iteration of a 100×100 grid sandpile, with z states in grey and an example of avalanche in white. (c) Distribution of avalanche sizes S with power-law fit represented by a dashed line.

4.1.1.2 Extension to Agent-Based Models (ABM)

An **agent-based model** is an algorithmic method that simulates both the actions and the interactions of autonomous agents. One of the earliest examples is the urban residential segregation model of Schelling (1971) where agents relocate on a grid if a certain fraction of their neighbours is not from their own group. A segregation pattern emerges when a critical fraction threshold is crossed. This model is actually a special case of the checkerboard model of Sakoda (1971), which included other types of social interactions. By the mid-1990s, combinations of ABMs and cellular automata already led to the modelling of complex 'artificial societies' and of their interactions with perils such as conflicts, epidemics, and pollution (Epstein and Axtell, 1996). Similarly to cellular automata, ABM interactions follow a set of simple local rules that can lead to complexity, including properties such as emergence and chaos (Clarke, 2014). An ABM provides more flexibility than a cellular automaton by having local rules of interactions applied to moving agents instead of to a fixed lattice. Their uses are, however, complementary, with an ABM employed when a behavioural unit must be considered such as people (for epidemics, social unrest, and conflicts), pests (for pest infestations), or vehicles (e.g., for perils related to transport dynamics). ABMs additionally combine concepts of game theory and evolutionary programming by defining a set of characteristics per agent that evolves over time and makes the agent take different sets of actions. Applications will include stampedes (Section 4.1.6) and social unrest (Section 4.1.7).

4.1.1.3 Applicability to CAT Modelling

Since their usage remains limited to a few perils, there is so far no standard method defined to implement the outputs of a cellular automaton or ABM in a CAT model. Yet, taking the sandpile model as an illustrative example (with avalanches equivalent to potentially

(a)

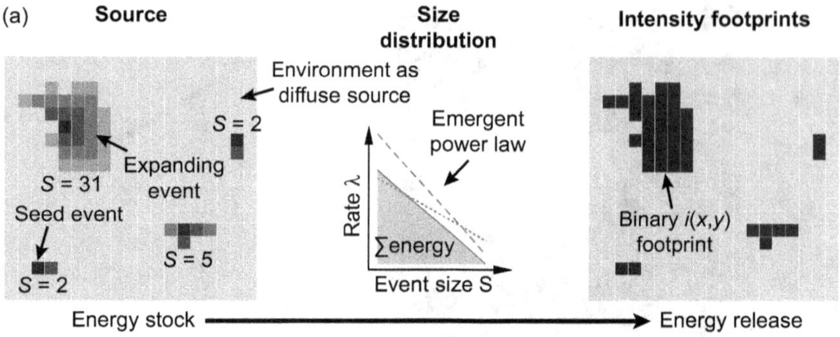

(b) **Event table**

ID	S	Loc.	λ
ID1	2	1	$1/\Delta T$
ID2	2	2	$1/\Delta T$
ID3	2	3	$1/\Delta T$
...	...	...	...
ID12	3	12	$1/\Delta T$
...	...	...	...
...	...	...	...
ID75	13	75	$1/\Delta T$
...	...	...	...

Fig. 4.3 Cellular automata in the CAT modelling framework. (a) Hazard physical process in the context of a cellular automaton for a diffuse source and dynamic event propagation (to be compared to Figures 2.3 and 2.11). (b) Implementation of the results of a cellular automaton in an event table. The parameter ΔT defines the duration of the cellular automaton simulation in years, which means that the cellular automaton's iterations must be related to physical time.

damaging events), a comparison with the hazard modelling framework described in Chapter 2 can be made. In physical terms, the cellular automaton rules (Eq. (4.1)) can be interpreted as an energy transfer between neighbouring geographical cells, with long-term loading $z(x, y) \rightarrow z(x, y) + 1$ leading to the stochastic occurrence of seed events $z(x, y) \geq z_c$ and the triggering of event propagation up to size S. The power-law size distribution of events, observed for most perils (Section 2.3.1), is here an emergent property of SOC. The hazard intensity footprint $I(x, y)$ is simply the area covered by the sandpile, which represents the static version of the dynamic footprint $\max_t I(x, y, t)$. Figure 2.3 on energy transfer and Figure 2.11 on diffuse sources can therefore be adapted for the cellular automaton context, as shown in Figure 4.3a. The intensity value I can be binary, such as burnt/not burnt for a wildfire or electricity served/not served for a blackout. It can also be a scalar value, such as landslide depth. Figure 4.3b finally illustrates how a stochastic event set can be generated from a cellular automaton for implementation in the hazard module of a CAT model, one of the main issues being the ability to calibrate the event rates to past observations.

It is important to note that cellular automata and ABMs are considered here because they represent some of the simplest possible approaches to model the complex behaviour of catastrophes. They can also capture local heterogeneities, which is not possible with the system dynamics approach based on aggregate differential equations (Bonabeau, 2002; Section 4.2). Moreover, since they are based on local rules, cellular automaton and ABM algorithms are well suited to parallel computation (Cagigas-Muñiz et al., 2022). *The applications that follow must be understood by the reader as toy models, deliberately simplified to illustrate how selected perils can be represented by different cellular automaton and ABM rules.* More sophisticated cellular automaton models that consider additional parameters, as well as physical time, are necessary for a realistic modelling of hazard footprints. ABMs, even the simplest ones, remain very rarely used in practice for hazard footprint modelling, but are commonly employed in the development of crisis management scenarios.

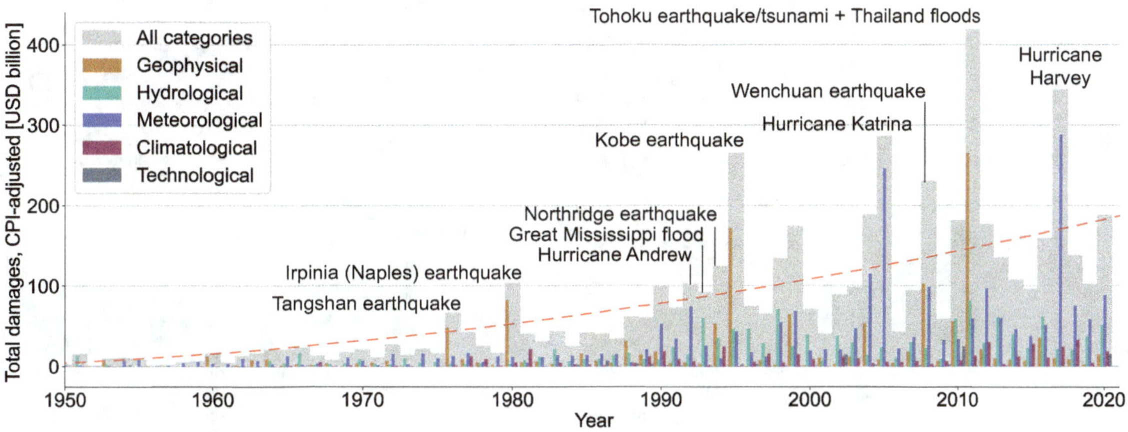

Fig. 1.1 Annual damage losses attributed to natural and human-made disasters during the period 1950–2020, as estimated from the international disaster database, or EMergency events DATabase (EM-DAT) (Guha-Sapir et al., 2022) (data used with permission from the Centre for Research on the Epidemiology of Disasters (CRED)). The year 2011, recording the highest losses, earned the epithet *annus horribilis* by The Geneva Association (Courbage and Stahel, 2012). Accelerating trend in economic losses highlighted by the dashed curve.

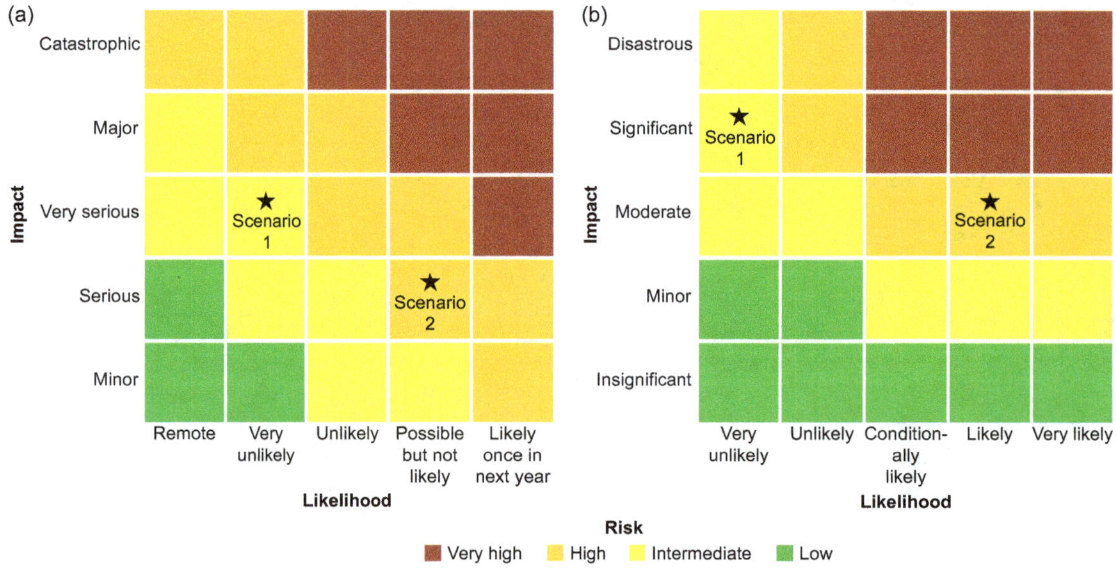

Fig. 1.2 Typical scheme of a risk matrix where scenarios are ranked as a function of likelihood and impact for comparative risk judgements. Risk categories according to (a) Cox (1998, fig. III.1.1) and (b) Komendantova et al. (2014, fig. 3).

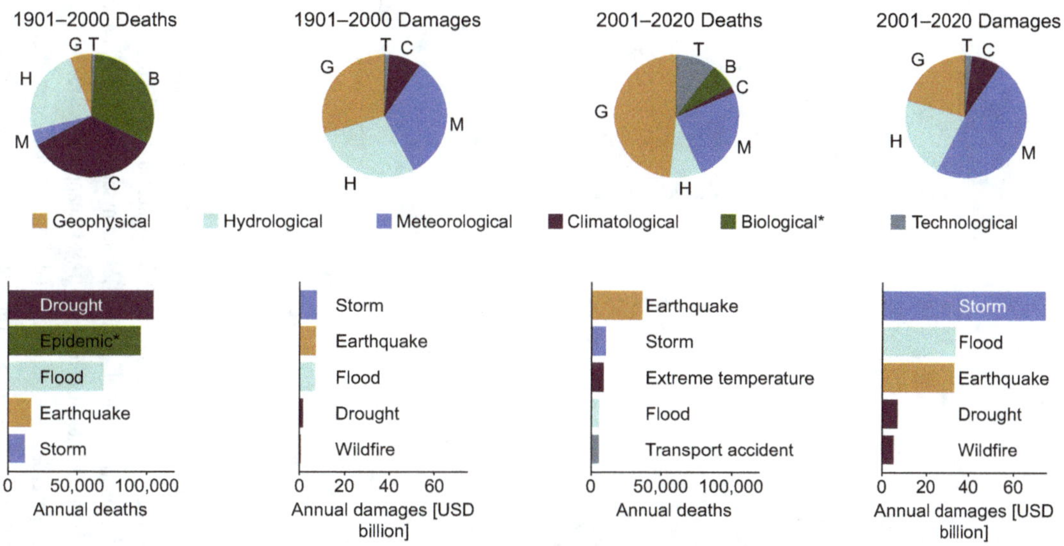

Fig. 1.4 Ranking of catastrophic losses per peril category in terms of deaths and damage for the twentieth and early twenty-first centuries, inferred from EM-DAT (Guha-Sapir et al., 2022) (data used with permission from the CRED). * Epidemic data incomplete in EM-DAT (pers. comm., R. Below, February 2023), with COVID-19 deaths not yet counted and the 1918 Spanish Flu, responsible for 32 million deaths (Marani et al., 2021a), missing due to early-twentieth-century data incompleteness.

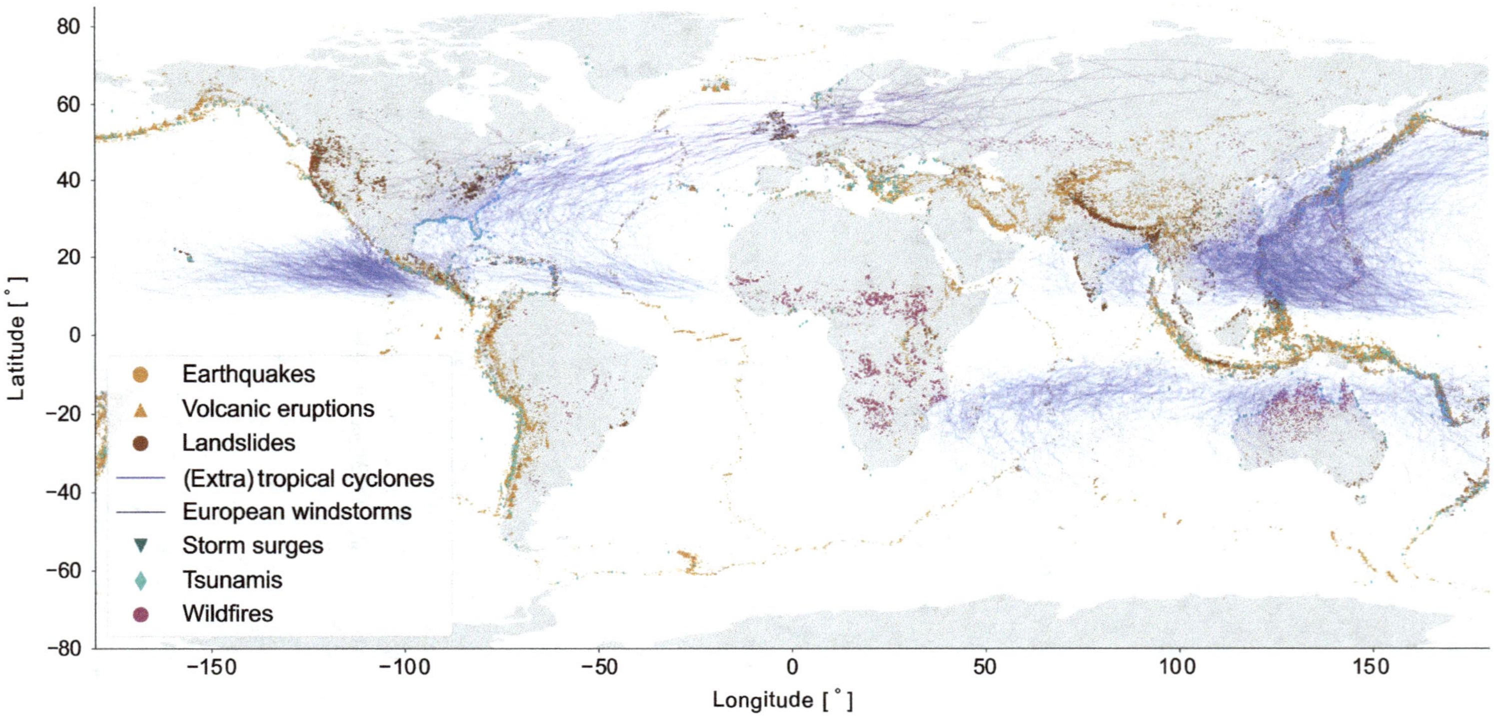

Fig. 1.8 Geographical distribution of some of the main natural perils on Earth. Earthquakes from the International Seismological Centre-Global Earthquake Model (ISC-GEM) Global Instrumental Earthquake Catalogue (ISC, 2023; Storchak et al., 2013, 2015; Di Giacomo et al., 2018) (CC-BY-SA, with permission); landslides (rain-induced) from the Global (rain-induced) Landslide Catalog (GLC) (Kirschbaum et al., 2010, 2015); storm surges from SURGEDAT (Needham and Keim, 2012); tropical and extratropical cyclones from the International Best Track Archive for Climate Stewardship (IBTrACS) (Knapp et al., 2010, 2018); tsunamis from the Global Historical Tsunami Database (NGDC/WDS, 2022); volcanic eruptions from the global database Large Magnitude Explosive Volcanic Eruptions (LaMEVE) (Crosweller et al., 2012); wildfires from the FRY global database of fire patches (Laurent et al., 2018a,b) (CC-BY 3.0); European windstorms from the eXtreme WindStorms (XWS) catalogue (Met Office, University of Reading and University of Exeter, 2023; Roberts et al., 2014) (CC-BY 4.0, https://creativecommons.org/licenses/by/4.0/).

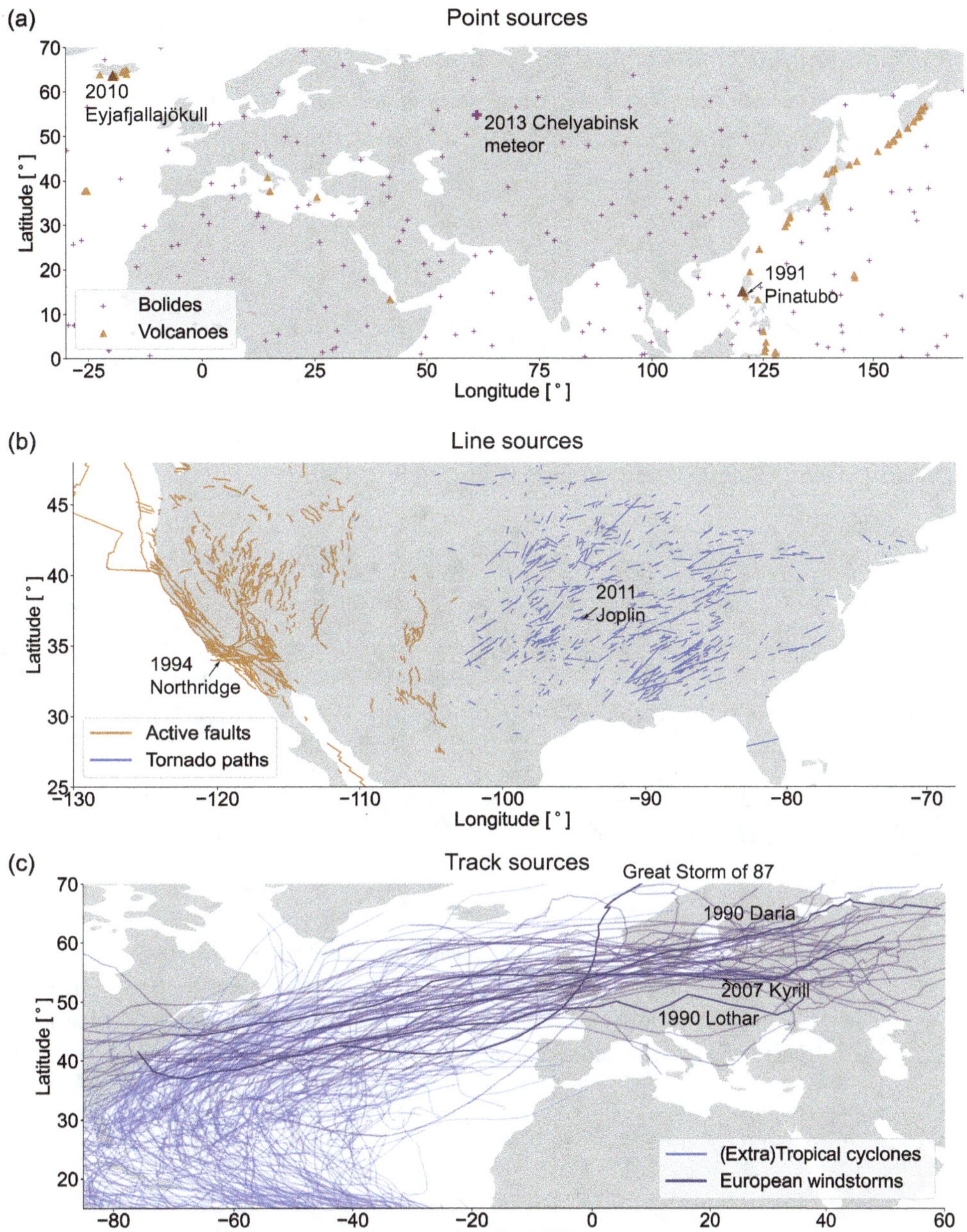

Fig. 2.4 Examples of localized sources: (a) Point sources: bolides in the 1988–2022 period (CNEOS, 2023) and volcanoes (LaMEVE; Crosweller et al., 2012). (b) Line sources: active faults (GEM Global Active Faults Database; Styron and Pagani, 2020 – CC-BY-SA, published with permission) and F4–F5 tornado paths in the period 1950–2021 (NOAA's Storm Prediction Center, 2023). (c) Track sources: Atlantic hurricanes and extratropical cyclones (IBTrACS; Knapp et al., 2010, 2018) and European windstorms (XWS; Met Office, University of Reading and University of Exeter, 2023; Roberts et al., 2014; CC-BY 4.0, https://creativecommons.org/licenses/by/4.0/) for the same period 1979–2012.

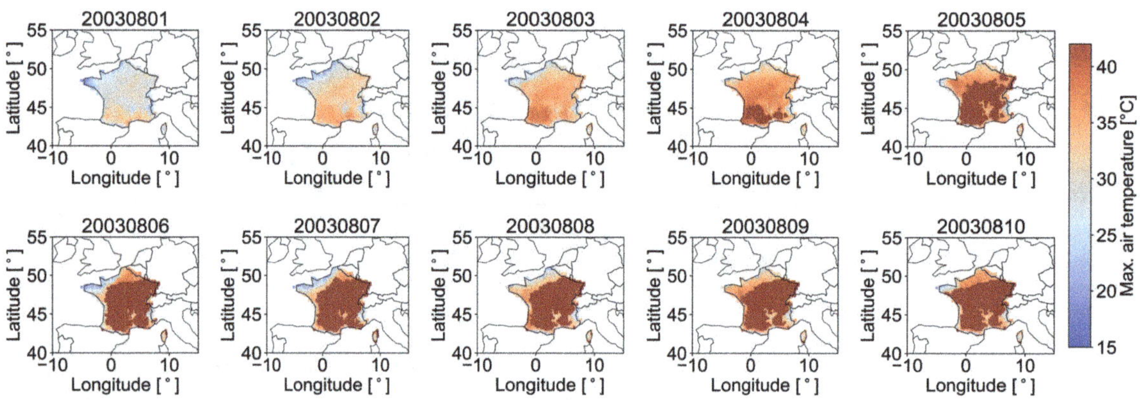

Fig. 2.20 The 2003 European heatwave footprint defined with a threshold method. The darker points delineate the heatwave and represent the locations (here in France) where the maximum daily air temperature exceeds the threshold $T_{th} = 35°C$ for at least 3 consecutive days. Data from the Gridded Agro-Meteorological Data in Europe (Toreti, 2014), from the Agri4Cast Resources DATAPORTAL (CC-BY 4.0, https://creativecommons.org/licenses/by/4.0/).

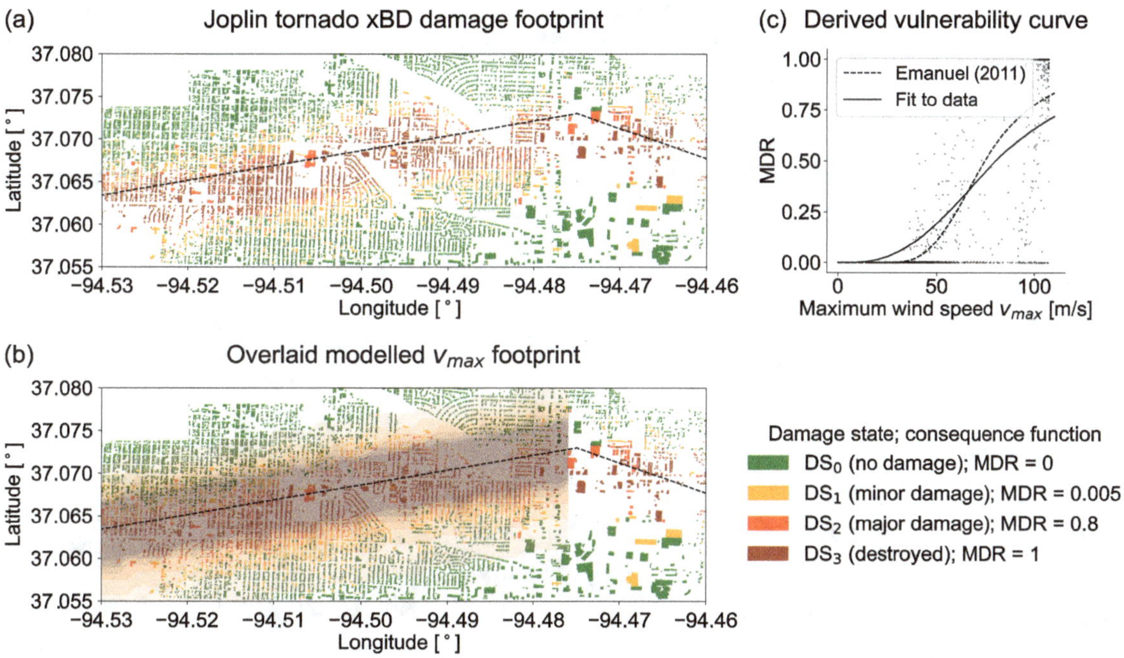

Fig. 3.5 Damage footprint of the 2011 Joplin tornado and vulnerability curve fit. (a) Damage footprint from the xBD data set (Gupta et al., 2019) (CC-BY-NC – published with permission). (b) Overlaid maximum wind speed v_{max} intensity footprint (Eq. (2.59)), with $r_{max} = 300$ m, $v_f = 11$ m/s, $v_{rad} = 86$ m/s and $v_{tan} = 43$ m/s from Karstens et al. (2013) and track coordinates from NOAA's Storm Prediction Center (2023) with mid-point fixed to $(-94.475°, 37.073°)$. Intensity modelled only on the track subsection where the tornado was at its maximum intensity (EF5). (c) Pairs of (v_{max}, D) data points with D estimated from Eq. (3.6), and fit $D = (cv_{max})^3/(1 + (cv_{max})^3)$ with $c = 0.012$ compared to the Emanuel (2011) model (Eq. (3.14)).

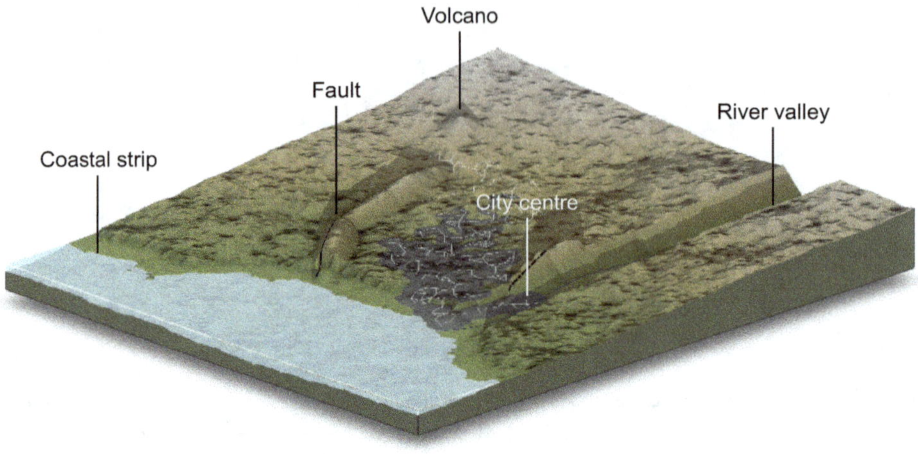

Fig. 3.10 One instance of the GenMR digital template (default parameterization) with topography layer and land-use layer (here consisting of a water mass, built area, and road network). Simulation rendered with Rayshader (Morgan-Wall, 2023).

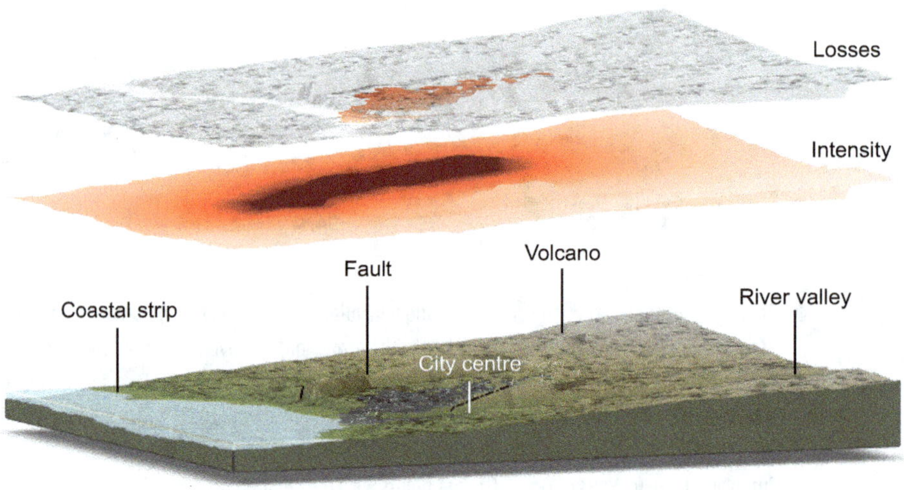

Fig. 3.11 Intensity footprint and loss footprint in the GenMR digital template (default parameterization), here for a large earthquake. Simulation rendered with Rayshader (Morgan-Wall, 2023).

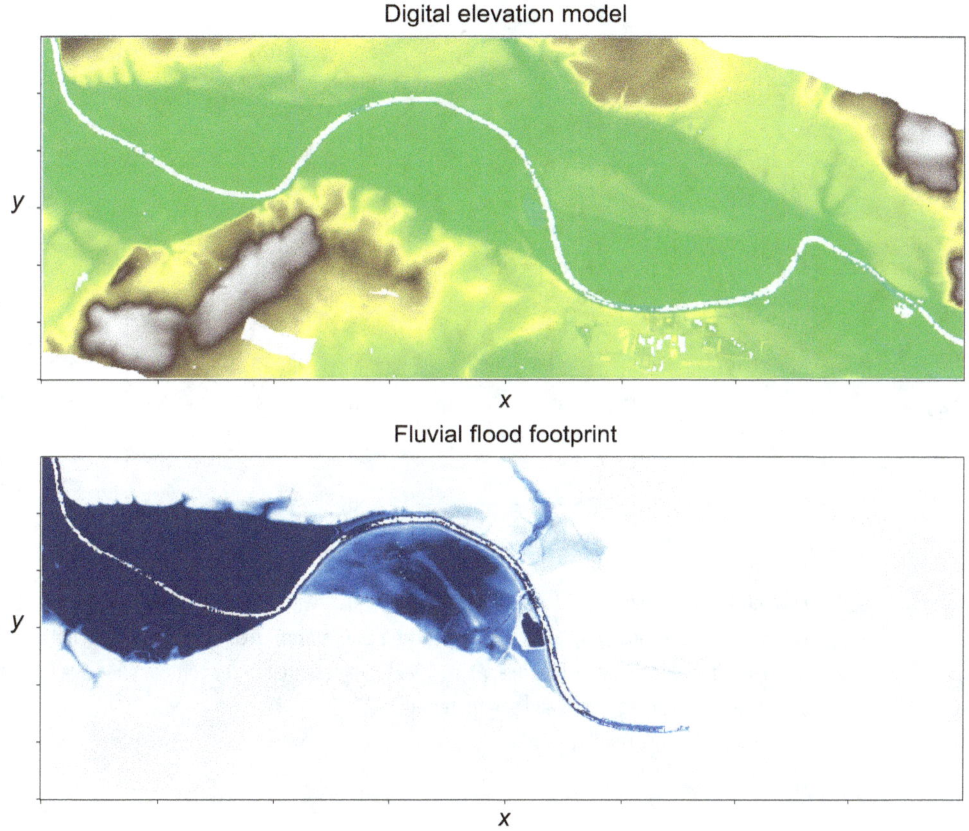

Fig. 4.6 Example of a fluvial flood footprint modelled with a simple cellular automaton, here using the Upton-upon-Severn test-site digital elevation model (test 7) of '2D benchmarking: evaluating the latest generation of the hydraulic models for FCRM' available from the UK Environment Agency (2021) and described in Néelz and Pender (2013). A darker colour on the footprint map represents a greater flood depth h_w.

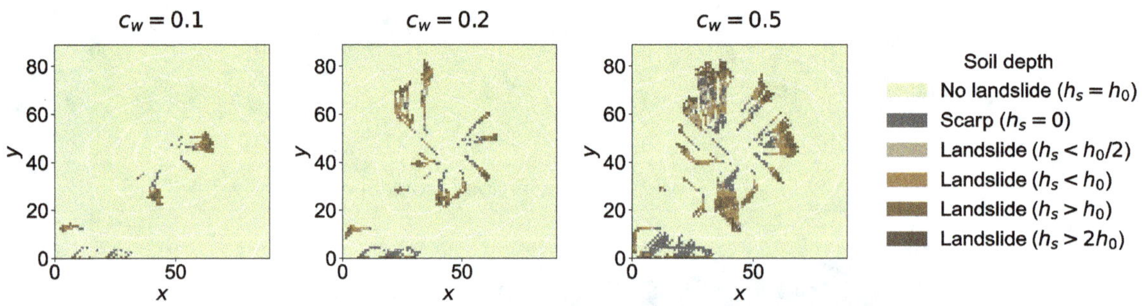

Fig. 4.8 Examples of simulated landslides triggered by rainstorms – with relative wetness $c_W = 0.1$ (large rainstorm), $c_W = 0.2$ (severe rainstorm), and $c_W = 0.5$ (extreme rainstorm), according to the toy cellular automaton. The simulated fractal topography (Fournier et al., 1982) is represented by the contour lines. See Exercise 4.2 for details about the modelling approach and parameterization.

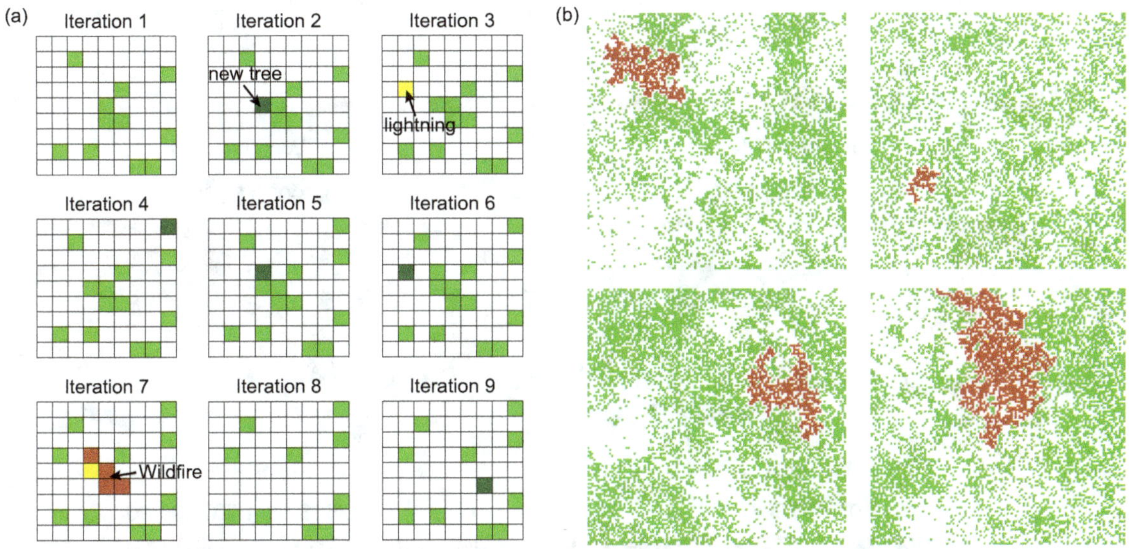

Fig. 4.9 The forest-fire model. (a) Cellular automaton rules applied for 9 iterations on a 9×9 grid. White cells are unoccupied sites; other cell states are tree, lightning strike, and burning tree. The fire at iteration 7 consumes 5 trees (event of size $S = 5$). (b) Examples of wildfire footprints (here in red) simulated with the forest-fire toy model on a 128×128 lattice (event sizes $S = 493$, 103, 1458, and 527 clockwise from top left). See also Malamud and Turcotte (2000, figs. 1 and 2).

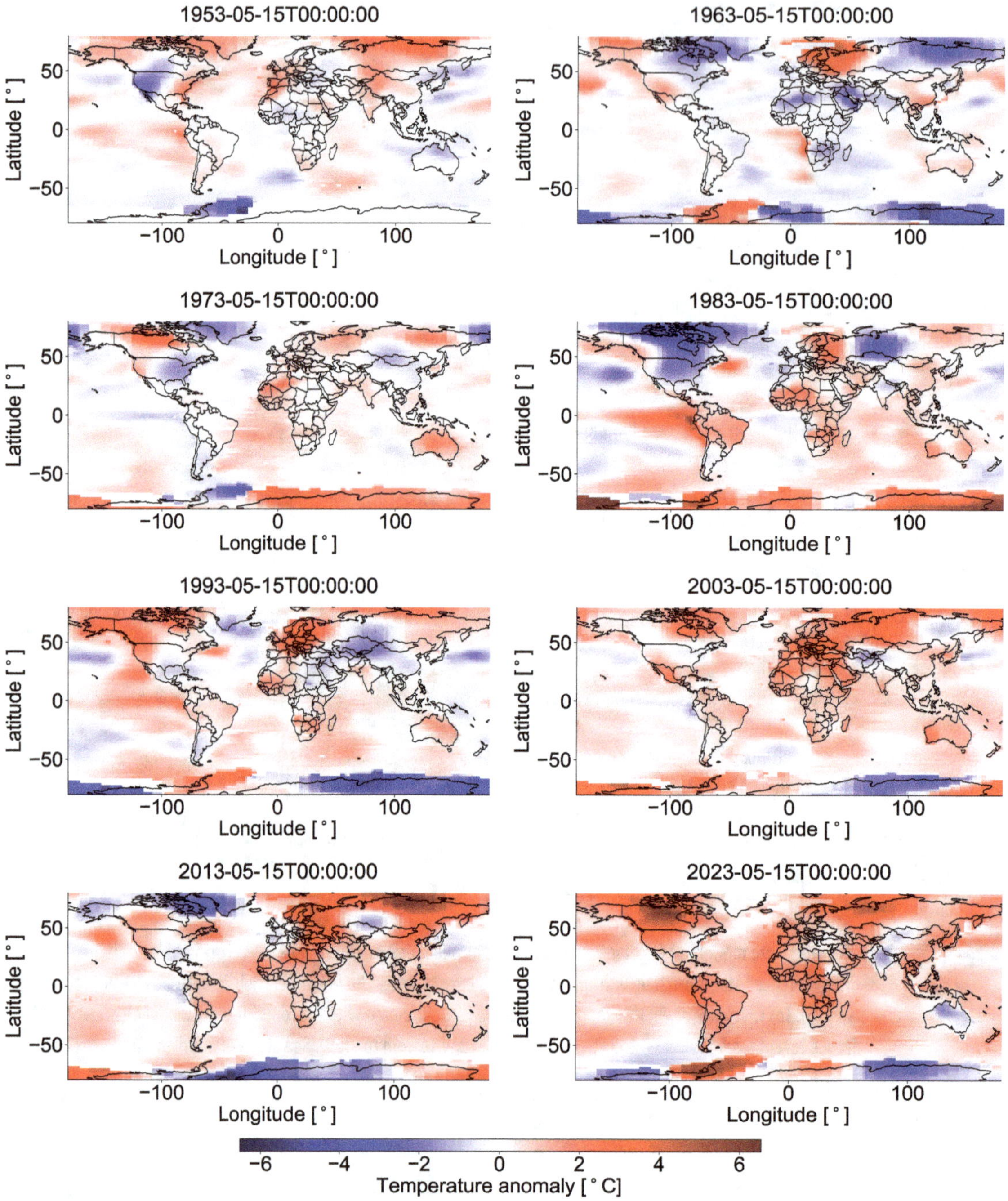

Fig. 5.13 Maps of GMST changes observed from May 1953 to May 2023, with temperature anomalies from the GISTEMP v4, with 1951–1980 baseline (GISTEMP Team, 2023; Lenssen et al., 2019).

4.1.2 Blackout (BO) Cellular Automaton

A power grid is an example of an engineered complex system, and the dynamics of blackouts can be modelled by the variant of a sandpile. As already explained in Section 1.3.7.4, a blackout is the loss of electrical power affecting end users in large parts of a power grid. This is caused by power outages cascading through the network with elements (nodes or lines) failing due to power overload.

One simple model proposed by Sachtjen et al. (2000) considers a ring network consisting of N nodes, each linked to the k nearest nodes (subjected to some random link reconnections; Fig. 4.4a). Here, a node (or vertex) represents any component of the power grid, including both nodes (or buses in power grid terminology, i.e., generation, transmission, and distribution nodes) and edges (transmission lines, transformers). Instead of using a Cartesian grid system as in the case of the standard sandpile model, connections between nodes are defined here in an adjacency matrix, which is a square matrix with its elements indicating whether pairs of nodes are linked or not. Network theory and related terms will be formally introduced in Chapter 5. No additional detail is needed here for the comprehension of the model. At any given time, nodes are characterized by a load Z. Random fluctuations occur over time consisting of the transfer of one unit load from (or to) node i to (or from) one of its neighbours j:

$$\begin{cases} Z_i \to Z_i \pm 1 \\ Z_j \to Z_j \mp 1 \end{cases} \tag{4.2}$$

Here, j identifies the location of a randomly chosen node among the k neighbours of node i. Each element of the power grid has a region of safe operation. When the power load exceeds the safety margin Z_c, the element trips and the power redistributes itself throughout the network according to the impedances of the surrounding lines. Therefore, if for any element i, $Z_i \geq Z_c$, then N_f units of load are transferred to its neighbours j:

$$\begin{cases} Z_i \to Z_i - N_f \\ Z_j \to Z_j + n_{f,j} \end{cases} \tag{4.3}$$

(e.g., with $Z_c = 105$ and $N_f = 10$ in this illustrative example). The load amounts $n_{f,j}$ transferred are chosen randomly with the constraint that $\sum_j n_{f,j} = N_f$ (Sachtjen et al., 2000). An event size S is then defined as the number of nodes failing during one sandpile avalanche.

When the grid is run with low loading of power (which is economically impractical), a single event, such as a line tripping out, is unlikely to cause others. Events are independent, and the distribution of event sizes has an exponential tail. Once loading is high, however, the system self-organizes, with cascading effects potentially leading to large-size events (Fig. 4.4a; Ex. 4.1b). The size distribution of outages follows a power law with exponent $\alpha \approx 1$, which is attenuated by an exponential cut-off when approaching the size of the network. Sachtjen et al. (2000) proposed the following functional form for the size distribution:

$$f(S) = c\frac{\exp(-S/S_0)}{1 + S/S_1}, \tag{4.4}$$

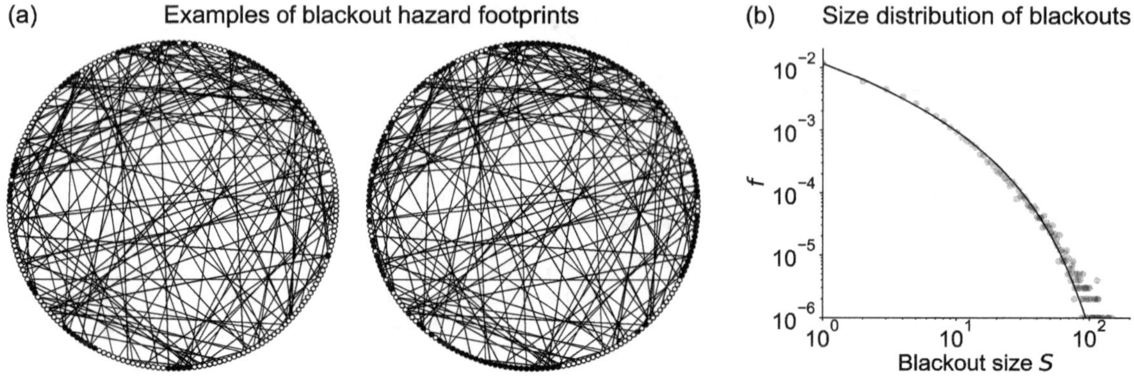

Fig. 4.4 Example of a blackout sandpile model, here based on the method of Sachtjen et al. (2000). (a) Examples of hazard footprints for a ring network composed of $N = 200$ nodes connected to $k = 4$ nearest nodes with 20% of links randomly reconnected. Failed nodes are represented in black and nodes in operation in white. Sizes $S = 57$ and 119, respectively. (b) Size distribution of simulated blackouts (for 1 million iterations). The solid curve represents the fit to Eq. (4.4) ($c = 0.04$, $S_0 = 18.0$, and $S_1 = 0.42$).

with f the PDF, S_0 and S_1 fitting parameters, and an S^{-1} dependence combined with an exponential cut-off (Fig. 4.4b). Note that S could be defined in terms of energy unserved (MWh) in real electricity networks.

Blackout models for real power grids are more complex since they need to follow electrical circuit rules to realistically describe power flows. Yet, cascading failures are based on the same basic SOC principle as shown above (Dobson et al., 2007). A number of open-source initiatives provide the tools necessary for power system analysis (Brown et al., 2018; Thurner et al., 2018). Simple approximations of parts of the American Electric Power system (e.g., IEEE 118-bus system) are also freely available for tests (Christie, 1993). Note finally the possible use of network topology metrics to model blackouts (Dueñas-Osorio and Vemuru, 2009) (see Chapter 5).

4.1.3 Flood (FF, PF, ...) Cellular Automaton

The physical modelling of floods typically relies on the **shallow-water equations** (SWE), which are derived from the Navier–Stokes equations (example of partial differential equations). In the simplest setting, the SWE are:

$$\begin{cases} \dfrac{\partial(\rho h)}{\partial t} + \dfrac{\partial(\rho h v_x)}{\partial x} + \dfrac{\partial(\rho h v_y)}{\partial y} = 0 \\ \dfrac{\partial(\rho h v_x)}{\partial t} + \dfrac{\partial}{\partial x}\left(\rho h v_x^2 + \dfrac{1}{2}\rho g h^2\right) + \dfrac{\partial(\rho h v_x v_y)}{\partial y} = 0 \\ \dfrac{\partial(\rho h v_y)}{\partial t} + \dfrac{\partial(\rho h v_x v_y)}{\partial x} + \dfrac{\partial}{\partial y}\left(\rho h v_y^2 + \dfrac{1}{2}\rho g h^2\right) = 0 \end{cases} \quad (4.5)$$

where h is the total fluid column height (instantaneous fluid depth as a function of x, y, and t) and (v_x, v_y) is the fluid's horizontal flow velocity, averaged across the vertical column. Parameter g is the acceleration due to gravity and ρ is the fluid density. The first equation is derived from mass conservation, and the next two from momentum conservation.

Shallow water flows include fluvial floods (FF) and pluvial floods (PF), but also storm surges (SS), tsunamis (Ts), and dam floods (DF, including levee failures) (Vreugdenhil, 1994). Physically based two-dimensional flood models require high computational resources to solve the SWE (Néelz and Pender, 2013). The cellular automaton is a computationally efficient alternative to approximate the dynamic process for rapid flood analysis (Guidolin et al., 2016). Fast computing, via a reduced complexity approach, is especially important in probabilistic flood assessment, which requires the modelling of many flood footprints at a regional or national level (Jamali et al., 2019).

Various flood cellular automaton models are available in the literature (Guidolin et al., 2016; Jamali et al., 2019; Issermann et al., 2020), which are all based on the same basic loop:

(1) Determine the level difference $\Delta l_{0i} = l_0 - l_i$ between the central cell of index 0 and the neighbouring cells i, with $l = z + h_w$ the water level, z the ground elevation, and h_w the water depth.
(2) Determine the volume of water $V \leq h_{w0}A$ to be transferred to neighbours of lower level $l_i < l_0$ constrained by the available water depth h_{w0} in the central cell of area A.
(3) Determine the weight of the water volume V to be kept in the central cell w_0 and to be distributed to the neighbours w_i.
(4) Update the water depth h_w in the respective cells, considering possible additional inflows (e.g., local river discharge at point source, distributed rainfall on diffuse source) and outflows (e.g., soil infiltration, evaporation, grid exit).

The process is illustrated in Figure 4.5 for the von Neumann neighbourhood, noting that other neighbourhood types may also be used. The water depth in a cell can be represented by the equation

$$h_{w0}^{t+\Delta t} = h_{w0}^{t} + \frac{\Delta t}{A} \sum_i (Q_{i0}^t - Q_{0i}^t) + (G_0^{t+\Delta t} - L_0^{t+\Delta t})\Delta t \tag{4.6}$$

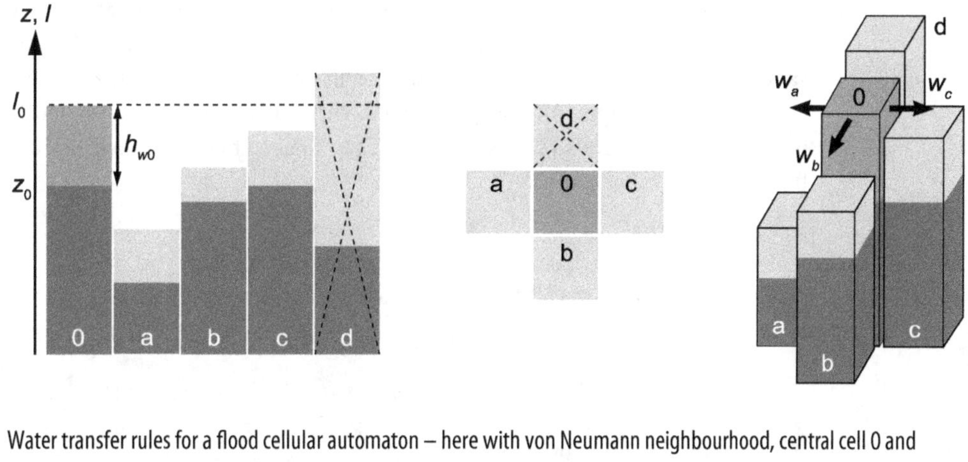

Fig. 4.5 Water transfer rules for a flood cellular automaton – here with von Neumann neighbourhood, central cell 0 and neighbour cells *a*–*d*. Water is distributed only to cells with a lower level l_i, conditional on the presence of water $h_{w0} > 0$. Water distribution weights w_i depend on the defined cellular automaton. See also Issermann et al. (2020, figs. 1 and 2).

where $h_{w0}^{t+\Delta t}$ is the water depth at $t + \Delta t$ in the central cell [L], Q_{i0}^{t} is the inflow discharge from the neighbouring cells to the central cell [L^3 T^{-1}], Q_{0i}^{t} is the outflow discharge from the central cell to the neighbouring cells [L^3 T^{-1}], $G_0^{t+\Delta t}$ is the water gain rate [L T^{-1}] (e.g., rainfall intensity), $L_0^{t+\Delta t}$ is the water loss rate [L T^{-1}] (e.g., soil infiltration rate), A is the cell area [L^2], and Δt is the time interval between two iterations [T].

The main difference between different cellular automaton models is the way the available water volume V and weights w are computed (Guidolin et al., 2016; Jamali et al., 2019; Issermann et al., 2020, and references therein). The simplest option is to define $w_i = c\Delta l_{0i} / \sum \Delta l_{0i}$ ignoring neighbour cells with $l_i \geq l_0$. If $c = 1$, water-level oscillations are observed between time steps. Lowering c can help reduce the algorithm artefact although further normalization of the weights is needed for stability over many iterations. This problem is discussed in, for example, Guidolin et al. (2016). Some cellular automata integrate physical time and flow velocities (Guidolin et al., 2016, and references therein) while the simplest cellular automata model only spatial extents and depths for rapid footprint definition (Jamali et al., 2019), which is enough for first-order damage assessment (Chapter 3). Models can be validated against the results expected from solving the SWE in benchmark tests (Néelz and Pender, 2013). Application of a simple cellular automaton, which models the spatial extent and depth $h_w(x, y)$ only, is shown in Figure 4.6 in the case of a fluvial flood (FF). Other applications could include pluvial flood (PF) and water system failure (Wf) during a rainstorm (RS) (e.g., overflowing manholes as point sources). Similarly, such a cellular automaton could be used to model the propagation of a toxic spill (TS). However, more sophisticated cellular automata or full SWE models are required to model the flow velocity in levee failures, dam floods (DF), and coastal floods (SS, Ts).

4.1.4 Landslide (LS) Cellular Automaton

The initiation of a slope failure was quantified by the factor of safety F_S in Section 2.2.4.1 (Eq. (2.29)). The mass movement that fully characterizes a landslide is a dynamic process highly dependent on local interactions with the terrain. While this process can be modelled via differential equations (Densmore et al., 1998; Hergarten and Neugebauer, 1998), a sandpile model variant provides a simple and natural approach to describe landslides.

A modified sandpile model is described below that is specific to the landslide slope instability process. For cells (x, y) with $F_S(x, y) < 1$ (unstable slope) and soil depth $h_s(x, y) > 0$, the following rules apply:

$$\begin{cases} z(x,y) \rightarrow z(x,y) - \Delta h_s \\ h_s(x,y) \rightarrow h_s(x,y) - \Delta h_s \\ z(\text{dir}_{Moore}(\theta_{max}(x,y))) \rightarrow z(\text{dir}_{Moore}(\theta_{max}(x,y))) + \Delta h_s \\ h_s(\text{dir}_{Moore}(\theta_{max}(x,y))) \rightarrow h_s(\text{dir}_{Moore}(\theta_{max}(x,y))) + \Delta h_s \end{cases} \tag{4.7}$$

where the mass movement is defined by

$$\Delta h_s = \frac{1}{2}\left[z(x,y) - z(dir_{Moore}(\theta_{max}(x,y))) - W\tan(\theta_{stable})\right] \tag{4.8}$$

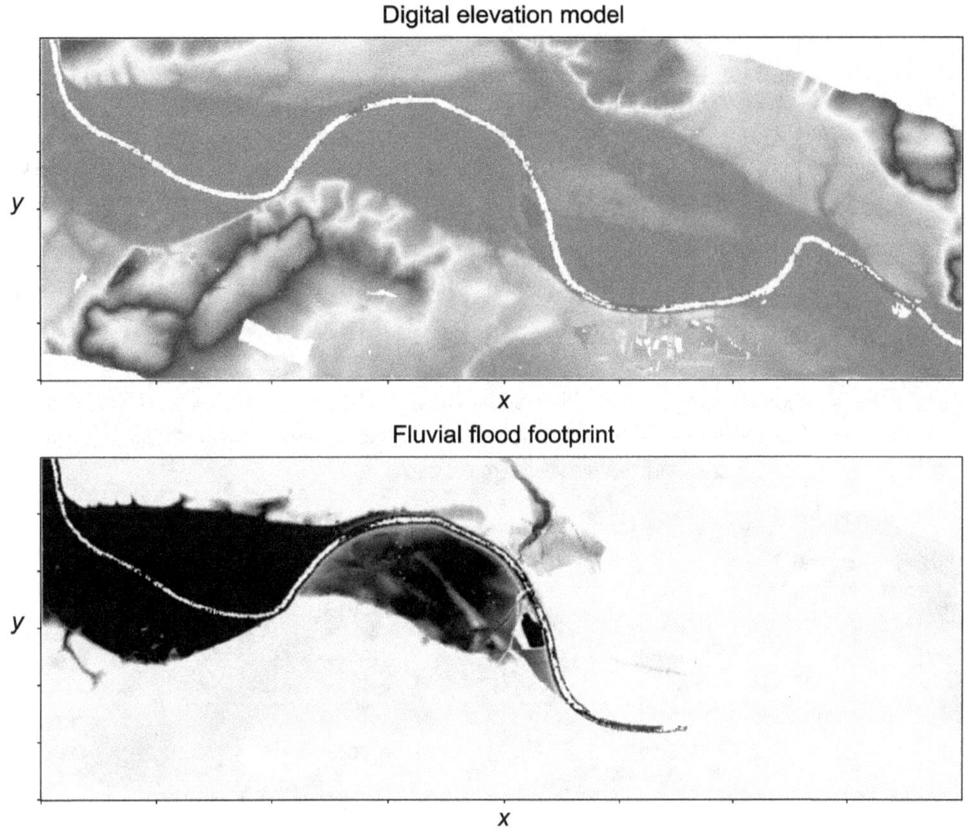

Fig. 4.6 Example of a fluvial flood footprint modelled with a simple cellular automaton, here using the Upton-upon-Severn test-site digital elevation model (test 7) of '2D benchmarking: evaluating the latest generation of the hydraulic models for FCRM' available from the UK Environment Agency (2021) and described in Néelz and Pender (2013). A darker colour on the footprint map represents a greater flood depth h_W. A black and white version of this figure will appear in some formats. For the colour version, please refer to the plate section.

with W the cell width (Mignan, 2022a). The mass is progressively transferred from cell to cell by Δh_s such that a stable slope θ_{stable}, if h_s is high enough, is reached (Fig. 4.7). If h_s is too low, that is, $h_s < \Delta h_s$, a scarp forms equivalent to $h_s = 0$ once the sandpile rule is applied. The mass movement occurs in the direction of maximum slope θ_{max}, here following the Moore neighbourhood nomenclature. The process continues until the above condition fails at all cells (or after a fixed number of iterations). θ_{stable} depends on the triggering mechanism, which is usually a rainstorm (Iverson, 2000) or an earthquake (Jibson, 1993).

4.1.4.1 Rainstorm (RS)-Triggered Landslide

In the case of landslides triggered by a rainstorm, the factor of safety F_S may be reduced by increasing the relative wetness c_w in the soil due to rainwater infiltration (Eq. (2.29)).

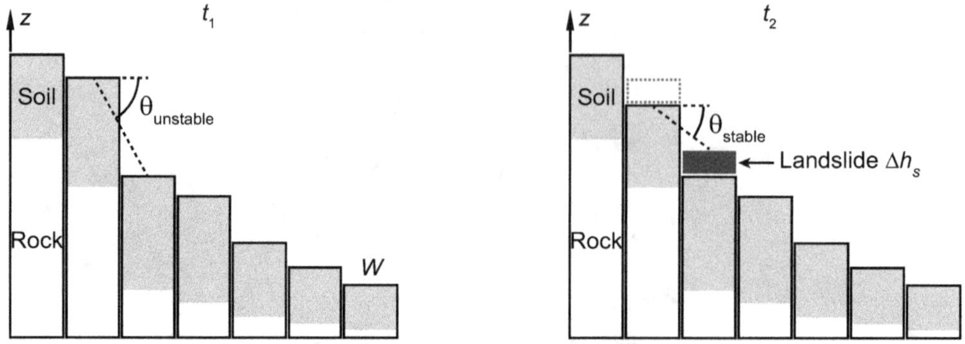

Fig. 4.7 Mass transfer rules for a landslide cellular automaton. Soil is transferred downward to the cell of highest gradient, from cells with unstable slope at the initiation phase or with critical slope at the propagation phase. The amount of displaced soil is such that the slope becomes stable.

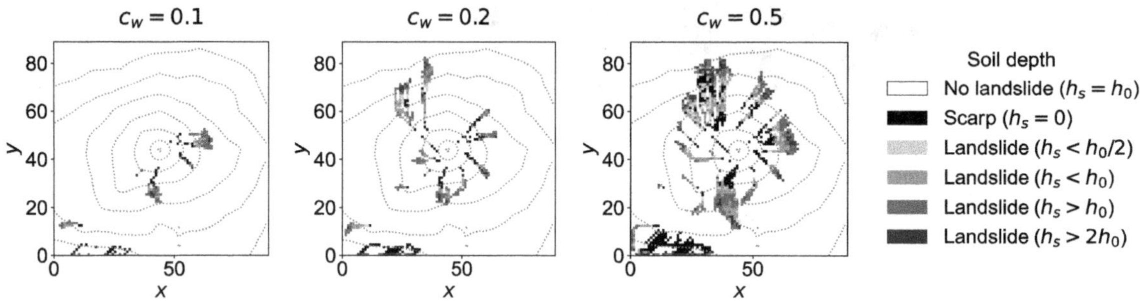

Fig. 4.8 Examples of simulated landslides triggered by rainstorms – with relative wetness $c_w = 0.1$ (large rainstorm), $c_w = 0.2$ (severe rainstorm), and $c_w = 0.5$ (extreme rainstorm), according to the toy cellular automaton. The simulated fractal topography (Fournier et al., 1982) is represented by the contour lines. See Exercise 4.2 for details about the modelling approach and parameterization. A black and white version of this figure will appear in some formats. For the colour version, please refer to the plate section.

The propagation rules previously defined are initiated for the cells that reach $F_S(x, y) < 1$ at the first time step $t = 0$ (see Figure 2.12a for the role of c_w on θ_{stable}). Any cell which is critical (i.e., $1 \leq F_S(x, y) < 1.5$) can participate to mass movement until θ_{stable} is reached everywhere. This parameter is estimated from Eq. (2.29) for the condition $F_S(x, y) = 1.5$. Examples of rainstorm-triggered landslides are shown in Figure 4.8 for the toy model defined by Eqs (4.7) and (4.8) (Ex. 4.2).

4.1.4.2 Earthquake (EQ)-Triggered Landslide

In the case of landslides triggered by an earthquake, the terrain slope stability is not only a function of the factor of safety F_S, but also of ground-shaking. This can be quantified with the **Newmark stability method** (Newmark, 1965; Jibson, 1993). One possible empirical relationship is given by Jibson (2007):

$$\log_{10} d_N(x,y) = -2.710 + \log_{10}\left[\left(1 - \frac{I_c(x,y)}{I_{EQ}(x,y)}\right)^{2.335}\left(\frac{I_c(x,y)}{I_{EQ}(x,y)}\right)^{-1.478}\right] + 0.424M, \quad (4.9)$$

with d_N [m] the Newmark displacement, M the earthquake magnitude, and I_{EQ} the matching hazard intensity footprint, defined as PGA [m/s^2] (Section 2.4.1). The parameter I_c is the critical acceleration function of the terrain characteristics, such that

$$I_c(x,y) = (F_S(x,y) - 1)g \sin\theta, \quad (4.10)$$

θ being the slope angle. The initial landslide footprint depends on d_N crossing a fixed threshold. The stable slope (leading to rest) could be calculated for the same d_N threshold but by replacing I_c with, for instance, what could be termed a stable acceleration $(F_S(x,y) - 1.5)g \sin\theta$, by making an analogy with the difference between critical and stable F_S values.

4.1.5 Wildfire (WF) Cellular Automaton

The **forest-fire model** is another classic cellular automaton model that, similarly to the sandpile model, displays a SOC behaviour (with power-law size distribution emergence). Here, a cell can be empty, occupied by a tree, or burning. The model of Drossel and Schwabl (1992), for example, is defined by four rules, which are executed simultaneously:

(1) an empty space fills with a tree with probability p_{tree} (i.e., tree growth);
(2) a tree ignites due to a lightning strike of probability p_{Li};
(3) a tree will burn if at least one neighbour is burning (i.e., fire spread);
(4) a burning cell turns into an empty cell.

An event footprint is defined by the cells that are part of a burning cluster, which consists of all the cells filled with a tree that can reach each other via nearest neighbour-relations, such as the von Neumann neighbourhood (Fig. 4.9a). The term forest fire is generic and applies to any type of fire propagation, from wildfires (WF) to urban fires (Fi). The energy transfer principle is again observed with vegetation growth (i.e., fuel accumulation) representing some energy loading, and fires representing the release of the stocked energy in the environment. The forest-fire model represents a classic example of a **percolation system**: as loading increases, clusters merge leading to events of greater size.

Although the shape of the footprints obtained from a basic forest-fire model (Fig. 4.9b) can resemble real wildfires, realistic modelling of the process requires the consideration of physical parameters including wind direction, topography, and fuel characteristics (e.g., elliptical fire spread model; Eq. (2.30); Rothermel, 1972; Anderson et al., 1982). This means that a set of local rules of interactions must be defined instead of using the simpler concept of instantaneous cell cluster. One of the simplest physics-based cellular automaton models for fire propagation consists of defining a weight for each Moore neighbour as a function of the fuel load, terrain slope, terrain aspect, wind direction, and wind intensity. After fire ignition, the probability that a firelet moves to a neighbouring cell depends on

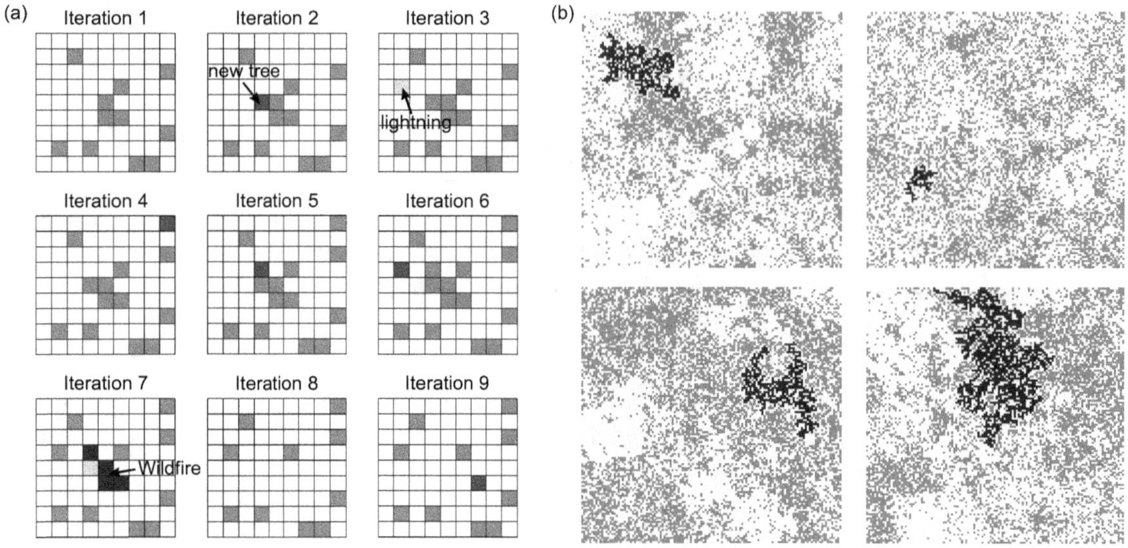

Fig. 4.9 The forest-fire model. (a) Cellular automaton rules applied for 9 iterations on a 9×9 grid. White cells are unoccupied sites; other cell states are tree, lightning strike, and burning tree. The fire at iteration 7 consumes 5 trees (event of size $S = 5$). (b) Examples of wildfire footprints (here in dark grey, in red in the colour version) simulated with the forest-fire toy model on a 128×128 lattice (event sizes $S = 493$, 103, 1458, and 527 clockwise from top left). See also Malamud and Turcotte (2000, figs. 1 and 2). A black and white version of this figure will appear in some formats. For the colour version, please refer to the plate section.

the estimated weight distribution (Clarke et al., 1994). The elliptical fire spread model may also be used to constrain the preferred direction of fire propagation in a cellular automaton (Trunfio et al., 2011).

4.1.6 Stampede (St) Agent-Based Model

Crowd dynamics, or the study of the movement of many people in respect of time, can naturally be implemented by an ABM. The process consists of simulating the trajectories of agents controlled by their decisions to move toward a goal (e.g., exit door) or to avoid a collision (e.g., with a wall or with other agents), constrained by the laws of Newtonian mechanics. In a life-threatening situation that leads to panic, such as a fire (Fi) or a terrorist attack (TA), crowd movements can provoke collisions due to many people moving fast toward the same direction or moving irrationally. The seminal model of Helbing et al. (2000) describes the dynamics of the process by the sum of socio-psychological and physical forces that drive the overall behaviour of the crowd. This is described by **Newton's second law of motion**,

$$\sum \vec{F}_i = m_i \vec{a}_i = m_i \frac{\mathrm{d}\vec{v}_i}{\mathrm{d}t} = m_i \frac{v_{0,i}\vec{e}_{0,i} - \vec{v}_i(t)}{\tau} + \sum_{j(\neq i)} \vec{F}_{ij} + \sum_{\mathcal{W}} \vec{F}_{i\mathcal{W}}, \tag{4.11}$$

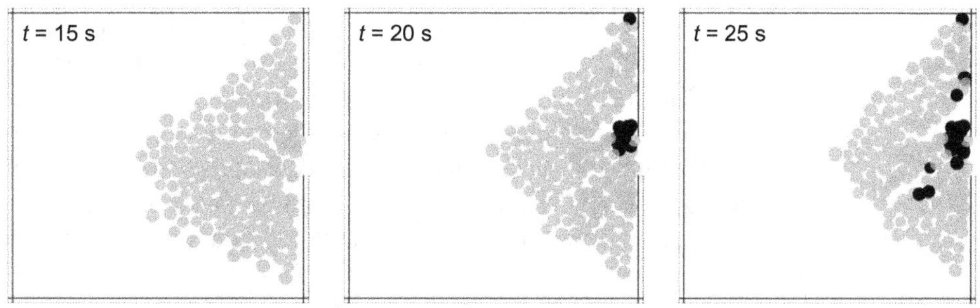

Fig. 4.10 Crowd dynamics ABM for stampede simulation, here based on the model of Helbing et al. (2000). The room is 15 m large with walls of thickness $h = 0.5$ m and a 2-m- wide exit on the right side of the room. $N = 200$ agents (in grey) panic and try to escape with velocity $v_{0,i} = 6$ m/s. Fatalities (agents who experienced a load of 6,227 N or more for at least 15 s) are represented in black. Considering the terminology of Chapter 2, hazard intensity footprints could be defined in terms of applied pressure $P(x, y)$. The images shown here are examples of loss footprints $L(x, y)$.

where m_i is the mass of the agent i, $\vec{a}_i$ is its acceleration, $\vec{v}_i$ is its velocity, and $\tau = 0.5$ s is the characteristic time for the agent to adapt its speed. The forces include a motivational force with desired velocity $v_{0,i}$ in the direction $\vec{e}_{0,i}$, interaction forces $\vec{F}_{ij}$ between agent i and the other agents j, and interaction forces $\vec{F}_{i\mathcal{W}}$ between the agent and walls $\mathcal{W}$. These interaction forces comprise a repulsion force from other agents and walls to keep some distance, as well as a body force and a sliding friction force,

$$\vec{F}_{ij} = \{c_A \exp[(r_{ij} - d_{ij})/c_B] + c_k f_{\mathcal{H}}(r_{ij} - d_{ij})\}\vec{n}_{ij} + c_\kappa f_{\mathcal{H}}(r_{ij} - d_{ij})\Delta\vec{v}^t_{ji}\vec{t}_{ij}, \tag{4.12}$$

where $c_A = 2 \times 10^3$ N, $c_B = 0.08$ m, $c_k = 1.2 \times 10^5$ kg/s, $c_\kappa = 2.4 \times 10^5$ kg/m/s, r_{ij} is the contact distance between i and j, d_{ij} is the distance between i and j, $\vec{n}_{ij} = (n^1_{ij}, n^2_{ij}) = (\vec{r}_i - \vec{r}_j)/d_{ij}$, $\vec{t}_{ij} = (-n^2_{ij}, n^1_{ij})$ the tangential direction, $\Delta\vec{v}^t_{ji} = (\vec{v}_j - \vec{v}_i) \cdot \vec{t}_{ij}$, and $f_{\mathcal{H}}(x) = x$ if $d_{ij} \le r_{ij}$ and 0 otherwise (similar rules apply for $\vec{f}_{i\mathcal{W}}$ by replacing j by $\mathcal{W}$, $v_{\mathcal{W}} = 0$, and $r_{i\mathcal{W}} = (d_i + h)/2$ with h the wall thickness) (Helbing et al., 2000). More details and model variations can be found in Helbing et al. (2002, 2005). The location of an agent at time t is defined from $d\vec{x}_i/dt = \vec{v}_i$. Results from simulations realized with this model are shown in Figure 4.10.

During a stampede, injuries and fatalities are due to crushing and trampling (Rutty et al., 2017). The level of damage can be estimated from the pressure P applied on the agent circumference by the sum of radial forces, with injuries expected for $P_{th} \ge 1{,}600$ N/m (e.g., Helbing et al., 2000). It is known that crowd movement can bend steel barriers, suggesting a pressure potentially reaching 4,450 N/m in extreme cases (e.g., Elliott and Smith, 1993). In the example shown in Figure 4.10, fatalities, represented in black, are agents who have experienced a load of 6,227 N or more for at least 15 s, causing cardiorespiratory arrest (Rutty et al., 2017). These agents become immobile, and here, further obstruct the way to the exit. In other ABMs where stampedes are combined with another peril, such as a terrorist attack, an agent's injury level can be described in terms of blood loss, which incrementally increases at each contact with another agent, and with the intensity increment increasing if collided, trampled, or attacked (Lu et al., 2020). The laws of motion described by Eq. (4.11) can model the movement of agents in any physical environment. Since such an ABM is computationally demanding, much simpler approaches may be used to model agent interactions, as shown below in the case of social unrest and later on for epidemics and pest infestations. Stampede simulations provide hazard and damage footprints for small areas and for specific contexts, such as mass gatherings. They are mostly used for risk reduction planning and not for CAT risk modelling since once such a catastrophe has occurred, the conditions for a similar event to repeat at the same place is, in most cases, mitigated.

4.1.7 Social Unrest (SU) Agent-Based Model

Social conflicts are extremely heterogeneous and varied. In terms of intensity scale, they go from simple protests and strikes, to riots, insurgence, and terrorism, to war (civil or international) (Section 1.3.8). Attention will be given here to social unrest emergence, which can be illustrated with the simple ABM proposed by Epstein (2002). Riots, which correspond to aggregate acts of violence against individuals and property, including looting and starting fires (e.g., McPhail and Wohlstein, 1983), result from the interactions between malicious agents who inflict damage and law enforcement agents who stop them. This is again an example of a complex system in which a set of simple local rules of interactions leads to an emergent behaviour.

The seminal civil-violence model of Epstein (2002) consists of two types of participants: population agents and 'cops'. Population agents can be in one of three states (quiet $\mathcal{S}_Q$, active $\mathcal{S}_A$, or jailed $\mathcal{S}_J$). Their behaviour depends on their fixed degree of grievance c_G, risk aversion c_{RA} (on the arbitrarily defined range of 0 to 1), and a certain threshold for rebellion c_{th}. Their other attributes are their location on a grid and their vision radius r_v. Cops are defined by their location and vision radius r_{v*} only. Grievance is defined as $c_G = c_H(1-c_L)$, where c_H is the perceived hardship (on the arbitrarily defined range of 0 to 1) and c_L the perceived legitimacy of the authority. Another parameter is the agent's net risk $c_{NR} = pc_{RA}$, where

$$p = 1 - \exp\left(-c\left\lfloor\frac{N_C}{N_A + 1}\right\rfloor\right) \tag{4.13}$$

is the arrest probability, which depends on the number of cops N_C and of active agents N_A (i.e., rioters) observed by the agent (c is a normalization constant). The rules are then defined as follows:

(1) move rule: move to an empty cell (or where someone is jailed);
(2) agent rule: if $(c_G - c_{NR}) > c_{th}$, become active ($\mathcal{S}_Q \rightarrow \mathcal{S}_A$), otherwise stay quiet ($\mathcal{S}_Q$);
(3) cop rule: inspect all cells within vision field and arrest a random active agent ($\mathcal{S}_A \rightarrow \mathcal{S}_J$).

For each jailed agent, their jail term is a random uniform value between 0 and a fixed number of simulation iterations $t_{J,max}$. Once released from custody, the agent becomes quiet again.

Riots, defined as clusters of active (i.e., malevolent) agents, emerge in the complex system after long periods of quiescence. A riot is triggered locally by the rule $(c_G - c_{NR}) > c_{th}$, requiring that the first individuals who act have a high level of grievance and/or a strong risk neutrality (Epstein, 2002). As quiet agents see the proportion of rioters increase compared to the number of cops, their net risk of being arrested decreases and they become active if their level of grievance is relatively high and risk aversion relatively low. This phenomenon of contagion is reminiscent of epidemics (e.g., Davies et al., 2013) (Section 4.2.2). As the riot develops, the number of agents being jailed increases, thereby stopping the riot after several iterations (Fig. 4.11a).

The process of agent interactions is marked by periods and regions of both stability (i.e., quiescence) and chaos (i.e, riots), representing a typical case of **punctuated equilibrium** in a complex system (Epstein, 2002). This is illustrated in Figure 4.11b (Ex. 4.3). Such a model can reproduce characteristics of real riots under the condition that additional variables are included, such as distance travelled to riot locations (Davies et al., 2013). Since it is challenging to estimate the levels of grievance and risk aversion for a given population,

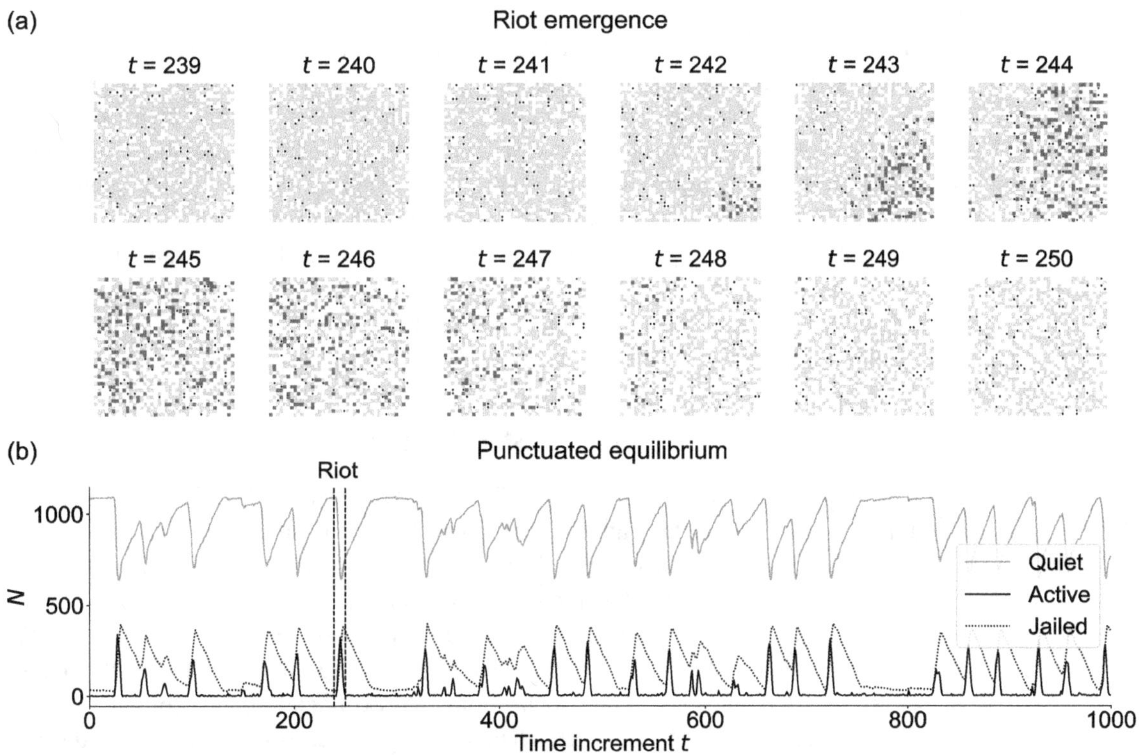

Fig. 4.11 Example of a civil-violence ABM for riot simulation, here based on the model of Epstein (2002). (a) Action maps for successive time increments showing the emergence of a riot (agents are light grey if quiet, dark grey if active; cops as smaller black points). Parameters are $c = 2.3$, $c_L = 0.82$, $r_v = r_{v*} = 7$, $c_{th} = 0.1$, $t_{J,max} = 30$, and a density of 70% for agents and 5% for cops, following Epstein (2002). (b) Punctuated equilibrium of riots, described as long periods of relative stability punctuated by outbursts of rebellious activity.

an ABM here mostly offers scenarios for crisis management strategies and not inputs for a standard CAT risk model. For the latter, historical data on riots may be used to estimate associated losses due to fires (Fi) and business interruptions (BI) (e.g, Launie, 1969; Baade et al., 2007).

4.2 System Dynamics

4.2.1 Concepts

System dynamics is a computational approach that models the non-linear behaviour of complex systems over time by using stocks, flows, and internal feedback loops (Forrester, 1961, 1969, 1973). It is grounded in **calculus**, which is the mathematical study of continuous change. In contrast to cellular automata where changes are algorithmic, system dynamics is concerned with differential and integral calculus. Clarke (2014) noted that the algorithmic approach had been favoured over system dynamics since both cellular automata and ABMs offer micro-level control of the parameters. However, models of system dynamics remain very useful to assess the overall behaviour of a system without going into the details of each interacting element. The choice of a modelling approach acting at a system level or at a disaggregated level should be guided by a cost–benefit analysis depending on the needs of the considered risk management policy (Rahmandad and Sterman, 2008). Three perils will be described in this section in the context of system dynamics: epidemics (Section 4.2.2), pest infestations (Section 4.2.3), and armed conflicts (Section 4.2.4). The first two perils will also be described by ABMs for comparison. Global warming and societal collapse models, far more complex and involving multiple perils, will be described in Chapter 5.

4.2.1.1 Stock Evolution

Stocks and **flows** are the building blocks of system dynamics models. Stocks can change only through their flows, which are the quantities added to or subtracted from a stock over time and defined as inflows and outflows, respectively (Fig. 4.12). When the total sum of all inflows to a stock is greater than the total sum of all outflows, the stock rises; when the total sum of all inflows to a stock is less than the total sum of all outflows, the stock falls; when the total sum of inflows to a stock equals the total sum of outflows, the stock remains unchanged, a state known as dynamic equilibrium. A **feedback loop** is a defining element of system dynamics, defined as a closed chain of causal connections from a stock, through a set of rules that are dependent on the level of the stock, and back again through a flow to change the stock (Meadows et al., 2004). An epidemic will be shown to be a good example of a feedback loop in Section 4.2.2. Note that the **order of a system** typically refers to the number of ODEs necessary to describe the dynamic flows within the system. For instance, it will be demonstrated shortly that the system order of a basic epidemiological model is three. In contrast, the order for complex Earth systems can be significantly higher, reaching,

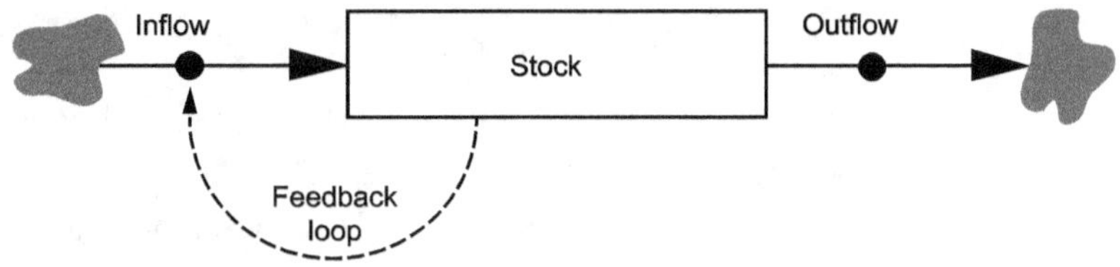

Fig. 4.12 Stocks, flows, and feedback loops in system dynamics. This type of visualization, showing how different variables in the system are causally related, is called a **causal loop diagram**.

for instance, up to 50 in the case of a low-complexity global warming model (Randers et al., 2016). These models, also including World2 (Forrester, 1973) and World3 (Meadows et al., 1972) and covered in Chapter 5, are called **Forrester-type models** after Jay Wright Forrester (see Box 5.3).

4.2.1.2 Numerical Integration

Integration is the mathematical process of calculating the area under the net flow curve, between initial and final times. Exact analytical solutions are unavailable for most dynamical systems; therefore, a numerical approach must be used instead to approximate the solution (discover some exceptions in Exercise 4.4). The simplest numerical algorithm, known as **Euler's method**, estimates the area under the net flow curve through a sequence of rectangles of identical width. The rectangle height is the opening value of the net flow applied over the interval (or time step) Δt. As the time step gets smaller, the overall numerical solution becomes more accurate. Euler's equation accumulates the successive areas of these rectangles (Eq. (4.14)) by assuming that the net flow is constant within each time interval:

$$\text{Stock}_t = \text{Stock}_{t-\Delta t} + (\text{Inflow}_{t-\Delta t} - \text{Outflow}_{t-\Delta t}) \times \Delta t \tag{4.14}$$

(Ex. 4.5). More sophisticated integration techniques exist to make smaller errors for a given step size Δt, such as the fourth-order Runge–Kutta routine (Press and Teukolsky, 1992).

4.2.2 Epidemic (Ep) Dynamics

A basic illustration of system dynamics is given by the **compartmental models** in epidemiology, where people are separated into compartments (stocks) representative of different states. The simplest model, described below, is the **susceptible–infectious–recovered (SIR) model** in which people flow between stocks as they become infected and then recover from the infectious disease (Kermack and McKendrick, 1927). Extensions of the SIR model are described by other acronyms, such as the SIS, SEIR, MSEIR, etc., which reflect the flow patterns between stocks. The additional labels E and M are for exposed (but not contagious) and maternally derived immunity, respectively (Hethcote, 2000) (see Box 4.2 for models with the fictitious label Z). The occurrence of epidemics has been

Box 4.2 **Zombie Apocalypse Modelled with a Modified SIR Model**

An apocalyptic variant of SIR is the susceptible–zombie–removed (SZR) model with its underlying assumptions based on popular zombie movies (Munz et al., 2009). The SZR model consists of susceptible people turning into zombies when bitten by one (S ⟶ Z; i.e., zombification) or dying of a natural cause (S ⟶ R), zombies which can be killed by people (Z ⟶ R), and the deceased who can be turned to zombies too (R ⟶ Z). The dead coming back to life and the removal of zombies by the healthy are key differences with real epidemiological models, making the system more complex than the SIR model. Zombie stories captivate the imagination and are, as such, ideal as instructional material (Ex. 4.5a). Many zombie analyses have followed the Munz et al. (2009) study with a list given in Alemi et al. (2015).

shown to be both cyclical and chaotic, which is again characteristic of the behaviour of a complex system (Earn et al., 2000).

4.2.2.1 The SIR Model

A widely used model of infectious disease transmission is the SIR model (Fig. 4.13), which divides a population composed of N individuals into three stocks:

- **susceptible** (S): people who have not acquired a contagious disease but who are at risk of being infected, should they encounter an infected person;
- **infectious** (I): people who have the disease and can spread it to others through contact;
- **recovered** (R): people who have become immune to the disease, and will not transmit it through contacts with other members of the population.

The model is subject to a feedback loop, with one infected individual potentially leading to a chain reaction of infections. Indeed, the **infection rate** is defined as $\beta \mathrm{SI}/N$, with β the transmission rate per infectious individual. If the number of infected people (I) increases, the infection rate also increases (at least at an early stage), which in turn leads to higher numbers of infectious cases (I). This process leads to an exponential growth process, an outbreak becoming an epidemic (and an epidemic potentially expanding into a pandemic). The parameter β depends on a number of factors – related to the probability of disease transmission per contact and to the contact rate – that include age and geographic setting. The recovery rate is defined as $\gamma \mathrm{I}$ with γ the recovery rate per infectious individual (the reciprocal $1/\gamma$ being the average infectious period). This flow transfers individuals from the infectious stock to the recovered one.

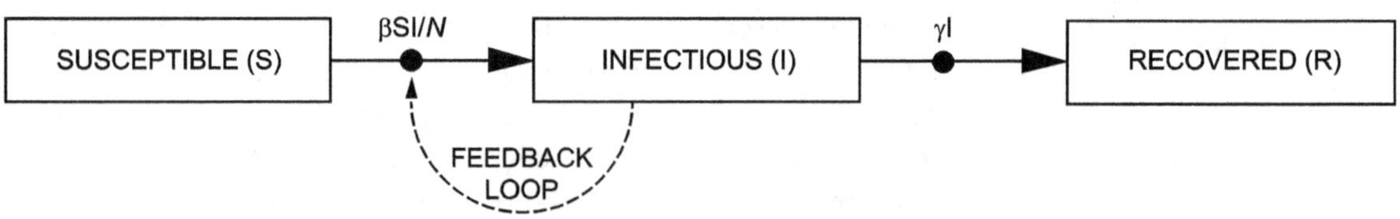

Fig. 4.13 Causal loop diagram of the SIR model.

The dynamics of the SIR model is described by the following set of ODEs:

$$\begin{cases} \frac{\mathrm{dS}}{\mathrm{d}t} = -\beta \mathrm{SI}/N \\ \frac{\mathrm{dI}}{\mathrm{d}t} = \beta \mathrm{SI}/N - \gamma \mathrm{I} \\ \frac{\mathrm{dR}}{\mathrm{d}t} = \gamma \mathrm{I} \end{cases}, \tag{4.15}$$

with $\mathrm{S}(t) + \mathrm{I}(t) + \mathrm{R}(t) = N$ (note that to distinguish stock variables from other variables in this textbook, upright capital letters (S and I) are used instead of italics (such as size S and intensity I) for clarity and ease of identification). Finally, the presence of a feedback loop is estimated from the **basic reproduction number** R_0, which is the expected number of secondary infections resulting from one infectious person in a totally susceptible population, with

$$R_0 = \frac{\beta}{\gamma}. \tag{4.16}$$

If $R_0 < 1$, the rate of infected people who recover is greater than the rate of new infections, leading to the depletion of the I stock; if $R_0 > 1$, the number of infectious people increases, leading to a self-sustained increase of the infection rate, which represents the **epidemic regime**.

The SIR model described by Eq. (4.15) can be solved by numerical integration, as described in the following pseudocode:

```
# ODE
diff_eq_SIR = func(S, I, par):
    IR = par[beta]*S*I/par[N] #infection rate
    RR = par[gamma]*I #recovery rate
    dS_dt = -IR
    dI_dt = IR - RR
    dR_dt = RR
    return (dS_dt, dI_dt, dR_dt)

# initialization
par = {N=10000, beta=2, gamma=1/2}
dt = 0.001      # tends to zero
S[0] = par[N] - 1
I[0] = 1
R[0] = 0

# Euler's method
for t in sequence(1, tmax, dt):
    dSIR = diff_eq_SIR(S[t-1], I[t-1], par)
    S[t] = S[t-1] + dSIR[1]*dt
    I[t] = I[t-1] + dSIR[2]*dt
    R[t] = R[t-1] + dSIR[3]*dt
```

The present example considers a total population $N = 10,000$ and one individual being infectious at $t = 0$. The model parameters β and γ are such that $R_0 > 1$. The results,

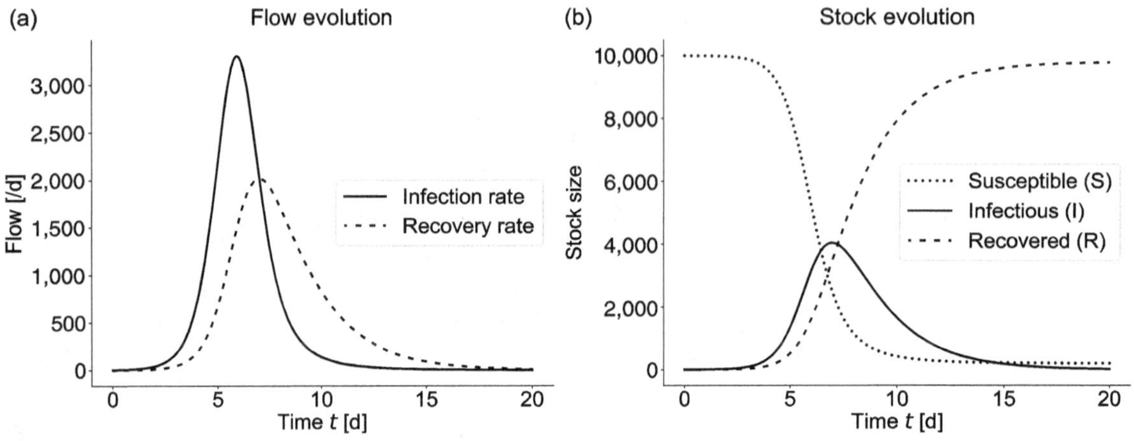

Fig. 4.14 SIR model of the spread of an infectious disease, according to Eq. (4.15) and modelled using the Euler numerical integration method (see pseudocode above).

shown in Figure 4.14, illustrate how the infection rate increases as the contagion spreads to more individuals, and eventually drops as the contagion diminishes with fewer and fewer susceptible people being available for further infection. The temporal evolution of the three stocks also gives a clear indication of the epidemic dynamics, with the larger stock shifting from the susceptible people, to the infectious people, to, finally, the people who have recovered.

The simulation of epidemics based on differential equations has some drawbacks in that it neglects the local characteristics of the spreading process. In particular, it fails to simulate in a proper way the spatial constraints linked to individual contacts (e.g., White et al., 2007). Instead, it assumes homogeneity and perfect mixing within compartments, which is unrealistic (Rahmandad and Sterman, 2008). The algorithmic approach has been proposed to improve the simulation of an epidemic, using cellular automata (Fuentes and Kuperman, 1999; White et al., 2007) and ABMs (Rahmandad and Sterman, 2008; Epstein, 2009). These models can capture different irrational behaviours and network topologies (Danon et al., 2011). A toy SIR ABM could be as simple as having N agents located in a unit grid, moving in a random uniform fashion in the range $\pm\Delta r = 0.05$ at each iteration. Each agent is in one of the three possible states S, I, or R. Two agents i and j are in contact range if within the ad hoc distance $d_{ij} = 0.03$ of each other. The dynamics are then controlled by the same β and γ values as above. Results, comparable at first-order to the system dynamics approach, are shown in Figure 4.15a. The temporal evolution of an epidemic footprint is illustrated in Figure 4.15b where several foci of infections are present at $t = 0$.

4.2.2.2 Connection to CAT Modelling

The SIR models and their variants are typically used to forecast the number of infections in the near future, for example in the next days and weeks, which is a shorter temporal horizon than the one of CAT models (see Part III on short-term forecasting). Yet a probabilistic pandemic risk model, similar in structure to the CAT models developed for

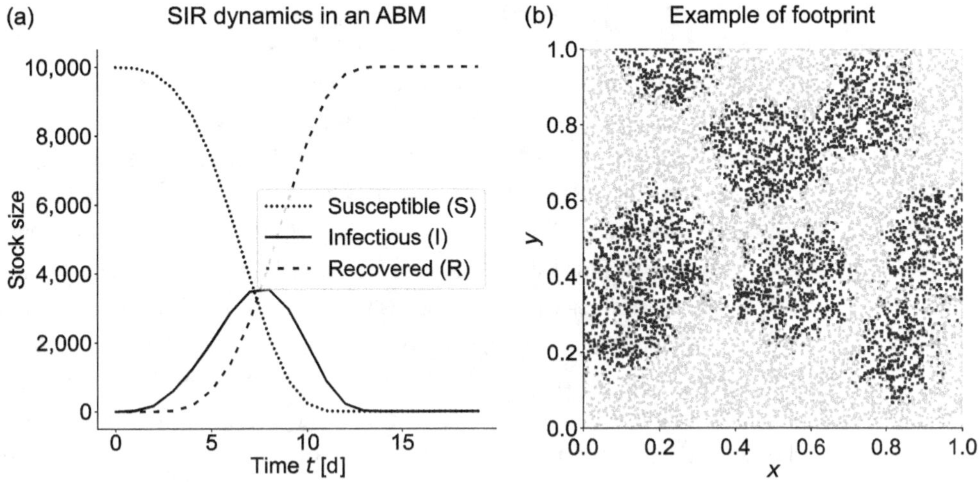

Fig. 4.15 A SIR process simulated with a toy ABM. (a) Results similar at first-order to the ones obtained by numerical integration of Eq. (4.15) (Fig. 4.14). (b) An example of hazard intensity footprint at a given time increment t, with infections in dark grey considering multiple foci of propagation.

other perils, can be built (Smith, 2017; Qiu, 2020). It first requires the modelling of a stochastic event set. Although historical data could be used to develop a size distribution, as done in Section 2.3.2 (Fig. 2.13b), the reliability of the results is degraded by the paucity of observations, and the evolution over time of both human population and its interactions with the environment (Woo, 2015). For instance, only the period 1600–1945 was previously used to fit a generalized Pareto distribution to the 1500–2020 Global Epidemics Dataset in order to limit biases due to the recent role of vaccination (Marani et al., 2021a). Moreover, in this case, the size S is equivalent to the event loss L with the size distribution directly representing an EP curve (in terms of fatalities). For improved epidemic risk modelling, a distribution of the basic reproduction number R_0 may be required to allow SIR-type simulations. By analogy to the standard CAT modelling pipeline, R_0 could be referred to as the event size S with the size distribution developed by expert judgement (Risk Management Solutions, 2007), with R_0 depending on the pathogen type, among other factors (Delamater et al., 2019). The hazard intensity footprint could then be the number of infections $\mathrm{I}(x, y)$ obtained from a SIR-like ABM. Accordingly, the exposure component would define population **cohorts** to represent groups of individuals who share similar characteristics, such as age, sex, and presence or not of comorbidities. To continue the parallel with the standard CAT risk framework, a cost function could finally be defined to estimate the total fatalities (i.e., losses L) from the **case fatality rate**, which gives the ratio of deaths to total infections.

4.2.3 Pest Infestation (PI) Dynamics

Pest infestations, such as insect outbreaks, can have dramatic consequences on crop production and cause considerable economic losses (Tonnang et al., 2017). The dynamics of

Box 4.3 Logistic Growth

The **logistic differential equation** of logistic growth is $dN/dt = RN(1 - N/K)$ with N the population count or population density, R the growth rate, and K the carrying capacity. It was originally derived by Verhulst (1838) to describe the self-limiting growth of a biological population in response to the (unconstrained) Malthusian exponential growth model $dN/dt = RN$ (Malthus, 1798). The analytical solution is the logistic function, or $N(t) = K/[1 + \{(K - N_0)/N_0\} \exp(-Rt)]$, with N_0 the initial population and $\lim_{t\to\infty} N(t) = K$. This model is at the basis of most models of population dynamics. Interestingly, the SI (susceptible–infected) epidemiological model (a simplification of the SIR model; Ex. 4.4a) is analogous to the logistic equation with $I(t) = N(t)$, $I_0 = N_0$, $N_{tot} = S + I = K$, and $\beta = R$. All solutions of first-order systems either settle down to equilibrium (i.e., **fixed points**) or diverge toward $\pm\infty$ (Strogatz, 2018). It is, however, simple in appearance only, since for different values of R, different equilibria are reached, which is illustrated by the **logistic map** described by $N_{i+1} = RN_i(1 - N_i)$ with $K = 1$ (May, 1976) (a description remains outside the scope of this textbook).

insect population growth is a function of an insect's characteristics and its interactions with the environment. Two models will be described: (1) the Lotka–Volterra predator–prey model, which is the simplest set of ODEs to describe the process of cycles of pest disturbances (Lotka, 1910; Volterra, 1928), and (2) a 'cusp catastrophe' model, which can explain potentially unanticipated extreme insect outbreaks (Ludwig et al., 1978). The logistic growth function, nested in population dynamics models, is described in Box 4.3.

4.2.3.1 The Lotka–Volterra Model

The **Lotka–Volterra model** describes the dynamics of a two-species system in which one is a predator and the other is its prey. In the context of catastrophe risk, the prey can correspond to a pest that damages crops. The set of ODEs governing the dynamics is:

$$\begin{cases} \frac{dN_1}{dt} = R_\alpha N_1 - R_\beta N_1 N_2 \\ \frac{dN_2}{dt} = R_\delta N_1 N_2 - R_\gamma N_2 \end{cases}, \tag{4.17}$$

where N_1 is the prey density (in terms of number per unit area), N_2 is the predator density, R_α is the net birth rate of the prey population, R_β is the rate at which the prey are killed due to the presence of the predator, R_δ is the birth rate of the predator population due to the presence of the prey, and R_γ is the death rate of the predator in the absence of the prey. Each equation is basically the logistic equation of population growth under resource constraint (Box 4.3). It was initially proposed by Alfred Lotka in 1910 (Lotka, 1910) and developed independently by Vito Volterra in 1926 (Volterra, 1928). Numerical integration can again be used to calculate the evolution of N_1 and N_2 over time (Ex. 4.5b). Figure 4.16a shows the resulting cyclical pattern of the two populations. Such oscillations are expected by previously noting that Eq. (4.17) describes a two-variable system (Strogatz, 2018, fig. 1.3.1). The prey population initially increases, leading to an abundance of food for the predators.

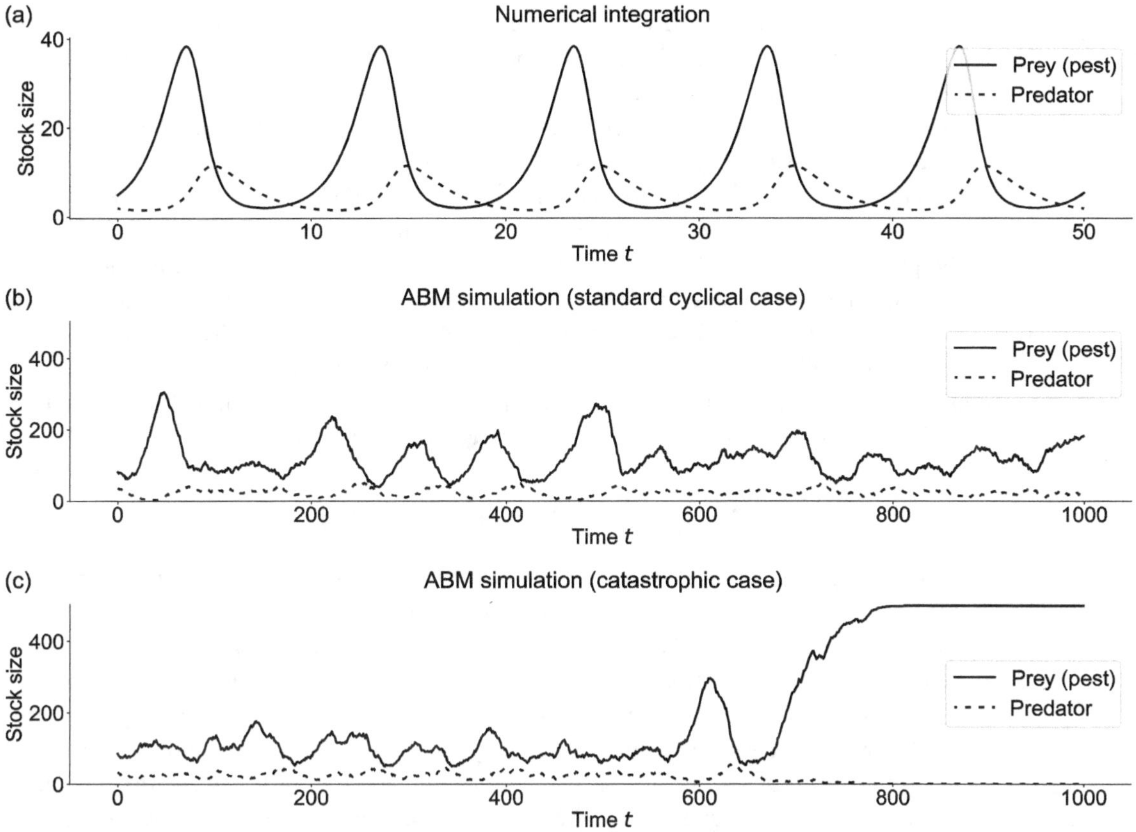

Fig. 4.16 Prey–predator population dynamics following the Lotka–Volterra ODEs (Eq. (4.17)). (a) Numerical integration with parameters $R_\alpha = 1, R_\beta = 0.2, R_\delta = 0.04, R_\gamma = 0.5$, and initial populations $N_1(t = 0) = 5$ and $N_2(t = 0) = 2$. (b) Toy ABM results with $R_\alpha = 0.1, R_\beta = 1, R_\delta = 0.5, R_\gamma = 0.1, N_1(t = 0) = 100$, and $N_2(t = 0) = 30$ (see text for further details). Notice the irregular cyclical behaviour in comparison to the results of (a). (c) Similar to (b) but showing an extreme behaviour, with the pest reaching maximum capacity when predators become extinct.

The number of predators increases in response, lagging the prey population but eventually overwhelming the prey population, which crashes. This in turn causes the predator population to crash due to lack of food, and the cycle repeats as the prey can reproduce freely. An outbreak can be said to occur when the density of prey N_1 crosses a relatively high threshold. Time series analyses of real insect populations have shown the validity of the Lotka–Volterra equations (Tonnang et al., 2009). A related model was developed by Nicholson and Bailey (1935) for the host–parasitoid system.

The insect outbreak scenario showcases the potential limits of the system dynamics approach compared to the algorithmic one in which heterogeneous, and therefore more realistic, spatial interactions are implemented. For instance, a catastrophic event, such as the abrupt increase of prey (i.e., pest) due to the extinction of predators is not expected by the simple Lotka–Volterra equations, but is possible in a cellular automaton. For example, an abnormal increase of prey may come from the formation of spatial clusters, with predators unable to easily reach the prey (Tainaka, 1994). Cellular automata and ABMs for insect pest prediction are reviewed in Tonnang et al. (2017). A very simple ABM is again sketched for illustration purposes. Agents, both prey and predators, are located in a unit grid and move in a random uniform fashion in the range $\pm\Delta r_{N_1} = 0.03$ and $\pm\Delta r_{N_2} = 0.05$ at each iteration. Two agents i and j are in contact range if within $d_{ij} = 0.05$ of each other. The parameters of Eq. (4.17) are fixed to $R_\alpha = 0.1$, $R_\beta = 1$, $R_\delta = 0.5$, and $R_\gamma = 0.1$. Initial populations are fixed to $N_1(t = 0) = 100$ and $N_2(t = 0) = 30$. To avoid an infinite growth of the prey in the case of predator extinction, a carrying capacity $K = 500$ is added for N_1. Figures 4.16b and c show the results of two ABM simulations, the first showing the typical cyclical behaviour of the Lotka–Volterra equation system and the second the extreme case of pest explosive growth after predator extinction.

4.2.3.2 Extreme Insect Outbreak (Cusp Catastrophe Example)

A more complex dynamical system, described in the seminal study by Ludwig et al. (1978), is the collapse of a forest system due to spruce budworm outbreaks. These insect infestations recur in eastern Canadian forests, leading to major financial disasters in forestry. The system dynamics is reminiscent of a predator–prey model but with fast and slow variables, and multiple levels of predation. This still remains a very simple dynamical system compared to any real ecosystem composed of many interlinked species. The model is described by a set of three ODEs that describes the temporal evolution of the budworm population density N_B, tree size S, and energy reserve E (representing foliage and tree health). During an outbreak, S and E decline sharply. For the sake of simplicity, only the first ODE will be investigated, which is

$$\frac{dN_B}{dt} = R_B N_B \left(1 - \frac{N_B}{K_B}\right) - c_\beta \frac{N_B^2}{c_\alpha^2 + N_B^2}, \tag{4.18}$$

with R_B the growth rate, K_B the carrying capacity, and c_α and c_β some factors. The first term corresponds to the logistic growth function for the budworm population (in the absence of predation) and the second term to the effect of avian predation, which has an S-shaped functional response (Fig. 4.17a). This second term indicates that predation starts slowly as the birds show little interest in a scarce resource of budworms while predation saturates when birds cannot keep up with high numbers of budworms. This leads to a highly nonlinear behaviour with a bifurcation point where the budworm density suddenly jumps to a significantly higher value. Note that a **bifurcation** is defined as an abrupt change in the system from a fixed point to another one (or to divergence) (Strogatz, 2018). This process, described below, was termed '**cusp catastrophe**' in the catastrophe theory of the 1970s (Box 4.4). The equilibrium solutions (fixed points) of Eq. (4.18) are obtained for the condition $dN_B/dt = 0$, which yields

$$R^* \left(1 - \frac{N^*}{K^*}\right) = \frac{N^*}{1 + N^{*2}} \tag{4.19}$$

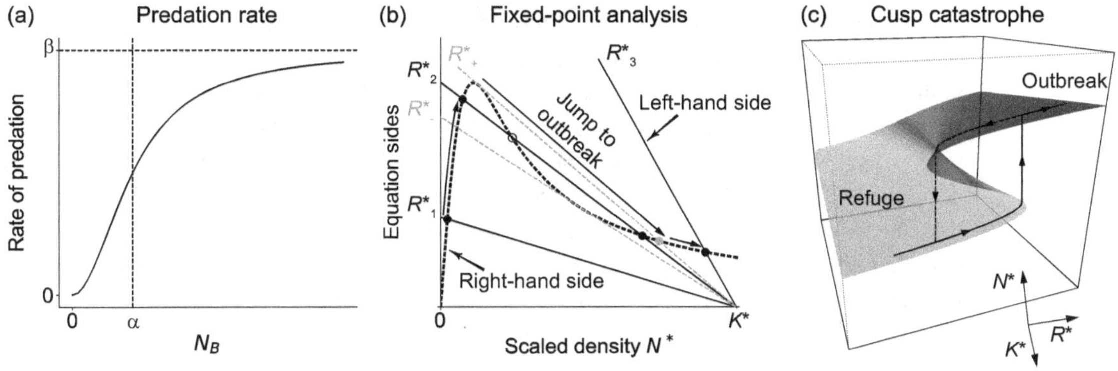

Fig. 4.17 Spruce budworm outbreak as a cusp catastrophe, according to the seminal Ludwig et al. (1978) study. (a) Avian predation sigmoid function of the budworm population density N_B (second term of Eq. 4.18)). (b) An analysis of the fixed points in Eq. (4.19) with N^* the scaled budworm density. The scaled carrying capacity K^* is fixed to a value high enough to observe the bifurcation, while R^* increases from R_1^* to R_3^*. Solutions to Eq. (4.19) correspond to the points at which the line and the curve intersect. The system has three roots for $R_-^* < R^* < R_+^*$ (read text for details). (c) Cusp catastrophe with the trajectory of N^* over time illustrating the phenomenon of **hysteresis** where the system's path depends on its history, leading to non-reversibility (for R^* first increasing and then decreasing, and K^* constant). Jumps occur at R_+^* and R_-^*. Cusp surface generated using the function `cusp3d.surface()` of the cusp R package (Grasman et al., 2009).

Box 4.4 Catastrophe Theory

Catastrophe theory (Thom, 1975) constitutes an attempt to classify bifurcation patterns, such as the cusp singularity (Whitney, 1955; Fig. 4.17c), into different function families, the term catastrophe referring to the abrupt jump of state variables at singular points. Very popular in the 1970s, it was originally considered a promising technique for the modelling of complex systems, including stock market crashes, fishery collapses, and prison riots (Rosser, 2007, and references therein). However, the theory's proponents made exaggerated claims about its universality and applicability (Zeeman, 1976), which led to a backlash and its rapid demise (Zahler and Sussmann, 1977). It later only survived within mathematics as a special case of bifurcation theory. Despite this, the term 'cusp catastrophe' remained in use, for example to illustrate the dynamics of insect outbreaks (e.g., Strogatz, 2018, following Ludwig et al., 1978). In a review, Rosser (2007) showed that many recent models actually resemble catastrophe theory, showing that it still has some potential.

in dimensionless form, with $N^* = N_B/c_\alpha$ the scaled budworm density, $R^* = c_\alpha R_B/c_\beta$, and $K^* = K_B/c_\alpha$. Figure 4.17b shows the dynamics of the system by plotting the right- and left-hand sides of Eq. (4.19) as a function of N^*. The solutions, or roots, are simply their points of intersection. As R^* increases from zero, there is at first only one root monotonically increasing until R_-^* is reached. An increase of R^* could be due to a long-term increase of c_α, representing the idea that birds would need to see more budworms to eat them as the overall surface of the forest increases. Above R_-^*, the system has three roots with the middle one an unstable one (read, e.g., Strogatz (2018, p. 77) on how to distinguish stable fixed points from unstable ones). As R^* continues to increase and reaches R^*_+, the first two roots collide and annihilate (i.e., **saddle-node bifurcation**), sending the scaled insect density N^* to the third fixed point, which represents the catastrophic outbreak. A characteristic of the cusp catastrophe is the irreversibility of the process due to the presence of a hysteresis loop (Fig. 4.17c). Once the system is at the high equilibrium, reducing R^* would only slightly decrease N^*, and R^*_- must be reached to jump back to the lower equilibrium. This example illustrates the complexity of ecological systems and the potential challenges in mitigating potential ecological risks. A detailed analysis of the Ludwig et al. (1978) model can be found in Strogatz (2018, section 3.7).

4.2.3.3 Connection to CAT Modelling

A framework for calculating economic losses in crop production is linking a pest model (as defined above) to a crop productivity model to estimate the impact of pest damage on the production yield (the loss L) (e.g., van Wijk et al., 2009; Reynolds et al., 2018; Rasche and Taylor, 2019). This ideally requires knowledge of the different levels of interactions in the complex agricultural ecosystem. This is, however, very challenging, as proven by the previous forest–budworm–bird model. Such a system-level approach has yet to be fully developed, with existing pest–crop interaction models limited to parts of an ecosystem (Tonnang et al., 2017; Reynolds et al., 2018; Rasche and Taylor, 2019). Note that damage is estimated by the ratio of biomass eaten to available plant biomass, with feeding rates depending on the species involved. Another challenge in CAT risk modelling is the calibration of pest model results to historical observations of crop damage for proper probabilistic assessment (Chivers et al., 2014).

4.2.4 Armed Conflict (AC) Dynamics

A war is a cumulation of attacks and counterattacks, whose dynamics can be explained with game theory (e.g., Kress, 2012). Although highly complex and heterogeneous in nature, some basic rules will be mentioned here for completeness. The simplest model of attrition warfare is a set of ODEs, called **Lanchester equations** after Frederick William Lanchester (Lanchester, 1916). It is defined as

$$\begin{cases} \frac{dN_A}{dt} = -c_\beta N_B; \quad N_A > 0 \\ \frac{dN_B}{dt} = -c_\alpha N_A; \quad N_B > 0 \end{cases}, \tag{4.20}$$

where N_A is the number of soldiers in army A, each with offensive fire power c_α (i.e., number of enemy soldiers killed per soldier from A), and N_B is the number of enemy soldiers in army B, each with offensive fire power c_β. This represents an **attrition** strategy since each army aims to gradually wear down the enemy's number of soldiers through sustained engagement. Note that M. Osipov had independently developed a similar model by 1915 (Helmbold and Rehm, 1995). A common modification made to Eq. (4.20) is the addition of a positive term representing reinforcement troops being introduced over time (Engel, 1954). Solving Eq. (4.20) indicates that the effectiveness of an army rises proportionally to the square of the number of its soldiers, but only linearly with their fighting ability (the so-called **N-square Law**; Lanchester, 1916; Ex. 4.4b). The Lanchester model has been validated from actual combat data (Engel, 1954), while variants have been proposed to describe the dynamics of additional types of battles (MacKay, 2015; Kress, 2020). Compared to ODEs, ABMs provide the advantage of being able to include spatial variations of forces, as well as decision-making and psychology (Ilachinski, 2000). Agents are here combat units with a mission and situational awareness, among other characteristics. Their possible states are alive, injured, or killed. The battlefield is the geographical lattice. However, the rules would be too numerous to list here (Epstein and Axtell, 1996; Ilachinski, 2000) – for a related model, see Section 4.1.7. The hazard footprint of an armed conflict would, in theory, be the sum of a heterogeneous set of sub-footprints (e.g., explosions, agents' individual acts of violence, fires, critical infrastructure sabotages).

To the best of the author's knowledge, there is so far no CAT model per se for armed conflicts, primarily due to the war exclusion clause in insurance policies. Such a clause is defined when insurance firms cannot properly assess the risk (e.g., complexity of a war's dynamics) and when the expected damage is potentially colossal (e.g., total destruction of cities) (Simon, 1981). War games are however common in military strategy developments (Kress, 2012). Some aspects of game theory for decision-making will be covered in Part III.

4.3 Concluding Remarks

In this chapter, we have explored two central computational methods for modelling the dynamics of catastrophes: the algorithmic approach, which encompasses cellular automata and agent-based models (Section 4.1), and the system dynamics method, involving the numerical integration of ordinary differential equations (Section 4.2). While both of these methods possess the capability to address similar challenges, their distinct purposes have become evident through our exploration. Within the context of CAT modelling, we have demonstrated how cellular automata excel at simulating dynamic hazard intensity footprints, denoted as $I(x, y, t)$. ABMs, while also capable, come with a higher computational cost due to their high-resolution nature. In contrast, system dynamics offers a different perspective by exclusively modelling the temporal evolution of an intensity metric for the entire system. This approach has nevertheless been proven to be sufficient for loss assessment in specific cases (e.g., the fatality count can be derived from the number of infections during an epidemic even in the absence of detailed spatial information). The present chapter was complementary to Chapter 2 in terms of perils covered, in order of appearance: blackouts, floods, landslides, wildfires, stampedes, social unrest, epidemics, pest infestations, and armed conflicts. As we progress to the next chapter, we will expand our understanding to extreme catastrophes resulting from event interactions and environmental feedback loops.

Further recommended reading

- *How Nature Works* (Bak, 1996)

- *Growing Artificial Societies* (Epstein and Axtell, 1996)
- *Nonlinear Dynamics and Chaos* (2nd ed.) (Strogatz, 2018)
- *Introduction to the Theory of Complex Systems* (Thurner et al., 2018)

4.4 Exercises

Most of this chapter's exercises require the use of the codes provided in the online exercise material folder. Go to `exercise_notebook.ipynb` to access the main codes, listed by exercise number.

Exercise 4.1

(a) The sketch depicted in Figure 4.18 shows the current state of a sandpile (Eq. (4.1); Fig. 4.2).

First, as a preliminary step, draw the next iteration(s) until the 'avalanche' stops. Use the von Neumann neighbourhood and $z_c = 4$. If you cannot make a three-dimensional sketch, simply write the height values $z(x, y)$ (as shown at the time step $t = 0$).

Second, complete the class `gen_CA_sandpile` to automatize the process. Then run it to generate a generic sandpile. You should observe examples of avalanche footprints in the folder named sandpile.

(b) The modified sandpile of Sachtjen et al. (2000) to model generic blackouts is now considered (Section 4.1.2). First, complete the loop by using Eq. (4.3). Second, find a simple way to prove that very large cascades (i.e., leading to a blackout in more than half of the power grid) require the system to be highly loaded.

Exercise 4.2

The toy landslide cellular automaton described in Section 4.1.4 is now considered. First, complete the `# PROPAGATION PHASE` loop by using Eqs (4.7) and (4.8), then run the cellular automaton by following these three steps for the rain-induced landslide case, and, finally, summarize your observations.

(1) Generate a fractal topography with constant soil depth $h_{cst} = 5$ m with the `diamondsquare()` algorithm (Fournier et al., 1982).
(2) Run the landslide cellular automaton a first time to erode the pristine soil. The new topography $z(x, y)$ and soil depth $h(x, y)$ will automatically be saved.
(3) Rerun the landslide cellular automaton on the eroded topography and test increasing values $w = (0.1, 0.2, 0.5)$ to represent storms of increasing rainfall.

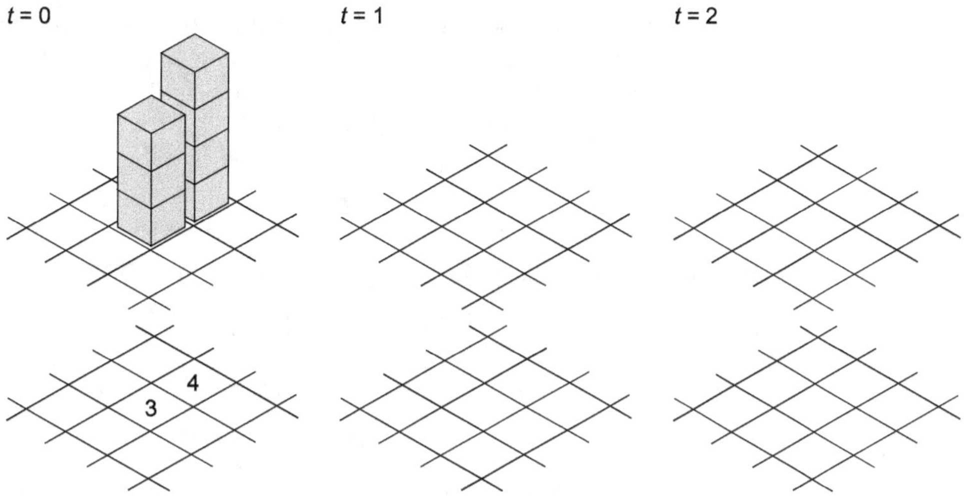

Fig. 4.18 Exercise 4.1a: Sandpile iterations. Initial state $z(x, y)$ at $t = 0$ and empty lattices to be filled for the next iteration(s) of a sandpile.

Exercise 4.3

In this exercise, the social unrest ABM model of Epstein (2002) is considered. First, fill the list of input parameters and complete the loop by using the information contained in Section 4.1.7 (*Hint:* the operator $\lfloor \cdot \rfloor$ is the floor function). Then, run the code and observe the punctuated equilibrium of riots by defining a riot as at least 100 active agents. Finally, change one of the input parameters so that no riot can occur anymore. Verify on a plot that your choice has worked and explain the feasibility of this choice in a realistic situation.

Exercise 4.4

(a) Demonstrate that the number of infections increases exponentially in the SI model described by the ODE

$$\frac{\mathrm{dI}}{\mathrm{d}t} = \beta \mathrm{IS}/N,$$

where I is the number of infections (I_0 at time $t = 0$), S the number of susceptible people, β the transmission rate per infectious individual, and N the total number of people. Prove also that everyone gets infected in this model.

Hints: Compared to the SIR model, nobody recovers (i.e., R $= 0$). For the integration, consider using the **partial fraction expansion method** by finding A and B in $1/(XY) = A/X + B/Y$. Your result should be equivalent to the solution of the logistic equation (Box 4.3).

(b) Solve the Lanchester ODE set (Eq. (4.20)) to prove that the effectiveness of an army rises proportionally to the square of the number of its soldiers, but only linearly with their fighting ability. What is the condition for army B to lose? From that relationship, finally estimate how many $N_{A,0}$ soldiers army A must contain to win the battle against army B composed of $N_{B,0} = 1{,}000$ soldiers, knowing that their respective offensive firepowers are $c_\alpha = 2$ and $c_\beta = 1$ kill(s)/soldier.

Exercise 4.5

(a) The zombie SZR model is defined by the following set of ODEs:

$$\begin{cases} \frac{\mathrm{dS}}{\mathrm{d}_t} = c_\Pi - c_\beta \mathrm{SZ} - c_\delta \mathrm{S} \\ \frac{\mathrm{dZ}}{\mathrm{d}_t} = c_\beta \mathrm{SZ} + c_\zeta \mathrm{R} - c_\alpha \mathrm{SZ} \\ \frac{\mathrm{dR}}{\mathrm{d}_t} = c_\delta \mathrm{S} + c_\alpha \mathrm{SZ} - c_\zeta \mathrm{R} \end{cases},$$

where S, Z and R stand for susceptible (people), zombie (living dead), and removed (dead), respectively. People have a birth rate c_Π, die from natural causes (parameter c_δ), or become a zombie by being bitten by one (parameter c_β). Zombies can be killed (by having their brain destroyed, parameter c_α). Finally, the dead can resurrect as zombies (parameter c_ζ) (Munz et al., 2009).

First, draw the SZR causal loop diagram.

Second, simulate and plot the dynamics of the model assuming the following parameters: $c_\alpha = 0.005$, $c_\beta = 0.0095$, $c_\zeta = 0.0001$, $c_\delta = 0.0001$, and, for population replacement, $c_\Pi = c_\delta \mathrm{S}$ (Munz et al., 2009). What can you conclude?

Hint: Use Euler's method by following a structure similar to the one shown in the pseudocode provided for the SIR model (Section 4.2.2).

(b) Simulate and plot the dynamics of a predator and prey system described by the Lotka–Volterra model (Eq. (4.17); Section 4.2.3) using Euler's method to reproduce Figure 4.16a.

5 Modelling of Cascading Effects

Extreme events were previously categorized as low-probability–high-consequence occurrences, and classified in Chapter 1 into three primary categories: tail events, meta-events, and speculative events (the last falls beyond the scope of this introductory textbook). Tail events represent the largest and rarest disasters found within the power-law size distributions discussed in Chapter 2. While they are examined in the stress testing of highly critical infrastructures (Argyroudis et al., 2020), they are generally regarded as too infrequent for inclusion in standard CAT models, not to mention mega-events, which can potentially lead to societal collapse. Extreme consequences can also arise from more common, and initially less severe, events when they cluster together. These temporally correlated events often exhibit overlapping footprints, and could be merged into meta-events representative of greater catastrophes due to the aggregation and amplification of losses. This phenomenon of event escalation is also sometimes referred to as a '**super-catastrophe**' (or 'super-cat') in CAT risk terminology (Grossi and Muir-Wood, 2006; Mignan et al., 2022b). Chapter 1 already reviewed numerous historical examples that illustrated a wide range of interaction patterns. Although secondary perils such as storm surges, tsunamis, and landslides triggered by rainstorms or earthquakes were modelled in Chapters 2 and 4, these events have so far been treated as individual entities. The purpose of this chapter is to explore the dynamics of hazard and risk interdependencies within CAT models.

While random co-occurrences of multiple events are infrequent, the triggering of one event by another is relatively commonplace. These chains of events can cross through the natural, technological, and socio-economic systems, posing a formidable challenge for their full characterization (Mignan and Wang, 2020). For instance, consider the 2011 Tohoku earthquake, which not only generated a devastating tsunami but also precipitated the Fukushima nuclear disaster (Norio et al., 2011). Remarkably, this event, on the opposite side of the globe, influenced Germany's nuclear energy policy (Wittneben, 2012). Other infamous occurrences categorized as super-cats are the 1906 San Francisco earthquake and 2005's Hurricane Katrina (Grossi and Muir-Wood, 2006). The dynamic interplay of these events can result in loss amplification, where losses accumulate non-linearly, leading to unexpected and far-reaching consequences. Moreover, current trends in the complex Earth system, including global warming and resource depletion, serve as additional regional to global stressors that increase the frequency, intensity, and clustering of catastrophes (Knutson et al., 2020; Fan et al., 2021; Brozovic, 2023).

The various approaches available for incorporating interdependencies within the CAT modelling pipeline will be covered in Section 5.1. Section 5.2 will focus on exploring domino effects, while Section 5.3 will examine the impact of complex Earth system feedback loops, and more generally, compound effects. The catastrophes considered in this chapter are storm clustering, critical network failures resulting from natural events, and other relevant scenarios. Several complex Earth system models will also be employed, including the Earth System Climate Interpretable Model (ESCIMO) global warming model (Randers et al., 2016), as well as the World2 and World3 models that address societal collapse (Meadows et al., 2004). It is essential to recognize that while the processes discussed in this chapter can indeed lead to extreme losses, the consequences may also be much milder but still relatively high compared to the events previously covered in this textbook.

5.1 Interdependencies in CAT Models

The implementation of interdependencies in CAT models remains relatively rare, which can be due to a number of factors: lack of understanding of interacting processes, insufficient number of data for robust calibration, or simply problems related to lack of insurability (e.g., extremes beyond the capacity of reinsurance firms, meta-events too blurry to be defined in insurance policies). Yet, *it is known that indirect losses caused by cascading effects can be greater than the direct losses induced by the initial event* (e.g., Mignan, 2022b). Recent Sigma reports, for example, indicate the important role of secondary perils in total loss estimates (Swiss Re Institute, 2021) and losses being amplified by time-dependent drivers such as climate change (Swiss Re Institute, 2020), but also ageing structures and inflation (Swiss Re Institute, 2023).

Three types of interdependencies will now be covered: statistical clustering (Section 5.1.1), chains of events due to cascading effects (Section 5.1.2), and other time-dependent processes (Section 5.1.3).

5.1.1 Alternative to the Poisson Distribution

Let's recall that the Poisson distribution (Eq. (2.1)) is the statistical distribution used to model counts of discrete events when events are assumed to be independent. It is found nested in both the hazard curve (Eq. (2.3)) and EP curve (Eq. (3.22)), and is used to sample the number of events per year in the simulation table (Section 2.1.1). Since the probability of observing both events i and j in the same year is $p_{i,j} = \mathbb{P}(i \cap j) = p_i \times p_j$, and since hazardous events remain relatively rare in a given region ($p_i \ll 1$), the likelihood $p_{i,j}$ of a random co-occurrence of independent events is slight at best. In a simulation table, this translates to very few simulation years containing more than one event. A recent example of two coinciding events in a narrow spatiotemporal window is the case of the 2021 Haiti earthquake and Tropical Storm Grace, which occurred within 2 days of each other. In order to model event dependency in a CAT model, the simplest approach is to replace the Poisson distribution with the negative binomial distribution, such as in the case of storm clustering (Mailier et al., 2006) (Fig. 5.1).

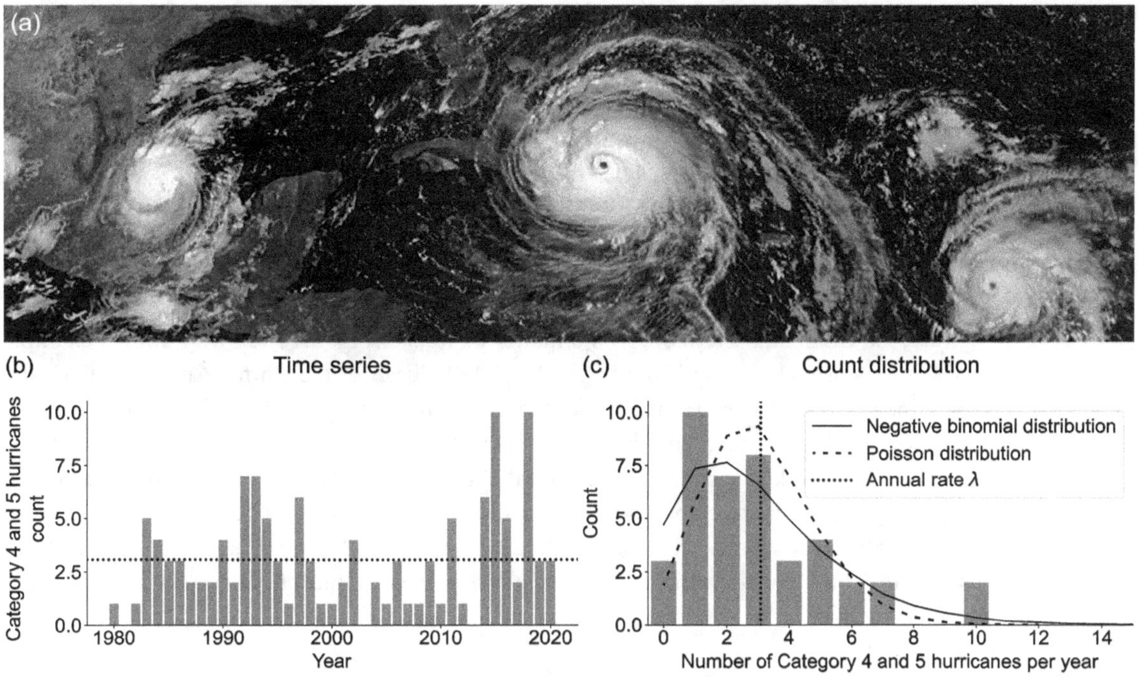

Fig. 5.1 Statistical clustering modelled with a negative binomial distribution. (a) A glimpse into the intense 2017 North Atlantic hurricane season, revealing the clustering of Tropical Storm Katia, Hurricane Irma, and the impending Hurricane Jose in early September. Hurricane Harvey had already made landfall in August, while Hurricane Maria had yet to form in late September (NOAA, public domain). (b) Time series depicting Category 4 and 5 hurricanes observed in the eastern Pacific basin from 1980 to 2020, using data from IBTrACS (Knapp et al., 2010, 2018) (events with a maximum windspeed $v_{max} \geq 58$ m/s). (c) Event count distribution and the corresponding Poisson and negative binomial distributions with parameters $\lambda = 3.1$ and $\sigma^2 = 5.9$. The negative binomial distribution provides a better fit to the data than the Poisson distribution, as demonstrated in Exercise 5.1b.

5.1.1.1 Negative Binomial Distribution

The **negative binomial distribution**, of which the Poisson distribution is a special case (Ex. 5.1a), is a discrete probability distribution of the form

$$\mathbb{P}(X = k) = \frac{(k + r - 1)!}{(r - 1)!k!}(1 - p)^k p^r, \tag{5.1}$$

where X is the random variable of count values. The parameters $r > 0$ (not to be confused with the radius in other sections) and $0 \leq p \leq 1$ are related to the mean rate and variance with $\lambda = r(1-p)/p$ and $\sigma^2 = r(1-p)/p^2$, respectively. The difference between the Poisson distribution and the negative binomial distribution is illustrated in Figure 5.1 in the case of storms. The implementation of clustering in a CAT model is straightforward when using the simulation approach. It consists of replacing the Poisson distribution with the negative binomial distribution when estimating the number of events per year k in the YLT. In the pseudocode of Box 3.5, one simply changes `k = Poisson_sample(rate_tot, Nsim)` to `k = NegativeBinomial_sample(rate_tot, variance, Nsim)` (Ex. 5.1b).

5.1.1.2 Application to Storm (TC, xC) Clustering

Statistical analysis of storm occurrences shows that they are not random, but clustered in time, with periods of elevated activity followed by relative quiescence. This has been demonstrated for both tropical and extratropical cyclones (Mumby et al., 2011; Mailier et al., 2006). In 2017, a cluster of Category 4 and 5 hurricanes, Harvey, Irma, and Maria, struck the US southeastern coast in a matter of weeks (Fig. 5.1a). A well-known example of winter storm clustering is the December 1999 cluster of the storms Anatol, Lothar, and Martin, which cost an estimated EUR 18.5 billion in combined economic damage (Munich Re, 2002). The two main possible mechanisms for serial clustering are individual cyclone tracks being similarly affected by temporal changes in the large-scale atmospheric flow or offspring being generated by one parent cyclone (Mailier et al., 2006). This clustering can be described statistically with the negative binomial distribution (Eq. (5.1)). A **dispersion statistic** ψ based on the variance-to-mean ratio of event counts provides a measure of storm clustering, with

$$\psi = \frac{\sigma^2}{\lambda} - 1 = \phi_d - 1, \tag{5.2}$$

where ϕ_d is the **index of dispersion** (Mailier et al., 2006; Vitolo et al., 2009). The condition $\psi > 0$, when the variance exceeds the mean, represents a clustered pattern (i.e., overdispersion), whereas $\psi < 0$ would represent a regular pattern (underdispersion) (Ex. 5.1a). For example, the Category 4 and 5 hurricanes shown in Figure 5.1b, which occurred in the eastern Pacific basin between 1980 and 2020, exhibit a notable clustering pattern with $\psi = 0.91$ (Ex. 5.1b). The concept of overdispersion becomes evident in Figure 5.1c as it illustrates the larger variance associated with the negative binomial distribution when compared to the Poisson distribution. It is important to emphasize that characterizing clustering using statistical distributions is a viable approach when dealing with phenomena such as storms, given the inherent stochastic nature of their sources (Section 2.2). However, this approach does not extend to phenomena such as earthquake clustering, where the location of significant aftershocks is physically constrained by the mainshock's location. This constraint also applies to scenarios involving multi-peril clusters, where the sequence, location, and order of events need to be precisely specified. To address such situations, the development of algorithms is necessary to model interconnections between events. These algorithms may involve triggering rules to define cascading effects (Section 5.1.2) or other rules aimed at adjusting hazard and risk parameters over time in response to the occurrence of preceding events (Section 5.1.3).

5.1.1.3 OEP and AEP Curves from the YLT

Event clustering has an impact on aggregate annual losses, which can be assessed by comparing the OEP and AEP curves, terms which have already been defined in Section 3.3.2.

Here, both risk metrics are calculated from the YLT, with the probability that a given loss level L_i is exceeded in a year being

$$\mathrm{EP}(L_i) = \mathbb{P}(X > L_i) = \frac{N_{sim} - i}{N_{sim}}, \tag{5.3}$$

with N_{sim} the number of simulations, and $1 \leq i \leq N_{sim}$ the simulation ID in a YLT ordered by ascending losses L_i (see Eq. (3.22) to calculate the EP (OEP) curve from an ELT). Recall that the OEP is the probability that the maximum event loss in a year exceeds a given level while the AEP is the probability that it is the sum of event losses that exceeds that level. It follows that, for the OEP, L_i is defined as the maximum event loss in year i. Conversely, for the AEP, L_i is the sum of all event losses in year i. Note also that the area under the AEP is equal to the AAL. Figure 5.2 presents a comparison of the OEP and AEP curves for two YLTs generated from the ELT of Chapter 3 (Table 3.9; Ex. 5.1c). These two YLTs differ in the nature of event occurrences: one follows the Poisson distribution with independent events, while the other exhibits clustering according to the negative binomial distribution, with the dispersion parameter $\psi = 0.91$. A comparison is also made with the basic EP curve directly calculated from the ELT. The figure illustrates the equivalence between the EP and Poisson OEP curves (Fig. 5.2a). In contrast, the AEP curves provide insights into the aggregated losses resulting from multiple events occurring within a single year. This accounts for the extension of the AEP curves towards higher loss values. In the case of the Poisson YLT, these losses reflect the effects of random coinciding events, whereas in the negative binomial YLT, they indicate clustered event occurrences. This distinction clarifies the observed **tail fattening** of the negative binomial AEP curve compared to the Poisson AEP curve, indicating a higher loss given an exceedance probability or a higher probability given a loss threshold for the clustered disaster scenario (Fig. 5.2b).

5.1.2 Chain-of-Events Encoding

To effectively model chains of events, it is first essential to grasp two crucial theories: Markov theory and graph theory.

5.1.2.1 Markov Theory

A chain-of-$(k+1)$ events $x_1 \to x_2 \to \cdots \to x_{k+1}$ represents a cascading effect where a first event x_1 occurs and triggers a second event x_2, which itself triggers a third event x_3, etc., up to x_{k+1}. For $k+1$ events, there are k links causing up to kth-order effects (e.g., Rinaldi et al., 2001). Considering n possible events (i.e., states), all the links potentially existing between these events can be encoded in an $n \times n$ matrix. This matrix can be formulated in different ways: the adjacency matrix **A** in graph theory, and the transition matrix **P** or transition-rate matrix **Q** in Markov theory (Box 5.1). In the simplest case where an event depends only on the previous one, the process is called a **Markov chain** (after Russian mathematician Andrey Markov, 1856–1922). The **Markov property** is the assumption that $\mathbb{P}(x_t|x_{t-1}, \cdots, x_1, x_0) = \mathbb{P}(x_t|x_{t-1})$. A Markov chain with memory (of jth order) satisfies $\mathbb{P}(x_t|x_{t-1}, \cdots, x_1, x_0) = \mathbb{P}(x_t|x_{t-1}, \cdots, x_{t-j})$ with $j < t$. A process that depends on all

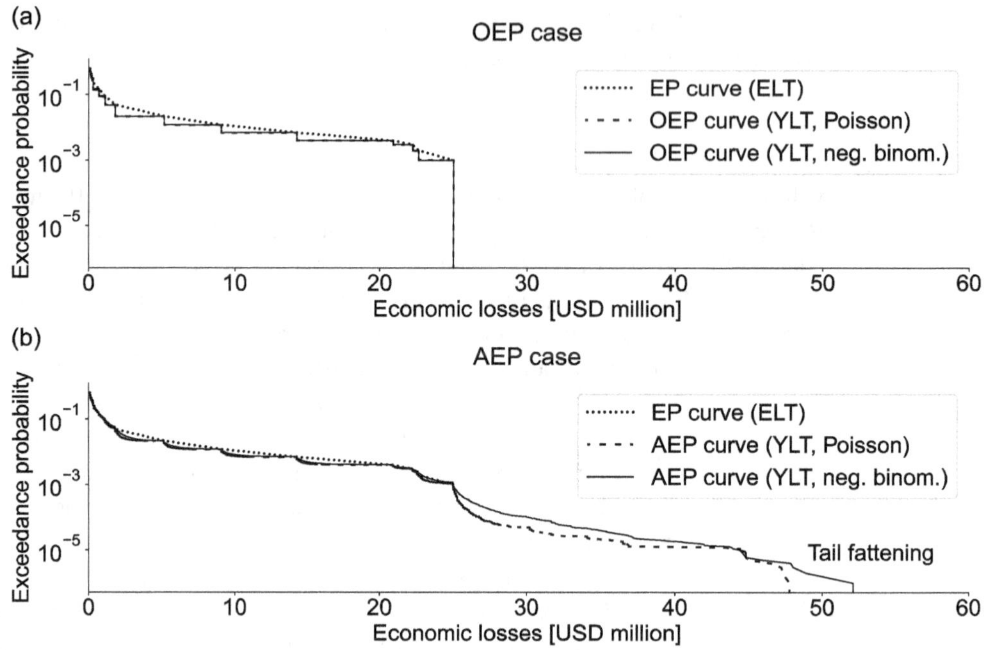

Fig. 5.2 Impact of clustering on different types of EP curves. (a) Semi-logarithmic graph of OEP curves (different cases almost indistinguishable from each other). (b) Semi-logarithmic graph of AEP curves. The EP curve was derived from the ELT shown in Table 3.9 while the OEP and AEP curves were calculated from two YLTs composed of $N_{sim} = 10^6$ simulations, one with the Poisson distribution and the other with the negative binomial distribution (for $\psi = 0.91$). Step-like behaviour and abrupt stop of OEP curves are due here to the fact that secondary aleatory uncertainties in losses are not included (see Part III).

Box 5.1 Square Matrix Definitions for Chains of Events

Interactions between a set of n events can be encoded in an $n \times n$ square matrix where rows i represent trigger events and columns j the triggered events (i.e., $i \longrightarrow j$). Depending on how the matrix is filled, different terms and rules apply. The **adjacency matrix A** defines the connections in a finite graph $\mathcal{G}$ (or network) with a non-zero element $a_{ij} \neq 0$ indicating an edge $\mathcal{E}_{ij}$ from vertex $\mathcal{V}_i$ to vertex $\mathcal{V}_j$ (Fig. 5.3). An **edge** is also called a link and a **vertex** a node. Each node represents an event or some other state. The **transition matrix P** describes the transitions of a time-discrete Markov chain with its elements p_{ij} probabilities satisfying the condition $\sum_{j=1}^{n} p_{ij} = 1$. Finally, the **transition-rate matrix Q** describes the transitions of a continuous time Markov chain with its elements q_{ij} instantaneous rates of change respecting the condition $\sum_{j=1}^{n} q_{ij} = 0$. The relationship between the transition matrix and the rate matrix is $d\mathbf{P}/dt = \mathbf{P}(t)\mathbf{Q}$, which yields $\mathbf{P}(t) = \exp(\mathbf{Q}t)$. It is important to note that the rate matrix **Q** is more fundamental than **P** as it captures the essential dynamics and interactions of the system. A static transition matrix allows us to directly describe simple chains of events (Section 5.2), but a rate matrix is necessary when considering feedback loops (Section 5.3), similar to the ones described by the ODEs of Section 4.2 (Ex. 5.2).

the past states is a **non-Markovian process**. The dynamics of the interacting system are described by a **master equation**, that is, a set of ODEs which takes the matrix form

$$\frac{\mathrm{d}\vec{p}}{\mathrm{d}t} = \vec{p}\mathbf{Q}, \tag{5.4}$$

with $\mathbf{Q}$ the rate matrix of connections, $\vec{p} = (\mathbb{P}(X = \mathrm{ID}_1), \mathbb{P}(X = \mathrm{ID}_2), \cdots, \mathbb{P}(X = \mathrm{ID}_n))$ the row vector of event probabilities, and $\vec{X} \sim \vec{p}$ the event state vector. The process is Markovian if the connections are time-independent rate constants. The application of Eq. (5.4) for the study of the temporal development of disasters has been referred to as **catastrophe dynamics** by Helbing and Kuehnert (2003) (not to be confused with catastrophe theory; Box 4.4). To make sense of Eq. (5.4) and of its rate matrix, one should note that the sets of ODEs defined in Chapter 4, such as in the case of epidemiological models, can be reformulated as a master equation (Wang et al., 2020; Ex. 5.2). Despite the apparent simplicity of Eq. (5.4), it is the $n \times n$ connections encoded in the rate matrix that govern the dynamics of $\vec{p}$ (see graph theory below). Further aspects of Markov theory will be developed in Section 5.2 to show how to model Markov chains and analyse specific chains, or trajectories $\vec{X}(t)$.

5.1.2.2 Graph Theory

The interactions defined in an adjacency matrix $\mathbf{A}$ can be represented in the form of a graph $\mathcal{G}$, as shown in Figure 5.3a and b. Each event i represents a vertex $\mathcal{V}_i$ (or node) and each element a_{ij} of $\mathbf{A}$ represents an edge $\mathcal{E}_{ij}$ (or link) from event i to event j. Since the action of j on i is rarely reciprocal, event interactions are represented in a directed graph where edges $\mathcal{E}_{ij}$ and $\mathcal{E}_{ji}$ link two vertices asymmetrically. Graph theory provides different types of centrality measures to assess the relative importance of nodes in a network. In catastrophe dynamics, they can help to classify events in terms of their relative roles in cascading effects. Note that graphs of event interactions represent abstract causal networks, but the failure of physical networks (e.g., power grids, transportation systems) can also be described by the same types of rules. This will be illustrated in Section 5.2. **Centrality measures** quantify the place that a node takes in a network, usually by counting the number of paths going through that node (Fig. 5.3c; Ex. 5.3a). The main centrality measures are degree centrality, closeness centrality, and betweenness centrality (Freeman, 1978). **Degree centrality** $C_D(\mathcal{V}_i)$ is the number of connections to the vertex $\mathcal{V}_i$. In-degree centrality is the number of incoming network links while out-degree centrality is the number of outgoing network links. For $a_{ij} = 1$ when a link is present and 0 otherwise, $C_{D,in}(\mathcal{V}_i) = \sum_{j=1,i \neq j}^{n} a_{ji}$ and $C_{D,out}(\mathcal{V}_i) = \sum_{j=1,i \neq j}^{n} a_{ij}$. A high in-degree centrality indicates that the event will be triggered in more cascades while a high out-degree centrality indicates that the event has more secondary consequences (Rocha et al., 2018; Tang et al., 2019). **Closeness centrality** of a vertex $C_C(\mathcal{V}_i) = k / \sum_j^k d(\mathcal{V}_i, \mathcal{V}_j)$ is the reciprocal of the average of the shortest path lengths d from the vertex to the k other vertices accessible in the network. A higher value indicates an event that plays a more important role in the spreading of a cascade. **Betweenness centrality** of a vertex $C_B(\mathcal{V}_i)$ is the number of the shortest paths that pass through that vertex. Events with a high betweenness act as intermediaries in the chain and can play a critical

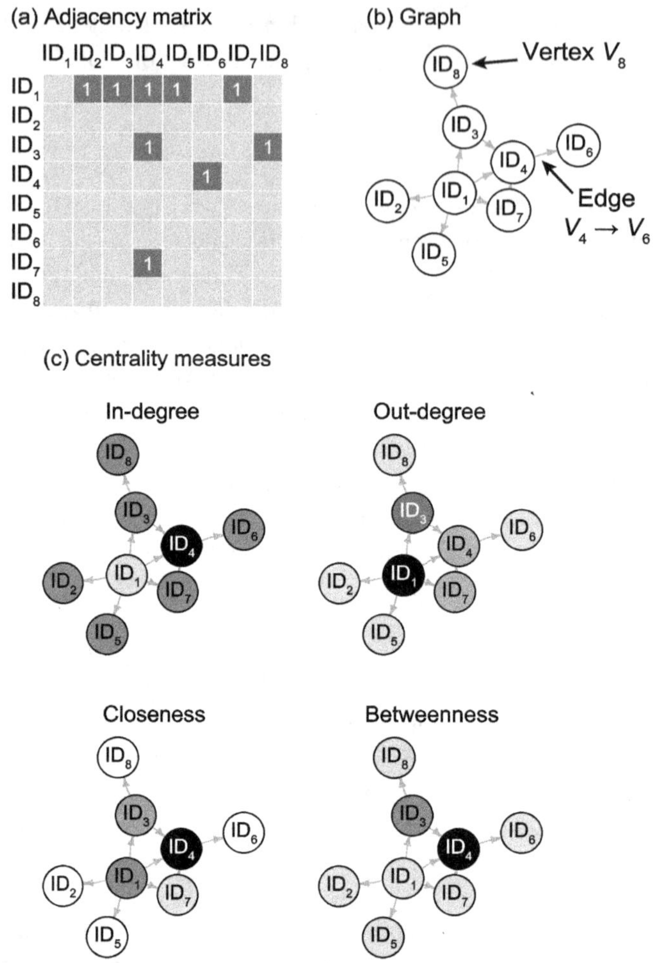

Fig. 5.3 Some concepts of graph theory to model chains of events. (a) Event interactions encoded in an adjacency matrix **A** with elements $a_{ij} = 1$ representing the triggering of event ID_j (columns) by event ID_i (rows). (b) Graph derived from the adjacency matrix. (c) Centrality measures for the same graph with values ordered by event ID: $C_{D,in} = (0, 1, 1, 3, \ldots)$, $C_{D,out} = (5, 0, 2, 1, \ldots)$, $C_C = (0.78, -, 0.75, 1, \ldots)$ and $C_B = (0, 0, 1, 3, \ldots)$ (complete the other values in Exercise 5.3a). Higher values are represented by darker nodes.

role in the development of the cascade (Rocha et al., 2018; Tang et al., 2019). Various applications will be described in Section 5.2.

5.1.2.3 Application to Earthquake (EQ) Clustering

The modelling of cross-peril cascading effects will be the subject matter of Section 5.2. Here, large earthquake clustering is considered (as a parallel to storm clustering) to illustrate the process. Doublets or even triplets of very large earthquakes are a rather common phenomenon. Examples include the 2023 Türkiye–Syria $M7.8$–7.6 doublet (Box 2.6), the

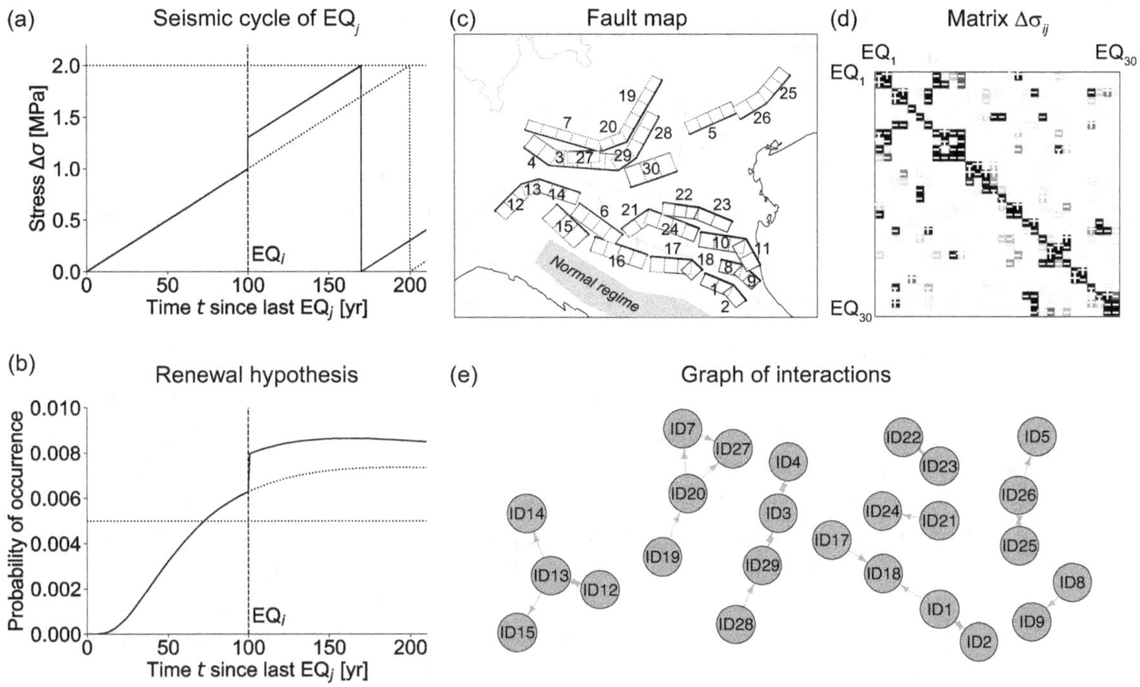

Fig. 5.4 Illustrating the process of earthquake clustering. (a) Concept of 'clock change': an idealized earthquake EQ_j recurs every 200 years if the tectonic loading is $\dot{\tau} = 0.01$ MPa/yr and the stress drop $\Delta\sigma_0 = 2$ MPa. The occurrence of a nearby earthquake EQ_i can however advance the next EQ_j if $\Delta\sigma_{ij} > 0$ (here 0.3 MPa). (b) Time-dependent probability of the occurrence of EQ_j according to the renewal hypothesis with log-normal distribution ($\mu = 200$ yr, $\sigma = 0.7$) and its increase due to EQ_i ($\mu = \mu - \Delta\sigma_{ij}/\dot{\tau}$). (c) Map of reverse faults in Northern Italy, simplified from the EDSF (Basili et al., 2013), with $n = 30$ possible events. (d) The $n \times n$ stress-change matrix $\Delta\sigma_{ij}$, from which the time change ΔT_{ij} is derived – with estimates from Mignan et al. (2018). (e) Graph of interactions with links highlighting the main potential clusters of earthquakes ($a_{ij} = 1$ if $\Delta\sigma_{ij} \geq 0.05$ MPa and 0 otherwise).

2004–2005 *M*9.0–8.7 Sunda megathrust doublet (Mignan et al., 2006), the 1999 *M*7.4–7.1 Izmit and Duzce North Anatolian doublet (Parsons et al., 2000), and the 1811–1812 *M*7.3–7.0–7.5 New Madrid Central US triplet (Mueller et al., 2004) where events occurred from within hours to months of each other. Earthquake clustering is caused by stress changes on faults, which is relatively well explained by the theory of **static stress transfer** (King et al., 1994; Lin and Stein, 2004). An earthquake is a mechanism that releases the stress accumulated by long-term tectonic loading. In an idealized case, if the tectonic stressing rate is $\dot{\tau}$ [MPa/yr] and an earthquake corresponds to a stress drop $\Delta\sigma_0$ [MPa] on the hosting fault, earthquakes would recur periodically on that fault with an annual rate $\lambda = \dot{\tau}/(\Delta\sigma_0)$ (concept of **seismic cycle**). In reality, heterogeneities in the stress history and in fault friction make the process stochastic. Moreover, the occurrence of an earthquake EQ_i changes the stress on neighbouring faults, with earthquakes EQ_j advanced if their host fault is located in an area of increased stress $\Delta\sigma_{ij} > 0$, and delayed if $\Delta\sigma_{ij} < 0$. The so-called **clock change model** calculates the time shift as $\Delta T_{ij} = \Delta\sigma_{ij}/\dot{\tau}$ (Fig. 5.4a). Since the likelihood of an earthquake changes over time, it is described by a time-dependent probability of the form:

$$\mathbb{P}(t_0 \leq t \leq t_0 + \Delta t) = \int_{t_0}^{t_0+\Delta t} f(t)dt, \tag{5.5}$$

with $f(t)$ a PDF such that the likelihood of earthquake EQ_i is lowest just after the last EQ_i occurrence (at $t = 0$) and then increases as tectonic stress accumulates (i.e., **renewal hypothesis**). The **log-normal distribution** is one such function with

$$f(t, \Delta T, \sigma) = \frac{1}{t\sigma\sqrt{2\pi}} \exp\left(-\frac{(\log t - \Delta T)^2}{2\sigma^2}\right), \tag{5.6}$$

where ΔT is the recurrence interval and σ is the standard deviation in log-space (σ is called aperiodicity and is estimated from past earthquake time series). If an earthquake EQ_i occurs at t, the probability of EQ_j is calculated from Eq. (5.5) with $\Delta T_j = \Delta T_j - \Delta T_{ij}$ derived from the stress change $\Delta\sigma_{ij}$ (Parsons, 2005; Fig. 5.4b). Softwares, such as Coulomb 3 by the USGS, compute $\Delta\sigma_{ij}$ as a function of rupture slip, mechanism, friction, and other parameters (Lin and Stein, 2004; Toda et al., 2011). An example of an earthquake–stress interaction system is shown in Figure 5.4c. Although the values in matrices $\Delta\sigma_{ij}$ (Fig. 5.4d) and ΔT_{ij} are constants, the conditional probability p_{ij} is not since all previous time changes are memorized in ΔT_j, making the process non-Markovian. Interactions can still be shown in a graph (Fig. 5.4e) but the modelling of cascades in a CAT model requires the development of some algorithm (Mignan et al., 2018).

5.1.2.4 Specialized Algorithms

Implementing chains of events and event-dependent hazard and risk parameters is more complicated and requires the development of some algorithms, as illustrated in Figure 5.5.

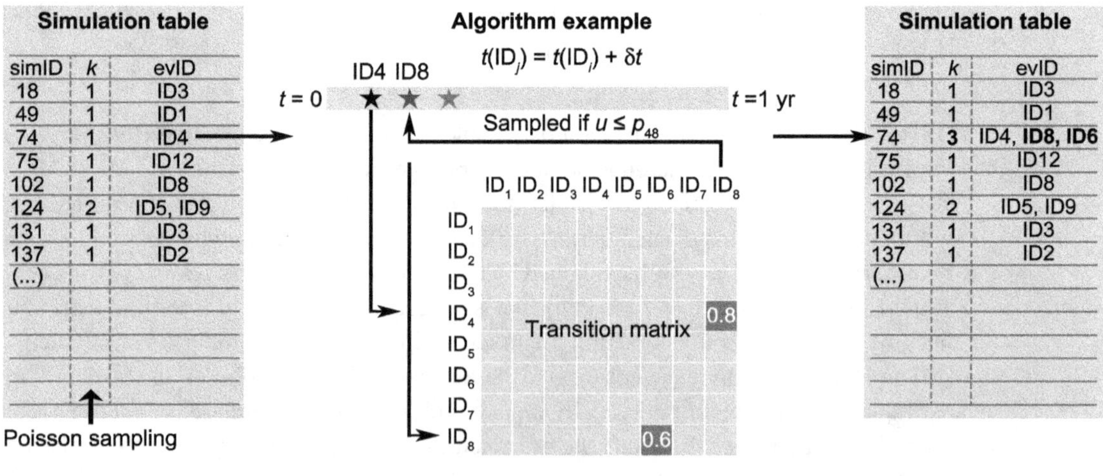

Fig. 5.5 Implementation of chains of events in a YET. A simple algorithm consists of sampling triggered events from a transition matrix. The occurrence time of the triggered event is fixed to the time of the triggering event plus a fixed δt interval.

In this example, simulation numbered 74 contains one event (ID4) before the application of any event interaction algorithm. The algorithm then consists of sampling events possibly triggered by event ID4 according to the transition matrix **P** (Section 5.1.2). Here, event ID8 is triggered, and the process repeats until no more events occur in the year-simulation. A deeper description goes beyond the scope of this textbook.

5.1.3 Non-stationarity of CAT Model Parameters

The CAT risk modelling pipeline described in Part I involved constant hazard and risk parameters to estimate event rates λ, hazard intensities I, damages D, and losses L (Fig. 3.1). Input parameters can, however, vary with time. Two categories of time-dependent processes are described below: (1) long-term exogenous drivers and (2) short-term endogenous changes caused by events.

5.1.3.1 Long-Term Risk Drivers

Long-term exogenous drivers of catastrophes include global warming and demographic growth coupled to urbanization, but also other aspects such as inflation and the ageing of structures (Fig. 5.6). Different drivers impact different components of a CAT model, but current trends seem to tend towards higher future losses. Global warming can increase the frequency and intensity of climate and weather perils (van Aalst, 2006; Knutson et al., 2020). Demographic growth and inflation lead to higher exposure values over time (Bilham, 2009; Swiss Re Institute, 2023), while urbanization leads to potential risk concentration in megacities (Bilham, 2009). The ageing of structures makes them more vulnerable as time passes (Ghosh and Padgett, 2010). Since changes occur on a time period relatively long compared to the standard 1-year time window used in CAT models, these long-term trends can be modelled as projected risk estimates for a given year in the future (e.g., year

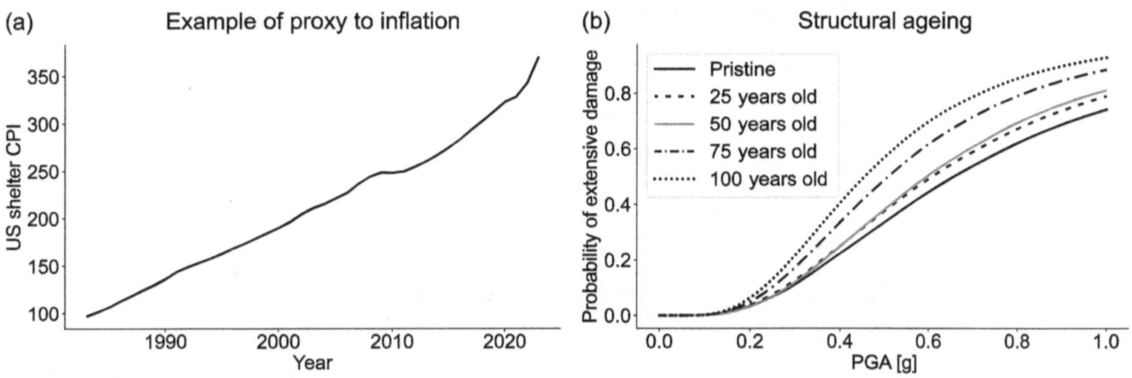

Fig. 5.6 Long-term risk drivers in exposure and vulnerability. (a) 1983–2023 US shelter consumer price index (CPI) from the US Bureau of Labor Statistics (2023), which correlates with house price evolution (with a time lag). (b) Impact of structural ageing on seismic fragility curves for extensive damage, with cumulative log-normal distribution parameters $\Theta_\nu(t) = (\mu(t), \sigma(t))$ from Ghosh and Padgett (2010, table 3).

2050 or 2100). Changes to exposure values (Fig. 5.6a) are relatively direct to implement since they consist of increasing, for instance, the nominal value of buildings and other assets (Swiss Re Institute, 2023). For ageing, the parameters of damage functions – which are part of the asset characteristics parameter set Θ_ν – can be defined as a function of time based on models of structural corrosion (Ghosh and Padgett, 2010) (e.g., $\mu(t)$ and $\sigma(t)$ in Eq. (3.4); Fig. 5.6b). The role of global warming on hazard, which is complex, will be the subject matter of Section 5.3.1.

5.1.3.2 Risk Dependency on Past Event Occurrences

With losses a function of the events expected to occur in a given year, the impact of an event on CAT model parameters is perceived only if more than one event happens per year. With storm clustering, earthquake clustering, and other cross-peril cascades potentially occurring from within minutes to months, such conditions can easily be met. Once an event has occurred, damage occurs with the exposure asset value ν reduced to $\nu(1-(1-\eta)D)$ with D defined as the MDR and $0 \leq \eta \leq 1$ the proportion of damage being repaired prior to the occurrence of the next event (Mignan et al., 2014). Losses can however increase due to a process of **vulnerability amplification**. When several events occur in a short spatiotemporal window, buildings may indeed experience accelerated fatigue. The combined role of two perils can be described by a **vulnerability surface** instead of the vulnerability curve defined in Chapter 3. Engineering models of the response of a structure to the interactions of several damage mechanisms remain rare but are already available for the following couplings: earthquake and tsunami (EQ+Ts), earthquake and landslide (EQ+LS), earthquake and windstorm (EQ+WS), windstorm and storm surge (WS+SS), and volcanic eruption and earthquake (VE+EQ) (Gentile et al., (2022) and references therein). For instance, for the VE+EQ case (Fig. 5.7), it has been evaluated that 600 kg/m^2 or 1,200 kg/m^2 of ash load on a roof of a weak masonry building is equivalent to the macroseismic intensity being increased by two or four units, respectively (Zuccaro et al., 2008). For the WS+EQ case, the probability of exceeding extensive damage for a PGA of 1 g goes from less than 20%

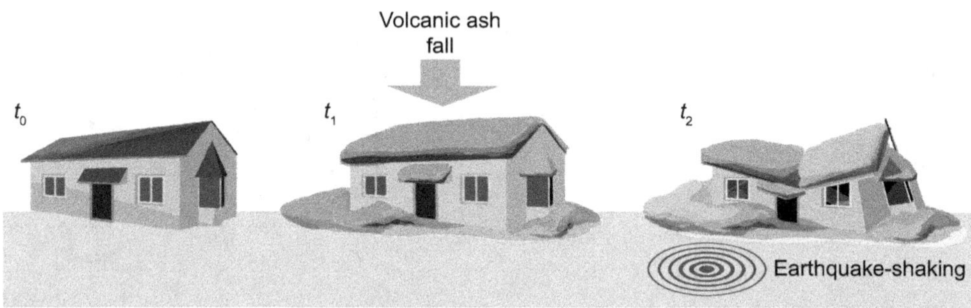

Fig. 5.7 Combined roles of volcanic eruptions and earthquakes in building damage (refer to the text for quantification of the impact). Building illustrations by Marion Deichmann, 2012 (published with permission), developed within the scope of the Multi-hazard and multi-risk assessment methods for Europe (MATRIX) project (e.g., Mignan et al., 2014), based on the author's sketch.

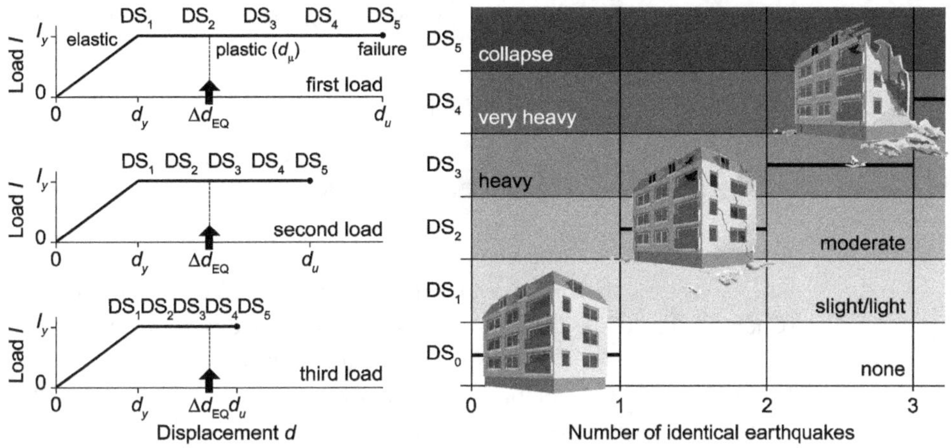

Fig. 5.8 Accelerated building fatigue from earthquake clustering. Ground-shaking yields a drift ratio Δd_{EQ}, which, if above the yield displacement capacity d_y of the building, will reduce its ultimate displacement capacity from d_u to $d_u - (\Delta d_{EQ} - d_y)$. The repeat of an earthquake will, each time, increase the vulnerability of the building until a collapse occurs (Mignan et al., 2018). Building illustrations by Marion Deichmann, 2012 (published with permission), developed within the scope of the MATRIX project (e.g., Mignan et al., 2014), based on the author's sketch.

for the marginal case to more than 60% when wind speed is 35 m/s taking one instance of building class (Li et al., 2020). The counteracting effects of exposure decrease and vulnerability increase make the dynamic risk process non-trivial. Note that short-term factors may also affect the hazard component. For example, successive heavy rains can increase the soil moisture and increase flood intensity over time. Possibilities are too numerous to list here (Leonard et al., 2014).

5.1.3.3 Application to Earthquake (EQ) Clustering

The earthquake clustering case is again illustrative. When a building experiences successive shaking episodes, its vulnerability can increase due to **accelerated fatigue**. Considering the capacity curve of a generic building (Fig. 3.7), the occurrence of a damaging event will reduce the ductility range d_μ of the structure by the earthquake drift ratio Δd_{EQ} (function of the ground-shaking) minus the yield displacement capacity d_y (Section 3.2.4). This can also be phrased in terms of the reduction of the ultimate displacement capacity d_u $(= d_y + d_\mu)$. Since a fragility curve depends on the length of the ductility range of the structure, its progressive reduction leads to a shift of fragility curves toward lower shaking intensities. This is illustrated in Figure 5.8. In this example, an ad hoc earthquake leads to moderate damage if it occurs. The impact of the reoccurrence of an identical event on the same – but now moderately damaged – building leads to heavy damage. Occurrence of the same event a third time may lead to the collapse of the building, which has now experienced three successive loads (Mignan et al., 2018). In reality, the earthquake triplet would occur along successive ruptures and not exactly on the same segment (see renewal hypothesis; Section 5.1.2). This shows a non-linear, amplifying damage process, via the principle of memory (a non-Markovian process).

5.2 Domino Effects

Domino effects describe a succession (a chain) of interconnected events triggered by an initial event that results in overall consequences greater than those of that primary event (Chen et al., 2020). Domino effects can be very diverse in nature. Many examples of secondary effects of disasters were already mentioned in Section 1.3 on peril taxonomy. Tertiary, quaternary, and longer chains of events can now be modelled thanks to the mathematical frameworks previously introduced. A few selected examples from the vast array of possible interaction patterns will be described to cover a wide variety of situations: domino effects occurring in refineries (Section 5.2.1); a natural event triggering cascading failures in a physical network (Section 5.2.2); and, finally, a generalization to any type of cascade across the natural, technological, and socio-economic systems (Section 5.2.3). This generalization will remain academic in nature since much more historical data combined with physical models of interactions will be needed for any future potential implementation in a CAT model. Note that although the terms '**cascading effects**' and 'domino effects' are often used interchangeably, *domino effects strictly represent one-way dependencies, while cascades would also include the role of feedback mechanisms, which are self-amplifying processes* (Rocha et al., 2018) (Section 5.3).

5.2.1 Domino Effects at Refineries (Ex, Fi, TS)

Technological accidents can be due to human error, technical failure, malicious acts, or natural events. For any of those initial triggers, further cascading effects may take place within and across critical infrastructures. Refineries are such complex objects, an example being the BP Texas City Refinery where a disaster killed 15 workers, injured 180, and caused losses of up to USD 1.5 billion in 2005 (Isimite and Rubini, 2016). A description of the chain of events is given in ENSAD (Hirschberg et al., 2004) as follows: due to an error, a raffinate splitter tower had been overfilled, causing the opening of the pressure relief devices, which resulted in a flammable liquid geyser from a blowdown stack that was not equipped with a flare. The release of flammables led to an explosion followed by fires (Mignan et al., 2022a; see also Khan and Amyotte, 2007). ENSAD lists more than 4,000 severe accidents and includes information on chains of events for 673 refinery disasters spanning from 1970 to 2016 (Mignan et al., 2022a). This provides enough data to develop a transition matrix, generate chains of events, and rank them by likelihood.

5.2.1.1 ENSAD Transition Matrix

ENSAD includes an attribute for chains-of-k events with k up to 6 for refineries. Events in ENSAD do not necessarily generate losses directly but are needed to define the causal links leading to a catastrophe (e.g., human error, corrosion, mechanical failure, rupture). The 24 events defined in ENSAD for 673 cases of severe accidents at refineries are listed in Figure 5.9. In critical infrastructure risk assessment, successive loss-generating accidents are referred to as domino effects. In the chemical process industry, they represent 'situations when a fire/explosion/missile/toxic load generated by an accident in one unit in an industry causes secondary and higher order accidents in other units' (Khan and Abbasi, 1998a). The inclusion of other types of events to deduce how the catastrophe unfolds is commonly called **fault tree analysis** (Khan and Abbasi, 1998b). To study potential chains of events at refineries, one-to-one interactions are first encoded in a transition matrix **P** with conditional probabilities $p_{ij} = n_{ij}/n_i$ empirically assessed, with n_i the number of triggers i and n_{ij} the number of interactions $i \rightarrow j$ observed in ENSAD (Mignan et al., 2022a). The result is shown in Figure 5.9a. Not shown here is the presence of an additional outflow event with conditional probability $p_{i,n+1} = 1 - \sum_j^n p_{ij}$ for conservation of probabilities, which represents an **absorbing state** that ends a chain (e.g., Mignan and Wang, 2020).

5.2.1.2 Chain-of-Events Modelling

If the Markov property is assumed, the probability of a chain transitioning from event i to event j over two steps is $\mathbb{P}(X_2 = j|X_0 = i) = \sum_{k=1}^n \mathbb{P}(X_2 = j|X_1 = k) \times \mathbb{P}(X_1 = k|X_0 = i)$. For higher-order interdependencies with $t-$step transitions (or chains-of-$t+1$ events),

$$\mathbb{P}(X_t = j|X_0 = i) = \mathbf{P}^t, \tag{5.7}$$

with t the number of time steps and **P** the transition matrix (Ex. 5.3b). The resulting matrix for $t = 4$ is shown in Figure 5.9b. It shows, for example, taking human error (ID_{13}) as initial trigger, that fire is the most likely event occurring at the end of chains-of-5 events. The initial state vector is in this case $\vec{X}(t=0) = (\varnothing, \varnothing, \cdots, \mathrm{ID}_{13}, \cdots, \varnothing)$ or $\vec{p}(t=0) = (\mathbb{P}(X_0 = \mathrm{ID}_1) = 0, \mathbb{P}(X_0 = \mathrm{ID}_2) = 0, \cdots, \mathbb{P}(X_0 = \mathrm{ID}_{13}) = 1, \cdots, \mathbb{P}(X_0 = \mathrm{ID}_n) = 0)$. At any given time t,

$$\vec{X}(t) \sim \vec{p}(t) = \vec{p}(t=0)^\intercal \mathbf{P}^t, \tag{5.8}$$

where $\intercal$ is the transpose operator. The probability of a specific trajectory, for example $\mathrm{ID}_5 \rightarrow \mathrm{ID}_2 \rightarrow \mathrm{ID}_3 \rightarrow \mathrm{ID}_2$, is $\mathbb{P}(X_0 = \mathrm{ID}_5) \times p_{52} \times p_{23} \times p_{32}$. The four most likely chains-of-5 events starting with human error in ENSAD (for refineries) are as follows: '→ explosion → fire → explosion → fire', '→ rupture → toxic release → explosion → fire', '→ corrosion → rupture → toxic release → gas poisoning', and '→ toxic release → fire → explosion → fire' (Fig. 5.9c).

One can note that the ENSAD transition matrix has a zero diagonal (Fig. 5.9a), suggesting that cases of successive identical events (in terms of identifiers) have been concatenated into one unique event. Successive explosions, for instance, are lacking from the database (an example of event aggregation bias – see Box 1.1). Exercise 5.3c will give the reader the opportunity to model a chain of explosions by combining tools from Part I and from the present chapter.

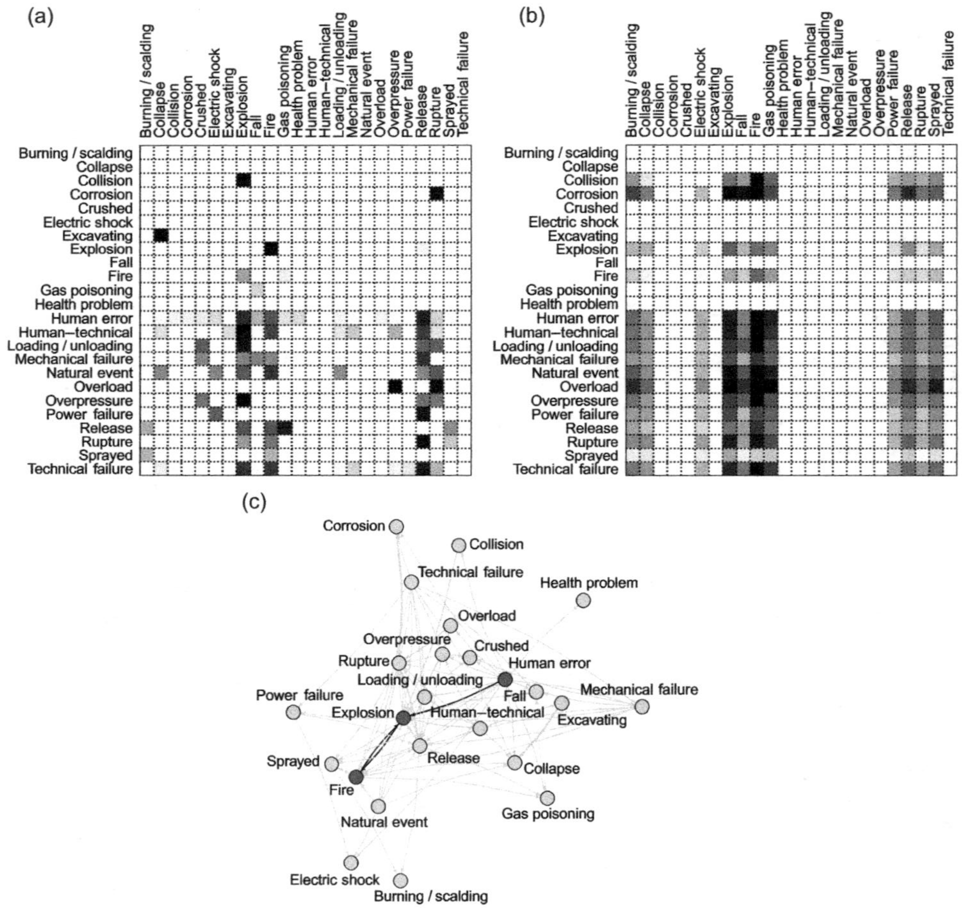

Fig. 5.9 Domino effects at refineries, derived from ENSAD. (a) Transition matrix $\mathbf{P} = (p_{ij})$ according to Mignan et al. (2022a) (CC-BY 4.0, https://creativecommons.org/licenses/by/4.0/). (b) Fourth-step transition matrix $\mathbf{P}^4$. (c) Graph of event interactions with the chain 'human error $\longrightarrow$ explosion (Ex) $\longrightarrow$ fire (Fi) $\longrightarrow$ explosion $\longrightarrow$ fire' highlighted – 24 events are defined here, including both loss-generating events and other events having a role in the causal chain ('release' here means a toxic spill, TS). Note that the reduced text size was necessary to accommodate the numerous event entries. Despite the small font, every effort has been made to ensure readability.

5.2.2 Cascading Failures in Physical Networks ($\cdots \rightarrow$ BO)

The failure of a physical network, such as of a power grid (referred to as a blackout, BO) or of a transportation network (abbreviated as Tf), is a complex process contingent upon the network's ability to redistribute the flow of physical quantities, such as electricity or vehicles, in response to component failures. Despite the introduction of redundancies aimed at mitigating system-level vulnerabilities, it is imperative to acknowledge that the risk can never be entirely eradicated, as emphasized by **normal accident theory** (Perrow, 2001). This was, for example, illustrated with the phenomenon of self-organized outages in Section 4.1.2. The complexities extend further, from the influence of natural events, such as earthquakes, storms, or floods, on the network's structural integrity, to the impact of network failures on higher-order cascades throughout technological and socio-economic systems. This type of super-cat is only starting to be implemented in CAT risk platforms (Mühlhofer et al., 2023). One approach to model these interdependencies revolves around **topology-based analyses**, which involve examining changes in centrality values when network elements fail. Case studies encompass a spectrum of scenarios, from tropical cyclones triggering blackouts (TC $\rightarrow$ BO; Winkler et al., 2010) to earthquakes and lightning causing blackouts (EQ, Li $\rightarrow$ BO; Dueñas-Osorio and Vemuru, 2009), and blackouts inducing water and gas supply failures (BO $\rightarrow$ Wf, Ef; Winkler et al., 2011). In the context of road networks, topological measures for connectivity loss have been used, often in tandem with origin-destination matrices of traffic volumes, to estimate the impact of various natural hazards on traffic flow, including earthquakes (EQ $\rightarrow$ Tf; Argyroudis et al., 2015), landslides (LS $\rightarrow$ Tf; Muriel-Villegas et al., 2016), and floods (FF, PF $\rightarrow$ Tf; Kermanshah and Derrible, 2017). To illustrate the modelling of a blackout caused by cascading failures triggered by a natural event, such as a tornado (To), we will employ the simple topology-based approach. Although more sophisticated techniques exist to model network interdependencies, ranging from system dynamics and agent-based models to physical flow analysis (Ouyang, 2014), they fall beyond the scope of this introductory textbook (for a generic cellular automaton approach, refer to Section 4.1.2).

5.2.2.1 Network Component Failures

A power grid, in its simplest form, is a graph $\mathcal{G}(\mathcal{V}, \mathcal{E})$ composed of substations (vertices $\mathcal{V}$) and transmission lines (edges $\mathcal{E}$). Substations include generators where electricity is produced, loads where electricity is consumed, and other distribution substations which relay electricity from generators to loads. An ad hoc power grid of $n = 12$ substations is shown in Figure 5.10a, with the adjacency matrix $\mathbf{A}_0$ representing the initial undamaged directed graph where electricity can only exit the generator, enter the loads, and enter or exit the distribution substations in both directions. A more realistic application could consist of, for example, using one of the IEEE bus networks (Christie, 1993). However, the basic principle of cascading failure would remain the same. A cascade can be initiated by the failure of the kth network element $\forall k \in \mathcal{V}, \mathcal{E}$. Failure $\mathcal{S}_{f,k} = 1$ occurs if $u \leq p_{f,k}$ with $p_{f,k}$ the probability of failure and u a random number. With a triggering event potentially affecting several elements in the network, different direct damage configurations can be reached at a system

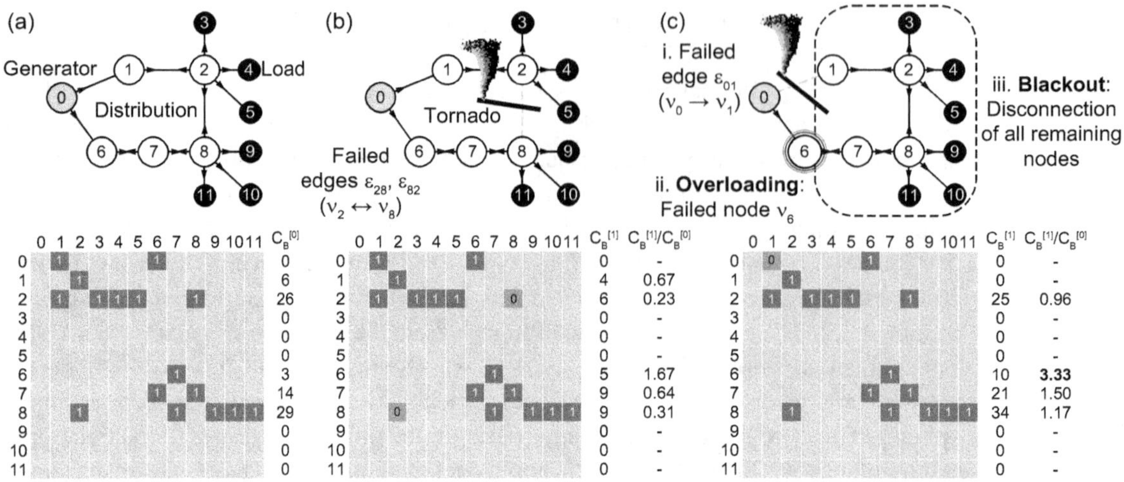

Fig. 5.10 Cascading failure in a power grid due to a natural event. (a) Undamaged ad hoc network and matching adjacency matrix $\mathbf{A_0}$. (b) A tornado makes the transmission line 2 $\leftrightarrow$ 8 fail, yielding $\mathbf{A_1}$ at iteration 1. There is no additional iteration since all nodes remain connected to the generator and no overloading occurs. (c) A tornado makes the transmission line 0 $\longrightarrow$ 1 fail, yielding a different $\mathbf{A_1}$. All nodes remain connected to the generator but overloading occurs at node $\mathcal{V}_6$ (i.e., $C_B^{[1]}/C_B^{[0]} \geq 3$ in our ad hoc scenario). It fails and all other nodes get disconnected from supply at iteration 2. This cascading failure leads to a system-level blackout. Tornado icon from SimCity's open source version Micropolis (General Public License).

level for different Monte Carlo simulations (leading to an altered adjacency matrix $\mathbf{A}_1$). The probability $p_{f,i}$ is estimated from component fragility functions (Section 3.2.1). In general, structural damage due to a given peril load is assessed with the cumulative log-normal distribution (Eq. (3.4)), with the input parameters $\Theta_\nu = (\mu, \sigma)$ a function of the substation type. Note that, by design, generators (e.g., hydro-dam, nuclear power plant) are usually far more resistant than other substations. Here, only transmission lines will be cut to illustrate the process of cascading failure. Tornadoes, because of their narrow line-like footprints (Eq. (2.59)), could act as 'cutters'. Transmission line fragility to wind load is commonly modelled by following the engineering-based approach (Kempner, 2007; Section 3.2.4).

5.2.2.2 Topology-Based Approach

Cascading failure in a power grid is a multi-step process that can be described by a simple algorithm:

1. update the adjacency matrix $\mathbf{A_l}$ with entries $a_{ij}^{[l]} = 0$ when the link $i \rightarrow j$, node i or node j has failed (i.e., if $\mathcal{S}_{f,k} = 1$ for $k \in \mathcal{V}, \mathcal{E}$);
2. fix $\mathcal{S}_{f,k} = 1$ for distribution and load nodes $\mathcal{V}_k$ disconnected from all generators and update $\mathbf{A_l}$ according to the previous rule;
3. calculate the load at each distribution node $\mathcal{V}_k$ using a proxy based on graph topology statistics (see text below) and fix $\mathcal{S}_{f,k} = 1$ for the nodes which are overloaded;

4. fix $l = l + 1$ and repeat from 1 until there is no more element-k failure (i.e., $\sum_k \mathcal{S}_{f,k}^{[l]} - \sum_k \mathcal{S}_{f,k}^{[l-1]} = 0$).

For an example of pseudocode based on a similar algorithm, see Winkler et al. (2010). At iteration $l = 0$, failures $\mathcal{S}_{f,k} = 1$ represent the direct damage inflicted by the exogenous event. Indirect failures due to these initial disconnections are the other nodes which become disconnected from all generators. Given the damaged network topology, nodes still connected to a generator do not necessarily get served (i.e., overloading failures). This is best assessed via power-flow analysis, but node betweenness centrality $C_B(\mathcal{V}_i)$ can be used as a proxy of the amount of flow passing through the node $\mathcal{V}_i$ (Motter and Lai, 2002; Dueñas-Osorio and Vemuru, 2009).

5.2.2.3 Betweenness Centrality as a Proxy for Node Load

Assuming that the quantity that flows in a network is transmitted along the shortest path connecting every pair of nodes, the load at a node is then the total number of shortest paths passing through the node (Motter and Lai, 2002). This is the definition of betweenness centrality $C_B(\mathcal{V}_k)$ for vertex $\mathcal{V}_k$ (Section 5.1.2.2). Values of C_B for $\mathbf{A_0}$ are given in Figure 5.10a. To illustrate how loads may redistribute when the network topology changes, the impact on the graph of two different tornadoes is shown in Figure 5.10b and c. Overload (algorithm step 3) is defined as the node capacity being exceeded and, for this ad hoc power grid, is represented by a positive change $C_B^{[l]}/C_B^{[0]} \geq 3$. Different metrics are available in the literature, which are calibrated to historical data (Winkler et al., 2010). Since power grids are tuned to operate close to their maximum capacity in realistic conditions, betweenness centralities derived from $\mathbf{A_0}$ must be close to the maximum capacity value of the substations. In the first scenario, removing the transmission line $2 \leftrightarrow 8$ that makes the ring-mesh does not lead to an outage since the two remaining branches still have a relatively direct connection to the generator. In the second scenario, removing the transmission line $0 \rightarrow 1$ requires all the electricity to journey through the lower nodes, which leads to overloading and failure of node $\mathcal{V}_6$. Applying the previously defined algorithm leads to a system-level blackout. The size of the event could then be defined as the total number of failed load substations. The dynamic process becomes more complex when more than one network component is targeted by an external event or when the network itself more becomes complex, for example in the case of a network of networks (Buldyrev et al., 2010).

5.2.3 Generalization to Any Peril Interaction

An investigation of historical catastrophes characterized by cascading effects illustrates how perils can be intertwined within and across the natural, technological, and socioeconomic systems. Such an exploration was realized by Mignan and Wang (2020), who scrutinized 29 such instances, encompassing renowned super-cats such as the 2011 Tohoku earthquake, the 1906 San Francisco earthquake, the 2010 Eyjafjallajökull volcanic eruption, and Hurricane Harvey of 2017, alongside more peculiar historical anomalies such as a Valley Fever outbreak triggered by landslides following the 1994 Northridge earthquake, a case involving airborne dust contamination. The primary objective was to explore the space of physically plausible interactions. Figure 5.11 visually presents the resulting

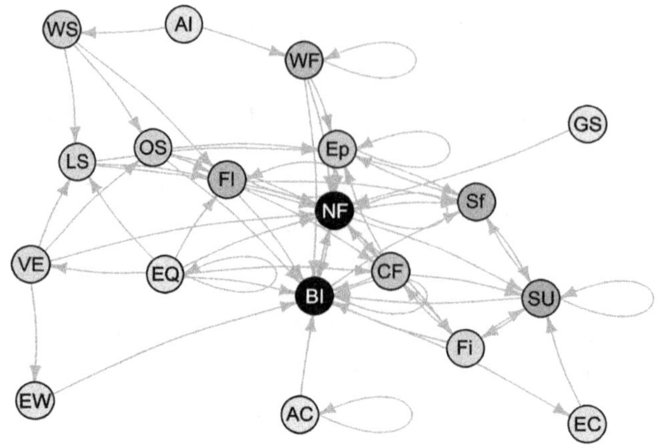

Fig. 5.11 General graph of peril interactions, within and across the natural, technological, and socio-economic systems according to Mignan and Wang (2020). Most peril identifiers can be found in Table 1.7. Others are generic flood (FI), other storm (OS, i.e., sub-perils beyond wind), extreme weather (EW, i.e., CW, Dr, HW), localized critical infrastructure failures (CF, i.e., DF, Ef, Ex, TS), network failure (NF, i.e., BO, Ef, If, Tf), and economic crisis (EC, i.e., Bf, FC). Higher betweenness centrality values are represented in darker shades of grey. Modified from Mignan and Wang (2020) (CC BY).

interaction graph, constructed from the encoding of one-to-one interactions in an empirical adjacency matrix (**A**) with $n = 19$ perils. While historical instances are too few to assess the transition matrix (**P**) necessary for quantitative evaluation and potential incorporation into a CAT model, the topology of the interaction graph already provides some invaluable insights. Notably, physical network failures like blackouts (BO) or transportation system failures (Tf), as well as business interruptions (BI), emerge as key events within chains of events, with the highest in-degree and betweenness centralities. This observation suggests that network failure and business interruption act as the main bridges for super-cats to cross from the natural system to the technological and socio-economic ones. A systematic exploration of all observed chains of events and modelling of all physical relationships have yet to be achieved (Ex. 5.4).

5.3 Compound Events in the Complex Earth System

The Earth is a highly complex system (commonly called the **complex Earth system**) composed of interacting subsystems, which are the lithosphere, hydrosphere, cryosphere, atmosphere, and biosphere, on top of which our techno- and socio-economic system is built. All of these subsystems interact by the exchange of matter, energy, and information (Donner et al., 2009). *Various phenomena can disrupt the equilibrium of these systems, potentially leading to* ***regime shifts****, which are critical transitions involving a fundamental and persistent change in the system's function and structure* (Rocha et al., 2018). *Moving away from an equilibrium state usually has negative consequences; therefore, regime shifts can be considered a type of extreme catastrophe*, although they appear to be sudden only at the timescale of the system's process. One example, at the scale of a three-species forest ecosystem, was the spruce budworm cusp catastrophe (Section 4.2.3). Larger ecosystems, the climate system, and our socio-economic system require far more variables to be realistically modelled.

The previous section focused mainly on one-way interactions (i.e., domino effects); the present section will instead cover feedback loops as another type of cascading effect. As already mentioned, one common approach to model feedback effects is system dynamics. It was previously employed to model epidemics, pest infestations, and armed conflicts (Section 4.2). Here, some parsimonious system-dynamics models will be presented to describe global warming (Section 5.3.1), ecosystem collapse (Section 5.3.2), and, finally, societal collapse (Section 5.3.3). All of these processes are interconnected: natural catastrophes and human activities can act as triggers of regime shifts, and regime shifts can trigger other regime shifts as well as various catastrophes in different subsystems. While it is possible to assess the impact of global warming trends in CAT models by projecting risk forecasts for a given year in the future, the impact of ecosystem and societal collapses (including existential threats) goes beyond the applications of current standard CAT models.

The term 'compound event' (used in the previous title), which represents an extreme event resulting from a combination of multiple physical processes, is primarily employed in the domain of climate change and will therefore be formally introduced in the following section.

5.3.1 Global Warming

In the context of a climate influenced by global warming (Meehl et al., 2003; Lenssen et al., 2019; IPCC, 2021), the occurrence of increasingly severe events aligns with the observation discussed in Chapter 2 that hazardous events involve the transfer of energy from a source to the environment. As the Earth's atmosphere warms, with temperature T defined as the global mean surface temperature (GMST) (Figs. 5.12 and 5.13), more energy becomes available for the generation of storms, coastal floods, rain-induced events such as floods and landslides, wildfires, droughts, and other perils. This energy load translates into an increase in the frequency and/or intensity of these natural disasters (van Aalst, 2006; Seneviratne et al., 2012; Knutson et al., 2020). The impact of global warming is most likely to be perceived by the occurrence of more extreme events, since a relatively small increase in mean temperature leads to the largest per-cent changes being observed in the upper tail of the temperature distribution (Fig. 5.12b; Trenberth, 2011).

While the basic concept in itself is simple and the process leading to hazard amplification is potentially a single variable in an extreme state, it is more often the case that the process involves a combination of variables that are not necessarily extreme when taken individually (Leonard et al., 2014). An extreme event resulting from a combination of multiple physical processes has been termed a **compound event** by the Intergovernmental Panel on Climate Change (IPCC) (Seneviratne et al., 2012, p. 118), from which Leonard et al. (2014) proposed the general definition: 'an extreme impact that depends on multiple statistically dependent variables or events' (see also Myers (1987) for a much earlier definition). Compound events include the simple one-to-one chain of events previously described (Section 5.2; linear system of Eq. (5.4)), but also more complex

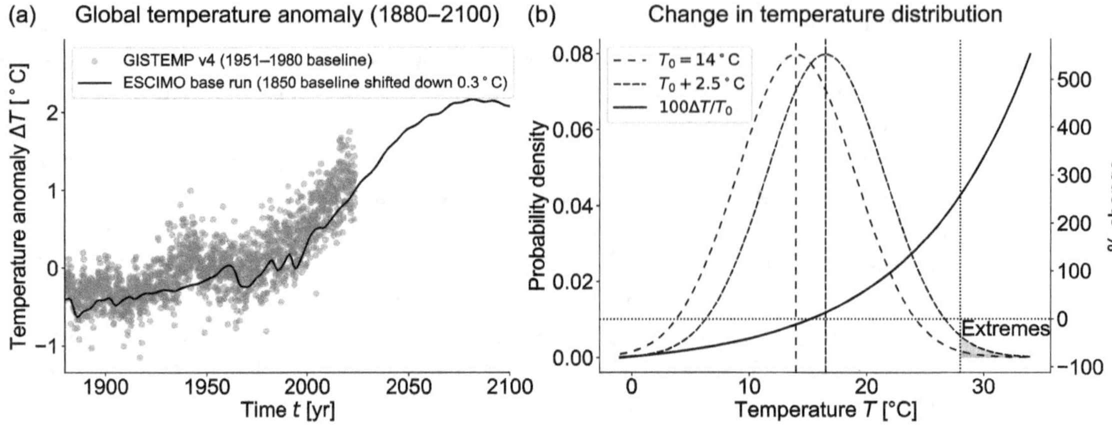

Fig. 5.12 Global warming and emergence of temperature extremes. (a) Temporal evolution of the temperature (GMST) anomaly with monthly data from the Goddard Institute for Space Studies Surface Temperature product version 4 (GISTEMP v4) (GISTEMP Team, 2023; Lenssen et al., 2019). Includes the ESCIMO model prediction up to 2100 (Randers et al., 2016) (ESCIMO source file in .mdl format courtesy of Professor Ulrich Golüke, June 2023). (b) Emergence of extreme temperatures conceptualized: 'A small change in the average temperature value can have a large effect on extremes' which is reflected by the increase of the per-cent change of T as T increases (Trenberth, 2011) (temperature distribution normally distributed with mean value T_0 or $T_0 + \Delta T$, and standard deviation $\sigma = 5$ kept constant – an increase of σ would lead to more extremes on both tails).

non-linear relationships, with feedback loops being a special case (Fig. 5.14). Various chain reactions can further result in **tipping points** (or critical thresholds) being crossed with potentially irreversible changes to the climate system (i.e., regime shifts) occurring and risks spilling into ecosystems and socioecological systems (Sections 5.3.2 and 5.3.3). In contrast to the simple domino effects described in Section 5.2, here both events and variables (temperature, rainfall, soil moisture, etc.) must be considered to describe the dynamic process. Although mainly discussed in the climate risk literature, the concept of compound events applies to other types of super-catastrophes (e.g., the 2011 Tohoku earthquake leading to the Fukushima nuclear accident can be defined as a compound geophysical/hydrological/technological event).

A relatively simple model of global warming based on system dynamics is presented below to illustrate some of the main interactions and feedback loops of the climate system (Randers et al., 2016). A plethora of more complex global climate models, not discussed here, form the basis of the IPCC assessment reports (Randall et al., 2007, IPCC, 2021, annex II) and of the related Coupled Model Intercomparison Project (CMIP) (Meehl et al., 2000; Touzé-Peiffer et al., 2020).

5.3.1.1 A Parsimonious Global Climate Model (ESCIMO)

Randers et al. (2016) proposed a low-complexity model called the Earth System Climate Interpretable Model (ESCIMO), which represents the main causal mechanisms at work in

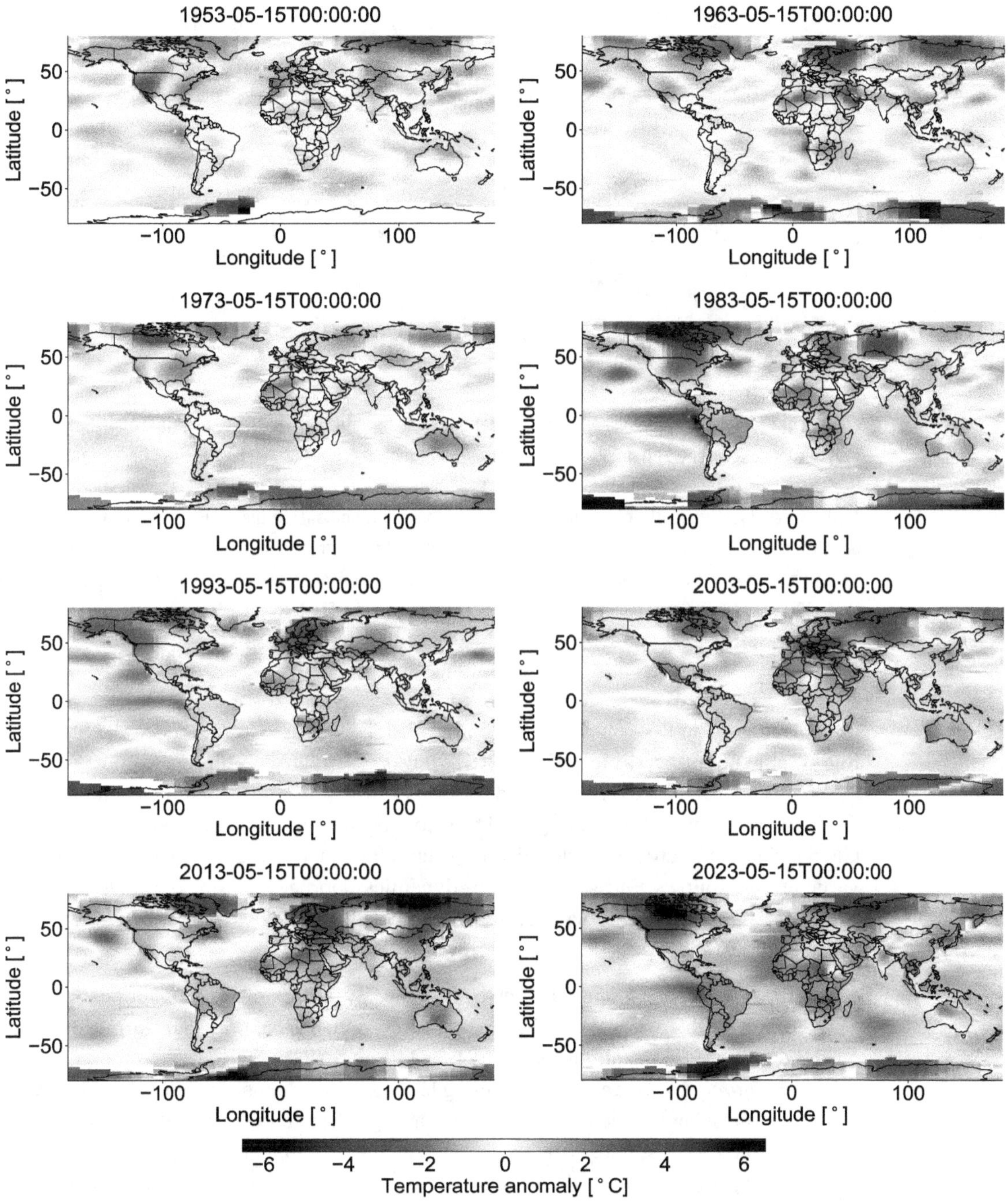

Fig. 5.13 Maps of GMST changes observed from May 1953 to May 2023, with temperature anomalies from the GISTEMP v4, with 1951–1980 baseline (GISTEMP Team, 2023; Lenssen et al., 2019). A black and white version of this figure will appear in some formats. For the colour version, please refer to the plate section.

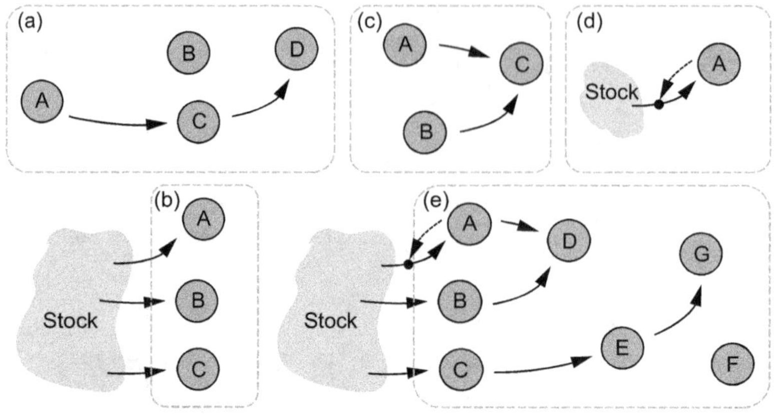

Fig. 5.14 Concept of a compound event following the IPCC definition (Seneviratne et al., 2012, p. 118). (a) Combinations of causally unrelated events and simple one-to-one chains of events (Section 5.2). (b) A common external forcing factor (e.g., warming) changing the probability or intensity of multiple events (e.g., tropical cyclone and heatwave; Matthews et al., 2019). (c) The occurrence or impact of one event conditional on the occurrence of more than one other event (e.g., extreme wildfire due to combination of drought and strong winds). (d) System feedback loops reinforcing an event (e.g., drought or heatwave amplification from soil moisture–temperature coupling; Seneviratne et al., 2010). (e) A complex combination of all the above (e.g., complex Earth system). A compound event is also called a meta-event in this book.

the climate system to explain changes in the GMST. It is able to reproduce the broad outline of climate history from 1850 to 2015 and is coherent with the IPCC forecasts up to 2100. It is here chosen over other models for its simplicity. ESCIMO models global carbon flows, global energy flows, and global albedo changes that influence the evolution of the GMST via a set of around 50 non-linear differential equations (and a total of almost 1,200 variables). It follows Forrester's system-dynamics approach of stocks, flows, and feedback loops (Section 4.2). There are seven major feedback loops in ESCIMO (Fig. 5.15), summarized below.

1. *Balancing feedback loop*: A higher GMST (i.e. warming) leads to higher heat radiation from the Earth's surface and higher heat flow into space, which in turn reduces the amount of heat in the atmosphere and thus lowers the GMST.
2. *Self-reinforcing feedback loop*: Warming leads to a higher concentration of water vapour in the atmosphere, which, in turn, blocks a larger fraction of the heat that escapes into space. The result is more heat in the atmosphere and a higher GMST.
3. *Balancing feedback loop*: Warming increases the area covered by low clouds, which in turn leads to higher reflection of the incoming sunlight, with a cooling effect (high clouds have a warming effect 10 times weaker than that of low clouds).
4. *Self-reinforcing feedback loop*: Warming leads to melting of ice and snow, exposing darker surfaces, which in turn reduces the albedo and leads to higher absorption of incoming sunlight and increased warming (this effect is reduced in the long term, with brighter deserts replacing dark forests).

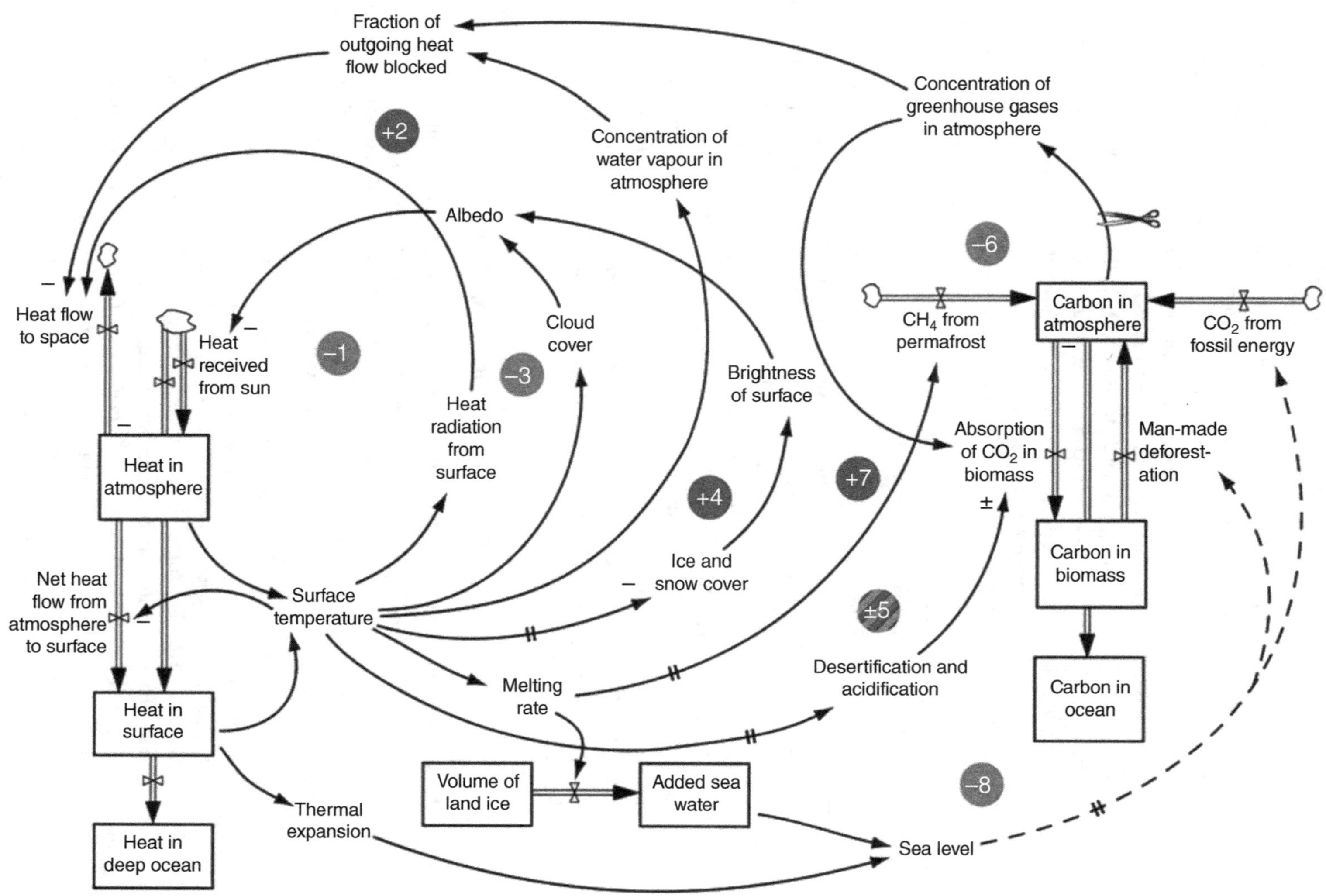

Fig. 5.15 Causal loop diagram of the ESCIMO global climate model, reproduced from Randers et al. (2016, fig. 2) (CC-BY 3.0).

5. *Balancing feedback loop*: Warming increases the rate of biomass growth through photosynthesis, which in turn reduces the amount of CO_2 in the atmosphere. This reduces the concentration of greenhouse gases, which in turn blocks a smaller fraction of heat that escapes into space, finally reducing warming.
6. *Balancing feedback loop*: More CO_2 in the atmosphere leads to increased absorption of CO_2 in terrestrial biomass (through fertilization), which, in turn, leads to less greenhouse gases in the atmosphere and less absorption in terrestrial biomass. In parallel, more carbon in the atmosphere leads to more CO_2 in the ocean surface layer (through chemical diffusion) and in turn to faster removal of CO_2 from ocean surface water through assimilation in aquatic biomass and sedimentation (reversed in the long term by the water becoming more acidic).
7. *Self-reinforcing feedback loop*: Warming melts permafrost, which releases CH_4 (methane) and CO_2 from thawed organic material. This increases the concentration of greenhouse gases in the atmosphere, which in turn blocks a bigger fraction of heat, which escapes into space. This results in increased warming.

One can note from the described processes that the climate system involves interactions between five major subsystems: the atmosphere, the hydrosphere, the cryosphere, the land surface, and the biosphere (e.g., Fan et al., 2021). The ESCIMO model evaluates the evolution of the GMST up to 2100. Figure 5.12a shows the model base run forecast in comparison to the GMST observations between 1880 and 2015. A reasonable match is observed between observations and model, with the temperature predicted to peak at $\approx +2.4$°C, compared to pre-industrial levels, around year 2085. Parameters of the model may also be changed to investigate the impact of different policies on future GMST trends (Ex. 5.5a). Notably, Randers et al. (2016) estimated that a significant increase in volcanic activity would lead to slight cooling while a significant increase in wildfire activity would lead to slight warming.

5.3.1.2 Consequences for Perils

Most categories of perils can be perturbed by global warming. An increase in GMST can have a direct impact on climatological, meteorological, and hydrological perils and a secondary impact on other perils via cascading effects. Although the dynamics of extremes is complex, with compound events potentially depending on a myriad of variables (Leonard et al., 2014), some general consequences are worth mentioning.

- *Climatological perils*: The frequency and intensity of heatwaves (HW), droughts (Dr), and wildfires (WF) are directly impacted by increasing temperatures and drier conditions (van Aalst, 2006; Seneviratne et al., 2010; Vicente-Serrano et al., 2014; Abatzoglou and Williams, 2016). For some regions, it could reflect a regime shift toward desertification due to a change in atmospheric circulation (e.g., Fan et al., 2021).
- *Meteorological perils*: Since tropical cyclones (TC) transform latent heat of condensation into mechanical wind energy, they become more intense, and the largest ones more frequent, with increasing temperatures (Knutson et al., 2020; Emanuel, 2021). Similarly, the likelihood of hurricane-force extratropical cyclones (xC) hitting western

Europe could become greater (Haarsma et al., 2013). Extreme precipitation events (rainstorms, RS) also become more common as the water-holding capacity of the atmosphere increases with rising temperature (e.g., Walsh and Pittock, 1998; Allen and Ingram, 2002; Held and Soden, 2006).

- *Hydrological perils*: More extreme precipitation events also naturally lead to more fluvial and pluvial floods (FF, PF) (Allen and Ingram, 2002; van Aalst, 2006). With sea levels rising due to ocean thermal expansion, land-ice melting, and larger storms, storm surges (SS) are expected to have greater consequences (e.g., Anthes et al., 2006).
- *Biological perils*: Climate change can enable pest proliferation and their poleward migration (pest infestations, PI) (Patterson et al., 1999). As a corollary, it can also significantly affect vector-borne diseases in humans (epidemics, Ep). Disease transmission would increase in regions becoming warmer and more humid, but decrease in warmer and drier conditions (Gage et al., 2008).
- *Geophysical perils*: Landslides (LS) become more likely as extreme precipitations become more frequent (van Aalst, 2006).
- *Technological perils*: Climatic extremes can more readily damage critical infrastructures, enhancing the overall risk of domino effects. Moreover, heatwaves (HW) and cold waves (CW) can trigger blackouts (BO) due to power grid congestion (Burillo et al., 2017). Water scarcity can stress water systems (Wf), as well as energy systems dependent on water cooling or storage (energy supply failures, Ef), such as nuclear power plants and hydro-dams (Mikellidou et al., 2018).
- *Socio-economic perils*: As the annual average temperature exceeds 13°C for a given country, its economic production starts to decline (Burke et al., 2015). Many businesses (e.g., agriculture, tourism) can also be disrupted (BI) by population migrations and land-use changes, which can act as stressors and be caused by new flood patterns (Kam et al., 2021), saltwater intrusion, and drought on crops and livestock (van Aalst, 2006), etc. There exists a multitude of transmission channels from climate risk drivers to the financial risks faced by the banking sector (Basel Committee on Banking Supervision, 2021). Economic deterioration combined with climatic stress can possibly lead to social unrest (SU) and ultimately to armed conflicts (AC) (Hsiang et al., 2013; Gleick, 2014). For a review of the impact of climate change on crops, income, health, and violence, see Carleton and Hsiang (2016).

Further cascading effects (too many to list) could occur according to the catastrophe network links defined in Section 5.2.3. Note that patterns of extremes due to global warming differ between different parts of the world. A discussion on regional patterns and their modelling via downscaling approaches is, however, beyond the scope of this textbook (Emanuel, 2021). The level of extreme events to be expected in the future due to global warming highly depends on the number of tipping points being crossed (Lenton, 2011; McKay et al., 2022). At the time of writing this textbook, there are concerns that the tipping point toward West Antarctic Ice Sheet (WAIS) collapse might have already been crossed (Rosier et al., 2021; Naughten et al., 2023) and that the critical transition to Amazon forest dieback may be fast approaching (Boulton et al., 2022). Relatively simple system-dynamics models are available to describe the general processes of tipping points, for example of ice sheet loss, monsoon disruption, and collapse of the Atlantic meridional overturning circulation (AMOC) (e.g., Ritchie et al., 2021). Particularly worrying is the concept of a **tipping cascade** in which tipping a threshold increases the likelihood of tipping another (Klose et al., 2021). Climate-related cascading effects spilling into ecosystems (e.g., forest dieback) and into the socio-economic system will be described in the next two sections.

5.3.1.3 Implementation in a CAT Model

The consequences of global warming on existing perils in the years and decades to come may be assessed from the long-term trends predicted by global climate models. Given one such model and some relationships linking warming to hazard variables, especially the occurrence rate $\lambda(t)$ and the intensity $I(t)$, it becomes trivial to update the CAT model parameterization for a time t in the future (Section 5.1.3). This is so far mainly tested for future tropical cyclone CAT risk (Knutson et al., 2020; Bloemendaal et al., 2022). However, outside statistical analyses, there is no simple way to quantitatively relate tropical cyclone rate and intensity to atmospheric conditions (Walsh et al., 2015). The student of risk would need to have access to the outputs of sophisticated numerical climate models to assess future hazard trends. For instance, considering a GMST increase of 2°C (i.e., year 2055 projection in CMIP5; Taylor et al., 2012), the maximum windspeed v_{max} is forecasted to increase by 5% on average across various climate models (Knutson et al., 2020). One of the most transparent approaches is the one of Bloemendaal et al. (2022) where the authors provide the list of climate model variables used as input to generate a stochastic set of tropical cyclones for the year 2050 (see their supplementary materials). The genesis of heavy precipitations and wildfires also depend on various subsystem couplings which are only captured in complex numerical models (Powers et al., 2017). Following the philosophy of this textbook, which is to allow the reader to make simple computations, analytical expressions linking temperature change to key hazard variables based on first physical principles would be ideal. One such law is the TC precipitation rate increasing by 7% per 1°C, according to the Clausius–Clapeyron equation (Held and Soden, 2006; Knutson et al., 2015). Considering the equations of Chapter 2, heatwave (HW) intensities can be directly calculated from projected future temperature and humidity values, and storm surge (SS) height derived from the projected windspeed $v_{max}(t)$ (Eq. (2.16)), corrected for sea-rise trends (e.g., +3 mm/yr; Church and White, 2011). Inland flood (FF, PF) and landslide (LS) amplification could similarly be inferred. Empirical relationships may also be useful at first order to assess other trends. For instance, it has been claimed that one standard deviation change toward warmer temperatures might increase the frequency of intergroup conflict (AC) by 14% (Hsiang et al., 2013) or that, for each degree-Celsius increase in GMST, global yields of wheat, rice, maize, and soybean may be reduced by 3–7% (Zhao et al., 2017) (going from business interruption, BI, to ultimately famine, Fa). However, all these relationships remain highly approximative due to hidden variable correlations and regional disparities (see, e.g., the debate on global run-off increase from global warming; Labat et al., 2004; Legates et al., 2005).

5.3.2 Ecosystem Collapse

'An **ecosystem** is a dynamic complex of plant, animal, and microorganism communities and the nonliving environment interacting as a functional unit' (Millennium Ecosystem Assessment, 2005, p. v). It has been argued that the Sixth Mass Extinction is underway (Barnosky et al., 2011), with the main factors driving ecosystem collapse being related to human activities such as land-use changes, overexploitation (e.g., deforestation, hunting), pollution, and global warming (Myers, 1987). This situation – in which society has a dominant influence on the Earth system – has been referred to as the **Anthropocene** (Verburg et al., 2016). Impacts on the biosphere can be difficult to evaluate due to the high complexity of ecosystems (Box 5.2). The spruce budworm outbreak case (Section 4.2.3), although only a three-species system, already included the complex dynamics of a regime shift caused by a cusp bifurcation. To consider the impact of humans on ecosystems, social–ecological models based on system dynamics have been developed to study specific regime shifts (e.g., Ritchie et al., 2021; Willcock et al., 2023). Considering the entire dynamics of the complex Earth system would require assessing a network of cascading regime shifts crossing between the climate system and the biosphere (Rocha et al., 2018). These drivers can accelerate tipping-point dates and, combined, render unstable other subsystems otherwise resilient to individual regime shifts (Willcock et al., 2023). Here, two reduced-complexity **social–ecological models** are described to illustrate the interactions between climate, ecosystems, and human activities. Similarly to the system dynamics section of Chapter 4, processes will be defined from simple ODEs. Similar types of models, but focused on human population dynamics, will be covered in Section 5.3.3.

5.3.2.1 Amazon Rainforest Dieback Model

The Amazon rainforest **dieback**, that is, a loss of biomass by transformation of the Amazon basin into savanna, is recognized as a potential tipping element in the climate system

Box 5.2 Trophic Cascades

Ecosystems involve interactions between multiple animal and plant species and their environment. Rules of cascading catastrophes in ecosystems can be generalized via the concept of **trophic cascades** (Beschta and Ripple, 2009), which are powerful indirect interactions that can control an entire ecosystem. In a top-down cascade, the top predator controls the primary consumer population. Displacing large predators can have a long-term impact on plant communities and potentially lead to a shift towards an alternative ecosystem state. This is due to the role of the top predator in regulating populations of herbivores. The impact on the vegetation can lead to landscape changes due to erosion and other difficult-to-anticipate consequences, such as beaver river dams disappearing, impacting the water system (Beschta and Ripple, 2009). The complexity of ecosystems can be modelled via ecological networks that consider the multiple interactions involving predation, herbivory, parasitism, pollination, etc., in order to explore some well-defined patterns (Montoya et al., 2006). This, however, has yet to be implemented in a CAT risk context.

(Cox et al., 2004; Lenton et al., 2008; Boulton et al., 2022). Such a regime shift would have catastrophic repercussions on biodiversity, carbon storage, and climate change. Megadroughts cause the Amazon carbon sink to decline (Yang et al., 2018) while they exacerbate the risk of wildfires, turning forests into carbon sources. One complex positive feedback loop is the role of droughts on forest evapotranspiration reduction, hampering moisture release and cooling; consequently, this leads to lower rainfall intensity and higher temperatures, further droughts, and further rainforest degradation. A simplified version of the Top-down Representation of Interactive Foliage and Flora Including Dynamics (TRIFFID) model (Cox et al., 2000), proposed by Ritchie et al. (2021), describes the rainforest dieback dynamics by a simple Lotka–Volterra equation (Section 4.2.3) where temperature increase is the external driver (Section 5.3.1). The evapotranspiration feedback loop is not included here. The following Lotka–Volterra equation describes the change in vegetation carbon density N_v:

$$\frac{\mathrm{d}N_v}{\mathrm{d}t} = R_g N_v (1 - N_v) - c_\gamma N_v, \tag{5.9}$$

where the parameter c_γ is a disturbance rate and R_g the growth rate defined as a parabolic function of the local temperature T_l:

$$R_g = R_{g,0}\left[1 - \left(\frac{T_l - T_{opt}}{c_\beta}\right)^2\right]. \tag{5.10}$$

There is an optimal temperature T_{opt} for which growth is maximal and equal to $R_{g,0}$ (note that $R_g < 0$ implies additional tree mortality). The parameter c_β determines the temperature half-width for which vegetation grows (Ritchie et al., 2021). There is an additional feedback on the local temperature, which increases as the vegetation coverage declines (i.e., climate–vegetation feedback):

$$T_l = T_f + c_\alpha (1 - N_v), \tag{5.11}$$

where c_α is the difference between surface temperature of bare soil and forest, and T_f is the temperature for total forest cover. The effect of global warming can be injected in T_f as external forcing. Figure 5.16a shows the dynamics of the system for the parameter values of Ritchie et al. (2021). One clearly observes the collapse of the forest system when the tipping point is crossed.

5.3.2.2 Lake Eutrophication Model

Pollution is another anthropogenic factor potentially leading to ecosystem collapse. An example is the diffusion of by-products of human activities, such as phosphorus from agricultural and urban lands into lakes. This can lead to a regime shift from an oligotrophic to an eutrophic lake, corresponding to a sudden change from a clear water system to one with excessive growth of algae (i.e., algal bloom) due to phosphorus acting as a fertilizer. This **eutrophication** has harmful consequences on the organisms populating the lake, due to oxygen depletion caused by bacterial decomposition of dead algae and some toxin release. In turn, it can have a detrimental impact on the local lake economy (with business interruptions, BI), such as on fisheries, with the collapse of fish populations, as well as on tourism and property value due to the toxicity of the lake and the unpleasant aspects of an algal bloom (murky greenish water, smell). The process can also occur in rivers, estuaries, and coastal oceans (e.g., Carpenter et al., 1999). However, the lake, as a more isolated mass of water, provides a simpler modelling set-up. The dynamics of the amount of phosphorus N_P in the water column can be described by the equation

$$\frac{dN_P}{dt} = R_{in} - R_{out}N_P + \frac{R_{max}N_P^\alpha}{N_m^\alpha + N_P^\alpha}, \tag{5.12}$$

where R_{in} is the phosphorus input rate (the controlling pollution factor), R_{max} is the maximum recycling rate, R_{out} is the phosphorus loss rate by sedimentation, outflow or sequestration in biomass, the power exponent α is the strength of the recycling response to phosphorus concentrations, and N_m is the phosphorus amount at which recycling reaches half the maximum rate (Carpenter et al., 1999). Notice that the third term on the right-hand side is a sigmoid function, which is reminiscent of the spruce budworm outbreak dynamic system of Section 4.2.3. Figure 5.16b shows the dynamics of the system for the parameter values of Wang et al. (2012). In addition to the clear regime shift, one can observe a hysteresis as one tries to reduce the pollution by decreasing R_{in} (similarly to what was presented in Section 4.2.3). For more details on the role of model parameterization, including the addition of an economic model, see Carpenter et al. (1999).

5.3.2.3 In the Context of a CAT Model

The two previous examples showed the impact of global warming and of pollution, respectively, on ecosystem collapse. In a CAT model in which risk is defined in economic terms, exposure could be defined in terms of the goods and services provided by ecosystems (Carpenter et al., 1999; Millennium Ecosystem Assessment, 2005). The collapse of some local or regional ecosystem could, for instance, be interpreted as a business interruption (BI event) due to a drastic reduction of the crop yield or livestock production. Similarly to the CAT models that provide risk projections of the impact of global warming on climatological and weather perils, one could, in theory, estimate the BI risk given the temperature $T_f(t)$ or pollution rate $R_{in}(t)$ expected at a future time t (Fig. 5.16). Due to the system switching back and forth between alternative states prior to a permanent change of regime (e.g., Wang et al., 2012), a rate of occurrence $\lambda(t)$ would be quantifiable in theory but subject to high uncertainties in practice. Moreover, many potential losses are **intangible**, meaning that they cannot be easily quantified. For instance, what is 'the worth of a songbird' in ecological economics? (Funtowicz and Ravetz, 1994). Similarly, what is the market value of ecosystem services such as water and air purification, crop pollination, climate regulation, or protection against various perils (including against floods, landslides, wildfires, or pests)? (Kousky and Light, 2019). Despite the many modelling obstacles and the challenges of insuring natural resources and ecosystem services (Kousky and Light, 2019), the CAT model formalism provides a standardized approach for the development of innovative risk transfer mechanisms to repair ecosystems after disasters, and of risk reduction strategies to prevent risks from reaching such extreme levels that they become uninsurable (e.g., post-regime shift, there is almost a certainty of total loss). Risk transfer and risk reduction tools will be presented in Part III.

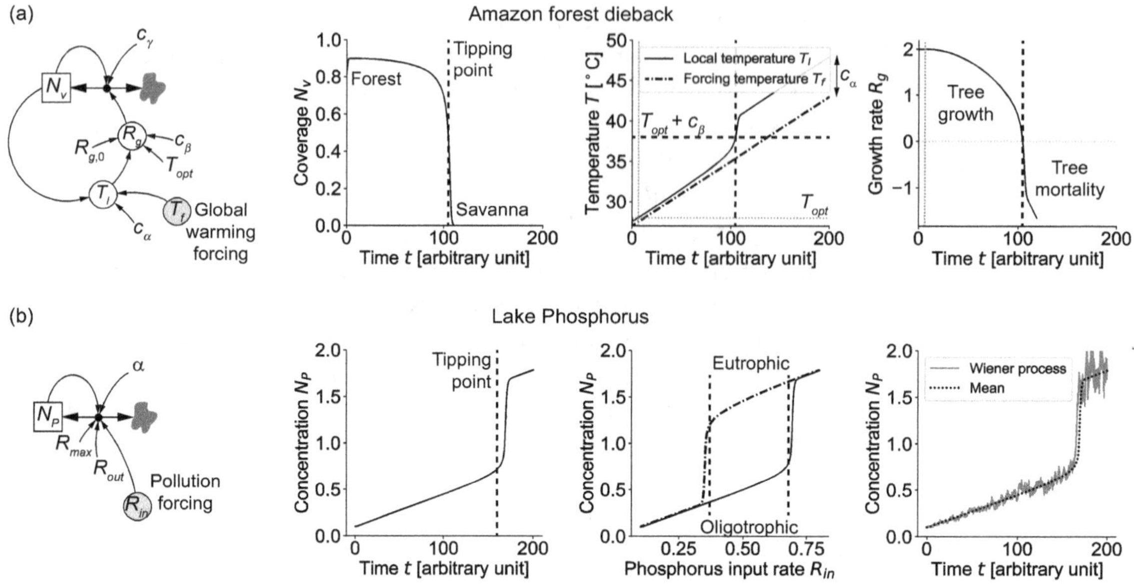

Fig. 5.16 Examples of models describing ecosystem collapse due to a regime shift. (a) Amazon forest dieback with the causal loop diagram representing Eqs (5.9)–(5.11) with parameters $c_\alpha = 5°C$, $c_\beta = 10°C$, $R_{g,0} = 2$ per yr, $c_\gamma = 0.2$ per yr, and $T_{opt} = 28°C$ (Ritchie et al., 2021). The graphs show the impact of the different parameters on the dynamics of the system, which abruptly shifts from a forest to a savanna when the local temperature T_l reaches $T_{opt} + c_\beta$ due to the constantly increasing T_f (global warming forcing). (b) Lake eutrophication by phosphorus pollution with the causal loop diagram representing Eq. (5.12) with parameters $R_{out} = 1$, $R_{max} = 1$, $\alpha = 8$, $N_m = 1$ from Wang et al. (2012), with R_{in} increasing linearly from 0.1 to 0.8 (notice the different path when decreasing from 0.8 to 0.1). Note that white noise is commonly added to take into account random fluctuations in a dynamic system (i.e., **Wiener process**), here $dN_P = f(N_P)dt + \sigma N_P \mathcal{N}(0, \Delta t)$ with $\mathcal{N}$ the normal distribution with mean 0 and standard deviation $\sqrt{\Delta t}$, Δt the time bin, and σ a scaling factor (here 0.1 for illustration purposes).

5.3.3 Societal and Civilizational Collapse

Societal collapse can be defined as a 'significant sociopolitical fragmentation and/or state failure along with the relatively rapid, enduring, and significant loss [of] capital, and systems identity; this can lead to large-scale increases in mortality and morbidity' (Kemp et al., 2022, table 1). A **civilizational collapse** represents the collapse of a more extensive and complex societal structure that spans a vast geographical area or even multiple continents, such as an empire or a cultural sphere, leading to the potential loss of shared knowledge, technologies, and institutions. There must be a combination of severe environmental and/or societal stressors for such an extreme catastrophe to occur. As Tainter (1988) postulated in his seminal work *Collapse of Complex Societies*, these stressors, such as conflicts, natural disasters, or resource depletion, require the society to have already undergone a process of economic decline to trigger the collapse. This hypothesis suggests that economic power and human ingenuity may act as counteracting forces, potentially finding solutions to mitigate the impact of such stressors (Nordhaus, 1992). According to

Diamond's (2011) *Collapse: How Societies Choose to Fail or Succeed*, it is the choices made by societies and their ability to adapt that ultimately determine their fate. The analysis of historical events by the scientific method, termed **'cliodynamics'** (Turchin, 2008), aims at modelling the dynamics of past societies, including catastrophic changes (examples include the falls of the Roman Empire, of the classic Maya civilization, and of Easter Island).

In the present day, there exists the alarming scenario of a worldwide societal collapse. The concept can be illustrated using world-dynamics models, such as those presented in the groundbreaking work *The Limits to Growth* (Meadows et al., 1972, 2004) and adopted from the system-dynamics approach developed by Forrester (1973). These models highlight resource depletion as a central driver of potential collapse. Additionally, the current concern of climate change and its potential cascading effects on a global scale (Sections 5.3.1 and 5.3.2) have also raised the spectre of worldwide collapse. Kemp et al. (2022) described the dynamics of the process with a causal diagram where state fragility and mortality (two metrics of societal collapse) are caused by resource scarcity, diseases, and conflicts, which can all be triggered by global warming (e.g., food and water shortage by ecosystem service collapse, vector-borne diseases by pest migrations, and conflicts by extreme weather events and linked population displacements). In the most extreme case, if positive feedback loops leading to regime shifts cascading from the climate system to the ecosystems that support our global society cannot be avoided, the situation might escalate into an existential threat (Kareiva and Carranza, 2018). The threshold has been set to approximatively a 6°C increase compared to pre-industrial levels. This so-called **'Hothouse Earth'** (Steffen et al., 2018) is plausible when considering potential tipping cascades in the complex Earth system.

In order to gain a deeper insight into these complex dynamics, the World2 model of Forrester (1973) will be introduced. This model will shed light on the interplay between environmental stressors, the socio-economic system, and the possible collapse of the human population. In this context, the primary driving factor will be resource depletion, as the ramifications of global warming have already been partially covered (for instance, an Amazon rainforest dieback, detailed in Section 5.3.2, would result in the collapse of ecosystem services, water and food scarcity, and substantial socio-economic repercussions; Lapola et al., 2018).

5.3.3.1 World Dynamics Models (World2, World3)

System-dynamics models were developed in the early 1970s to simulate the behaviour of global ecological and socio-economic systems and were originally designed to investigate potential solutions to a wide range of global issues. These models are World2, documented in Forrester (1973), and World3, presented in Meadows et al. (1972) (see also World1 in Box 5.3). Considered the first model formulation at the scale of the Anthropocene (Verburg et al., 2016), **World3** is composed of 150 governing equations describing the human population dynamics in relationship to food production, industrial production, pollution, and consumption of non-renewable resources. In comparison, World2 comprises a modest set of 43 governing equations. **World2** focuses on five main stocks: population (P), natural

Box 5.3 The Legacy of J.W. Forrester

System dynamics was founded by Jay Wright Forrester (1918–2016), an American computer engineer and systems scientist. He first applied the method to factory planning in *Industrial Dynamics* (Forrester, 1961), and then to city planning in *Urban Dynamics* (Forrester, 1969). Forrester's work on urban dynamics was noticed by the Club of Rome which led to discussions on global sustainability and the book *World Dynamics* in 1971 (Forrester, 1973). He originally sketched a model called World1 on his flight home from his meeting with the Club of Rome in Switzerland (Hayes, 1993). World2 was published in 1971 and World3, by the Meadows team, the following year. Interestingly, Forrester's work on city planning also had an impact on the video game industry, being an inspiration for SimCity. SimCity is worth mentioning here for its inclusion of various disasters: earthquakes, floods, tornados, fires, nuclear meltdowns, and some exotic peril (monster attacks), alongside smaller-scale events, such as criminality and traffic congestion. Additional perils were added in later versions, including asteroid impacts and (zombie) outbreaks among others (e.g., Mignan, 2022b).

resource (NR), capital investment (CI), capital-investment-in-agriculture fraction (CIAF), and pollution (POL). World3 retained the core elements of World2 while incorporating additional variables and feedback mechanisms to better represent real-world dynamics (e.g., more detailed sub-models for agriculture, industrial production, and pollution generation). World3 is renowned for being the underlying model of *The Limits to Growth* (Meadows et al., 1972, 2004), which is infamous for predicting a collapse of our socio-technological–natural system by the mid-twenty-first century. Although World2 lacks the complexity of its successor, this simpler model includes the same key mechanisms and yields comparable conclusions. It finds its relevance here for illustrative purposes, as its causal diagram fits neatly on one page (Fig. 5.17), unlike the more extensive World3 model. The World2 model algorithm (in the DYNAmic MOdels (DYNAMO) language) is provided in Forrester (1973) and translated into modern programming languages in the supplementary material of this textbook. The result of the model's standard run up to year 2100 is shown in Figure 5.18: one can observe that natural resources are progressively depleted; population peaks at about 2030 before collapsing; a similar pattern is observed for pollution and capital investment but with a lag and peak at about 2050. The capital-investment-in-agriculture fraction decreases from the 1950s. These 1970s models have been shown to be essentially correct in their forecasts so far, despite being crude representations of the real-world dynamics. Turner (2008) showed that 30 years of historical data compare favourably with key features of a business-as-usual scenario called the 'standard run' scenario, which results in a collapse of the global system midway through the twenty-first century (for other scenarios of the world's future and potential impacts of policy interventions, see Exercise 5.5b). 'Both [World2 and World3] models [were] intended to open research and discussion toward improved understanding of social systems' (Forrester et al., 1974, p. 170). Consequently, they serve as valuable foundations for the student of risk interested in exploring the potential of world-dynamics models.

Noteworthily, some economists have criticized the pessimistic predictions of the World2 and World3 models. Of all the criticisms (Nordhaus, 1973), while many have been deemed unwarranted (Forrester et al., 1974), one deserves particular mention. The influence of

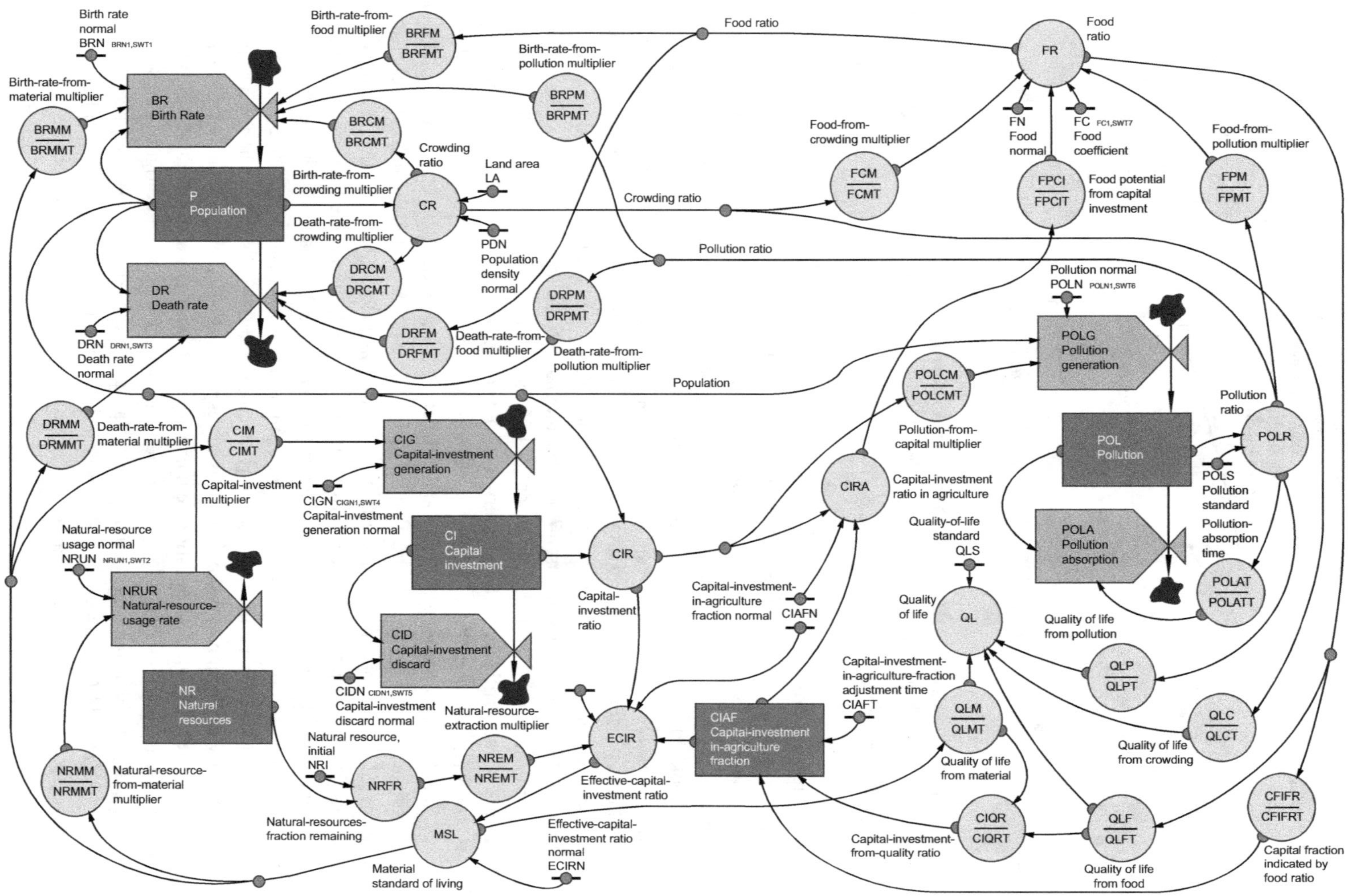

Fig. 5.17 Diagram of the World2 model interrelating the five variables: population, natural resource, capital investment, capital-investment-in-agriculture fraction, and pollution. Redrawn/modified from Forrester (1973, fig. 2-1). Original figure courtesy of the System Dynamics Society, https://systemdynamics.org.

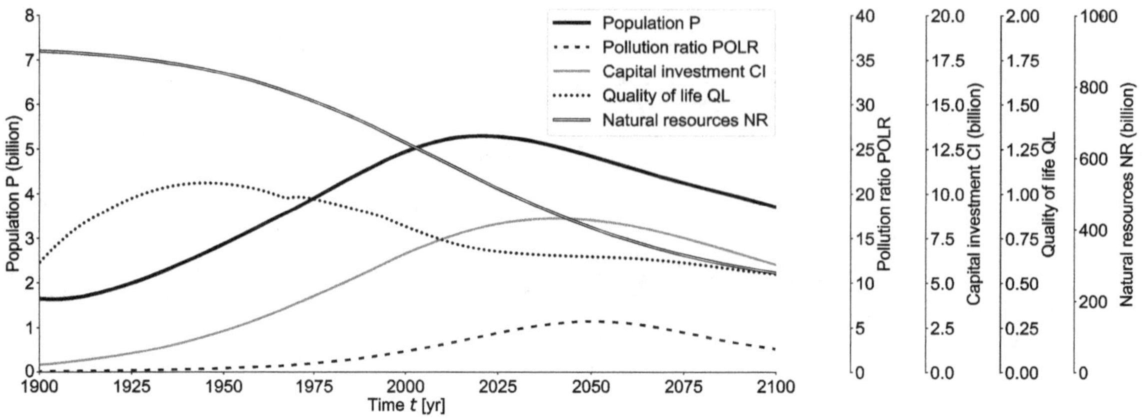

Fig. 5.18 Results of the World2 model's standard run, showing the evolution of population, pollution ratio, capital investment, quality of life, and natural resources. Notice the predicted reduction in population while resources get depleted (see other scenarios in Exercise 5.5b).

resources on production is defined in world dynamics as a long-term process that reflects greater required efforts as stocks are depleted. A different view is to assume that, due to technological progress, production can be sustained at any amount with a lower resource input. The debate between the two views has been fierce and is outside the scope of this textbook (Nordhaus, 1973; Forrester et al., 1974; Fiddaman, 2002; Hänsel et al., 2020; Keen, 2021). However, to avoid introducing a bias toward catastrophic thinking, the role of human ingenuity in addressing the aforementioned challenges will be discussed briefly for completeness. Historically, Malthus' concerns of a world population increasing faster than the food supply (Malthus, 1798) was alleviated to a large extent by the Second Agricultural Revolution, which took place in the nineteenth century and was linked to the use of chemical fertilizers such as superphosphates (Thompson, 1968). Concerns of overpopulation again emerged by the mid-twentieth century and climaxed with the publication of *The Population Bomb* (Ehrlich and Ehrlich, 1968). Predictions of worldwide famines fortunately did not materialize thanks to the Green Revolution, which brought significant advancements in world agriculture by the late 1960s, such as the introduction of high-yielding varieties of cereals (Borlaug, 1971). Simply put, the logistic growth underlying most of the previously described models (Box 4.3) is observed for organisms which cannot plan in advance, such as bacteria on a Petri dish (Thornton, 1922; Meyer et al., 1999). Innovations manifest as successive technological revolutions in the form of cascading S-curves, leading to radical changes to respond to new market demands (Christensen, 1992; Perez, 2010). The question remains whether innovation can endure indefinitely in response to finite resources (e.g., hypothetical nuclear fusion technological revolution). If one can hope that human ingenuity will play a critical role, considering scenarios *à la* World2/World3 (but also *à la* ESCIMO on the climate front) are crucial to better understand the world's dynamics based on current knowledge (Forrester et al., 1974) and to compel action (e.g., Kemp et al., 2022). If future environmental conditions are confirmed to be far more dangerous than currently believed, it would indeed become a challenge of avoiding a 'ghastly future' (Bradshaw et al., 2021).

5.3.3.2 In the Context of a CAT Model

The concept of societal collapse goes beyond the applications of current CAT models. Yet, similarly to the innovative methods being considered to insure ecosystems (Kousky and Light, 2019), the probabilistic assessment of a collapse using a holistic CAT modelling framework would be helpful to pose the problem, analyse the mechanisms, and inform decision-makers. This is in agreement with the 'integrated catastrophe assessment' of worldwide societal collapse due to climate change suggested by Kemp et al. (2022) and formalized by, for example, Pescaroli and Alexander (2018) or Simpson et al. (2021). Many of the perils contributing to societal collapse, such as armed conflicts, weather perils, and diseases, as well as cascading effects leading to food scarcity (basically a type of business interruption) and ultimately to famines, have all already been mentioned or modelled in this textbook (see, e.g., Maslow's hierarchy of needs; Fig. 1.21). Challenges remain regarding the high complexity of the Earth system and how to realistically model all these interdependencies. Clearly, the outcome of a societal collapse CAT model would find very limited use in risk transfer instruments but would be critical for risk reduction planing and worst-case scenario analysis. For this reason, the standard hazard/exposure/vulnerability/loss pipeline (Fig. 3.1) should be extended to include the role of response to risk within the modelling framework (Simpson et al., 2021).

5.4 Concluding Remarks

This chapter showed the richness of cascading effects from a CAT modelling perspective. The described processes were increasingly complex, from clustering to compound effects via domino effects. The aggregate effects of individual events led to low-probability–high-consequences meta-events, sometimes called super-cats or compound events depending on the domain of expertise. Feedback loops caused by the inclusion of environmental variables, described for individual perils in Chapter 4, were here described as regime shifts in the complex Earth system, including the impact of various worrying trends (global warming, pollution, resource depletion) on the future frequency and intensity of individual perils, and their subsequent impact on ecosystems and the global socio-economic system. Interactions were observed at different spatiotemporal scales, from cascading explosions at a refinery to the potential collapse of the Amazon rainforest at continental scale.

Despite the variety of the interaction processes involved, two main modelling approaches were used: the encoding of one-to-one interactions in an adjacency matrix for domino effects and physical network failures, and the definition of non-linear dynamical processes in sets of ODEs to model feedback loops leading to catastrophic regime shifts. Examples were kept as simple as possible to illustrate the main concepts while covering as many physical processes as possible. It was demonstrated that these complex interactions can, in theory, be implemented in the CAT risk modelling framework. The development of a holistic, all-encompassing CAT model remains at a very preliminary stage of research and it is the hope of the author that Parts I and II of this textbook, concluding with this chapter, can contribute to some further development by some enthusiastic students of risk or other cross-disciplinary professionals.

The concluding part of this textbook, Part III (Chapter 6) on catastrophe risk management, will offer a comprehensive exploration of risk management strategies rooted in CAT model outcomes. It will heavily draw upon the insights gained in Part I while also mentioning the challenges associated with dealing with the complex risks described in Part II.

Further recommended reading

- *Collapse of Complex Societies* (Tainter, 1988)
- *Limits to Growth – The 30-Year Update* (Meadows et al., 2004)
- *Climate Change 2021: The Physical Science Basis* (IPCC, 2021)

5.5 Exercises

Exercise 5.1

(a) Prove that the Poisson distribution (i.e., mean $\lambda =$ variance σ^2) is a special case of the negative binomial distribution. From your solution, prove that the negative binomial distribution is always overdispersed (i.e., $\psi > 0$).

(b) The annual counts of Category 4 and 5 hurricanes which occurred in the eastern Pacific basin between 1980 and 2020 are $n = (1, 0, 1, 5, 4, 3, 3, 2, 2, 2, 4, 2, 7, 7, 5,$ $3, 1, 6, 3, 1, 1, 2, 4, 0, 2, 1, 3, 1, 1, 3, 1, 5, 1, 0, 6, 10, 5, 2, 10, 3, 3)$ (see time series in Figure 5.1b).

First, calculate the dispersion statistic ψ to verify that the data are indeed overdispersed.

Second, reproduce the histogram of Figure 5.1c by showing the data count distribution alongside the distributions obtained from Poisson and negative binomial simulations (to be used in (c) also).

Hint: A code template is provided in `exercise_notebook.ipynb` for your convenience.

(c) Use the ELT and method of Exercise 3.3 to generate a YLT from the negative binomial distribution for the ψ estimate obtained in (b). Then calculate and plot the OEP and AEP curves. Finally, verify that the area under the AEP curve equates to the AAL.

Exercise 5.2

Your goal is to define the transition-rate matrix $\mathbf{Q}$ to represent the set of ODEs of the SIR epidemiological model (Eq. (4.15), reproduced below) as a master equation (Eq. (5.4)). This should help you better understand the link between the ODEs presented in this textbook and the concept of the master equation.

Hints: Use matrix multiplication and, instead of probability p, the rates S, I, and R giving the input vector (S, I, R). Verify that the elements of your $\mathbf{Q}$ matrix follow the condition $\sum_{j=1}^{n} q_{ij} = 0$ (Box 5.1).

$$\begin{cases} \frac{\mathrm{dS}}{\mathrm{d}_t} = -\beta \mathrm{SI}/N \\ \frac{\mathrm{dI}}{\mathrm{d}_t} = \beta \mathrm{SI}/N - \gamma \mathrm{I} \\ \frac{\mathrm{dR}}{\mathrm{d}_t} = \gamma \mathrm{I} \end{cases}$$

Exercise 5.3

(a) Fill in the missing values of the four centrality measures in the caption of Figure 5.3.

(b) Let's now consider the equation $\mathbb{P}(X_t = j|X_0 = i) = \mathbf{P}^t$ (Eq. (5.7)). Draw a logic tree of all possible paths from $i = 1$ to $j = 3$ for $t = 2$ in a 4×4 transition matrix. Demonstrate that it is equivalent to $\mathbf{P}^2$, that is, the multiplication of matrix $\mathbf{P}$ with itself.

(c) Figure 5.19 represents an oil storage facility where three atmospheric storage vessels are present (vessels with coordinates $x = (0, 50, 100)$ m and $y = 0$ m, diameter $D = 15$ m and height $h = 4$ m). A VCE explosion occurs in the first one on the left (ID = 1). What is the probability $\mathbb{P}(X_2 = 3|X_0 = 1)$ of the three vessels exploding in a domino effect with the chain $1 \rightarrow 2 \rightarrow 3$?

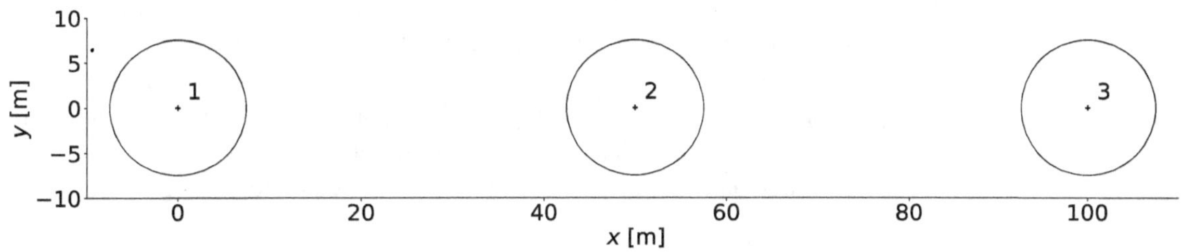

Fig. 5.19 Exercise 5.3c: Set-up of the critical infrastructure. Location of three atmospheric storage vessels at an oil storage facility. See text for details.

This exercise requires using models from Chapter 2 to estimate the size of an explosion assuming a point source (Eq. (2.7) with $\eta = 0.03$, $\Delta H_c = 46.4$ kJ/g, $\rho_{gasoline} = 742.9$ kg/m^3, and 1% of total mass participating in the explosion yield) and the overpressure field. An explosion is assumed to occur when there is failure of the tank. Use as generic fragility curve the probit function $\mathbb{P}(X) = \Phi(-20 + 2 \log P)$ where Φ is the cumulative normal distribution and P [Pa] the overpressure (compatible with the models of Salzano and Cozzani (2005); the code of the probit model is provided in `exercise_notebook.ipynb`).

Exercise 5.4

Choose a category of perils in Section 1.3 (e.g., meteorological perils, climatological perils) to define a list of possible chains of events. To do so, follow these three steps: (a) define an adjacency matrix for the n perils mentioned in the category of your choice and write $a_{ij} = 1$ when the interaction $i \to j$ is indicated as possible. (b) Draw the matching graph of interactions (by hand or with a code). (c) List some of the longest chains of events observed in the graph (up to, e.g., $k = 4$ links).

Exercise 5.5

Two complex Earth system models are showcased in `exercise_notebook.ipynb`, offering the student of risk a practical opportunity to engage with these types of sophisticated models.

(a) Explore the ESCIMO model by Randers et al. (2016), with the source file (in .mdl format) graciously provided by Professor Ulrich Golüke. Launch the standard run of the model to obtain the results depicted in Figure 5.12a. Then, alter the parameter values in `exercise_notebook.ipynb` to investigate the effects of stratospheric aerosol injection on potential temperature reduction (caused by a great volcanic eruption and by some radical geoengineering what-if scenario). Comment on the results.

(b) Explore the World2 model by Forrester (1973), here translated from DYNAMO to Python (also available in R in the supplementary material). Commence with the standard run of the model to retrieve the results illustrated in Figure 5.18. Then, modify the function `exercise_codes.World2_model()` to replicate a scenario developed by Forrester (1973): reduce the natural-resource-usage rate, pollution generation, and capital generation by 1970, transitioning NRUN1 from 1 to 0.25, POLN1 from 1 to 0.5, and CIGN1 from 0.05 to 0.0. Describe your observations and draw comparisons to the potential outcomes if these changes were introduced in 2025 instead of 1970.

PART III

CAT RISK MANAGEMENT

6 CAT Risk Management

The main tools of catastrophe risk modelling were described in Parts I and II to understand and quantify the potential impacts of catastrophic events. However, the essence of catastrophe risk modelling lies in its practical application with catastrophe risk management. This involves the identification of risk in a given societal context, the quantification of this risk (as demonstrated by the CAT models discussed in Parts I and II), and ultimately, the implementation of well-informed decisions aimed at reducing the risk (Jones et al., 2017, p. 2). How do the insights derived from CAT models translate into actionable strategies that effectively mitigate the threats posed by both natural and human-made disasters? This final chapter seeks to provide answers to this crucial question.

One strategy for managing catastrophe risk involves the transfer of risk from one entity to another, noting that the tolerance to risk (i.e., the extent to which an entity is willing to take on risk – also called risk appetite) varies significantly between entities. Section 6.1 will describe the mechanisms of risk transfer instruments, represented by insurance and catastrophe bonds, and their significance in a better distribution of risk among different parties. It will be complemented by a description of approaches seeking to directly reduce or altogether prevent the intrinsic risk through **mitigation measures**. While the probabilistic models that constitute the foundation of this textbook have not yet achieved widespread integration into risk management practices, there is a discernible trend toward their increasing adoption. This shift can be attributed to expanding computational capabilities and data resources that contribute to enhancing the efficacy of decision-making processes. As a consequence, the descriptions provided in this concluding chapter will be tailored to the CAT model formalism. Risk transfer strategies will be functions of the risk metrics derived from CAT models, such as the AAL and EP curves, while risk mitigation techniques will depend on the CAT model modules considered. The concept of **peril forecasting**, aimed at facilitating timely responses and preparedness against potential disasters by predicting their occurrence, will also be formalized within the CAT modelling framework and presented in Section 6.2.

Central to effective risk management is the recognition that different stakeholders can exhibit distinct behaviours. These **stakeholders** are individuals, organizations, governments, and any other entities that hold a vested interest in the outcomes of CAT models. Whether it is a homeowner looking to safeguard their property, an insurance company aiming to assess potential liabilities, or a government striving to enhance public safety, stakeholders collectively constitute a complex web of actors engaged in a system termed as risk governance (Renn, 2008). Section 6.3 will highlight the subtleties of risk governance, using a case study that examines anthropogenic seismic risk in the geothermal industry. This analysis will demonstrate how risk governance proves to be a formidable task, requiring the alignment of decision-making with collective goals and values. This challenge arises due to the unavoidable presence of uncertainty in CAT models and to the heterogeneities in risk perceptions held by various stakeholders. Here, these aspects will be quantified in terms of epistemic uncertainty and utility functions, respectively. The case study will serve as a platform to explore the heterogeneous nature of risk governance, revealing how insights derived from CAT models, while subject to debate, can ultimately translate into concrete action plans in the real world.

6.1 Risk Transfer and Risk Reduction Instruments

This section offers a concise overview of the main characteristics of risk transfer, encompassing insurance and CAT bonds. The descriptions will draw upon the basic actuarial concepts already provided in the CAT risk management literature (Grossi and Kunreuther, 2005a; Mitchell-Wallace et al., 2017). Additionally, this section briefly introduces risk reduction methodologies within the CAT modelling framework, keeping one form of risk mitigation strategy – peril forecasting – for detailed examination in Section 6.2.

6.1.1 Risk Transfer via Catastrophe Insurance

Catastrophe insurance involves the sharing of a potential future loss in exchange for an upfront payment of a **risk premium** to the stakeholder taking the extra risk. The risk is effectively transferred from the **policyholder**, who is the individual or company purchasing insurance, to an insurance company through an insurance policy. The aggregation of numerous insurance policies constitutes an **insurance portfolio**. Furthermore, insurers can offload a portion of their risk to reinsurers through risk cession; some of the remaining risk may also be transferred to the financial markets via CAT bonds by either insurers or reinsurers (Section 6.1.2). This establishes a cascading flow of risk transfer, starting from property owners to insurers, reinsurers, and, ultimately, capital markets. Notably, regulatory agencies and commissions overseeing insurance and financial markets also play critical roles (Grossi and Kunreuther, 2005b, fig. 1.3).

The primary objectives of an insurance company revolve around two key constraints: a safety-first constraint and a return-on-assets constraint. *The insurer's imperative is first to withstand catastrophic events and maintain viability. Risk insurability requires an estimate of risk (the CAT model output) and the ability to determine appropriate premiums.* After satisfying the survival constraint, insurers formulate risk management strategies to optimize expected profits. The provision of coverage depends on whether the premium structure allows the insurer to generate a net positive profit. When confronted with elevated uncertainty or ambiguity, insurers might opt for significantly higher premiums. By diversifying their portfolio, insurers can theoretically mitigate the **process risk** associated with stochastic processes (aleatory uncertainty), while retaining some residual exposure to the **parameter risk** subject to the potential bias introduced by incomplete knowledge (epistemic uncertainty). This is achievable when insurers issue a substantial number of

Box 6.1 Portfolio Diversification

The principle of **diversification** in catastrophe risk insurance consists of covering a wide variety of uncorrelated perils and establishing an extensive portfolio widely distributed in space. This serves to reduce uncertainties and therefore the risk load R_l (see Section 6.1.1.3) taken by the insurer. An illustration, similar to the one described by Jones et al. (2017, box 1.1), explains the process: a house worth \$1 million is located in a seismically active area where there is an annual probability $p = 0.01$ of total loss L due to earthquake shaking. The expected loss is $\mathbb{E}[X] = pL = 0.01 \times \$10^6 = \$10{,}000$. While this value is relatively low, the uncertainty is high regarding a total loss with probability p or no loss at all with probability $1 - p$. The variance is $\text{Var}[X] = \mathbb{E}[X^2] - \mathbb{E}[X]^2 = pL^2 - p^2L^2$, giving the standard deviation $\sigma = L\sqrt{p(1-p)} = \$10^6 \times \sqrt{0.01 \times (1 - 0.01)} = \$99{,}499$. Considering a group of $n = 1{,}000$ identical houses with identical earthquake risk, but in different locations, the expected loss per house remains the same. However, depending on the house locations being correlated or not, the standard deviation will differ. If only one house can be affected by each earthquake (i.e., the risk to each house is independent), then the standard deviations are combined by taking the square root of the sum of squares (i.e., sum of variances), which is equal to $\sigma_n = \sqrt{n \times \sigma^2} = \$\,3{,}146{,}427$, or $\sigma = \$3{,}146$ per house. If all houses were similarly affected by each earthquake (100% correlated), then the standard deviations would add up, meaning no reduction in σ. The reduction in loss uncertainty observed in this simple example is one of the fundamental reasons why insurance is a viable concept. Real catastrophe processes lie somewhere between complete independence and perfect correlation, which are quantified via the CAT risk models previously described. A problem occurs when hidden correlations are missed, for example due to cascading effects and other compound effects (Chapter 5).

policies with spatially distributed and independent losses (Dong and Grossi, 2005) (Box 6.1). The role of epistemic uncertainties on risk management will be covered in the final case study (Section 6.3.2).

The most common risk metrics derived from the ELT and YLT were comprehensively defined in Chapter 3. These are the AAL, EP curves, VaR, and TVaR (Section 3.3.2). These metrics serve as crucial tools in catastrophe insurance, as evidenced below. Notably, an array of additional metrics exists; however, their description falls beyond the scope of this introductory textbook. Examples of such metrics are discussed in works such as Jones et al. (2017). An important attribute of the loss tables that has so far remained unaddressed is the standard deviation σ_{ID} associated with the loss L_{ID} for each distinct stochastic event. This critical aspect, which quantifies the level of loss dispersion, will now be covered in this section. The rationale for this inclusion is rooted in the concept of risk and return in the context of investment portfolios where variability of returns is the 'risk' (Markowitz, 1952).

6.1.1.1 Standard Deviation of Event Loss

The variance of a compound random variable S_i is defined by

$$\text{Var}[S_i] = \mathbb{E}[N_i]\text{Var}[X_i] + \text{Var}[N_i]\mathbb{E}[X_i]^2 \tag{6.1}$$

when the event's frequency N_i and severity X_i distributions are assumed to be independent (e.g., Kaas et al., 2008, eq. (3.4)). The expectations (or mean values) $\mathbb{E}[N_i]$ and $\mathbb{E}[X_i]$ are the event rate λ_i and the event loss L_i, respectively. With event occurrence assumed to follow the Poisson distribution (Section 2.1.1), $\mathrm{Var}[N_i] = \lambda_i$. This yields

$$\mathrm{Var}[S_i] = \lambda_i \sigma_i^2 + \lambda_i L_i^2, \tag{6.2}$$

where σ_i ($= \sigma_{\mathrm{ID}}$) is the overall standard deviation for event i (encompassing both independent ($\sigma_{ind,i}$) and correlated ($\sigma_{corr,i}$) standard deviation terms). The value σ_{ID}, to be included as an additional event attribute in the ELT, represents the secondary uncertainty of the CAT model. Ground-motion random fields were an example of secondary uncertainty modelling in the hazard component (Section 2.5). In the vulnerability module, the beta distribution is well suited to model MDR secondary uncertainty since it is bounded in the [0, 1] range. This distribution is also used to directly describe the aggregated effects of secondary uncertainties on event losses (Homer and Li, 2017; Jones et al., 2017) (for more details about the beta distribution, see Exercise 6.1a). Note that σ_{corr} depends on the correlation of losses between coverages (e.g., between building and content value at a location) and between locations (e.g., adjacent buildings likely to experience similar hazard intensity variations, or buildings from the same builder likely to have similar vulnerability variations). These correlations are often estimated from claim data. When correlations are not well known, a CAT model could be run twice, once for full independence and once for full correlation to inspect the impact on the results. Simulating a 100% correlation simply consists of using the same random number for two variables (e.g., building and content damages) (Taylor and Carter, 2020). Finally, assuming event independence, the standard deviation σ around the AAL is, for the ELT,

$$\sigma_{ELT} = \sqrt{\sum_i \lambda_i (\sigma_i^2 + L_i^2)}. \tag{6.3}$$

For a YLT, it is

$$\sigma_{YLT} = \sqrt{\frac{\sum_{k=1}^{N_{sim}} \left(\sum_{j=1}^{n_k} L_{j,k}\right)^2}{N_{sim}} - \mathrm{AAL}^2}, \tag{6.4}$$

with N_{sim} the number of year-simulations, n_k the number of events per year-simulation, and $L_{j,k}$ the loss realized for the jth event in year k that directly includes the role of secondary uncertainties (e.g., Homer and Li, 2017) (Ex. 6.1b). The σ metric is of critical importance in insurance, as demonstrated below.

6.1.1.2 Financial Structure

Insurance policies are structured according to policy terms which include features applied on the loss values in the final module of a CAT model (previously termed loss module, also commonly known as financial module). These financial terms, the deductible L_d and the limit L_l, delineate the risk layer covered by the insurer. The **deductible**, also known as the excess or retention, is the amount of loss the policyholder has to bear before becoming eligible to reclaim compensation from the policy. Conversely, the **limit** defines the portion of

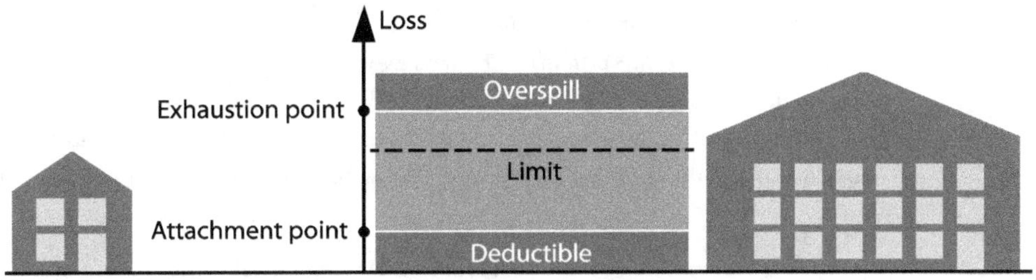

Fig. 6.1 Illustration of an insurance financial structure for some given policy terms. The limit area is covered by insurance, while the deductible and overspill areas remain the liability of the policyholder. The smaller property here is covered for the entire rebuild cost in excess of the deductible (no overspill; indicated by the dashed line).

the loss covered by the insurer. If the difference between the event ground-up loss L and the sum of the deductible and the limit is positive, this difference is known as the **overspill**, which also falls to the policyholder. Consequently, the insured loss, central to insurance calculations, is $L_{insured} = \min(\max(0, L - L_d), L_l)$ (e.g., Varghese and Rau-Chaplin, 2013). Figure 6.1 illustrates a standard insurance financial structure featuring deductible, limit, and overspill. For completeness, two additional terms are introduced: the **attachment point** (the loss value at which insurance coverage initiates) and the **exhaustion point** (the loss value at which coverage ends). A parallel layering is observed in reinsurance, where the reinsurer assumes the role of the insurer and the insurer becomes the **cedant** (who cedes the risk but remains liable for both deductible and overspill) (e.g., Mitchell-Wallace, 2017).

6.1.1.3 Pricing

Insurance rates are, like all economic products, the outcome of supply and demand forces. On the demand side, these rates must be sufficiently attractive concerning the insured's estimation of the expected loss, so that purchasing insurance emerges as an appealing choice. On the supply side, the premium needs to be adequately high for investors to anticipate an acceptable return on their invested capital, considering the risk characteristics of the insurer. CAT models provide the technical inputs for the formulation of insurance rate determination (Kuzak and Larsen, 2005). Although underwriting is an intricate exercise, the specifics of which lie beyond the scope of this textbook, the price or premium that the insurer should impose on policyholders is grounded in the summation of the AAL, the risk load, denoted as R_l, and the expense load:

$$\text{premium} = \text{AAL} + R_l + \text{expense load}. \tag{6.5}$$

Here, the AAL is calculated for the insured losses, which are constrained by the attachment point and the exhaustion point as defined in the policy terms. The **risk load** arises from the uncertainty surrounding the AAL and can be expressed as $R_l = c\sigma$, with σ the previously defined standard deviation (Eqs. (6.3) and (6.4)) and c a multiplier set at 0.5 as the upper limit for insurer charges (Kuzak and Larsen, 2005, tables 5.1 and 5.2) (Ex. 6.2a). Note that more sophisticated methods are usually employed to calculate the risk load (Kreps, 1990).

It is important to clarify that *the term risk in this context does not refer to the anticipated loss, but rather to the potential for loss beyond expectations*. To reduce R_l, it is crucial for the insurance portfolio to be diversified (Box 6.1). Lastly, the **expense load** encompasses the administrative costs associated with insurance contracts, including elements such as loss adjustment expenses, processing fees, premium taxes, commissions, and profits.

6.1.1.4 Survival Constraint

An insurer can successfully meet the survival constraint by meticulously constructing a portfolio of risks in such a way that the overall expected probability of insolvency falls below a predetermined threshold p. There are a number of approaches to verify if an insurance portfolio satisfies this condition. According to Grossi et al. (2005) for instance, the maximum number of policies n satisfying the survival constraint is governed by the inequality:

$$\mathbb{P}(L_{tot} > n \times \text{premium}_{policy} + \nu_C) < p, \tag{6.6}$$

where L_{tot} denotes the total loss due to a catastrophe and ν_C is the current currency surplus of the insurer. A similarly simple approach, given by Mitchell-Wallace (2017), is to define the necessary capital, referred to as the **risk-based capital** (RBC), as:

$$\text{RBC} = \text{VaR}_q[L] - \mathbb{E}[L], \tag{6.7}$$

where $\mathbb{E}[L]$ is the expected loss and $\text{VaR}_q[L]$ the loss amount at the specified percentile of the loss distribution q (e.g., q is 99.5% for **Solvency II**, the regulatory framework established by the European Union). The rationale for subtracting the mean loss is that this amount of loss is expected, and capital is needed only for the unexpected loss (Mitchell-Wallace, 2017, box 2.2). As a basic example, considering the ELT of Section 3.3.2 with an AAL of USD 524,450 (assuming losses to be insured losses instead of ground-up losses for the sake of illustration) and ν_C = USD 15 million for the current currency surplus, the existing portfolio of policies would be unsatisfactory. This stems from the fact that the $\text{VaR}_{99.5\%}$ for the portfolio is equal to USD 18.7 million, which means that a risk-based capital of USD 18.2 million is required (according to Eq. (6.7); Ex. 6.2b). As a consequence, the insurer would be compelled to explore avenues to reduce its portfolio's risk exposure. One approach could involve ceding more than USD 3 million of potential loss to a reinsurer or, alternatively, procuring a CAT bond to underwrite this amount. Another option would be to reduce the amount of coverage and reassess the level of solvency. These strategic actions would effectively maintain the probability of insolvency at an acceptable threshold (Grossi et al., 2005).

6.1.2 Risk Transfer via CAT Bonds

Once it is ascertained that the risk covered by insurance becomes unacceptably high, the company can transfer a portion of the risk associated with the tail of the EP curve (the part above a given VaR). This transfer can occur not only via traditional reinsurance, but also via **insurance-linked securities** (ILS). An ILS is a financial instrument, a type of derivative, whose price is determined by insurance losses. The most common type of ILS is the **catastrophe bond**, or CAT bond (Mitchell-Wallace, 2017, p. 158). The capital markets surpass the reinsurance market in size and undergo daily value fluctuations that exceed the magnitudes of the most significant observed catastrophes. Furthermore, catastrophe risk possesses an appealing potential for the capital markets, owing to its lack of correlation with other investments. Therefore, CAT bonds provide a means of diversification in capital market portfolios, theoretically enabling the capital markets to accept catastrophe risk with a reduced risk load compared to reinsurers (Lalonde, 2005). Other advantages include no counterparty-risk of default (since CAT bonds are fully collateralized) and the possibility of multi-year commitment (Polacek, 2018). Note that CAT bonds only made their debut in 1996–1997 (Litzenberger et al., 1996; Polacek, 2018). Like a regular bond, the issuer pays premium coupons (interest payments) while the principal serves as collateral, here to pay for potential disaster losses. The **prospectus**, a document which includes all the pertinent information related to the transaction, also outlines the specifics of the CAT model outcomes. It incorporates explanations regarding the constraints of the analysis and risk factors aimed at notifying investors about the inherent variability in the modelling procedure.

6.1.2.1 Bond Structure

The CAT modelling conducted to support the transaction quantifies the likelihood that investors have of recovering their principal fully, recovering it partially, or losing it entirely. The yield of the notes is a function of these probabilities, influencing their rating as well. *Any default on the notes is triggered by the actual occurrence of an event or events within the coverage period. The held-in-trust principal is subsequently employed to cover the losses of the issuer, or cedant.* Distinct transaction types exist for CAT bonds. They can adopt an indemnity-based approach, which relate payouts to actual losses suffered, or they can rely on predefined triggers to dictate payouts. In the latter scenario, there are index-based bonds based on external indices, potentially indirectly correlated with the catastrophe, while parametric bonds employ direct and quantifiable parameters directly linked to the event's impact. The choice between these approaches depends on the specific requirements and risk profiles of both issuers and investors (Lalonde, 2005). The example of the 2010 Mariah Re Ltd. CAT bond, described by Polacek (2018), is illustrative of the process: the deal consisted of up to USD 100 million in compensation if losses from severe thunderstorms and tornadoes across the US – estimated by an independent third-party CAT modeller – exceeded an USD 825 million attachment point (USD 100 million being the limit, as in Figure 6.1). For taking the risk, investors received a 6.25% annual coupon over the bond's 3-year term. With 2011 being the deadliest and costliest tornado season in US history (recall the Joplin tornado; Sections 1.3.3 and 3.2.3), modelled losses rose to USD 955 million, completely wiping out the investors' principal. For insights into the pricing of CAT bonds, the reader should refer to sources such as Mitchell-Wallace (2017, p. 163).

6.1.2.2 Parametric CAT Bonds

This type of transaction employs a CAT risk model to estimate the probability of an event reaching or exceeding a specific intensity within a designated location throughout the coverage period. The trigger for any payment to the cedant is rooted not in a particular loss amount, but rather in the physical parameters associated with the actual event, should it occur. More precisely, a **parametric index** corresponds to the measurement of an intensity indicator at multiple locations close to the portfolio's risk sites. Each recorded measurement is weighted by the values at risk and the vulnerabilities of each insured risk (Lalonde, 2005). The payout is contingent upon the intensity of an event, as objectively determined by an independent third party. For instance, earthquake magnitude could be provided by the USGS, while hurricane category might be assessed by NOAA. While parameterization may encompass potential loss modelling to structure and price the transaction, investors primarily need to assess the hazard probability. For example, the 2000 PRIME Capital bond was issued in such a way. It stated that a triggering earthquake's epicentre must fall within one of eight designated boxes or seismic source zones – four around the San Francisco Area and four around Los Angeles. The USGS served as the reporting agent for epicentral locations. Additionally, the moment magnitude of a covered earthquake had to equal or surpass certain predefined magnitudes within each source zone for payments to be made to the reinsurer from the notes. Concerning hurricane-related losses, triggers are functions of the central pressure within specific defined landfall zones, with the National Hurricane Center acting as the reporting agency. Similarly, for European windstorms, losses to the notes depend on a weighted parametric index computed from wind speeds recorded at stations across Western Europe, as reported by various governmental meteorological organizations. These transactions can exhibit considerable complexity in their structure – far more details can be found in Lalonde (2005).

6.1.3 Risk Reduction Instruments

Catastrophe risk models serve a critical role beyond their predictive capabilities – they extend their influence to governmental bodies, regulatory authorities, and emergency management agencies, for effective risk mitigation strategies. Mitigation strategies can also directly benefit property owners in the form of reduced losses and lower insurance rates (Kleindorfer et al., 2005). *In the context of a CAT model as input, risk reduction, often referred to as de-risking, necessitates a re-parameterization of the CAT model to explore the conditions that would ultimately reduce the expected risk in the future.* In every instance, the contours of risk management strategies are constrained by cost–benefit trade-offs. While the imperative to reduce risk may seem self-evident, it remains, in practice, dependent on the funds that can be allocated to such actions. Examples include evacuations, retrofitting old buildings, enhancing the structural designs of future buildings, and constructing defence structures. While the primary actions often involve reducing exposure or potential disaster damage, measures can also be taken to minimize the hazard. This is detailed in the next paragraphs, by model component.

6.1.3.1 Exposure and Vulnerability Components

The exposure data comprise the asset value, denoted as ν, along with its corresponding characteristics denoted as Θ_ν (Section 3.1). These data serve as the basis for assessing both damage (D) and loss (L), which can be determined with respect to a given hazard

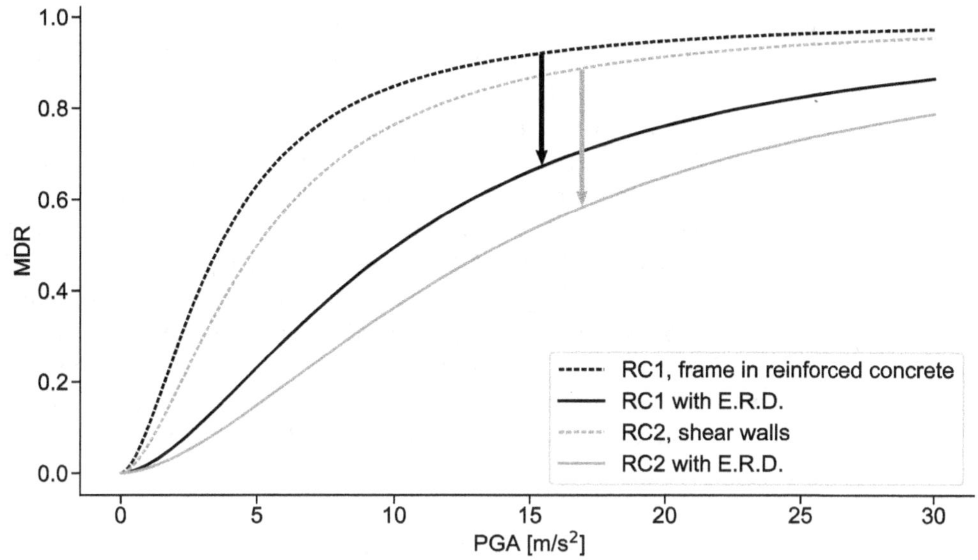

Fig. 6.2 Impact of earthquake reduction designs on building vulnerability, here denoted E.R.D., for two different reinforced concrete (RC) typologies – MDR defined as $D_\mu/5$ according to Eq. (3.8) with vulnerability index $\kappa_D = 0.644$ for RC1, 0.324 for RC1 with E.R.D., 0.544 for RC2, and 0.224 for RC2 with E.R.D. (Lagomarsino and Giovinazzi, 2006, table 3).

intensity (I) (recall the CAT modelling workflow of Figure 3.1). Initiating risk mitigation measures often involves strategies to limit asset concentration, thereby reducing the value of ν, within hazard-prone regions such as flood plains and fault-line areas. Land planning studies contribute to achieving this objective. This can however be problematic to apply in practice; for example, settlements around the world have continuously expanded into flood zones since 1985 (Rentschler et al., 2023). **Building codes** also mandate property developers to adhere to specific design standards. These standards ensure that structures possess the capability to withstand designated loads at specified probabilities of occurrence. For critical infrastructures, codes must account for the potential occurrence of extreme events, necessitating the implementation of **stress tests** that require structures to endure significantly higher loads (Argyroudis et al., 2020). These actions lead to adjustments in the Θ_ν parameters, resulting in the recalibration of vulnerability functions. This recalibration ultimately leads to lower damage under a given load I when assessed within the vulnerability module. The influence of earthquake-resistant designs on a vulnerability curve is depicted in Figure 6.2. Although the impact of these alterations is observable only over the course of years or decades, other measures yield more immediate effects. Regarding the built environment, Θ_ν can be modified to reflect building **retrofitting** (strengthening of a structure) or the implementation of defensive measures (e.g., flood-proofing). In terms of human exposure (where ν denotes the number of people), which is a dynamic process, relocations are feasible when a hazardous event is predicted for a particular area. Diverse **early-warning systems** exist to facilitate the evacuation of individuals prior to the arrival of severe weather phenomena (observed trajectory), tsunamis (following a tsunamigenic earthquake), floods or landslides (in the aftermath of heavy rainfall), and wildfires, among other perils. In regions susceptible to landslides, sensors can additionally monitor indications of ground slippage (Pei et al., 2011), while volcanic eruption forecasts can be informed by the monitoring of magma expansion (Aspinall et al., 2002). If earthquakes themselves remain unpredictable – except for aftershocks that follow well-known statistics (Section 6.2.2), early-warning systems can capitalize on the speed of seismic wave propagation as a mainshock unfolds (Kong et al., 2016). The course of action to be taken following an early warning depends on the time interval between the triggering event and the issuance of the alert to avoid panic (and the emergence of other hazards such as stampedes). If evacuation is viable, it is prioritized; however, in cases where evacuation is not feasible, automated steps can be initiated. It is important to acknowledge that there exists a trade-off due to the potential costs associated with false alarms. The subject of peril forecasting will be addressed in Section 6.2.

6.1.3.2 Hazard Component

The mitigation of technological perils focuses on reducing the impact of severe accidents while acknowledging that their complete avoidance may not always be achievable. The principle of **ALARP**, which stands for 'as low as reasonably practicable,' guides the regulation and management of critical infrastructure systems. Under the influence of operational constraints, the occurrence rate (λ) of human-made disasters typically needs to remain below 1 occurrence in 10,000 years. Moreover, concerted efforts are made to drive this occurrence rate even lower, aiming toward an acceptable rate of 1 in 1 million years (Melchers, 2001). Risk mitigation strategies encompass strict comprehensive safety measures and operational protocols to prevent a catastrophe. This involves rigorous infrastructure design and construction standards, routine inspections, and maintenance practices to avoid, for instance, reactor meltdowns and radioactive releases in the nuclear industry, dam failures that could lead to widespread flooding downstream in the hydrodam industry, or large explosions and toxic releases in the process industry. Whatever the industry, regulatory agencies play a critical role in enforcing safety standards and protocols. The framework for risk management in the process industry gained traction in Europe after the 1976 Seveso incident in Italy (the so-called Seveso directive; Jain et al., 2017). Following the Chernobyl disaster in 1986, the international community was prompted to revise safety standards and regulations for nuclear power plants (Ahearne, 1987) and lessons from the 2011 Fukushima disaster led to additional improvements. In the hydrodam industry, catastrophic dam failures have influenced risk management approaches (e.g., Pisaniello et al., 2023). These examples highlight the significant role that major incidents play in shaping risk management practices, resulting in regulatory reforms, technological advancements, and heightened industry awareness. A distinctive aspect of technological perils is the fact that both the frequency of events (λ) and their size (S) can be minimized through the reduction of vulnerability in critical infrastructure components. Notably, there are scenarios – such as induced seismicity during deep underground stimulations – where hazard reduction does not involve alterations to human-made structures (see case study;

Section 6.3). In contrast, managing natural hazards proves considerably more challenging, leading to a greater focus on risk mitigation measures within the exposure domain, as detailed earlier. Yet, efforts are made to mitigate natural hazards. For instance, controlled triggering of events can help regulate evacuation timing; a good example of this is the management of snow avalanches. Additionally, natural hazards can be reduced through actions such as forest management and fire suppression to combat wildfires, or pesticide application to control pest infestations. The STORMFURY project was an early initiative aimed at mitigating hurricanes (Gentry, 1969). Nonetheless, it is important to note that each solution carries potential negative side effects. It has been argued that avoiding medium-sized events in the short term could potentially channel unreleased energy toward more extreme, low-probability events (e.g., mega-fires caused by biomass accumulation because of smaller fires being extinguished). Similarly, proposals involving geoengineering to counteract global warming introduce the possibility of unintended dire consequences (Keith, 2000). These examples show the challenges in trying to reduce risk on the hazard front.

6.2 Peril Forecasting

Forecasting is the process of extrapolating future outcomes from historical and present data. The CAT models previously described in this textbook therefore 'forecast' the projected effects of future events across extended time spans. The term forecast is, however, usually more narrowly defined as a 'definite statement or statistical estimate of the likely occurrence of a future event or conditions for a specific area' (UNISDR, 2009).

The distinction between a CAT model and what is termed a forecast is conceptual. Catastrophe risk modelling is loosely tied to an average approach where the future $\hat{y}_{i+\delta t|\mathcal{D}}$ (e.g., the event rate $\hat{\lambda}$) is estimated as the average of the past data $\mathcal{D} = (y_1, y_2, ..., y_i)$ observed over interval ΔT. In this scenario, $\hat{y}_{i+1|\mathcal{D}} = (y_1 + y_2 + ... + y_i)/\Delta T$ with $\delta t = 1$ year and ΔT spanning from decades to millennia depending on the peril. A particularly rudimentary approach to forecasting is to assume that the future is identical to the most recent past observation, yielding $\hat{y}_{i+\delta t|\mathcal{D}} = y_i$. This sort of **persistence** rule might assert, for instance, that tomorrow's weather will mimic today's (with $\delta t = 1$ day). Although this method proves to be unpractical, it serves to emphasize the distinction between the models developed in previous chapters and the upcoming discussions. It introduces the concept of time-dependency, suggesting that recent data could hold more predictive value about impending events than the historical average. This distinction is important for short-term risk mitigation strategies based on event forecasting.

As noted by Woo (1999, p. 134), 'forecasting any natural hazard is itself a hazardous occupation'. Throughout history, a multitude of Cassandras have emerged, sharing doomsday predictions of comets, wars, plagues, and beyond. Only with the Scientific Revolution of the seventeenth and eighteenth centuries were data-based predictions made and subject to verification (Section 1.2). The recent convergence of machine learning and big data is now pushing the limits to unearth hidden signals, including the ones potentially preceding hazardous events (Sun et al., 2020). This section will explore both historical and methodological aspects of peril forecasting. The forecasting challenges will be shown to vary across perils. For instance, tropical cyclones offer observable cues before landfall (Section 6.2.1), while earthquakes tend to strike suddenly without evident precursory signals (Section 6.2.2).

6.2.1 Tropical Cyclone (TC) Forecasting

6.2.1.1 Historical Aspects

The first hurricane forecast models emerged in the 1950s, spurred by technological progress (upper-air network in the 1930s, aircraft reconnaissance in the 1940s, advent of computer modelling in the 1950s, and integration of satellite observations in the 1960s) (DeMaria, 1996). Early operational implementations were relatively basic. The Hurricane Analog (HURRAN) technique, introduced in 1969, employed a tropical cyclone database to identify storms with similar tracks, shifting their paths through the actual trajectory using location, direction, speed, and date for analogous matches (Hope and Neumann, 1970). Its successor, the Climatology and Persistence (CLIPER) statistical model, became a baseline for comparison with more advanced models (Knaff et al., 2003). The Statistical Hurricane Intensity Forecast (SHIFOR), established in 1979, predicted future hurricane intensity using climatology and persistence principles (Jarvinen and Neumann, 1979). Typhoon forecasts saw a parallel development in the form of STIFOR. Subsequent decades yielded a suite of increasingly sophisticated statistical and statistical-dynamical models. Escalating computing capacities promoted further progress, as numerical models of dynamic weather systems emerged. The Sanders barotropic (SANBAR) model, operational from 1970 to 1989, employed a vertically averaged wind field. Later models, termed baroclinic, would include multiple vertical layers. Three-dimensional hurricane models such as the Moveable Fine Mesh emerged by the late 1970s, limited in spatial domain due to computational constraints (DeMaria, 1996). The Geophysical Fluid Dynamics Laboratory (GFDL) model, introduced in 1992 and operational in 1995, outperformed contemporaries (Bender et al., 2007). The latest generation, the Weather Research and Forecasting (WRF) model, predicts both hurricane track and intensity. Its massively parallel computation system equips WRF for diverse Earth system predictions, spanning storms, air pollution, wildfires, and regional climate (Powers et al., 2017). Let us mention also the storm surge model Sea, Lake, and Overland Surges from Hurricanes (SLOSH), described in detail in Jelesnianski et al. (1992).

6.2.1.2 Storm Track Forecasting

While tropical cyclone forecasting typically demands advanced computational expertise, the basic concept of statistical storm track forecasting can be presented in a concise manner for the student of risk. This concept aligns with the principle of track persistence, using historical storm paths to reveal general patterns in storm behaviour within a specific region, similarly to the HURRAN technique previously mentioned. To test this method, IBTrACS

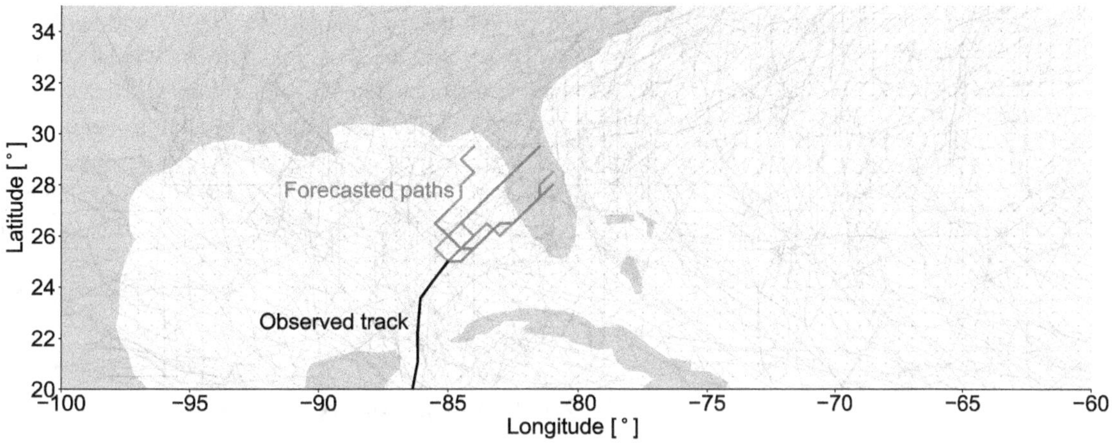

Fig. 6.3 Hurricane track forecasting using a Markov model. Historical observations in the Atlantic Basin from 1990 to 2019 are depicted in light grey and sourced from the IBTrACS data (Knapp et al., 2010, 2018). The distinctive irregular shape of the forecasted tracks arises from the 0.5° spatial resolution used to construct the transition matrix (small grey dots). Only the first 10 iterations are shown for each forecast.

data will be utilized (Knapp et al., 2010, 2018), with the historical set of tracks shown in grey in Figure 6.3. A statistical forecasting technique adopts the form of a Markov model, which, as previously defined in Section 5.1.2 on domino effects, operates on the assumption that the previous observation X_i contains all the relevant information for predicting the next occurrence. Assuming discrete time intervals, the joint distribution of a sequence of random variables $X_1, \cdots, X_i$ of arbitrary length k can be expressed as:

$$\mathbb{P}(X_{1:k}) = \mathbb{P}(X_1)\mathbb{P}(X_2 \mid X_1)\mathbb{P}(X_3 \mid X_2)\cdots = \mathbb{P}(X_1)\prod_{t=2}^{k}\mathbb{P}(X_t \mid X_{t-1}). \tag{6.8}$$

In the case where $\mathbb{P}(X_t \mid X_{t-1})$ remains independent of time, the chain is referred to as homogeneous stationary or time-invariant. For n discrete states, that is, $X_k \in 1, \cdots, n$, the conditional distribution $\mathbb{P}(X_t = j \mid X_{t-1} = i)$ can be expressed as the element p_{ij} of an $n \times n$ transition matrix $\mathbf{P}$, denoting the probability of transitioning from state i to state j. In the illustrative example depicted in Figure 6.3, storm tracks are initially discretized into a grid with a resolution of 0.5° by 0.5°. Each random variable X_k can take as its value one of the thousands of grid cells. For any cell (or state) traversed by at least one storm, the transition probability to the next cell is calculated by dividing the number of tracks passing through that state change by the total number of track state changes (including instances where the storm remains in the same cell). In this simplified scenario, as the transition matrix primarily contains non-zero values only for neighbouring cells, a more straightforward transition vector can be established as follows: $\vec{p}_i = (p_{i,i}, p_{i,N}, p_{i,S}, p_{i,E}, p_{i,W}, p_{i,NE}, p_{i,NW}, p_{i,SE}, p_{i,SW}, p_{i,stop})$. Here, $p_{i,i}$ denotes the probability of remaining in cell i while $p_{i,stop}$ represents the likelihood of the storm terminating at cell i. The other values represent the probability of moving from cell i to one of the four cardinal directions, north (N), east (E), south (S), and west (W), or one of the four intercardinal (or ordinal) directions: north-east (NE), south-east (SE), south-west (SW), and north-west (NW). The assumption that a storm trajectory depends solely on its current location, independent of its past path, can be attributed to the influence of cyclonic forces, driven by local Coriolis forces and wind direction. Given the current hurricane location, one can generate a set of stochastic projections for the anticipated future storm path, as demonstrated in Figure 6.3. While rudimentary in terms of storm forecasting, similar, albeit refined, methods are employed to generate stochastic sets of storm tracks for CAT models (Vickery et al., 2000; Bloemendaal et al., 2020).

6.2.2 Earthquake (EQ) Forecasting

6.2.2.1 Historical Aspects

Unlike storm forecasting, which relies on well-established governing equations and abundant accessible data, earthquake forecasting is fraught with challenges due to the lack of fundamental understanding of earthquake nucleation, the absence of direct observation of geological variables, and the rarity of large events. There are two primary approaches to the problem: large earthquake prediction using pattern recognition and physical inference, and earthquake operational forecasting, which employs statistical methods based on simple aftershock laws (Mignan, 2014). While the former holds the promise of high probability gains, it has consistently fallen short thus far. The latter approach offers limited probability gains but may suffice for basic risk mitigation measures. The history of earthquake prediction has been marked by periods of optimism followed by strong criticisms, citing a lack of generalization, conflicting results, and disputed statistical methodologies (Geller, 1996, 1997; Mulargia, 2001; Hardebeck et al., 2008; Mignan, 2011; Mignan and Broccardo, 2020; Mignan et al., 2021b). Modern earthquake prediction can be traced back to 1975 when the successful prediction of the magnitude 7.3 Haicheng, China, earthquake saved thousands of lives through an evacuation order (Wang et al., 2006). This prediction relied on various precursory signals, with a cluster of small earthquakes being the most significant (Jones and Molnar, 1976). Seismic precursors can manifest as changes in seismicity rates, shifts in the ratio of small to large earthquakes, spatiotemporal migration patterns, or combinations of these phenomena (Kanamori, 1981). While foreshocks were initially considered an obvious precursor, other seismicity patterns gained prominence, such as precursory quiescence in the 1980s (Wyss and Habermann, 1988) and precursory accelerating seismicity in the 1990s (Bowman et al., 1998; Mignan, 2011). In addition to seismic precursors, numerous reports have documented non-seismic precursory phenomena, including rock dilatancy, geo-electric and magnetic anomalies, radon activity, thermal infrared anomalies, groundwater changes, and abnormal animal behaviour (Scholz et al., 1973; Cicerone et al., 2009; Mignan et al., 2021b). Numerous algorithms with specific rule sets have been proposed for predicting large earthquakes (Varotsos et al., 1981; Huang et al., 2001; Keilis-Borok, 2002; Holliday et al., 2007; Mignan and Broccardo, 2020). While the most obvious earthquake precursor candidates have been identified, it is clear that if earthquake prediction is physically achievable, it will require new theoretical advancements in understanding the earthquake preparation process, innovative pattern recognition techniques, or improved data acquisition methods. The Collaboratory for the Study of Earthquake Predictability (CSEP) (Schorlemmer et al., 2018), a highly rigorous endeavour and successor to the Regional Earthquake Likelihood Models (RELM) experiment (Schorlemmer et al., 2010), showed that models based on basic aftershock statistics consistently outperform all other tested models. However, this enhanced performance comes at the expense of limited probability gains.

6.2.2.2 Precursory Pattern Recognition

Among the various pattern recognition techniques designed for discerning precursory signals from seemingly random noise, the region–time–length (RTL) algorithm stands out as an illustrative example due to its relative simplicity and its documented success in some historical cases (Di Giovambattista and Tyupkin, 2000; Huang, 2004; Sobolev, 2007; Mignan and Di Giovambattista, 2008). The RTL metric is calculated at a location x and time t and defined as the product of three functions, all derived from seismicity data:

$$\begin{cases} \mathrm{R}(x,t) = \left[\sum_{i=1}^{n} \exp(-r_i/r_0)\right] - \mathrm{R}_{bk}(x,t) \\ \mathrm{T}(x,t) = \left[\sum_{i=1}^{n} \exp(-(t-t_i)/t_0)\right] - \mathrm{T}_{bk}(x,t) \\ \mathrm{L}(x,t) = \left[\sum_{i=1}^{n} \exp(l_i/r_i)\right] - \mathrm{L}_{bk}(x,t). \end{cases} \tag{6.9}$$

Here, r_i represents the event distance from x, t_i signifies the event occurrence time, and $l_i = 10^{0.44M_i - 1.289}$ corresponds to the rupture dimension, which is a function of event magnitude M_i. The parameter n denotes the number of events i that meet the conditions $r_i \leq 2r_0$, $(t - t_i) \leq 2t_0$, and $M_{min} \leq M_i \leq M_{max}$. Additionally, r_0 and t_0 represent characteristic distance and time span. $\mathrm{R}_{bk}(x,t)$, $\mathrm{T}_{bk}(x,t)$, and $\mathrm{L}_{bk}(x,t)$ are the background trends of $\mathrm{R}(x,t)$, $\mathrm{T}(x,t)$, and $\mathrm{L}(x,t)$, respectively. These functions, $\mathrm{R}(x,t)$, $\mathrm{T}(x,t)$, and $\mathrm{L}(x,t)$, are dimensionless and further normalized by their standard deviations, σ_R, σ_T, and σ_L, respectively (Huang, 2004; Sobolev, 2007). The resulting RTL value quantifies the deviation from the background level of seismicity, with a negative RTL indicating quiescence and a positive RTL indicating activation. Figure 6.4a presents the RTL time series leading up to the 1997 $M = 6$ Umbria–Marche earthquake in Italy. The observed quiescence anomaly, well defined in time, is depicted in Figure 6.4b from the mean RTL value during this quiescence period for every location within a geographical grid of Central Italy. Remarkably, this anomaly was centred on the location of the impending mainshock (indicated by the white star). In terms of statistical and machine learning classification, this instance could be characterized as a **true positive** (TP, or hit), signifying a correct prediction of the target class. It is important to note that this conclusion arises from retrospective analyses (Di Giovambattista and Tyupkin, 2000; Mignan and Di Giovambattista, 2008), which may involve cherry-picking, rather than from a prospective or **pseudo-prospective** forecast (i.e., a forecast conducted afterward, although mimicking prospective conditions under blind circumstances). Such RTL anomalies may also occur without being followed by a major earthquake, resulting in a **false positive** (FP, or false alarm, type I error), and conversely, a significant earthquake may not necessarily be preceded by such an anomaly, leading to a **false negative** (FN, or a miss, type II error). The most common scenario is

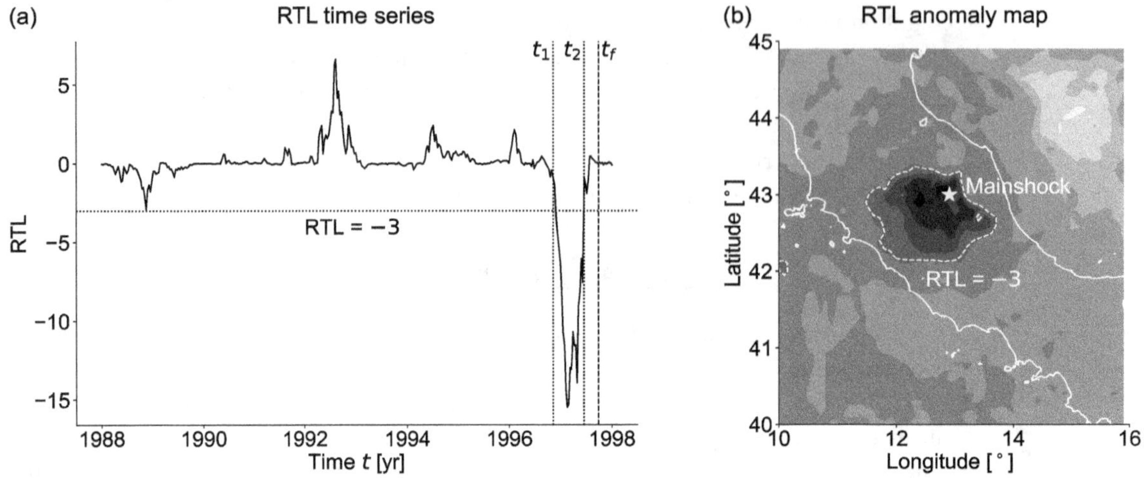

Fig. 6.4 Precursory pattern prior to the 1997 $M = 6$ Umbria and Marche earthquake, Italy, rerunning the analysis by Mignan and Di Giovambattista (2008) (see also Di Giovambattista and Tyupkin, 2000). (a) Region–time–length (RTL) time series at the location of the future mainshock with $r_0 = 50$ km, $t_0 = 1$ year, $M_{min} = 2.3$, and $M_{max} = 3.8$ (and mainshock occurrence time $t_f = 26$ September 1997). (b) Map of the RTL anomaly representing the mean RTL value in the time interval $[t_1 = 1996.85, t_2 = 1997.45]$.

observing no anomaly followed by no earthquake, known as a **true negative** (TN) or correct rejection. In a systematic approach, the accuracy of a forecast can be assessed based on the numbers of TP, TN, FP, and FN instances. An example of applying this approach to tornado forecasting is given in Box 6.2, where additional constraints on result interpretability are discussed. Although the pattern observed before the Umbria–Marche earthquake may seem evident in hindsight, conducting a blind test would require defining successful prediction criteria in precise terms. These criteria should encompass RTL thresholds below or above which a prediction is made, the duration of the anomaly, the time delay until the mainshock, and other **hyperparameters**, which are parameters defining the conditions under which a model is applied (Ex. 6.3).

6.2.2.3 Aftershock Forecasting

Earthquake statistics are grounded in two primary empirical laws: the Gutenberg–Richter law (Eq. (2.39)) and the Omori law of aftershocks (Omori, 1894). Remarkably, the Omori law pre-dates the Gutenberg–Richter law by half a century (e.g., Mignan, 2016a). Figure 6.5 shows the emblematic aftershock sequence of the 1891 Great Nobi earthquake in Japan, along with the original Omori fit expressed as $n(t) = K/(t+c)$, where n represents the rate of aftershocks, t is the time since the mainshock, and c_K and c_c are constants. The standard formulation, known as the **modified Omori law** or Omori–Utsu law, is given by:

$$n(t) = \frac{c_K}{(t+c_c)^{c_p}}, \tag{6.10}$$

Box 6.2 Measuring the Success of a Forecast: The 'Finley Affair'

'What constitutes a good forecast?' This fundamental question was posed by Murphy (1993) in his exploration of the nature of goodness in weather forecasting. The immediate response that often comes to mind is a high level of accuracy, but this can sometimes lead to a deceptive conclusion. The foundational issues regarding forecast verification began to surface in the 1880s with the publication of a groundbreaking paper on an experimental tornado forecasting program (Finley, 1884; Murphy, 1996). John Park Finley, now recognized as the pioneer of severe storm forecasting, embarked on a mission to demonstrate that tornadoes could be predicted (Galway, 1985). In his work, Finley (1884) applied a forecast system to the tornado-prone regions of North America. He provided predictions for both 8-hour and 16-hour time frames, relying on surface weather maps as the primary data source. It is worth noting that Finley's motivation might have been influenced by the devastating tornado outbreak of 19 February 1884, during which he recorded a staggering 60 tornadoes in the southern US, with an estimated death toll of 800 (Galway, 1985). Finley's study initially appeared remarkable due to its very high accuracy levels, surpassing 95%, in forecasting the occurrence or non-occurrence of tornadoes (i.e., a **binary classification**). However, the concept of **accuracy**, defined as ACC = (TP + TN)/*N* – the sum of true positives and true negatives divided by the total number of instances (*N*) – can be deceptive when the data set is **unbalanced**, meaning there is a significant disparity in the number of observations between different classes. The 'Finley Affair', as it later became known (Murphy, 1996), represents a classic example of such a pitfall, with his data set heavily skewed towards instances of no tornadoes (2,752) as opposed to tornado instances (51). To illustrate, consider a zero-information model that predicts no tornadoes ever. This model would yield ACC = (2752 + 0) / 2803 = 98.2% success, as highlighted by Gilbert (1884). This was characterized as a fallacy because it incorrectly assumed that forecasting tornadoes and forecasting no tornadoes were equally challenging. In the context of unbalanced data, alternative metrics must be employed. Following the Finley (1884) publication, various individuals promptly proposed alternative measures of forecasting performance (Gilbert, 1884; Murphy, 1996). Note that results of a classification problem can be summarized in a **confusion matrix**, an example of which is given below:

		Observation		
		Positive P	Negative N	Total
Prediction	Positive P	TP	FP	TP+FP
	Negative N	FN	TN	FN+TN
	Total	TP+FN	FP+TN	*N*

with c_p representing the power exponent (Utsu et al., 1995). Its generalization takes the form of the **epidemic-type aftershock sequence (ETAS) model**:

$$n(t) = R_\mu + \sum_{i:t_i<t} \frac{c_{K_0}}{(t - t_i + c_c)^{c_p}} e^{c_\alpha (M_i - M_c)}. \tag{6.11}$$

In this model, each earthquake has its own sequence of aftershocks, and additional parameters, namely c_{K_0} and c_α, replace the original c_K. Here, t_i and M_i represent the time and magnitude of the ith earthquake, with M_c denoting the completeness magnitude (Mignan,

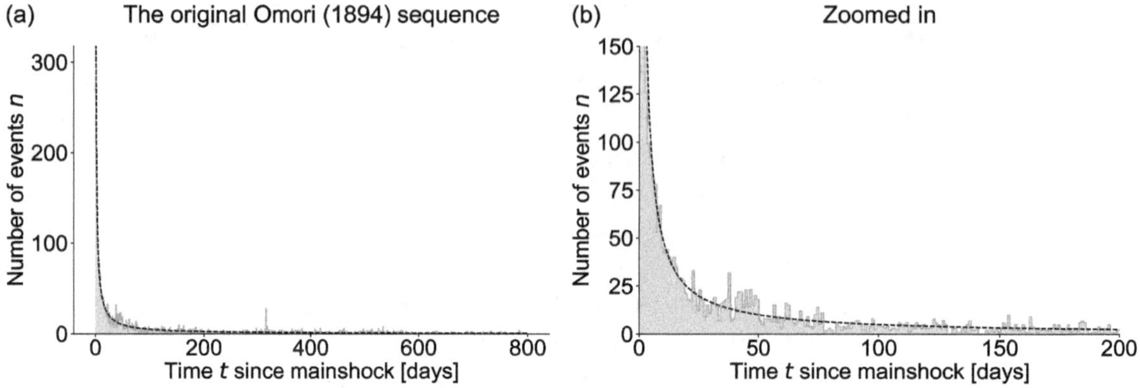

Fig. 6.5 The emblematic aftershock sequence of the 1891 Great Nobi earthquake, Japan, similar to Omori (1894, fig. 5) with the original Omori law $n = 16.9/(t + 0.397)$, where t denotes time in months since the mainshock and n the mean daily rate per month (dashed curve).

2012a), and R_μ indicating the background seismicity rate (Ogata, 1988). ETAS stands as the most effective model for capturing the temporal evolution of seismicity (Nandan et al., 2021), serving as the null hypothesis when evaluating different forecasts. Associated with the Omori law is **Båth's law**, which posits that the average difference in magnitude between a mainshock and its largest aftershock is approximately 1.2 (Båth, 1965). In 1989, the USGS initiated public forecasts of expected aftershock activity in California following major earthquakes by combining the Gutenberg–Richter and Omori laws, expressed as:

$$n(t, M) = 10^{a+b(M_m-M)}(t + c_c)^{-c_p}. \tag{6.12}$$

Here, n signifies the rate of aftershocks with a magnitude M or greater; t represents the time after the mainshock of magnitude M_m; and a, b, c_c, and c_p are parameters originating from the two empirical laws (Reasenberg and Jones, 1989, 1994). The probability of an 'aftershock' being larger than its 'mainshock' can be determined by assuming that any earthquake could be an aftershock, following the ETAS null hypothesis. This results in the equation:

$$\mathbb{P}(X \geq M, \Delta t = [t_1, t_2]) = 1 - \exp\left[-\int_{t_1}^{t_2} n(t, M)dt\right]. \tag{6.13}$$

Using the parameters $b = 0.91$, $a = -1.67$, $c_p = 1.08$, and $c_c = 0.05$ days according to Reasenberg and Jones (1994) yields the probability $\mathbb{P}(X \geq 6, \Delta t = [0, 7\text{days}]) \approx 0.0017$ of experiencing at least one earthquake with a magnitude $M = 6$ or greater within 7 days following an event of magnitude $M_m = 4$ (Ex. 6.4). This illustrates that aftershock statistics provide a practical and convenient approach to evaluate the probability of a mainshock occurring following any potential foreshock, regardless of potential precursory processes (which remain unknown). This approach was further refined by Gerstenberger et al. (2005). The 2009 $M = 6.3$ L'Aquila earthquake in Italy, which resulted in over 300 deaths, was preceded by significant foreshock activity, prompting questions about why no evacuation had been ordered (Hall, 2011). The application of aftershock statistics and a cost–benefit analysis subsequently revealed that an evacuation would not have been justified due to the low increase in probability in absolute terms (van Stiphout et al., 2010). However, this assessment assumed that the foreshocks were not indicative of a preparatory process leading to the mainshock. Various precursory patterns, including quiescence and accelerating seismicity, were observed in addition to the short-term foreshocks, but only in retrospect (Mignan, 2012b).

6.3 Risk Governance: A Case Study

Governance consists of the structures and processes necessary to facilitate decision-making involving a collaborative network of governmental and non-governmental actors – for additional, yet conceptually similar, definitions, refer to Rhodes (2007). Expanding upon this concept, risk governance applies the same overarching framework to decision-making in the realm of risk. This includes risk assessment, risk management, and risk communication. It necessitates a consideration of the legal, institutional, social, and economic contexts within which a specific risk is being evaluated, along with active engagement from the various actors and stakeholders representing these diverse sectors. In other words, risk governance considers the many interactions between actors, regulations, conventions, processes, and mechanisms governing the collection, analysis, and dissemination of pertinent risk information, as well as the formulation of management decisions (Renn, 2008). Figure 6.6 illustrates the position of CAT risk modelling within the broader framework of risk governance.

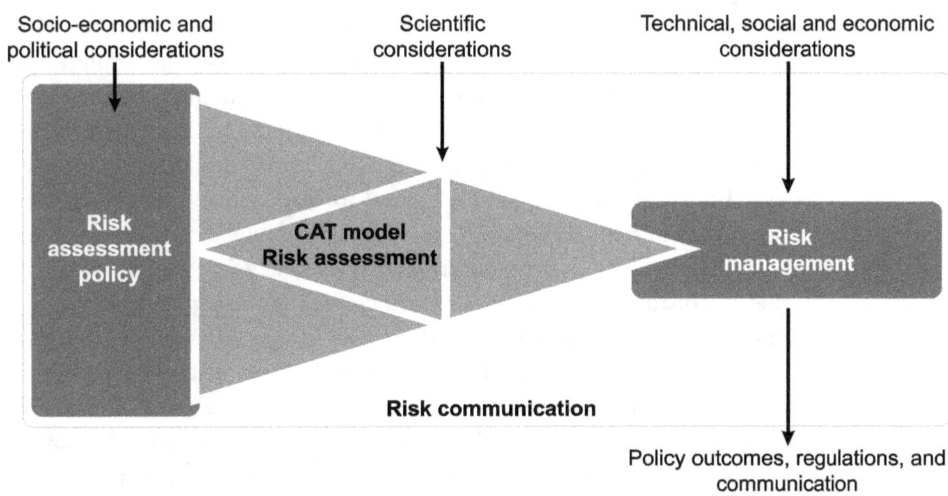

Fig. 6.6 CAT modelling in the general risk governance scheme, with governance aspects according to Renn (2008). Note that the final policy outcomes can be used as input in risk assessment policy, making CAT risk modelling (assessment phase) an ever-evolving process, but also the centre of the risk governance processes.

Risk governance, by its nature, extends its role beyond just CAT risk experts and risk managers. It encompasses a wide spectrum of stakeholders, including the government, various industries, and the general public. The **government** plays a pivotal role in risk governance by establishing agencies and committees responsible for decision-making, the enforcement of building codes, and the regulation of land use. These functions are executed through the executive, legislative, and judiciary systems. In fulfilling its role, the government must carefully weigh expert knowledge and the public's needs to ensure the optimal outcome for society. This outcome may be measured by various metrics, including public safety, energy security, economic growth, poverty reduction, or even more intangible factors such as happiness. Decisions are invariably constrained by a delicate balance between risks and benefits. **Industries**, representing the diverse economic activities involved in the production and supply of goods and services, also hold significance within the context of catastrophe risk. Three distinct categories of industries emerge: those that do not contribute to catastrophic events, those that have the potential to do so (e.g., critical infrastructures when they fail), and those actively engaged in catastrophe risk management, such as insurance firms. All of these sectors can be susceptible to the impact of disasters. The **public** consists of individuals, including private entities and property owners of both commercial and residential structures. It is widely understood that personal risk is an inherent part of life itself. Living is indeed enjoyable, but it is not devoid of danger. People acknowledge this reality and accept it. As noted by Starr (1980, p. 2), 'everyone reacts differently to risks they take voluntarily and to risks that are imposed by some outside group, the involuntary exposure'. The latter concept is called the **NIMBY** (not in my backyard) phenomenon, where individuals may resist certain developments or risks close to their immediate surroundings.

The challenges inherent in risk governance and the decision-making process can be vividly illustrated through the story of *The Lady, or the Tiger?* (Stockton, 1882), as revisited by Clark (1980) within the context of societal risk assessment. Here is a condensed version of the tale: three men are each faced with a choice between two doors. Behind one door lies a ferocious tiger that would surely tear them apart. Behind the other door stands a lady, the most suitable match for their respective lives. The first man, unwilling to take any risk, chooses not to open any door, living a safe but gloomy existence for the rest of his days. The second man, driven by a quest for knowledge, meticulously gathers data on the populations of both ladies and tigers. He deploys advanced sensors to deduce what might be behind each door. After careful analysis, he selects the door that appears to offer the best odds and opens it, only to be devoured by the unforeseen appearance of a low-probability tiger. The third man, adopting a different approach, embarks on a course to become a tiger tamer. He randomly picks a door and, to his fortune, discovers the lady. They later marry, but, tragically, she poisons him, leading to his demise. Clark (1980) interpreted this tale as a metaphor for the three possible responses to the unknown: withdrawal from it, seeking to comprehend it, or striving to control it. This narrative underscores the absence of a one-size-fits-all response and emphasizes the trade-off between costs and benefits that emerges due to the inevitable gaps in our understanding. It exemplifies the challenge of navigating uncertainty in a complex world, a fundamental objective of risk governance.

To guide the student of risk through these concepts, a single case study will be explored, carefully selected for its simplicity yet robust representation of a comprehensive risk governance framework. This case study will shed light on the interactions among an emerging industry and its risks, the role played by regulatory authorities, and the response of the public, in a context of highly uncertain risks and rewards. The chosen example centres on the seismic risks inherent in the stimulation of the deep underground to create a geothermal reservoir for the generation of electricity and heat. This case will follow the meta-model proposed by Mignan et al. (2019b), which offers a structured approach to evaluating the trade-off between ensuring public safety and maintaining energy security.

6.3.1 Case Study: Seismic Risk at a Geothermal Plant

Geothermal energy holds immense promise as a renewable resource, but its full potential remains largely untapped. The widespread adoption of geothermal technology is conditional on its environmental impact (Pan et al., 2019), with induced seismicity emerging as a major concern that could affect the growth and acceptance of this emerging industry (Lacirignola and Blanc, 2013). It is worth highlighting also that CO_2-storage initiatives can also trigger seismic events (White and Foxall, 2016), showing the urgency of addressing seismic risk mitigation strategies as we transition toward a carbon-neutral society. Conventional geothermal systems extract hot water from natural aquifers, and despite the ubiquitous geothermal gradient, such systems remain relatively scarce on Earth. In contrast, **enhanced geothermal systems** (EGS) involve the engineering of underground reservoirs through high-pressure water injection (Lu, 2018). In theory, EGS can be deployed worldwide, but its economic viability depends on local variations in the geothermal gradient and other geological, environmental, and technical factors (Mignan et al., 2019b). Some EGS projects have been abandoned due to concerns about the high seismic risk they pose. One illustrative case is the 2006 EGS stimulation in Basel, Switzerland, which triggered a magnitude $M = 3.2$ earthquake in the city, resulting in minor nonstructural damage and causing fear and frustration among the local population (Giardini, 2009). A more severe incident occurred at the Pohang, South Korea, EGS site in 2017, where an $M = 5.5$ earthquake injured over 100 residents and caused damage to tens of thousands of structures (Lee et al., 2019). Nevertheless, some EGS geo-energy production demonstrations have achieved success, offering a glimmer of hope in harnessing this valuable energy source (Lu, 2018; Pan et al., 2019).

Mignan et al. (2019b) introduced a comprehensive meta-model encompassing several key components: a geo-energy thermodynamic model for estimating electricity and heat production, a model for assessing induced seismicity risk associated with a **traffic light system** (TLS) designed to halt injection activities before safety thresholds are breached, and an economic model for electricity pricing. Notably, the seismic risk model serves as an example of a CAT model, while the operation of the TLS relies on outcomes derived from an earthquake forecast model. The essence of risk management within this context revolves around the dual objective of minimizing seismic risk while ensuring the viability of the geo-energy project. The complexity of this challenge is represented in Figure 6.7a,

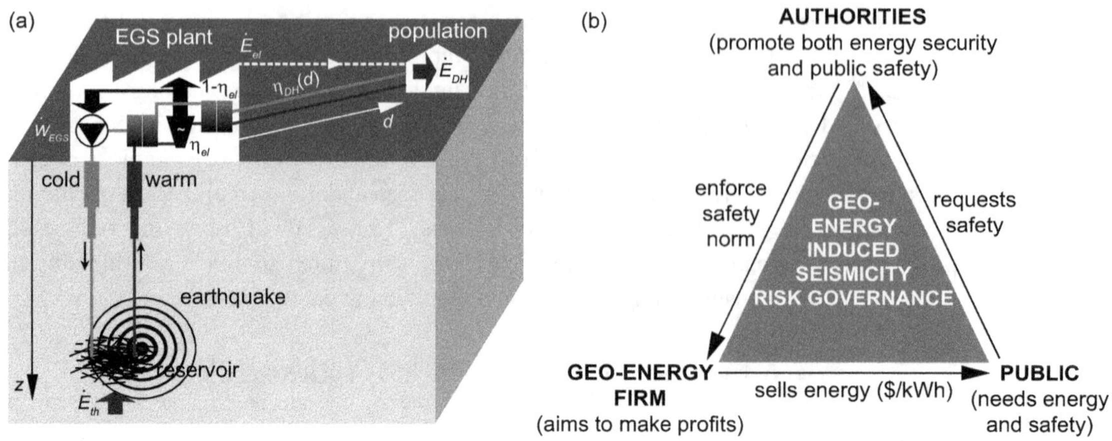

Fig. 6.7 Risk governance case study: seismic risk at a geothermal plant. (a) Depiction of a simplified EGS power plant featuring three energy cycles: an EGS cycle, a conversion cycle, and a district heating cycle. The potential for destructive earthquakes exists not only during the reservoir creation phase but also during subsequent circulation phases. Reprinted (minimally altered) from Mignan et al. (2019b), with permission from Elsevier. (b) Risk governance framework addressing induced seismicity within the geo-energy sector. Reproduced (minimally altered) from Mignan et al. (2019a), with permission from Springer Nature.

while the corresponding risk governance framework is illustrated in Figure 6.7b. In this landscape, governing bodies, which can include national, regional, or local authorities, face a dilemma. On one hand, they aspire to promote geo-energy projects for a multitude of commendable reasons, such as enhancing energy security, diversifying energy sources, advancing environmentally friendly energy solutions to combat global warming, and fostering the growth of the geo-energy industry. On the other hand, owing to the associated seismic risks, these authorities bear the responsibility of safeguarding local populations by enforcing safety protocols. Given the novelty of the challenge, various options for resolution remain the subject of ongoing debate. Since the risk is fundamentally tied to the geo-energy industry, it falls upon this industry to develop and implement appropriate safeguards. This means adhering to safety criteria, abiding by regulatory frameworks, and following safety guidelines mandated or recommended by the governing authorities. Additionally, there is a need to transfer a portion of the residual risk to environmental risk insurance providers. Given that the risk is imposed by the geo-energy industry, there exists the potential for a significant NIMBY effect, with the public seeking assurances of safety from the governing authorities. In the subsequent sections, each sub-model will be briefly introduced for a more detailed understanding of their respective contributions to the full story.

6.3.1.1 Energy Sub-Model

In this case study, a simplified energy system comprising three distinct cycles is examined: an EGS cycle, a conversion cycle, and a district heating cycle (Fig. 6.7a). The **EGS**

cycle's primary function is to transport geothermal heat to the surface, while the **conversion and district heating cycles** serve to deliver electrical and heating power to consumers, respectively. Throughout the process, a cold fluid with an enthalpy denoted as H_{inj} [J/kg] is consistently re-injected through a well, while a fluid with elevated enthalpy marked as H_{prod} [J/kg] is extracted from the reservoir. The geothermal heat flow upstreamed, denoted as $\dot{E}_{th}$ [J/s or W], can be defined as follows:

$$\dot{E}_{th} = \rho_{inj}\dot{V}_{inj}(H_{prod} - H_{inj}) = \rho_{inj}\dot{V}_{inj}c_w\left(T_0 + z\frac{dT}{dz} - T_{inj}\right). \tag{6.14}$$

Here, $\dot{V}_{inj}$ [m^3/s] represents the injected flow rate, ρ_{inj} [kg/m^3] corresponds to the fluid density at the injection temperature T_{inj} [°C], c_w is the specific heat capacity of water [J/(kg.K)], T_0 [°C] denotes the temperature at zero depth, and dT/dz [°C/km] characterizes the **geothermal gradient**. The obtained $\dot{E}_{th}$ is then converted into marketable electrical power $\dot{E}_{el}$ by expanding vapour within the conversion cycle, with an efficiency denoted as η_{el}. This electrical power is reduced by the converted electrical power $\dot{W}_{EGS}$ necessary to power the EGS pump:

$$\dot{E}_{el} = \eta_{el}\dot{E}_{th} - \dot{W}_{EGS}. \tag{6.15}$$

The EGS pump plays a critical role in compensating for pressure losses occurring within the reservoir and well systems. Finally, within the district heating cycle, a heat exchanger captures the waste heat generated in the conversion cycle. It subsequently directs this heat flow, denoted as $\dot{E}_{DH}$, towards an existing urban district heating facility. The heat flow delivered to consumers is determined by the efficiency factor $\eta_{DH}(d)$, which is a function of the distance d from the plant:

$$\dot{E}_{DH} = (1 - \eta_{el})\eta_{DH}(d)\dot{E}_{th}. \tag{6.16}$$

$\dot{E}_{el}$ and $\dot{E}_{DH}$ will later be used as inputs to the economic sub-model.

6.3.1.2 Seismic Risk Mitigation Sub-model

During the process of creating the EGS reservoir, the rate of induced seismicity, denoted as $\lambda(t)$, is directly proportional to the injected flow rate $\dot{V}_{inj}(t)$ at time t (Dinske and Shapiro, 2013; Mignan, 2016b; Mignan et al., 2017). Consequently, the overall seismic risk associated with the reservoir creation phase becomes a function of the total number N of earthquakes triggered during stimulation, given by the expression:

$$N(\geq M) = 10^{a_{fb} - bM}V. \tag{6.17}$$

This equation corresponds to the Gutenberg–Richter law, with a_{fb} representing the underground feedback activation or the normalized a-value, factored by the total volume of injected fluids, denoted as V (for the potential role of injection pressure $P(t)$ on $\lambda(t)$, see Feng et al., 2023). The PSHA can adopt a similar approach to the one outlined in Section 2.5. However, in this case, it is subject to very high epistemic uncertainties, which will be addressed in Section 6.3.2. The damage and loss assessment follows the standard CAT risk modelling procedure (Chapter 3). The risk mitigation strategy unfolds as follows. (1)

The governing authorities establish a safety standard that the geo-energy firm must adhere to, in the form of a risk threshold not to be breached. (2) In compliance with this safety standard, the injection operator employs a TLS that provides guidance on when to cease fluid injection if the forecasted risk approaches this threshold (Mignan et al., 2017). (3) In a simplified illustrative scenario, when the risk escalates beyond acceptable levels, the well is considered lost, and the stimulation process is abandoned. Prior to project commencement, the risk remains highly uncertain, making the loss of a well a probabilistic outcome (as detailed in Section 6.3.2 concerning epistemic uncertainty modelling).

6.3.1.3 Economic Sub-model

The final step consists of the computation of the **levelized cost of electricity** (LCOE) or energy price, denoted as ν_E [$/MWh]. This calculation involves dividing the total cost, represented as ν_C [$], required for constructing and operating the EGS plant by the total energy output, represented as E [MWh] produced over the plant's lifespan (excluding here any discount rate for simplicity) (Aldersey-Williams and Rubert, 2019). The energy output is determined by the plant's power, denoted as $\dot{E}$ [MW], multiplied by the number of operational hours during the plant's lifetime. It should be noted that the revenue generated from the sale of district heating contributes to the overall electricity pricing (as **heat credit**). Within the context of seismic risk mitigation measures, the probability p of well loss due to elevated seismic risk at a specific site location can be integrated into the LCOE (Mignan et al., 2019b). This integration serves as an illustration of decision-making under uncertainty in Section 6.3.3, where choices must be made among various alternatives. This integration can be formally represented as the null expectation of a **Bernoulli trial** (a random experiment with two possible outcomes, success and failure) where the outcomes are either gains or losses:

$$(1-p)(\nu_E E - \nu_C) + p(-\nu_{TLS}) = 0, \tag{6.18}$$

with p signifying the likelihood of well abandonment due to seismic risk. The value ν_{TLS} denotes the sunk costs associated with well drilling and fracturing. When $p = 0$, the original LCOE price remains unchanged. However, for $p > 0$, the price ν_E experiences an increase attributable to the additional costs incurred in potential early project terminations. Equation (6.18) encodes the two possible scenarios: either seismic risk remains low, the reservoir is successfully created, and the plant can operate (first term), or seismic risk proves to be too high, resulting in sunk costs (second term). Should a new project go ahead or not? It is expected that various stakeholders will exhibit diverse responses based on their individual risk profiles, as later described in Section 6.3.3.

6.3.2 Epistemic Uncertainties in CAT Models

The distinction between aleatory uncertainty and epistemic uncertainty was previously made in Section 1.1.4. Up to this point, this textbook has exclusively addressed aleatory uncertainty in the models developed, with the stochastic variability of their outputs encoded in probability distributions (refer to Chapter 2 for hazard and to Chapter 3 for risk. Recall the Poisson distribution (Eq. (2.1)) for primary uncertainty, and the log-normal distribution of intensity fields (Section 2.5.2) for secondary uncertainty in the hazard module. (See also Exercise 6.1 on the beta distribution for secondary aleatory uncertainty encoding in the loss module). In contrast, epistemic uncertainties, which represent a lack of perfect knowledge, were *assumed to be negligible*; it was presumed that each model used for hazard and risk assessment, given a peril, was entirely accurate. While simplifying the understanding of the CAT risk modelling process, *this assumption is unrealistic in practical applications* of risk management. For every physical process, different models exist, and so far, the simplest, most illustrative, or most commonly used models have been favoured. It is however worth noting that there is no strict boundary between the two types of uncertainties. 'Epistemic and aleatory uncertainties are fixed neither in space... nor in time. What is aleatory uncertainty in one model can be epistemic uncertainty in another model, at least in part. And what appears to be aleatory uncertainty at the present time may be cast, at least in part, into epistemic uncertainty at a later date' (Hanks and Cornell, 1994).

Epistemic uncertainties, as previously mentioned, encompass model errors, parameter uncertainty, and measurement errors, among others (Section 1.1.4; see Kennedy and O'Hagan (2001) for the different types of uncertainties in computer models). *Regarding model uncertainty and given a set of data, quantifying epistemic uncertainties involves evaluating the parameterization of each alternative model, comparing the ranges of values obtained, and subsequently combining them using a designated weighting strategy according to their adequacy.* It has been shown that the actuarial (or statistical prediction) approach is superior to the so-called clinical approach of informal judgement (Dawes et al., 1989; Grove, 2005). When observations are clearly lacking, then some judgement in determining the relative importance of different models may rely on the subjectivity of expert knowledge and experience. Philosophically, it is worth noting that even when model selection can be done 'objectively' from data alone (Zucchini, 2000), the process remains inherently subjective since any model is a modeller's creation (Frigg, 2010), with its functional form having an impact on the result (Pitt et al., 2002).

6.3.2.1 Frequentist versus Bayesian Inference

Up to this point, the frequentist approach has been employed for probability inference, with the probability of an event defined as the proportion of times it occurs in a sequence of hypothetical trials, relying solely on past data as evidence for inference (see, for instance, event size distributions in Section 2.3). However, this view oversimplifies the nature of probability. As Woo (1999, p. 156) aptly put it, 'probability is not an observable physical quantity, like wind speed or ground acceleration. It is invisible, inaudible, and intangible.' Bruno de Finetti went even further, asserting that there is no objective probability, boldly stating that 'probability does not exist' (de Finetti, 1974; Nau, 2001). Knight (1921), quoted in Section 1.1.4, stated the limits of inference from data alone when these data are heterogeneous, which is clearly the case of past disasters. This absence of an objective measure implies that probability estimates must be subjective, a concept formalized through the Bayesian approach to probability inference. In **Bayesian inference**, a prior distribution, representing a priori beliefs, is multiplied by a data-based likelihood distribution to obtain

Box 6.3 Bayesian Inference

We succinctly introduced Bayes' theorem, expressed as $\mathbb{P}(A|B) = \mathbb{P}(B|A)\mathbb{P}(A)/\mathbb{P}(B)$, in Section 2.3.2. For convenience, it can be reformulated as $\mathbb{P}(\mathcal{M}|\mathcal{D}) = \mathbb{P}(\mathcal{D}|\mathcal{M})\mathbb{P}(\mathcal{M})/\mathbb{P}(\mathcal{D})$ where $\mathcal{M}$ is a model (or hypothesis) and $\mathcal{D}$ the data (or evidence). $\mathbb{P}(\mathcal{M})$ is called the **prior**, $\mathbb{P}(\mathcal{D})$ the **marginal** (a normalization factor denoted c), $\mathbb{P}(\mathcal{D}|\mathcal{M})$ the **likelihood**, and $\mathbb{P}(\mathcal{M}|\mathcal{D})$ the **posterior**. In the context of forecasting induced seismicity using the model described by Eq. (6.17), Bayesian inference consists of updating the model parameter set $\Theta_{\mathcal{M}} = (a_{fb}, b)$ from new earthquake data $\mathcal{D}$. For this specific context, Bayes' theorem can be expressed in terms of PDFs with $f_{post}(\Theta_{\mathcal{M}}|\mathcal{D}) = cf_{lik}(\mathcal{D}|\Theta_{\mathcal{M}})f_{prior}(\Theta_{\mathcal{M}}) = c\mathcal{L}(\mathcal{D}|\Theta_{\mathcal{M}})\mathcal{B}(\Theta_{\mathcal{M}})$ (Broccardo et al., 2017). We have $\mathcal{B}$ representing the beta distribution (see Exercise 6.1), which is chosen subjectively and calculated from the prior $\Theta_{\mathcal{M}}$ set (the set of known parameter values from prior EGS stimulations). Since the model consists of counting the number of earthquakes N, the likelihood distribution $\mathcal{L}(\mathcal{D}|\Theta_{\mathcal{M}})$ must be described by a non-homogeneous Poisson process, the description of which remains outside the scope of this textbook. Importantly, the introduced subjective prior improves accuracy at the early stages of an underground stimulation (when observations remain limited) to predict the future risk (related to the predicted N), which will need to remain below a fixed safety threshold.

an updated distribution (Box 6.3). The debate between Frequentists and Bayesians has been fierce (Efron, 1986), but it lies outside the immediate scope of this textbook. These two approaches represent distinct philosophical and methodological views on statistical inference, and their suitability depends on the specific problem at hand. While some Bayesian priors (such as non-informative or flat priors) attempt to align with frequentist properties, their justification may not always be straightforward (Gelman et al., 2017). In his theory of decision under uncertainty, Gilboa (2009) emphasizes that classical and Bayesian statistics serve different purposes: 'Classical statistics is a collection of tools for a society to agree on what can be stated as officially established, with the understanding that this society includes different people with different beliefs and different goals. Bayesian statistics is the tool for individual reasoners or decision makers, who attempt to sort out their knowledge, belief, evidence and intuition.' The former seeks objectivity, while the latter aims to accommodate all beliefs, including both objective and subjective statements. This will be illustrated in the next paragraphs with our case study.

6.3.2.2 Logic Trees

A **logic tree** is a systematic representation of various alternative scenarios or models designed to capture diverse sources of epistemic uncertainties. It serves as a graphical structure for exploring and accounting for different hypotheses, assumptions, or parameterizations that can affect the outcomes of the CAT model. Although uncertainties may be sampled from continuous distributions in Monte Carlo simulations (e.g., from a parameter value distribution; Box 6.3), the logic tree approach offers practical advantages due to its tractability and its effectiveness to communicate risk to stakeholders. This approach has been employed since the early days of seismic hazard assessment (Kulkarni et al.,

1984). Figure 6.8a depicts a possible logic tree for the induced seismicity risk model of the case study. While different models may be tested at each step of the CAT modelling pipeline, here epistemic uncertainties are encoded in the hazard component only for the sake of illustration. Specifically, these uncertainties manifest in the event size distribution model $\lambda = f_S(S)$, where λ is the occurrence rate of event of size S (Section 2.3). For earthquakes, S corresponds to the event magnitude M. Notably, epistemic uncertainties related to the event rate model are typically not explicitly represented in current-generation CAT models (Foote et al., 2017, p. 372). In the context of earthquakes as a peril, the Gutenberg–Richter law is the standard model (Eq. (2.39)). However, alternative models exist, such as the characteristic earthquake model (Youngs and Coppersmith, 1985) or a hybrid of both. Earthquake rate models can also depend on different interpretations of fault slip rates (Danciu et al., 2018). In the case of induced seismicity, epistemic uncertainties are notably high due to limited historical data (Fig. 6.8b; Mignan et al., 2021a) and the absence of correlation between induced seismic responses to fluid injection (a_{fb}, b) (Eq. (6.17)) and other known parameters in a given region (Evans et al., 2012). These uncertainties are depicted by 3×3 branches in Figure 6.8a. The determination of the maximum magnitude M_{max} at which the Gutenberg–Richter law is truncated is also a subject of debate in the context of underground fluid stimulations. Historically, the **McGarr model** was commonly employed, assuming that the maximum magnitude is a function of the volume (V) of fluids injected, with

$$M_{max}(V) = \frac{2}{3} \log_{10}(V \cdot c_G \cdot 10^7) - 10.7, \tag{6.19}$$

where $c_G = 3 \times 10^{10}$ Pa, the modulus of rigidity (McGarr, 1976, 2014). While the predicted hard cap appears to be in agreement with historical data (Fig. 6.8c), a statistical study suggests that an unbounded Gutenberg–Richter law hypothesis cannot be rejected (van der Elst et al., 2016), implying that the tectonic M_{max} should be used in the size distribution model (leading to two further branches in Figure 6.8a). Another prevalent source of epistemic uncertainty is associated with the selection of intensity footprint models. For the purpose of illustration, a single **intensity prediction equation** (IPE) is employed, calibrated to global macroseismic intensity data (Allen et al., 2012) (refer to Exercise 6.5 for an exploration of IPE uncertainty. In contrast to a GMPE, an IPE uses macroseismic intensity instead of a physical measure of ground-shaking, such as PGA). It is essential to note that the illustrated logic tree structure serves for illustrative purposes, and other choices could have been made (Mignan et al., 2015a; Broccardo et al., 2020). Weights (w) are here defined to represent various approaches to addressing uncertainty: they are roughly based on historical data ranges for both a_{fb} and b while constraining the possible extrema (Fig. 6.8b; Mignan et al., 2021a). They are fixed at 0.8 for the tectonic M_{max} model (and thus 0.2 for the McGarr model) as an expert subjective choice derived from ambiguous observations (see Section 6.3.3 for approaches to deal with ambiguity). Alternatively, these weights could have been set as equal, guided by the principle of indifference (or of insufficient reason – read details in Section 6.3.3). In summary, various subjective choices are plausible.

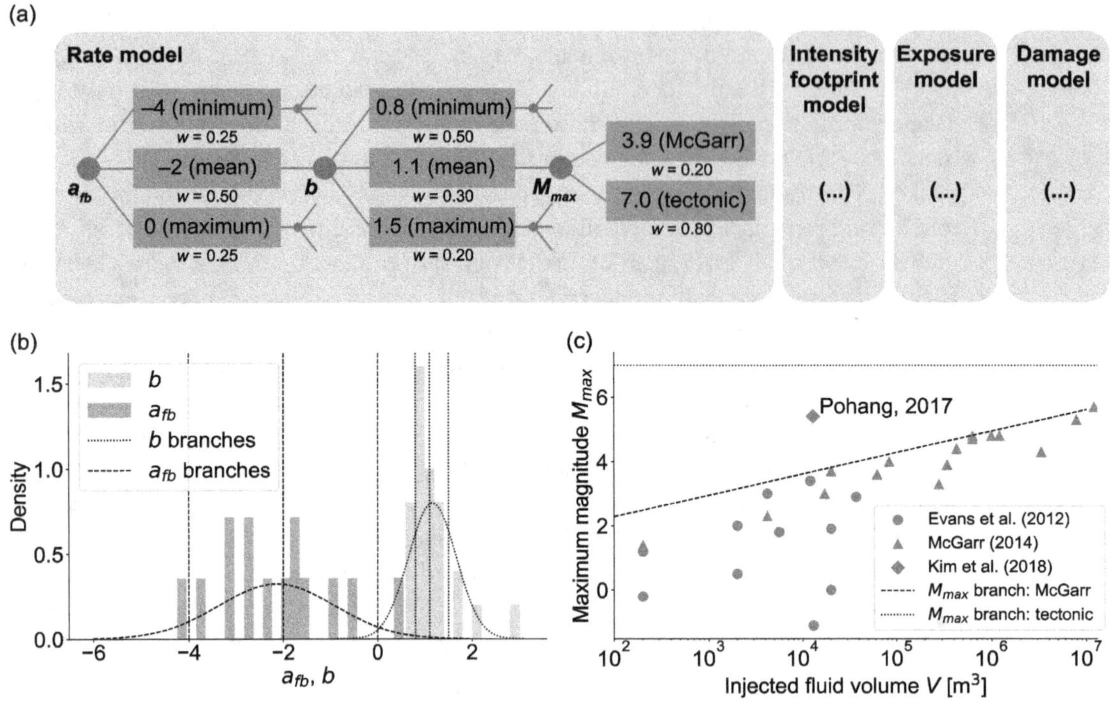

Fig. 6.8 Epistemic uncertainties considered in the case study. (a) Logic tree for the rate model of the hazard module. Weights (w) assigned to a_{fb} and b branches (Eq. (6.17)) are approximately based on empirical data – see (b) (the subjective extrema could also be incorporated in a Bayesian approach; Gelman et al., 2017). Weights for M_{max} determined by expert opinion – see (c). (b) Empirical distribution of a_{fb} and b based on data from Mignan et al. (2021a) (14 estimates for a_{fb} and 25 for b at geothermal sites). (c) Relationship between M_{max} and total injected fluid volume (V) using data from Evans et al. (2012), McGarr (2014), and Kim et al. (2018). In the considered area, $M_{max} = 7$ fixed for the tectonic setting while $M_{max} = 3.9$ for an injected volume of $V = 30,000$ m^3 with the McGarr model (Eq. (6.19)).

6.3.2.3 'Spaghetti' Plots of EP Curves

For each specific model combination within a designated logic tree path, an EP curve can be generated. These EP curves facilitate the comparison of various scenarios, and the results are typically visualized using 'spaghetti' plots, as exemplified in prior works (e.g., Mitchell-Wallace, 2017, p. 177). Figure 6.9 presents the outcomes of the seismic risk model designed for an underground fluid stimulation involving an injected fluid volume of $V = 30,000$ m^3. This amount influences both the seismicity rate (Eq. (6.17)) and the maximum magnitude for one branch (Eq. (6.19)). The hazard details, previously discussed in Figure 6.8, are part of this analysis. The calculation of building damage employs the RISK-UE method (Lagomarsino and Giovinazzi, 2006), with a vulnerability index (κ_D) set at 0.75, representing structures of simple stone masonry (see Eqs (3.7) and (3.8)). In the case study, losses (L) are defined in terms of fatalities using the consequence function of Eq. (3.9) with $L = N_{pop} \times \mathbb{P}(\text{death})$, where N_{pop} is the number of people exposed. Fixing $N_{pop} = 1$ yields the **individual risk** (IR), which signifies the probability of a statistically

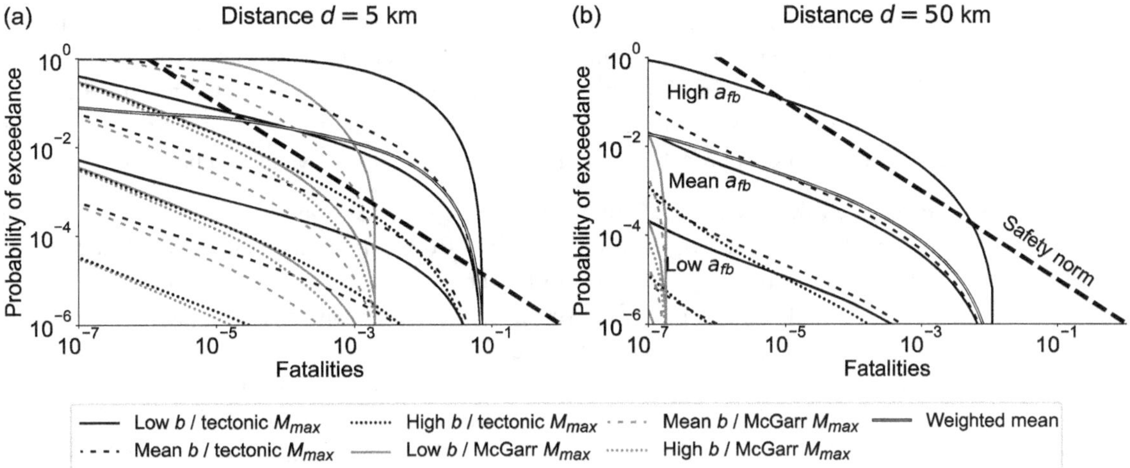

Fig. 6.9 EP-curve epistemic uncertainties visualized in a 'spaghetti' plot, as encoded in the logic tree outlined in Figure 6.8. Fatalities computed for $N_{pop} = 1$ (i.e., individual risk). (a) Results for the nearest populated area being situated at a distance of $d = 5$ km from the underground stimulation site (and stimulation taking place at a depth of $z = 6$ km). (b) Distance increased to $d = 50$ km. The dashed line delineates the safety threshold within the IR space, defined here as the iso-risk level of 1 μmt.

representative individual suffering a fatal outcome (within a given time frame and location). In the domain of catastrophe risk management, such as in reinsurance firms, each EP curve may correspond to the final output of a CAT model from different vendors. Epistemic uncertainty arises in that case because various modelling companies employ distinct models, data sets, and assumptions. Even within a single firm, new model releases are prone to variation from previous iterations due to the integration of novel research and data (Mitchell-Wallace, 2017, p. 123). Returning to the present case study, epistemic uncertainties are best understood in the context of the prescribed safety standard, here defined as a very low IR = 10^{-6}, or 1 micromort [μmt], for the duration of the stimulation. This is represented in Figure 6.9 by an **iso-risk line** where the ratio of the probability of loss to the magnitude of loss remains constant across different points on the line. Note that the **micromort** is a unit of risk equivalent to a one-in-a-million chance of death, which is convenient for comparisons with other risks; for example, the daily population risk of dying due to non-natural causes is approximately 1 μmt (Walker et al., 2014). The patterns observed in Figure 6.9 will serve as the basis for the next discussions.

6.3.3 Decision-Making under Uncertainties

Decision theory, the study of choices in the presence of uncertainty, draws upon concepts from game theory, psychology, and economics to determine optimal courses of action. It involves conducting a risk–benefit analysis, mapping out probabilities, consequences, and the societal value of various alternatives. The ultimate goal is to construct a kind of societal 'balance sheet' that can offer guidance to decision-makers (Lowrance, 1980, p. 10). As revealed in the previous section, the uncertain outcomes were implicitly influenced by the assigned weights at each branch within the logic tree. The weighted mean EP curves, depicted in Figure 6.9, were derived based on the selected weights illustrated in Figure 6.8a. Three distinct approaches to assigning probabilities to events can be identified.

1. ***Principle of indifference***: This approach suggests assigning equal probabilities to all potential outcomes. Originating from Jacob Bernoulli, it states that when faced with n possible outcomes and no reason to favour one over the others, each outcome should receive a probability of $1/n$. However, this principle was primarily applied to simple games of chance such as dice and card games. In more complex scenarios, where the state space can be partitioned in various ways, the principle of indifference can lead to arbitrary probability assignments (e.g., Woodward and Bishop, 1997).
2. *Frequentist approach*: This approach asserts that the probability of an event should be based on the relative frequency with which it has been observed in the past. The **law of large numbers** supports this by stating that when independent and identically distributed (i.i.d) random variables X_i are observed, their average $\bar{X} = \frac{1}{n}\sum_i X_i$ approaches their expected value as the number of observations (n) becomes large (Bernoulli, 1713). Classical statistics offers tools such as confidence intervals and hypothesis testing (Neyman and Pearson, 1933; Fisher, 1955; Lehmann, 1993). Nevertheless, a significant limitation is that real-world events are often not i.i.d., requiring the consideration of subjective choices when assessing probabilities. Smaller sample sizes (lower n) can lead to less reliable estimates, and for many types of catastrophes, finding a sufficiently large n is practically impossible. Furthermore, real systems often exhibit complexity, with inherent interactions and time-dependent trends. The frequentist approach thus simplifies these complexities. The CAT models described in this textbook followed this approach, but modelling choices were implicitly made, which were subjective.
3. *Subjective approach*: The subjective approach employs the mathematical framework of probability theory to model our uncertain beliefs and intuitions (Gilboa, 2009). This approach is the focus of this section and allows for the incorporation of individual judgements and qualitative assessments in the modelling process.

6.3.3.1 Risk Perception

Risk perception refers to how individuals or groups subjectively perceive and evaluate risks. It is a psychological and cognitive process through which people assess the likelihood and severity of potential risks or hazards in their environment, activities, or decisions. Risk perception can vary significantly among individuals and is influenced by a wide range of factors, including personal experiences, cognitive biases, and emotional responses. Cognitive biases were brought to the attention of the general reader in *Thinking, Fast and Slow* (Kahneman, 2011) and described in a disaster risk perspective by Meyer and Kunreuther (2017). Biases include the following.

- ***Assessment bias***: This cognitive bias frequently leads to the overestimation of risks that are objectively low while simultaneously underestimating risks that are objectively high.

For instance, individuals may exhibit a heightened fear of terrorist attacks compared to more prevalent risks such as traffic accidents, primarily because they perceive the potential consequences of the former as more severe, despite the significantly lower probability involved (Lichtenstein et al., 1978).

- ***Availability bias***: This cognitive bias is characterized by an individual's tendency to place undue significance on a recent experience, even if it was a rare occurrence. Such experiences can lead individuals to perceive the associated risk as particularly high. While this bias can be advantageous for preparing against risks similar to the recent experience – a concept related to memory bias – it can potentially be detrimental when allocating limited resources among various types of perils (de Guttry and Ratter, 2022). Other related terms include myopia bias (short-term horizon) and amnesia bias (fading memory of past events) (Meyer and Kunreuther, 2017).
- ***Optimism bias***: This cognitive bias is marked by individuals underestimating their personal risk in comparison to others (Weinstein, 1984). In the context of disasters, this bias is evident as individuals often believe that they are less likely to be adversely affected by a disaster than their neighbours or other community members. This phenomenon has been observed in the case of various perils, including earthquakes, floods, and tropical cyclones (e.g., Burger and Palmer, 1992; Royal and Walls, 2019; Wu et al., 2024). The negative consequence of optimism bias is that it can impede effective risk prevention measures, as exemplified during the COVID-19 pandemic (Druică et al., 2020).
- ***Anchoring effect***: This cognitive bias occurs when an individual, having been exposed to a specific numerical value (the anchor), uses that anchor as a reference point to evaluate an unrelated process. For example, when individuals are asked to estimate the frequency of various hazards, such as floods, tornadoes, and wildfires, their estimates can vary significantly depending on the anchor provided. For instance, if two groups of individuals (Group A and Group B) are presented with different independent anchors, their estimates for hazard frequencies will diverge. If Group A is informed that the annual death rate in the US due to car accidents is 50,000, and Group B is told that the annual death rate due to electrocution is 1,000, Group A is likely to overestimate the frequencies of other unrelated hazards compared to Group B (e.g., Lichtenstein et al., 1978; Nagaya and Nakayachi, 2015).

Other biases, including inertia bias, simplification bias, and herding bias, have been recognized as important factors in disaster risk perception (Meyer and Kunreuther, 2017). Cognitive biases have the potential to distort the results of risk assessments and must be carefully considered in CAT risk management. There are two contrasting approaches to address the influence of risk perceptions: one approach advocates relying solely on scientific risk concepts and correcting for erroneous risk perceptions through effective risk communication and education (Cross, 1998). Another perspective acknowledges that collective decisions are necessary, with science playing a role but not holding exclusive privilege in decision-making processes (Jasanoff, 1998). It is important to note that even scientific risk concepts are, to some extent, influenced by subjective modelling choices. Subjectivity permeates the very definition of research questions, the design of experiments used to gather evidence, and the assessment of the societal significance of risks (Lowrance,

1980, p. 7). Hence, dealing with subjective probabilities remains an integral aspect of risk management.

6.3.3.2 Expected Utility Theory

Various approaches have been developed to define, assign, and work with subjective probabilities in a rational manner, all grounded in the maximization of utility. **Utility**, in this context, represents the usefulness, worth, or value of a given quantity. A utility function denoted as f_u represents an individual's or agent's preferences. In this framework, an alternative preferred over another is assigned a higher utility value. As a consequence, selecting the most preferred alternative corresponds to maximizing the associated utility function (Gilboa, 2009). Decision-making under risk can be conceptualized as a process of evaluating preferences between simple lotteries or gambles. A **prospect** denoted as $(x_1, p_1; \cdots ;x_n, p_n)$ represents a contract that yields outcome x_i with a corresponding probability p_i, where the sum of all probabilities equals 1. It has long been recognized, dating back to Bernoulli's St. Petersburg paradox, that individuals do not maximize the expected monetary value of a game but rather its expected utility (Bernoulli, 1738/1954) (Box 6.4). The **expected utility** of these outcomes can be expressed as

$$\mathbb{E}[f_u(X)] = p_1 f_u(x_1) + \cdots + p_n f_u(x_n). \tag{6.20}$$

Here, the function f_u is concave, representing risk aversion. **Risk aversion** implies that when presented with a choice between a random outcome X and a sure pay-off equal to the expected value of X, a risk-averse agent consistently prefers the latter. The concept of

Box 6.4 St. Petersburg Paradox

The **St. Petersburg paradox** is a classic problem in probability theory and decision theory that raises the question about how people evaluate risks and make decisions. It is named after the city of St. Petersburg and was solved by Daniel Bernoulli in 1738 (Bernoulli, 1738/1954). Here is how the paradox works: imagine a gambling game where a fair coin (a coin with equal chances of coming up heads or tails) is flipped repeatedly until it comes up heads. Winnings are determined by how many coin flips it took to get that first heads with 2^k gains where k is the number of tosses. So, if heads comes up on the first flip, one wins $\$2^1 = \$2$. If it comes up on the second flip, one wins $\$2^2 = \$4$, and so on. The average payout, or expected value, is $\mathbb{E}[X] = \frac{1}{2} \cdot 2 + \frac{1}{4} \cdot 4 + \frac{1}{8} \cdot 8 + \cdots = 1 + 1 + 1 + \cdots = \infty$. According to traditional expected value calculations, the potential player should be willing to pay any finite amount to participate in this game because the expected value of winnings is infinite. This is the paradox: in reality, most people would not be willing to pay very much to enter the game. Bernoulli (1738/1954) addressed the St. Petersburg paradox by shifting the focus from monetary value to utility. He introduced a utility function, denoted as $f_u(X) = \log(X)$, which assigned values based on the satisfaction or utility derived from wealth. This adjustment leads to the finite series $\frac{1}{2} \log 2 + \frac{2}{4} \log 2 + \frac{3}{8} \log 2 + \cdots = 2 \log 2$ for $x_i = 2^i$. To estimate the cost to enter the game, x_i must represent the gambler's total wealth (Ex. 6.6). This paradox has triggered numerous discussions and debates within the world of decision theory (Samuelson, 1977).

expected utility remained relatively dormant for about two centuries until it was axiomatized in **expected utility theory** (EUT) by von Neumann and Morgenstern (1944) in their seminal work on game theory. A von Neumann–Morgenstern utility f_u is any function that satisfies

$$X \prec Y \text{ if and only if } \mathbb{E}[f_u(X)] < \mathbb{E}[f_u(Y)], \tag{6.21}$$

with $X \prec Y$, or equivalently $Y \succ X$, signifying that Y is preferred to X (i.e., Y is the rational choice). Savage (1954) introduced the **subjective expected utility theory** (SEUT), which replaces objective probabilities with subjective or 'personal probabilities', aligning with Bayesian priors. SEUT is similar to Eq. (6.20), but it replaces p_i with personal probabilities that can vary among individuals. While EUT, which explains phenomena such as insurance purchase and returns on risky assets, dominated as the primary model for decision-making under uncertainty for several decades, it faced significant scrutiny by the 1980s. It is now widely accepted that EUT does not adequately describe individual choices, yet it remains the standard theoretical tool for cost–benefit analysis under conditions of risk (Buchholz and Schymura, 2012).

6.3.3.3 Cumulative Prospect Theory

Prospect theory, introduced by Kahneman and Tversky (1979), emerged as an alternative model aimed at addressing the limitations of EUT. At its core, prospect theory employs two fundamental components: a **value function** denoted as $f_v(x)$ – instead of $f_u(x)$ – and a non-linear transformation of the probability scale $f_w(p)$. **Cumulative prospect theory** (CPT) expands upon the original concept, accommodating a variable number of outcomes, allowing for distinct weighting functions for gains (f_{w^+}) and for losses (f_{w^-}), and providing a unified approach to both risk and uncertainty (Tversky and Kahneman, 1992). Among the range of alternatives to EUT, CPT stands out as one of the most sophisticated and accurate in mimicking individual behaviour (Barberis, 2013; Grechuk and Zabarankin, 2014). Its value function, which exhibits concavity for gains, convexity for losses, and greater sensitivity to losses than gains, distinguishes CPT from EUT. Notably, CPT operates on the premise that individuals exhibit risk aversion concerning gains while displaying risk-seeking behaviour when facing losses. This concept is mathematically represented as a two-part power function,

$$f_v(x) = \begin{cases} x^{\alpha_1} \text{ if } x \geq 0 \\ -c_\lambda(-x)^{\alpha_2} \text{ if } x < 0 \end{cases}, \tag{6.22}$$

where $0 < \alpha_1 \leq \alpha_2 < 1$ and $c_\lambda > 1$ is the loss aversion coefficient. **Loss aversion** is a psychological concept that highlights the human tendency to be significantly more responsive to losses, even when they are relatively minor, compared to equivalent gains. The weighting function then assigns disproportionate importance to small probabilities while diminishing the significance of moderate and high probabilities (see assessment bias):

$$f_{w^+}(p) = \frac{p^{\alpha_3}}{(p^{\alpha_3} + (1-p)^{\alpha_3})^{1/\alpha_3}}, f_{w^-}(p) = \frac{p^{\alpha_4}}{(p^{\alpha_4} + (1-p)^{\alpha_4})^{1/\alpha_4}}, \tag{6.23}$$

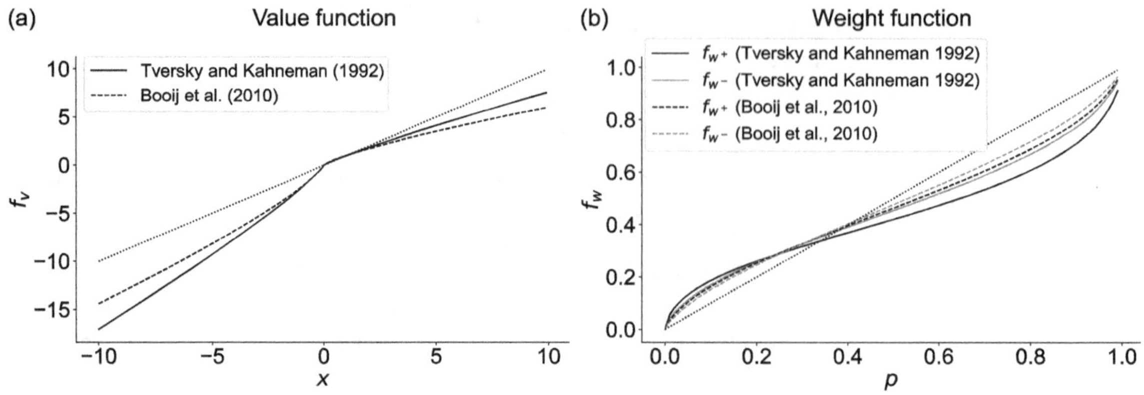

Fig. 6.10 Cumulative prospect theory parameterization. (a) Value function (Eq. (6.22)) with $\alpha_1 = 0.88, \alpha_2 = 0.88$, and $c_\lambda = 2.25$ for Tversky and Kahneman (1992) and $\alpha_1 = 0.78, \alpha_2 = 0.82$, and $c_\lambda = 2.18$ for Booij et al. (2010). (b) Weighting functions for gains (f_{w+}) and for losses (f_{w-}; Eq. (6.23)) with $\alpha_3 = 0.61$ and $\alpha_4 = 0.69$ for Tversky and Kahneman (1992) and $\alpha_3 = 0.72$ and $\alpha_4 = 0.77$ for Booij et al. (2010).

with the coefficients $\alpha_3 < 1$ and $\alpha_4 < 1$. These equations map objective probabilities p onto subjective probabilities $f_w(p)$, which also relates to SEUT. The combination of Eqs (6.22) and (6.23) results in a **fourfold pattern of risk attitude**: individuals exhibit risk-seeking behaviour over low-probability gains and high-probability losses, while displaying risk aversion over low-probability losses and high-probability gains (Tversky and Kahneman, 1992). There are four behaviours instead of two due to a reversal caused by the probability distortion. The four behaviours explain, respectively, the attractiveness of lottery tickets (hope of large gain), desperate gambles (hope of avoiding loss), the attractiveness of insurance (fear of large loss), and conservative investment strategies (fear of disappointment, e.g., low-risk, low-return investments such as bonds preferred over higher returns in the stock market). In the realm of insurance, CPT is now employed to estimate deductible levels and other policy preferences, as highlighted by Barberis (2013). The original CPT parameterization was $\alpha_1 = 0.88, \alpha_2 = 0.88, c_\lambda = 2.25, \alpha_3 = 0.61$, and $\alpha_4 = 0.69$ in Tversky and Kahneman (1992). The parameterization for the general population has been found to be $\alpha_1 = 0.78, \alpha_2 = 0.82, c_\lambda = 2.18, \alpha_3 = 0.72$, and $\alpha_4 = 0.77$ (Booij et al., 2010) (Fig. 6.10). It is worth noting that the choice of the coefficients in the value function and probability weighting function can lead to distinct behavioural outcomes. For instance, Blavatskyy (2005) demonstrated that α_1 must be lower than α_3 to avoid the St. Petersburg paradox (Box 6.4). This underscores the limitations of extrapolating findings from empirical studies that often involve moderate outcomes, possibly reflecting an intrinsic boundary of CPT. The interplay between utility functions and probability distributions in catastrophe risk management remains relatively unexplored (Barberis, 2013). While CPT captures human behaviour more accurately than EUT, it still relies on individual choices involving limited gains and losses. The applicability of existing decision theories, with EUT as a special case of CPT, to the context of CAT risk management is currently under test. This field is characterized by extreme events, where an event with unbounded loss ($x = -\infty, p > 0$)

challenges standard approaches aimed at minimizing the probability of such catastrophic events following the safety-first principle (Grechuk and Zabarankin, 2014). This issue has been referred to as the 'tyranny of catastrophe risks' by Buchholz and Schymura (2012) who demonstrated that EUT struggles to address low-probability–high-impact events in an ethically acceptable manner, as the cost of avoiding such extreme events with very low likelihood can be exorbitantly high, potentially driving society into poverty.

6.3.3.4 CPT Application to the Case Study

While CPT parameterization introduces its own set of uncertainties, its basic principles remain useful for illustrating risk aversion, as demonstrated in this chapter's case study. Eq. (6.18) defined the LCOE as a null expectation (representing the fair price of electricity, now denoted as $\nu_{E,fair}$) and accounted for uncertainty in pricing (due to the risk of a future geothermal reservoir creation not adhering to seismic safety norms). This uncertainty leads to two potential outcomes: the success of the project with electricity generation, or the incurring of sunk drilling costs on an abandoned site. When faced with the decision of whether to invest in a new project, an investor's perception of risk, as discussed earlier, is influenced by various biases. The LCOE can be re-evaluated using CPT, resulting in the following equation:

$$f_{w^+}(1-p)f_v(\nu_{E,averse}E-\nu_C)+f_{w^-}(p)f_v(-\nu_{TLS})=0. \tag{6.24}$$

The new LCOE ($\nu_{E,averse}$) accounts for both risk and loss aversions, with the result $\nu_{E,averse} > \nu_{E,fair}$ obtained (Mignan et al., 2019b). Note that the likelihood of well abandonment p can be defined as the proportion of EP curves crossing the IR = 1 μmt iso-risk line in Figure 6.9. Depending on factors such as the assessed seismic risk, adherence to specific safety norms, expected costs, and projected energy production, an investor's level of risk aversion may lead them to decide against participating in a new geothermal project, perceiving the risk of failure as too substantial because an LCOE appears to be non-competitive. Over the course of this chapter, this example has demonstrated how decision-making under uncertainty can unfold in a real-world scenario where both energy security and human safety are significant concerns. It underscores the central role of CAT risk modelling (as shown in Figure 6.9) in effective risk management.

6.3.3.5 Ambiguity

Before concluding this textbook, it is crucial to emphasize a key point: the results previously discussed are contingent on the probability p of safety norm failure, which, in turn, depends on the selection of the weights w in the logic tree of Figure 6.8. The pros and cons of various approaches to weight selection were shown to be inherently subjective. One particular challenge arises when weighting the M_{max} models, which basically controls the outcomes. In the McGarr branch, the safety norm is likely to always be met, while in the tectonic branch, it may or may not be met. Stakeholders, driven by their optimism or pessimism regarding project success, may assign high weights to the former or latter model, respectively. Ethical issues would emerge if preferences were implicitly concealed within the decision process. In decision-making under **ambiguity**, that is, when decisions must be taken in situations where probabilities of outcomes are unknown, the conventional expected utility hypothesis fails. In the presence of pure uncertainty, of complete ignorance, when there is no prior information available, rational decision criteria shift their focus away from weighted means and toward the extremes of all potential prospects, as demonstrated by Arrow and Hurwicz (1977). The quest for consensus among experts in complex risk scenarios, such as ecosystem collapse risk (Ludwig et al., 1993) and global warming risk (Woodward and Bishop, 1997), has been shown to be futile. Instead, the primary goal becomes minimizing the potential maximum loss, leading to a decision rule known as **minimax**. When dealing with gains, this criterion is referred to as **maximin**, aiming to maximize the potential minimum gains (e.g., Kelsey, 1993). It is the pessimistic models that tend to dominate due to the influence of worst-case scenarios on overall willingness to pay for mitigation, reflecting a tendency toward ambiguity aversion. However, the decision-maker can take a different stance, by having a penchant for optimism (ambiguity seeking) and striving to maximize the maximum possible gains (**maximax** criterion; Kelsey, 1993). The maximax criterion is frequently (and often implicitly) applied in situations where prospects for significant and immediate gains are evident, such as resource exploitation (Ludwig et al., 1993). To mitigate the future costs of climatic disasters, as suggested by Woodward and Bishop (1997), early and aggressive measures against global warming should be pursued according to the minimax criterion. Formally, a set of n actions denoted as $\mathcal{A} = (\text{action}_1, \cdots, \text{action}_i, \cdots, \text{action}_n)$ is defined by its set of outcomes (or payoffs) $\mathcal{O} = (\text{action}_i(\text{state}_j) \rightarrow \text{outcome}_{ij})$ under each alternative state of nature $\mathcal{S} = (\text{state}_1, \cdots, \text{state}_j, \cdots, \text{state}_m)$. Returning to our case study, there exists a set of two possible actions, where $\mathcal{A} = (\text{action}_1\text{:go}, \text{action}_2\text{:no go})$, and a set of two possible states, where $\mathcal{S} = (\text{state}_1\text{:acceptable risk}, \text{state}_2\text{:risk too high})$. In terms of gains, the outcomes are as follows: $\text{outcome}_{1,1} = PE - C \geq 0$ (breaking even at fair price or making profits if priced above the LCOE), $\text{outcome}_{1,2} = -C_{TLS} < 0$ (representing sunk costs), $\text{outcome}_{2,1} = 0$ and $\text{outcome}_{2,2} = 0$ (no project meaning no gains or losses) (Mignan et al., 2015a). Applying the maximin criterion advises the potential investor not to initiate the project, following this decision rule:

$$\text{action}_{maximin} = \max_{\mathcal{A}}\{\min_{\mathcal{S}}[\text{action}(\text{state})]\}. \quad (6.25)$$

This is the simplest and least demanding rule. For other decision criteria, the reader is referred to the review by Heal and Millner (2013). While this toy example may appear trivial, it underscores that maximin-type criteria represent common-sense logic. However, it is important to note that this scenario assumes complete ignorance about the underground process, which is primarily influenced by the M_{max} model. If, assuming a conservative tectonic M_{max} and with all other uncertainties informed by data and bounded, the safety norm is met under any risk configuration, then launching the project becomes the rational choice. This is especially true when the project site is far removed from populated areas (Fig. 6.9). However, in cases where ambiguity persists, the various criteria just discussed offer rational approaches to decision-making. Yet, quantifying costs and gains remains an essential aspect of the process and a very challenging one. In environmental risk management, for example, when anthropogenic activities may pollute, costs and gains can become intangible. As eloquently posed by Funtowicz and Ravetz (1994), in the end, what is the 'worth of a songbird'?

6.4 Final General Conclusions

The present journey through the convoluted landscape of catastrophe risk modelling is drawing to a close, and it is now time to reflect upon the insights and lessons gathered from this exploration. Most aspects of the field have been covered: foundational concepts were laid in Part I, advanced modelling techniques were presented in Part II, and all these insights were applied to the domain of risk management in Part III. It is hoped that this textbook can, at its level, help support the implementation of the Sendai Framework for Disaster Risk Reduction 2015–2030, priority number one of which is an understanding of disaster risk. It should indeed help '(j) to strengthen technical and scientific capacity to capitalize on and consolidate existing knowledge [on vulnerability, exposure, and all hazards]' and '(l) to promote the incorporation of disaster risk knowledge [...] in formal and non-formal education, as well as in civil education at all levels, as well as in professional education and training' (United Nations, 2015, pp. 14–15).

In Part I, a voyage into the basics of CAT risk modelling was embarked upon. An overview of catastrophe risks was given in Chapter 1, presenting the language of this scientific domain, its history, and – in an encyclopaedic manner – a comprehensive description of 42 perils. A quantitative understanding was then achieved in Chapters 2 and 3 by the exploration of probabilistic hazard and risk assessment, respectively. By the end of Part I, the student of risk already had all the necessary tools to develop a simple CAT model from scratch, including stochastic event set sampling, hazard intensity and damage footprint modelling, as well as the standard hazard and risk metrics used by professionals in this field.

In Part II, through cellular automata, agent-based models and system dynamics, doors were opened to understand complex, interconnected systems and to conceptualize them within the CAT modelling framework. This was the topic of advanced CAT risk modelling with individual perils covered in Chapter 4 and interacting perils in the complex Earth system in Chapter 5. This part was hopefully an eye-opener on the risks associated with the Earth machinery and making the student of risk humble in view of the challenges to be faced in fully comprehending the potential of extreme events.

In Part III (Chapter 6), the societal and economic utility of CAT modelling was proven with the investigation of risk management strategies. An examination of risk transfer and reduction instruments was carried out, revealing how catastrophe insurance and peril forecasting can help better control some of the adverse outcomes of disasters. The complexities of risk governance and decision-making in the face of uncertainties were highlighted in a concrete case study on seismic risk at a geothermal plant.

It is evident that catastrophe risk modelling is not only a scientific endeavour but a critical tool for the safeguarding of societies and environments. The insights gained within these pages are designed to empower the reader to face the ever-present threats with knowledge and foresight. In an increasingly interconnected world, the importance of this field cannot be overstated. May the knowledge acquired here guide the students of risk in their quest to mitigate catastrophe risks of the future.

Further recommended reading

- *Natural Catastrophe Risk Management and Modelling* (Mitchell-Wallace et al., 2017, chapter 2)
- *Risk Governance – Coping with Uncertainty in a Complex World* (Renn, 2008)
- *Theory of Decision under Uncertainty* (Gilboa, 2009)
- *Thinking, Fast and Slow* (Kahneman, 2011)

6.5 Exercises

Exercise 6.1

The **beta distribution** $\mathcal{B}(x, c_\alpha, c_\beta)$, with $0 \leq x \leq 1$ and c_α and c_β parameters, is commonly used to describe the uncertainties around the mean losses (Fig. 6.11). The parameters are related to the mean $\mathbb{E}[X]$ and variance $\text{Var}[X]$ of the loss random variable X as follows:

$$\begin{cases} c_\alpha = \mathbb{E}[X]^2(1 - \mathbb{E}[X])/\text{Var}[X] \\ c_\beta = c_\alpha(1 - \mathbb{E}[X])/\mathbb{E}[X] \end{cases} \tag{6.26}$$

(a) In this exercise, you will first generate stochastic loss estimates drawn from the beta distribution for a YLT defined from the ELT already used in previous exercises (Exs. 3.2, 3.3 and 5.1, Poisson case). It is indeed common to use the parameters from an ELT to simulate individual events and then work with the simulated data directly (Homer and Li, 2017). To do so, consider a total exposure v_{tot} = USD 30,000,000 to represent

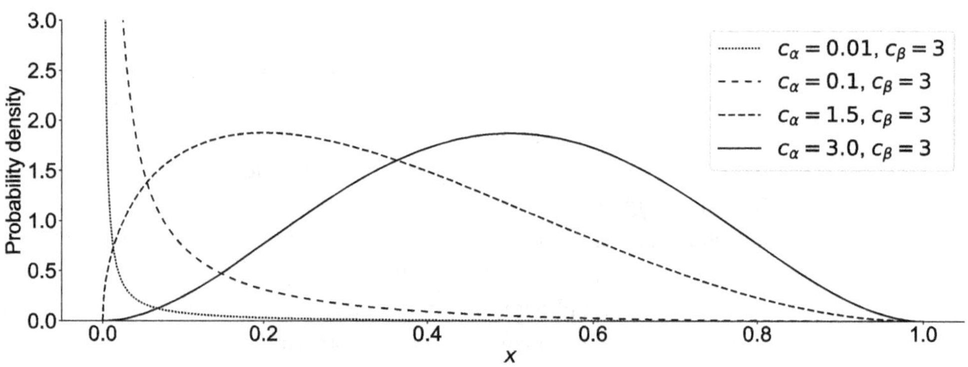

Fig. 6.11 Exercise 6.1: The beta distribution $\mathcal{B}(x, c_\alpha, c_\beta)$.

the maximum possible loss equal to $x = 1$ once losses are normalized in the beta distribution. Use an ad hoc parameter $c_\beta = 3$ (a beta distribution sampling function is provided in `exercise_notebook.ipynb`).

Hints: First generate a YLT as already done in Exercise 3.3. Then, estimate c_α for each event mean loss using Eq. (6.26).

(b) After calculating the standard deviation for each event of the ELT, show that the standard deviations calculated for the ELT (Eq. (6.3)) and for the YLT (Eq. (6.4)) are approximately the same.

Exercise 6.2

(a) Considering the previous ELT to be representative of an insurer's portfolio with the loss estimates defined as the insured losses instead of ground-up losses, and assuming no expense load in this ad hoc example, what is the upper bound on the premium that the insurer should charge?

(b) The insurer's surplus is $\nu_C = \$10$ million. Does the insurance respect Solvency II and does it survive the maximum-size event incurring \$15 million insured losses?

Exercise 6.3

The heat map shown in Figure 6.12 represents the spatial forecast results of an ad hoc model (bins and contours) for some ad hoc observations (stars). More precisely, the model provides the probability p that an event (such as a disaster) occurs. For the sake of simplicity, the heat map is binned into 25 spatial cells with each threshold represented by a different level of grey. Given a p threshold, a successful prediction (true positive TP) is obtained when an event is observed within the matching box. Commonly, the threshold $p \geq 0.5$ is used to define forecasting outcomes and to calculate the model accuracy (Box. 6.2). However, a more informative metric is the **receiver operating characteristic (ROC) curve**. It provides a graphical illustration of the performance of a binary classifier model for different threshold values by plotting the **true positive rate** TPR = TP/(TP+FN) against the **false positive rate** FPR = FP/(FP+TN) at each threshold. For $p \geq 0$, the forecast covers the entire map by construction and leads to the point (1, 1) on the ROC diagram. Similarly, we have (0, 0) systematically for $p = 1$. Random guesses populate the diagonal line while a perfect model would be represented by the point (0, 1). Any model with a ROC curve in the upper left triangle is better than a random classifier.

(a) First, calculate the counts of true positives TP, true negatives TN, false positives FP, and false negatives FN for the four thresholds $p \geq (0, 0.25, 0.5, 0.75)$. Then, calculate the model accuracy.

(b) Draw the ROC curve of the model by calculating the TPR and FPR for the different thresholds. Include the (0,0) point and explain its origin. What can you conclude?

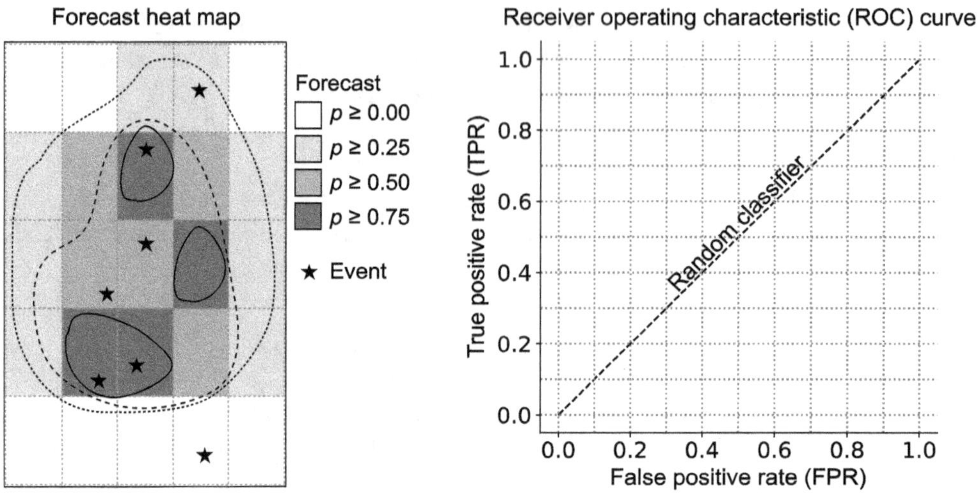

Fig. 6.12 Exercise 6.3: Ad hoc forecast heat map and ROC curve (to add). See text for details.

Exercise 6.4

This exercise is concerned with earthquake forecasting based on aftershock statistics.

(a) Fill in Table 6.1 using Eqs (6.12) and (6.13) and the parameters $b = 0.91$, $a = -1.67$, $c_p = 1.08$, and $c_c = 0.05$ days for the case $M = M_m$ and $t_1 = 0.01$ days (Reasenberg and Jones, 1994, table 1). Write down the equation that allows you to estimate $\mathbb{P}(X \geq M, \Delta t = [t_1, t_2])$ from the provided parameters.

Table 6.1 Exercise 6.4: Aftershock probabilities to be completed

$t_2 - t_1$	1	3	7	30	60	90	365	1000
$\mathbb{P}(X)$	.	.	.	..	..	..	...	....

(b) Is this model compatible with Båth's law?

Exercise 6.5

The intensity prediction equations (IPE) of Atkinson and Wald (2007) (derived from US data) and of Allen et al. (2012) (derived from global data) are provided below:

$$\begin{cases} I_{AW07}(M, R_{hyp}) = c_1 + c_2(M-6) + c_3(M-6)^2 + c_4 \log_{10} R_{hyp} + \\ \quad c_5 R_{hyp} + c_6 B + c_7 M \log_{10} R_{hyp} \\ I_{A12}(M, R_{hyp}) = c_0 + c_1 M + c_2 \log \sqrt{R_{hyp}^2 + R_M^2} + c_4 \log(R_{hyp}/50) \end{cases}$$

where I is the macroseismic intensity, M the earthquake magnitude, and R_{hyp} [km] the hypocentral distance. For I_{AW07}, $B = 0$ for $R_{hyp} \leq R_t$ and $B = \log_{10}(R_{hyp}/R_t)$ otherwise; $c_1 = 12.27$, $c_2 = 2.27$, $c_3 = 0.1304$, $c_4 = -1.3$, $c_5 = -0.000707$, $c_6 = 1.95$, $c_7 =$

-0.577, and $R_t = 30$ km for California and $c_1 = 11.72$, $c_2 = 2.36$, $c_3 = 0.1155$, $c_4 = -0.44$, $c_5 = -0.002044$, $c_6 = 2.31$, $c_7 = -0.479$, and $R_t = 80$ km for Central and Eastern US (Atkinson and Wald, 2007). For I_{A12}, $c_0 = 2.085$, $c_1 = 1.428$, $c_2 = -1.402$, and $c_4 = 0.078$ if $R_{hyp} > 50$ km (otherwise $c_4 = 0$), and $R_M = m_1 + m_2 \exp(M - 5)$ with $m_1 = -0.209$ and $m_2 = 2.042$ (Allen et al., 2012).

(a) For the three IPE models, plot the intensity I as a function of magnitude $4 \leq M \leq 8$ observed at an epicentral distance of 10 km away from the source, which is located at a 5-km depth.

(b) Discuss the level of epistemic uncertainty observed in terms of expected damage assuming the MMI as macroseismic scale.

Hint: You may need to refresh your memory on MMI by going back to Chapter 3.

Exercise 6.6

After reading the details of the St. Petersburg paradox gambling game outlined in Box 6.4, determine the maximum acceptable cost of entry, denoted as ν_C, for participants with varying levels of wealth: one possessing limited resources with $\nu_w = \$1{,}000$, and another with substantial wealth amounting to $\nu_w = \$10^6$.

Hints: Consider the two prospects of playing versus not playing. Identify the critical threshold for ν_C at which the change in expected utility becomes positive (i.e., rational choice according to Eq. (6.21)).

References

Abatzoglou, JT, and Williams, AP. 2016. Impact of anthropogenic climate change on wildfire across western US forests. *PNAS*, **113**(42), 11,770–11,775.

Abbasi, T, and Abbasi, SA. 2007. The boiling liquid expanding vapour explosion (BLEVE): Mechanism, consequence assessment, management. *Journal of Hazardous Materials*, **141**, 489–519.

Abe, M, and Ye, L. 2013. Building resilient supply chains against natural disasters: The cases of Japan and Thailand. *Global Business Review*, **14**(4), 567–586.

Acevedo, AB, Yepes-Estrada, C, González, D, Silva, V, Mora, M, et al. 2020. Seismic risk assessment for the residential buildings of the major three cities in Colombia: Bogotá, Medellín, and Cali. *Earthquake Spectra*, **36**, 298–320.

Adachi, T, and Ellingwood, BR. 2008. Serviceability of earthquake-damaged water systems: Effects of electrical power availability and power backup systems on system vulnerability. *Reliability Engineering and System Safety*, **93**, 78–88.

Adams, JM, Gasparini, NM, Hobley, DEJ, Tucker, GE, Hutton, EWH, et al. 2017. The Landlab v1.0 OverlandFlow component: A Python tool for computing shallow-water flow across watersheds. *Geoscientific Model Development*, **10**, 1645–1663.

Adamsky, V, and Smirnov, Y. 1994. Moscow's biggest bomb: The 50-megaton test of October 1961. *Cold War International History Project Bulletin*, **4**, 3.

Ager, AA, Day, MA, Alcasena, FJ, Evers, CR, Short, KC, and Grenfell, I. 2021. Predicting paradise: Modeling future wildfire disasters in the western US. *Science of the Total Environment*, **784**, 147057.

Ahearne, JF. 1987. Nuclear power after Chernobyl. *Science*, **236**(4802), 673–679.

Akkar, S, and Bommer, JJ. 2010. Empirical equations for the prediction of PGA, PGV and spectral accelerations in Europe, the Mediterranean Region, and the Middle East. *Seismological Research Letters*, **81**(2), 195–206.

Aldersey-Williams, J, and Rubert, T. 2019. Levelised cost of energy – A theoretical justification and critical assessment. *Energy Policy*, **124**, 169–179.

Alemi, AA, Bierbaum, M, Myers, CR, and Sethna, JP. 2015. You can run, you can hide: The epidemiology and statistical mechanics of zombies. *Physical Review E*, **92**, 052801.

Alexander, D. 1989. Urban landslides. *Progress in Physical Geography*, **13**, 157–191.

Allen, MR, and Ingram, WJ. 2002. Constraints on future changes in climate and the hydrologic cycle. *Nature*, **419**, 224–232.

Allen, TI, Wald, DJ, Hotovec, AJ, Lin, K, Earle, PS, and Marano, KD. 2008. *An Atlas of ShakeMaps for Selected Global Earthquakes*. USGS Open-File Report 2008–1236.

Allen, TI, Wald, DJ, and Worden, CB. 2012. Intensity attenuation for active crustal regions. *Journal of Seismology*, **16**, 409–433.

Alvarez, L, Alvarez, W, Asaro, F, and Michel, H. 1980. Extraterrestrial cause for the cretaceous-tertiary extinction. *Science*, **208**, 1095–1108.

Ambraseys, NN, and Bommer, JJ. 1991. The attenuation of ground accelerations in Europe. *Earthquake Engineering and Structural Dynamics*, **20**, 1179–1202.

Ambraseys, NN, Simpson, KA, and Bommer, JJ. 1996. Prediction of horizontal response spectra in Europe. *Earthquake Engineering and Structural Dynamics*, **25**, 371–400.

Anderson, CW, Santos, JR, and Haimes, YY. 2007. A risk-based input–output methodology for measuring the effects of the August 2003 Northeast Blackout. *Economic Systems Research*, **19**(2), 183–204.

Anderson, DH, Catchpole, EA, de Mestre, NJ, and Parkes, T. 1982. Modelling the spread of grass fires. *Journal of the Australian Mathematical Society (Series B)*, **23**(4), 451–466.

Anderson, JR, Hardy, EE, Roach, JT, and Witmer, RE. 1976. *A Land Use and Land Cover Classification System for Use with Remote Sensor Data*. Washington, DC: Geological Survey Professional Paper 964. US Government Printing Office.

Andersson, G, Donalek, P, Farmer, R, Hatziargyriou, N, Kamwa, I, et al. 2005. Causes of the 2003 major grid blackouts in North America and Europe, and recommended means to improve system dynamic performance. *IEEE Transactions on Power Systems*, **20**(4), 1922–1928.

Androulidakis, Y, Makris, C, Mallios, Z, Pytharoulis, I, Baltikas, V, and Krestenitis, Y. 2023. Storm surges and coastal inundation during extreme events in the Mediterranean Sea: The IANOS Medicane. *Natural Hazards*, **117**, 939–978.

Anthes, RA, Corell, RW, Holland, G, Hurrell, JW, MacCracken, MC, and Trenberth, KE. 2006. Hurricanes and global warming – Potential linkages and consequences. *Bulletin of the American Meteorological Society*, **87**(5), 623–628.

Apel, H, Aronica, GT, Kreibich, H, and Thieken, AH. 2009. Flood risk analyses – How detailed do we need to be? *Natural Hazards*, **49**, 79–98.

Arago, M. 1832. Tract on comets, and particularly on the comet that is to intersect the earth's path in October, 1832. By M. Arago. Translated from the French, by John Farrar, Boston, 1832. **14**(5), 315–320.

Argyroudis, S, Selva, J, Gehl, P, and Pitilakis, K. 2015. Systemic seismic risk assessment of road networks considering interactions with the built environment. *Computer-Aided Civil and Infrastructure Engineering*, **30**, 524–540.

Argyroudis, SA, Mitoulis, S, Winter, MG, and Kaynia, AM. 2019. Fragility of transport assets exposed to multiple hazards: State-of-the-art review toward infrastructural resilience. *Reliability Engineering and System Safety*, **191**, 106567.

Argyroudis, SA, Fotopoulou, S, Karafagka, S, Pitilakis, K, Selva, J, et al. 2020. A risk-based multi-level stress test methodology: Application to six critical non-nuclear infrastructures in Europe. *Natural Hazards*, **100**, 595–633.

Arrow, KJ, and Hurwicz, L. 1977. An optimality criterion for decision-making under ignorance. Appendix, pages 461–472 of: Arrow, KJ, and Hurwicz, L (eds), *Studies in Resource Allocation Processes*. Cambridge: Cambridge University Press.

Ashley, WS, and Gilson, CW. 2009. A reassessment of US lightning mortality. *Bulletin of the American Meteorological Society*, **90**(10), 1501–1518.

Aspinall, WP, Loughlin, SC, Michael, FV, Miller, AD, Norton, GE, et al. 2002. The Montserrat Volcano Observatory: Its evolution, organization, role and activities. Chapter 21, pages 71–91 of: Druitt, TH, and Kokelaar, BP (eds), *The Eruption of Soufridre Hills Volcano, Montserrat, from 1995 to 1999*. London: The Geological Society of London.

Atkinson, GM, and Wald, DJ. 2007. 'Did you feel it?' Intensity data: A surprisingly good measure of earthquake ground motion. *Seismological Research Letters*, **78**(3), 362–368.

Atlas, RM, and Hazen, TC. 2011. Oil biodegradation and bioremediation: A tale of the two worst spills in US history. *Environmental Science & Technology*, **45**, 6709–6715.

Aznar-Siguan, G, and Bresch, DN. 2019. CLIMADA v1: A global weather and climate risk assessment platform. *Geoscientific Model Development*, **12**, 3085–3097.

Baade, RA, Baumann, R, and Matheson, V. 2007. Estimating the economic impact of natural and social disasters, with an application to Hurricane Katrina. *Urban Studies*, **44**(11), 2061–2076.

Babbage, C. 1847. *Observations on the Temple of Serapis at Pozzuoli near Naples*. London: Privately printed.

Bacaër, N. 2011. *A Short History of Mathematical Population Dynamics*. Springer.

Bak, P. 1996. *How Nature Works*. New York, NY: Springer Science+Business Media.

Bak, P, and Tang, C. 1989. Earthquakes as a self-organized critical phenomenon. *Journal of Geophysical Research*, **94**(B11), 15,635–15,637.

Bak, P, Tang, C, and Wiesenfeld, K. 1987. Self-organized criticality: An explanation of $1/f$ noise. *Physical Review Letters*, **59**(4), 381–384.

Bak, P, Chen, K, and Tang, C. 1990. A forest-fire model and some thoughts on turbulence. *Physics Letters A*, **147**(5,6), 297–300.

Baker, JW, and Cornell, CA. 2006. Which spectral acceleration are you using? *Earthquake Spectra*, **22**(2), 293–312.

Balaei, B, Wilkinson, S, Potangaroa, R, and McFarlane, P. 2020. Investigating the technical dimension of water supply resilience to disasters. *Sustainable Cities and Society*, **56**, 102077.

Banks, S. 2003. The Erin Brockovich effect: How media shapes toxics policy. *Environs: Environmental Law and Policy Journal*, **26**(2), 219–251.

Barbeito, MS, Mathews, CT, and Taylor, LA. 1967. Microbiological laboratory hazard of bearded men. *Applied Microbiology*, **15**(4), 899–906.

Barberis, NC. 2013. Thirty years of prospect theory in economics: A review and assessment. *Journal of Economic Perspectives*, **27**(1), 173–196.

Bard, P-Y, Campillo, M, Chavez-Garcia, FJ, and Sanchez-Sesma, F. 1988. The Mexico Earthquake of September 19,1985 — A theoretical investigation of large- and small-scale amplification effects in the Mexico City Valley. *Earthquake Spectra*, **4**, 609–633.

Barnosky, AD, Matzke, N, Tomiya, S, Wogan, GOU, Swartz, B, et al. 2011. Has the Earth's sixth mass extinction already arrived? *Nature*, **471**, 51–57.

Basel Committee on Banking Supervision. 2021. *Climate-related Risk Drivers and their Transmission Channels*. Online; accessed 20 December 2023.

Basili, R, Kastelic, V, Demircioglu, MB, Moreno, DG, Nemser, ES, et al. 2013. The European Database of Seismogenic Faults (EDSF) compiled in the framework of the Project SHARE. Istituto Nazionale di Geofisica e Vulcanologia (INGV). Online; accessed 19 July 2023.

Bass, B, Irza, JN, Proft, J, Bedient, P, and Dawson, C. 2017. Fidelity of the integrated kinetic energy factor as an indicator of storm surge impacts. *Natural Hazards*, **85**, 575–595.

Bates, PD, Quinn, N, Sampson, C, Smith, A, Wing, O, et al. 2021. Combined modeling of US fluvial, pluvial, and coastal flood hazard under current and future climates. *Water Resources Research*, **57**, e2020WR028673.

Båth, M. 1965. Lateral inhomogeneities of the upper mantle. *Tectonophysics*, **2**(6), 483–514.

Båth, M. 1966. Earthquake energy and magnitude. Chapter 2, pages 115–165 of: Benioff, ME, Howell, BF, and Press, F (eds), *Contributions in Geophysics: In Honor of Beno Gutenberg*. New York, NY: Pergamon Press.

Batty, M. 1991. Cities as fractals: Simulating growth and form. Chapter 3, pages 43–69 of: Crilly, AJ, Earnshaw, RA, and Jones, H (eds), *Fractals and Chaos*. New York: Springer-Verlag.

BCSF-Rénass. 2022. *Instrumental Seismicity in Mainland France*. EOST UAR830, Université de Strasbourg, CNRS. (Collection). Online; accessed 26 November 2023.

Beauval, C, and Scotti, O. 2004. Quantifying sensitivities of PSHA for France to earthquake catalog uncertainties, truncation of ground-motion variability, and magnitude limits. *Bulletin of the Seismological Society of America*, **94**(5), 1579–1594.

Becker, GS. 1968. Crime and punishment: An economic approach. *Journal of Political Economy*, **76**, 169–217.

Belanger, JI, Curry, JA, and Hoyos, CD. 2009. Variability in tornado frequency associated with US landfalling tropical cyclones. *Geophysical Research Letters*, **36**, L17805.

Ben-Menahem, A. 1995. A concise history of mainstream seismology: Origins, legacy, and perspectives. *Bulletin of the Seismological Society of America*, **85**(4), 1202–1225.

Bender, MA, Ginis, I, Tuleya, R, Thomas, B, and Marchok, T. 2007. The operational GFDL coupled hurricane-ocean prediction system and a summary of its performance. *Monthly Weather Review*, **135**, 3965–3989.

Bentz, BJ, Régnière, J, Fettig, CJ, Hansen, EM, Hayes, JL, et al. 2010. Climate change and bark beetles of the Western United States and Canada: Direct and indirect effects. *BioScience*, **60**(8), 602–613.

Bernoulli, D. 1738/1954. Exposition of a new theory on the measurement of risk (originally published in 1738; translated by L Sommer). *Econometrica*, **22**(1), 23–36.

Bernoulli, D. 1766. Essai d'une nouvelle analyse de la mortalité causée par la petite vérole et des avantages de l'inoculation pour la prévenir. *Mémoires de Mathématique et de Physique de l'Académie Royale des Sciences*.

Bernoulli, J. 1713. *Ars Conjectandi: Usum & Applicationem Praecedentis Doctrinae in Civilibus, Moralibus & Oeconomicis*.

Beroggi, GEG, and Wallace, WA. 1994. Operational risk management: A new paradigm for decision making. *IEEE Transactions on Systems, Man, and Cybernetics*, **24**(10), 1450–1457.

Beschta, RL, and Ripple, WJ. 2009. Large predators and trophic cascades in terrestrial ecosystems of the western United States. *Biological Conservation*, **142**, 2401–2414.

Beven, KJ. 2020. A history of the concept of time of concentration. *Hydrology and Earth System Sciences*, **24**, 2655–2670.

Bilham, R. 2009. The seismic future of cities. *Bulletin of Earthquake Engineering*, **7**, 839–887.

Binzel, RP. 1997. A Near-Earth Object Hazard Index. *Annals of the New York Academy of Sciences*, **822**(1), 545–551.

Bird, JF, and Bommer, JJ. 2004. Earthquake losses due to ground failure. *Engineering Geology*, **75**, 147–179.

Bland, PA, and Artemieva, NA. 2006. The rate of small impacts on Earth. *Meteoritics & Planetary Science*, **41**(4), 607–631.

Blavatskyy, PR. 2005. Back to the St. Petersburg paradox? *Management Science*, **51**(4), 677–678.

Bloemendaal, N, Haigh, ID, de Moel, H, Muis, S, Haarsma, RJ, and Aerts, JCJH. 2020. Generation of a global synthetic tropical cyclone hazard dataset using STORM. *Scientific Data*, **7**, 40.

Bloemendaal, N, de Moel, H, Martinez, AB, Muis, S, Haigh, ID, et al. 2022. A globally consistent local-scale assessment of future tropical cyclone risk. *Science Advances*, **8**, eabm8438.

Blong, R. 2003a. A new Damage Index. *Natural Hazards*, **30**, 1–23.

Blong, R. 2003b. A review of damage intensity scales. *Natural Hazards*, **29**, 57–76.

Blong, RJ, Grasso, P, Jenkins, SF, Magill, CR, Wilson, TM, et al. 2017. Estimating building vulnerability to volcanic ash fall for insurance and other purposes. *Journal of Applied Volcanology*, **6**, 2.

Bogen, KT, and Jones, ED. 2006. Risks of mortality and morbidity from worldwide terrorism: 1968–2004. *Risk Analysis*, **26**(1), 45–59.

Boissay, F, Collard, F, and Smets, F. 2016. Booms and banking crises. *Journal of Political Economy*, **124**(2), 489–538.

Bolinger, RA, Brown, VM, Fuhrmann, CM, Gleason, KL, Joyner, TA, et al. 2022. An assessment of the extremes and impacts of the February 2021 South-Central U.S. Arctic outbreak, and how climate services can help. *Weather and Climate Extremes*, **36**, 100461.

Bolton, P, Kacperczyk, M, Hong, H, and Vives, X. 2021. *Resilience of the Financial System to Natural Disasters. The Future of Banking 3*. London: Centre for Economic Policy Research.

Bonabeau, E. 2002. Agent-based modeling: Methods and techniques for simulating human systems. *PNAS*, **99**, 7280–7287.

Bonadonna, C, Costa, A, Folch, A, and Koyaguchi, T. 2015. Tephra dispersal and sedimentation. Chapter 33, pages 587–597 of: Sigurdsson, H (ed), *The Encyclopedia of Volcanoes (2nd ed.)* London: Academic Press.

Bondarenko, M, Kerr, D, Sorichetta, A, and Tatem, AJ. 2020. *Census/projection-disaggregated gridded population datasets, adjusted to match the corresponding UNPD 2020 estimates, for 183 countries in 2020 using Built-Settlement Growth Model (BSGM) outputs*. WorldPop, University of Southampton. Online; accessed 8 March 2023.

Booij, AS, van Praag, BMS, and van de Kuilen, G. 2010. A parametric analysis of prospect theory's functionals for the general population. *Theory Decision*, **68**, 115–148.

Borlaug, NE. 1971. The Green Revolution: For bread and peace. *Bulletin of the Atomic Scientists*, **27**(6), 6–48.

Boslough, MBE, and Crawford, DA. 2008. Low-altitude airbursts and the impact threat. *International Journal of Impact Engineering*, **35**, 1441–1448.

Bostrom, N. 2002. Existential risks: Analyzing human extinction scenarios and related risks. *Journal of Evolution and Technology*, **9**, 1–36.

Bostrom, N. 2019. The vulnerable world hypothesis. *Global Policy*, **10**(4), 455–476.

Boteler, DH. 2003. Geomagnetic hazards to conducting networks. *Natural Hazards*, **28**, 537–561.

Boulton, CA, Lenton, TM, and Boers, N. 2022. Pronounced loss of Amazon rainforest resilience since the early 2000s. *Nature Climate Change*, **12**, 271–278.

Bowman, DD, Ouillon, G, Sammis, CG, Sornette, A, and Sornette, D. 1998. An observational test of the critical earthquake concept. *Journal of Geophysical Research*, **103**(B10), 24,359–24,372.

Bradshaw, CJA, Leroy, B, Bellard, C, Roiz, D, Albert, C, et al. 2016. Massive yet grossly underestimated global costs of invasive insects. *Nature Communications*, **7**, 12986.

Bradshaw, CJA, Ehrlich, PR, Beattie, A, Ceballos, G, Crist, E, et al. 2021. Underestimating the challenges of avoiding a ghastly future. *Frontiers in Conservation Science*, **1**, 615419.

Braga, F, Dolce, M, and Liberatore, D. 1982. A statistical study on damaged buildings and ensuing review of the M.S.K.-76 Scale. Pages 1–11 of: *Proceedings of the 7th European Conference on Earthquake Engineering*.

Brei, M, Mohan, P, and Strobl, E. 2019. The impact of natural disasters on the banking sector: Evidence from hurricane strikes in the Caribbean. *The Quarterly Review of Economics and Finance*, **72**, 232–239.

Broccardo, M, Mignan, A, Wiemer, S, Stojadinovic, B, and Giardini, D. 2017. Hierarchical Bayesian modeling of fluid-induced seismicity. *Geophysical Research Letters*, **44**, 11,357–11,367.

Broccardo, M, Mignan, A, Grigoli, F, Karvounis, D, Rinaldi, AP, et al. 2020. Induced seismicity risk analysis of the hydraulic stimulation of a geothermal well on Geldinganes, Iceland. *Natural Hazards and Earth System Sciences*, **20**, 1573–1593.

Brode, HL. 1959. Blast wave from a spherical charge. *Physics of Fluids*, **2**(2), 217–229.

Broughton, E. 2005. The Bhopal disaster and its aftermath: A review. *Environmental Health: A Global Access Science Source*, **4**, 6.

Brown, P, Spalding, RE, ReVelle, DO, Tagliaferri, E, and Worden, SP. 2002. The flux of small near-Earth objects colliding with the Earth. *Nature*, **420**, 294–296.

Brown, T, Hörsch, J, and Schlachtberger, D. 2018. PyPSA: Python for Power System Analysis. *Journal of Open Research Software*, **6**(4).

Brozovic, D. 2023. Societal collapse: A literature review. *Futures*, **145**, 103075.

Bryant, E, Walsh, G, and Abbott, D. 2007. Cosmogenic mega-tsunami in the Australia region: Are they supported by Aboriginal and Maori legends? *Geological Society, London, Special Publications*, **273**, 203–214.

Brzev, S, Scawthorn, C, Charleson, AW, Allen, L, Greene, M, et al. 2013. *GEM Building Taxonomy Version 2.0*. Pavia: GEM Technical Report 2013-02 V1.0.0, GEM Foundation.

Buchholz, W, and Schymura, M. 2012. Expected utility theory and the tyranny of catastrophic risks. *Ecological Economics*, **77**, 234–239.

Buldyrev, SV, Parshani, R, Paul, G, Stanley, H.E, and Havlin, S. 2010. Catastrophic cascade of failures in interdependent networks. *Nature*, **464**, 1025–1028.

Bulfin, A. 2015. The natural catastrophe in Late Victorian popular fiction: 'How will the world end?' *Critical Survey*, **27**(2), 81–101.

Burger, JM, and Palmer, ML. 1992. Changes in and generalization of unrealistic optimism following experiences with stressful events: Reactions to the 1989 California Earthquake. *Personality and Social Psychology Bulletin*, **18**(1), 3–110.

Burgherr, P, and Hirschberg, S. 2008. A comparative analysis of accident risks in fossil, hydro, and nuclear energy chains. *Human and Ecological Risk Assessment*, **14**, 947–973.

Burillo, D, Chester, MV, Ruddell, B, and Johnson, N. 2017. Electricity demand planning forecasts should consider climate non-stationarity to maintain reserve margins during heat waves. *Applied Energy*, **206**, 267–277.

Burke, M, Hsiang, SM, and Miguel, E. 2015. Global non-linear effect of temperature on economic production. *Nature*, **527**, 235–239.

Burow, D, Herrero, HV, and Ellis, KN. 2020. Damage analysis of three long-track tornadoes using high-resolution satellite imagery. *Atmosphere*, **11**, 613.

Burroughs, SM, and Tebbens, SF. 2005. Power-law scaling and probabilistic forecasting of tsunami runup heights. *Pure and Applied Geophysics*, **162**, 331–342.

Busby, JW, Baker, K, Bazilian, MD, Gilbert, AQ, Grubert, E, et al. 2021. Cascading risks: Understanding the 2021 winter blackout in Texas. *Energy Research & Social Science*, **77**, 102106.

Cagigas-Muñiz, D, Diaz del Rio, F, Sevillano-Ramos, JL, and Guisado-Lizar, J-L. 2022. Efficient simulation execution of cellular automata on GPU. *Simulation Modelling Practice and Theory*, **118**, 102519.

Cameron, J (director). 1984. *The Terminator* [Motion picture]. Orion Pictures.

Camus, A. 1947. *La Peste*. Gallimard.

Candau, JT. 2002. *Temporal Calibration Sensitivity of the SLEUTH Urban Growth Model*. PhD thesis, University of California Santa Barbara.

Canup, RM, and Asphaug, E. 2001. Origin of the Moon in a giant impact near the end of the Earth's formation. *Nature*, **412**, 708–712.

Cara, M, Cansi, Y, Schlupp, A, Arroucau, P, Béthoux, N, et al. 2015. SI-Hex: A new catalogue of instrumental seismicity for metropolitan France. *Bulletin de la Société Géologique de France*, **186**, 3–19.

Cardona, OD, Ordaz, MG, Reinoso, E, Yamín, LE, and Barbat, AH. 2012. CAPRA – Comprehensive Approach to Probabilistic Risk Assessment: International Initiative for Risk Management Effectiveness. *15 WCEE Lisboa*.

Cardone, VJ, Jensen, RE, Resio, DT, Swail, VR, and Cox, AT. 1996. Evaluation of contemporary ocean wave models in rare extreme events: The 'Halloween Storm' of October 1991 and the 'Storm of the Century' of March 1993. *Journal of Atmospheric and Oceanic Technology*, **13**(1), 198–230.

Carleton, TA, and Hsiang, SM. 2016. Social and economic impacts of climate. *Science*, **353**(6304), aad9837.

Carpenter, SR, Ludwig, D, and Brock, WA. 1999. Management of eutrophication for lakes subject to potentially irreversible change. *Ecological Applications*, **9**(3), 751–771.

Casadevall, TJ, Reyes, PJD, and Schneider, DJ. 1995. The 1991 Pinatubo eruptions and their effects on aircraft operations. In: Newhall, CG, and Punongbayan, RS (eds), *Fire and Mud*. Reston, VA: US Geological Survey.

Casal, J. 2008. *Evaluation of the Effects and Consequences of Major Accidents in Industrial Plants*. Elsevier.

Cebula, JJ, Popeck, ME, and Young, LR. 2014. *A Taxonomy of Operational Cyber Security Risks Version 2*. Carnegie Mellon University.

Cederman, L-E. 2003. Modeling the size of wars: From billiard balls to sandpiles. *The American Political Science Review*, **97**(1), 135–150.

Cerveny, RS, Bessemoulin, P, Burt, CC, Cooper, MA, Cunjie, Z, et al. 2017. WMO assessment of weather and climate mortality extremes: Lightning, tropical cyclones, tornadoes, and hail. *Weather, Climate, and Society*, **9**(3), 487–497.

Cetorelli, N, and Goldberg, LS. 2011. Global banks and international shock transmission: Evidence from the crisis. *IMF Economic Review*, **59**(1), 41–76.

Chang, SE. 2000. Disasters and transport systems: Loss, recovery and competition at the Port of Kobe after the 1995 earthquake. *Journal of Transport Geography*, **8**, 53–65.

Chang, SE, and Nojima, N. 2001. Measuring post-disaster transportation system performance: The 1995 Kobe earthquake in comparative perspective. *Transportation Research Part A*, **35**, 475–494.

Chang, W-Y, Lantz, VA, Hennigar, CR, and MacLean, DA. 2012. Economic impacts of forest pests: A case study of spruce budworm outbreaks and control in New Brunswick, Canada. *Canadian Journal of Forest Research*, **42**, 490–505.

Changnon, SA, Jr. 1977. The scales of hail. *Journal of Applied Meteorology (1962–1982)*, **16**(6), 626–648.

Chapman, CR, and Morrison, D. 1994. Impacts on the Earth by asteroids and comets: Assessing the hazard. *Nature*, **367**, 33–40.

Chen, C, Reniers, G, and Khakzad, N. 2020. A thorough classification and discussion of approaches for modeling and managing domino effects in the process industries. *Safety Science*, **125**, 104618.

Chen, S-C, Chen, M, Zhao, N, Hamid, S, Chatterjee, K, and Armella, M. 2009. Florida public hurricane loss model: Research in multi-disciplinary system integration assisting government policy making. *Government Information Quarterly*, **26**, 285–294.

Cheng, E, and Yeung, C. 2002. Generalized extreme gust wind speeds distributions. *Journal of Wind Engineering and Industrial Aerodynamics*, **90**, 1657–1669.

Cherry, JD. 2004. The chronology of the 2002–2003 SARS mini pandemic. *Paediatric Respiratory Reviews*, **5**, 262–269.

Child, L. 2007. *Deep Storm*. Doubleday.

Chiou, BS-J, and Youngs, RR. 2014. Update of the Chiou and Youngs NGA model for the average horizontal component of peak ground motion and response spectra. *Earthquake Spectra*, **30**(3), 1117–1153.

Chivers, C, Leung, B, and Yan, ND. 2014. Validation and calibration of probabilistic predictions in ecology. *Methods in Ecology and Evolution*, **5**, 1023–1032.

Christensen, CM. 1992. Exploring the limits of the technology S-curve. Part I. Component technologies. *Production and Operations Management*, **1**(4), 334–357.

Christie, R. 1993. Power systems test case archive. Online; accessed 12 April 2023.

Church, JA, and White, NJ. 2011. Sea-level rise from the late 19th to the early 21st century. *Surveys in Geophysics*, **32**, 585–602.

Cicerone, RD, Ebel, JE, and Britton, J. 2009. A systematic compilation of earthquake precursors. *Tectonophysics*, **476**, 371–396.

Cixin, L. 2014. *The Three-Body Problem*. New York: Tor Books.

Clark, WC. 1980. Witches, floods, and wonder drugs: Historical perspectives on risk management. Pages 287–313 of: Schwing, RC, and WA Albers, Jr (eds), *Societal Risk Assessment, How Safe is Safe Enough?* New York, NY: Springer Science+Business Media, LLC.

Clarke, KC. 2014. Cellular automata and agent-based models. Chapter 62, pages 1217–1233 of: Fischer, MM, and Nijkamp, P (eds), *Handbook of Regional Science*. Berlin Heidelberg: Springer-Verlag.

Clarke, KC, Brass, JA, and Riggan, PJ. 1994. A cellular automaton model of wildfire propagation and extinction. *Photogrammetric Engineering & Remote Sensing*, **60**(11), 1355–1367.

Clarke, KC, Hoppen, S, and Gaydos, L. 1997. A self-modifying cellular automaton model of historical urbanization in the San Francisco Bay area. *Environment and Planning B: Planning and Design*, **24**, 247–261.

Clauset, A, Young, M, and Gleditsch, KS. 2007. On the frequency of severe terrorist events. *Journal of Conflict Resolution*, **51**(1), 58–87.

Clauset, A, Shalizi, CR, and Newman, MEJ. 2009. Power-law distributions in empirical data. *SIAM Review*, **51**(4), 661–703.

CNEOS (Center for Near Earth Object Studies). 2023. Fireball and Bolide Data. Online; accessed 29 August 2022.

Coburn, A, Leverett, É, and Woo, G. 2019. *Solving Cyber Risk*. Hoboken, New Jersey: Wiley.

Coles, S. 2001. *An Introduction to Statistical Modeling of Extreme Values*. London: Springer-Verlag.

Collins, DA, Della-Marta, PM, Plummer, N, and Trewin, BC. 2000. Trends in annual frequencies of extreme temperature events in Australia. *Australian Meteorological Magazine*, **49**, 277–292.

Comfort, LK. 2006. Cities at risk: Hurricane Katrina and the drowning of New Orleans. *Urban Affairs Review*, **41**(4), 501–516.

Congleton, RD. 2006. The story of Katrina: New Orleans and the political economy of catastrophe. *Public Choice*, **127**(1/2), 5–30.

Cook, BI, Seager, R, and Smerdon, JE. 2014. The worst North American drought year of the last millennium: 1934. *Geophysical Research Letters*, **41**, 7298–7305.

Cook, ER, Seager, R, Cane, MA, and Stahle, DW. 2007. North American drought: Reconstructions, causes, and consequences. *Earth-Science Reviews*, **81**, 93–134.

Copernicus Climate Change Service. 2022. Synthetic windstorm events for Europe from 1986 to 2011. Copernicus Climate Change Service (C3S) Climate Data Store (CDS). Online; accessed 5 March 2023.

Cornell, CA. 1968. Engineering seismic risk analysis. *Bulletin of the Seismological Society of America*, **58**(5), 1583–1606.

Corral, A, and González, A. 2019. Power law size distributions in geoscience revisited. *Earth and Space Science*, **6**, 673–697.

Costa, JE, and O'Connor, JE. 1995. Geomorphically effective floods. Pages 45–56 of: Costa, JE, Miller, AJ, Potter, KW, and Wilcock, PR (eds), *Natural and Anthropogenic Influences in Fluvial Geomorphology*. Washington, DC: American Geophysical Union.

Courbage, C, and Stahel, WR. 2012. *Extreme Events and Insurance: 2011 Annus Horribilis*. The Geneva Reports Risk and Insurance Research.

Courbage, C, Orie, M, and Stahel, WR. 2012. 2011 Thai floods and insurance. Chapter 9, pages 121–132 of: Courbage, C, and Stahel, WR (eds), *Extreme Events and Insurance: 2011 Annus Horribilis*. The Geneva Reports Risk and Insurance Research.

Cox, LA. 1998. Risk integration and decision-making. Pages 277–311 of: Kirchssteiger, C, Christou, MD, and Papadakis, GA (eds), *Risk Assessment and Management in the Context of the Seveso II Directive*. Amsterdam: Elsevier.

Cox, LA. 2008. What's wrong with risk matrices? *Risk Analysis*, **28**(2), 497–512.

Cox, PM, Betts, RA, Jones, CD, Spall, SA, and Totterdell, IJ. 2000. Acceleration of global warming due to carbon-cycle feedbacks in a coupled climate model. *Nature*, **408**, 184–187.

Cox, PM, Betts, RA, Collins, M, Harris, PP, Huntingford, C, and Jones, CD. 2004. Amazonian forest dieback under climate-carbon cycle projections for the 21st century. *Theoretical and Applied Climatology*, **78**, 137–156.

CRESTA. 2023. Methodology and Definitions. Online; accessed 7 March 2023.

Crichton, D. 1999. The Risk Triangle. Pages 102–103 of: Ingleton, J (ed), *Natural Disaster Management*. London: Tudor Rose Holdings.

Cross, FB. 1998. Facts and values in risk assessment. *Reliability Engineering and System Safety*, **59**, 27–40.

Crosta, G. 1998. Regionalization of rainfall thresholds: An aid to landslide hazard evaluation. *Environmental Geology*, **35**(2–3), 131–145.

Crosweller, HS, Arora, B, Brown, SK, Cottrell, E, Deligne, NI, et al. 2012. Global database on large magnitude explosive volcanic eruptions (LaMEVE). *Journal of Applied Volcanology*, **1**, 4.

Crowl, DA. 2003. *Understanding Explosions*. New York, NY: American Institute of Chemical Engineers.

Crowley, H, and Pinho, R. 2009. Revisiting Eurocode 8 formulae for periods of vibration and their employment in linear seismic analysis. *Earthquake Engineering and Structural Dynamcis*, **39**, 223–235.

Crowley, H, Despotaki, V, Rodrigues, D, Silva, V, Costa, C, et al. 2021a. European Exposure Model Data Repository [Data set]. Online; accessed 8 March 2023.

Crowley, H, Dabbeek, J, Despotaki, V, Rodrigues, D, Martins, L, et al. 2021b. European Seismic Risk Model (ESRM20). Online; accessed 8 March 2023.

Cui, P, Zhu, Y-y, Han, Y-s, Chen, X-q, and Zhuang, J-q. 2009. The 12 May Wenchuan earthquake-induced landslide lakes: Distribution and preliminary risk evaluation. *Landslides*, **6**, 209–223.

Cui, W, and Perera, AH. 2008. What do we know about forest fire size distribution, and why is this knowledge useful for forest management? *International Journal of Wildland Fire*, **17**, 234–244.

Dal Zilio, L, and Ampuero, J-P. 2023. Earthquake doublet in Turkey and Syria. *Nature*, **4**, 71.

d'Alembert, J. 1746. *Réflexions sur la Cause Générale des Vents*. Paris, France: Chez David l'aîné, libraire.

Danciu, L, Sesetyan, K, Demircioglu, M, Gülen, L, Zare, M, et al. 2018. The 2014 earthquake model of the Middle East: Seismogenic sources. *Bulletin of Earthquake Engineering*, **16**, 3465–3496.

Danciu, L, Nandan, S, Reyes, C, Basili, R, Weatherill, G, et al. 2021. *The 2020 Update of the European Seismic Hazard Model: Model Overview*. EFEHR Technical Report 001, v1.0.0.

Daniell, JE, Khazai, B, Wenzel, F, and Vervaeck, A. 2011. The CATDAT damaging earthquakes database. *Natural Hazards and Earth System Sciences*, **11**, 2235–2251.

Daniell, JE, Schaefer, AM, and Wenzel, F. 2017. Losses associated with secondary effects in earthquakes. *Frontiers in Built Environment*, **3**, 1–14.

Danon, L, Ford, AP, House, T, Jewell, CP, Keeling, MJ, et al. 2011. Networks and the epidemiology of infectious disease. *Interdisciplinary Perspectives on Infectious Diseases*, **2011**, 284909.

Dasgupta, S, Laplante, B, Murray, S, and Wheeler, D. 2011. Exposure of developing countries to sea-level rise and storm surges. *Climatic Change*, **106**, 567–579.

Davies, TP, Fry, HM, Wilson, AG, and Bishop, SR. 2013. A mathematical model of the London riots and their policing. *Scientific Reports*, **3**, 1303.

Dawes, RM, Faust, D, and Meehl, PE. 1989. Clinical versus actuarial judgement. *Science*, **243**, 1668–1673.

De Bono, A, et al., and United Nations Environment Programme. 2004. Impacts of summer 2003 heat wave in Europe. *Environment Alert Bulleting*, **2**, 4.

de Finetti, B. 1974. *Theory of Probability, Vol. 1*. New York: John Wiley and Sons.

de Guttry, C, and Ratter, B. 2022. Expiry date of a disaster: Memory anchoring and the storm surge 1962 in Hamburg, Germany. *International Journal of Disaster Risk Reduction*, **70**, 102719.

Defoe, D. 1704. *The Storm: or, a Collection of the most Remarkable Casualties and Disasters Which happen'd in the Late Dreadful Tempoest, both by Sea and Land*. London: Printed for G. Sawbridge in Little Britain.

Delamater, PL, Street, EJ, Leslie, TF, Yang, YT, and Jacobsen, KH. 2019. Complexity of the Basic Reproduction Number (R_0). *Emerging Infectious Diseases*, **25**(1), 1–4.

Della-Marta, PM, Mathis, H, Frei, C, Liniger, MA, Kleinn, J, and Appenzeller, C. 2009. The return period of wind storms over Europe. *International Journal of Climatology*, **29**, 437–459.

DeMaria, M. 1996. A history of hurricane forecasting for the Atlantic Basin, 1920–1995. Chapter 9, page 263 of: Fleming, JR (ed), *Historical Essays on Meteorology 1919–1995*. American Meteorological Society.

Densmore, AL, Ellis, MA, and Anderson, RS. 1998. Landsliding and the evolution of normal-fault-bounded mountain. *Journal of Geophysical Research*, **103**(B7), 15,203–15,219.

DesRoches, R, Comerio, M, Eberhard, M, Mooney, W, and Rix, GJ. 2011. Overview of the 2010 Haiti earthquake. *Earthquake Spectra*, **27**(S1), S1–S21.

Devereux, S. 2009. Why does famine persist in Africa? *Food Security*, **1**, 25–35.

Devi, S. 2020. Locust swarms in east Africa could be 'a catastrophe'. *The Lancet*, **395**, 547.

Dhakal, N, Fang, X, Cleveland, TG, Thompson, DB, Asquith, WH, and Marzen, LJ. 2012. Estimation of volumetric runoff coefficients for Texas watersheds using land-use and rainfall-runoff data. *Journal of Irrigation and Drainage Engineering*, **138**(1), 43–54.

Dhiman, S, and Patra, KC. 2019. Studies of dam disaster in India and equations for breach parameter. *Natural Hazards*, **98**, 783–807.

Di Giacomo, D, Engdahl, ER, and Storchak, DA. 2018. The ISC-GEM Earthquake Catalogue (1904–2014): Status after the Extension Project. *Earth System Science Data*, **10**, 1877–1899.

Di Giovambattista, R, and Tyupkin, YS. 2000. Spatial and temporal distribution of seismicity before the Umbria–Marche September 26, 1997 earthquakes. *Journal of Seismology*, **4**, 589–598.

Diamond, J. 2011. *Collapse: How Societies Choose to Fail or Succeed*. London: Penguin Books.

Dinske, C, and Shapiro, SA. 2013. Seismotectonic state of reservoirs inferred from magnitude distributions of fluid-induced seismicity. *Journal of Seismology*, **17**, 13–25.

Dixon, PG, Brommer, DM, Hedquist, BC, Kalstein, AJ, Goodrich, GB, et al. 2005. Heat mortality versus cold mortality. A study of conflicting databases in the United States. *Bulletin of the American Meteorological Society*, **86**(7), 937–944.

Dobson, I, Carreras, BA, Lynch, VE, and Newman, DE. 2007. Complex systems analysis of series of blackouts: Cascading failure, critical points, and self-organization. *Chaos*, **17**, 026103.

Dolce, M, Speranza, E, Giordano, F, Borzi, B, Bocchi, F, Conte, C, et al. 2019. Observed damage database of past Italian earthquakes the Da.D.O. Webgis. *Bollettino di Geofisica Teorica e Applicata*, **60**(2), 141–164.

Dong, E, Du, H, and Gardner, L. 2020. An interactive web-based dashboard to track COVID-19 in real time. *The Lancet Infectious Diseases*, **20**(5), 533–534.

Dong, W, and Grossi, P. 2005. Insurance portfolio management. Chapter 6, pages 119–133 of: Grossi, P, and Kunreuther, H (eds), *Catastrophe Modeling: A New Approach to Managing Risk*. New York, NY: Springer.

Donner, R, Barbosa, S, Kurths, J, and Marwan, N. 2009. Understanding the Earth as a complex system – Recent advances in data analysis and modelling in Earth sciences. *The European Physical Journal Special Topics*, **174**, 1–9.

Doswell, CA, III. 2001. Severe convective storms – An overview. Chapter 1, pages 1–26 of: III, CA Doswell (ed), *Severe Convective Storms*. American Meteorological Society.

Dotzek, N, Groenemeijer, P, Feuerstein, B, and Holzer, AM. 2009. Overview of ESSL's severe convective storms research using the European Severe Weather Database ESWD. *Atmospheric Research*, **93**, 575–586.

Dowdy, AJ, Ye, H, Pepler, A, Thatcher, M, Osbrough, SL, et al. 2019. Future changes in extreme weather and pyroconvection risk factors for Australian wildfires. *Scientific Reports*, **9**, 10073.

Drossel, B, and Schwabl, F. 1992. Self-organized critical forest-fire model. *Physical Review Letters*, **69**(11), 1629–1632.

Druică, E, Musso, F, and Ianole-Călin, R. 2020. Optimism bias during the Covid-19 pandemic: Empirical evidence from Romania and Italy. *Games*, **11**, 39.

Dueñas-Osorio, L, and Vemuru, SM. 2009. Cascading failures in complex infrastructure systems. *Structural Safety*, **31**, 157–167.

Dunnavan, GM, and Diercks, JW. 1980. An analysis of Super Typhoon Tip (October 1979). *Monthly Weather Review*, **108**, 1915–1923.

Dunning, SA, Mitchell, WA, Rosser, NJ, and Petley, DN. 2007. The Hattian Bala rock avalanche and associated landslides triggered by the Kashmir Earthquake of 8 October 2005. *Engineering Geology*, **93**, 130–144.

Earn, DJD, Rohani, P, Bolker, BM, and Grenfell, BT. 2000. A simple model for complex dynamical transitions in epidemics. *Science*, **287**, 667–670.

Eastwood, JP, Biffis, E, Hapgood, MA, Green, L, Bisi, MM, et al. 2017. The economic impact of space weather: Where do we stand? *Risk Analysis*, **37**(2), 206–218.

Ebel, JE, and Kafka, AL. 1999. A Monte Carlo approach to seismic hazard analysis. *Bulletin of the Seismological Society of America*, **89**(4), 854–866.

Ebener, S, Murray, C, Tandon, A, and Elvidge, CC. 2005. From wealth to health: Modelling the distribution of income per capita at the sub-national level using night-time light imagery. *International Journal of Health Geographics*, **4**, 5.

Eberenz, S, Stocker, D, Röösli, T, and Bresch, DN. 2020. Asset exposure data for global physical risk assessment. *Earth System Science Data*, **12**, 817–833.

Edwards, R, LaDue, JG, Ferree, JT, Scharfenberg, K, Maier, C, and Coulbourne, WL. 2013. Tornado intensity estimation. Past, present, and future. *Bulletin of the American Meteorological Society*, **94**(5), 641–653.

Efron, B. 1986. Why isn't everyone a Bayesian? *The American Statistician*, **40**(1), 1–5.

Ehrlich, PR, and Ehrlich, AH. 1968. *The Population Bomb*. San Francisco, CA, USA: Sierra Club/Ballantine Books.

Ekström, G, and Stark, CP. 2013. Simple scaling of catastrophic landslide dynamics. *Science*, **339**, 1416–1419.

Eling, M, and Schnell, W. 2016. What do we know about cyber risk and cyber risk insurance? *The Journal of Risk Finance*, **17**(5), 474–491.

Elleder, L, Kašpárek, L, Šírová, J, and Kabelka, T. 2020. Low water stage marks on hunger stones: Verification for the Elbe from 1616 to 2015. *Climate Past*, **16**, 1821–1846.

Elliott, D, and Smith, D. 1993. Football stadia disasters in the United Kingdom: Learning from tragedy? *Industrial & Environmental Crisis Quarterly*, **7**(3), 205–229.

Elsayed, NM. 1997. Toxicology of blast overpressure. *Toxicology*, **121**, 1–15.

Elyoseph, Z, Hadar-Shoval, D, Angert, T, Yitshaki, N, Hol, E, et al. 2023. Mental health volunteers after the Oct 7 Gaza border crisis in Israel: silent warriors. *Lancet Psychiatry*, published online 9 November. https://doi.org/10.1016/S2215-0366(23)00369-3.

Emanuel, K. 2011. Global warming effects on US hurricane damage. *Weather, Climate, and Society*, **3**, 261–268.

Emanuel, K. 2021. Response of global tropical cyclone activity to increasing CO_2: Results from downscaling CMIP6 models. *Journal of Climate*, **34**(1), 57–70.

Emanuel, KA. 1986. An air–sea interaction theory for tropical cyclones. Part I: Steady-state maintenance. *Journal of the Atmospheric Sciences*, **43**(6), 585–605.

Emanuel, KA. 1999. The power of a hurricane: An example of reckless driving on the information superhighway. *Weather*, **54**(4), 107–108.

Emanuel, KA. 2005. Increasing destructiveness of tropical cyclones over the past 30 years. *Nature*, **436**, 686–688.

Embrechts, P, and Frei, M. 2009. Panjer recursion versus FFT for compound distributions. *Mathematical Methods of Operations Research*, **69**, 497–508.

Engel, JH. 1954. A verification of Lanchester's law. *Journal of the Operations Research Society of America*, **2**(2), 163–171.

Engman, ET. 1986. Roughness coefficients for routing surface runoff. *Journal of Irrigation and Drainage Engineering*, **112**(1), 1–78.

Epstein, JM. 2002. Modeling civil violence: An agent-based computational approach. *Proceedings of the National Academy of Sciences*, **99**(3), 7243–7250.

Epstein, JM. 2009. Modelling to contain pandemics. *Nature*, **460**, 687.

Epstein, JM, and Axtell, R. 1996. *Growing Artificial Societies: Social Science from the Bottom Up*. Washington, DC: The Brookings Institution.

Ericson, R, and Doyle, A. 2004. Catastrophe risk, insurance and terrorism. *Economy and Society*, **33**(2), 135–173.

Evans, KF, Zappone, A, Kraft, T, Deichmann, N, and Moia, F. 2012. A survey of the induced seismic responses to fluid injection in geothermal and CO_2 reservoirs in Europe. *Geothermics*, **41**, 30–54.

Faccioli, E, and Cauzzi, C. 2006. Macroseismic intensities for seismic scenarios, estimated from instrumentally based correlations. In: *First European Conference on Earthquake Engineering and Seismology*. Geneva: 13th ECEE and 30th General Assembly of the ESC.

Fan, J, Meng, J, Ludescher, J, Chen, X, Ashkenazy, Y, et al. 2021. Statistical physics approaches to the complex Earth system. *Physics Reports*, **896**, 1–84.

Farwell, JP, and Rohozinski, R. 2011. Stuxnet and the future of cyber war. *Survival*, **53**(1), 23–40.

Faturechi, R, and Miller-Hooks, E. 2015. Measuring the performance of transportation infrastructure systems in disasters: A comprehensive review. *Journal of Infrastructure Systems*, **21**(1), 04014025.

FEMA (Federal Emergency Management Agency). 2021. *Hazus Inventory Technical Manual*. Hazus 4.2 Service Pack 3. Washington, DC: FEMA.

FEMA. 2022. *Hazus Earthquake Model Technical Manual. Hazus 5.1*. Washington, DC: FEMA.

FEMA. 2023. OpenFEMA Dataset: FIMA NFIP Redacted Claims – v1. Online; accessed 15 March 2023.

Feng, Y, Mignan, A, Sornette, D, and Gao, K. 2023. Investigating injection pressure as a predictor to enhance real-time forecasting of fluid-induced seismicity: A Bayesian model comparison. *Seismological Research Letters*, **94**(2A), 708–719.

Feynman, RP. 1988. An outsider's inside view of the Challenger inquiry. *Physics Today*, **41**(2), 26–37.

Fiddaman, TS. 2002. Exploring policy options with a behavioral climate–economy model. *System Dynamics Review*, **18**(2), 243–267.

Field, EH, Arrowsmith, RJ, Biasi, GP, Bird, P, Dawson, TE, et al. 2014. Uniform California Earthquake Rupture Forecast, Version 3 (UCERF3) – The time-independent model. *Bulletin of the Seismological Society of America*, **104**, 1122–1180.

Finley, JP. 1884. Tornado predictions. *American Meteorological Journal*, **1**, 85–88.

Fisher, R. 1955. Statistical methods and scientific induction. *Journal of the Royal Statistical Society B*, **17**, 69–78.

Flynn, B (producer), and Peyton, B (director). 2015. *San Andreas* [Motion picture]. Warner Bros Pictures.

Foden, J. 1986. The petrology of Tambora Volcano, Indonesia: A model for the 1815 eruption. *Journal of Volcanology and Geothermal Research*, **27**, 1–41.

Foote, M, Mitchell-Wallace, K, Jones, M, and Hillier, J. 2017. Building catastrophe models. Chapter 4, pages 297–388 of: Mitchell-Wallace, K, Jones, M, Hillier, J, and Foote, M (eds), *Natural Catastrophe Risk Management and Modelling, A Practitioner's Guide*. Chichester: Wiley Blackwell.

Forrester, JW. 1961. *Industrial Dynamics*. Cambridge, MA: MIT Press.

Forrester, JW. 1969. *Urban Dynamics*. Waltham, MA: Pegasus Communications.

Forrester, JW. 1973. *World Dynamics (2nd ed.)*. Waltham, MA: Pegasus Communications.

Forrester, JW, Low, GW, and Mass, NJ. 1974. The debate on world dynamics: A response to Nordhaus. *Policy Sciences*, **5**, 169–190.

Foster, ZJ. 2015. The 1915 locust attack in Syria and Palestine and its role in the famine during the First World War. *Middle Eastern Studies*, **51**(3), 370–394.

Foulger, GR, Wilson, MP, Gluyas, JG, Julian, BR, and Davies, RJ. 2018. Global review of human-induced earthquakes. *Earth-Science Reviews*, **178**, 438–514.

Fournier, A, Fussell, D, and Carpenter, L. 1982. Computer rendering of stochastic models. *Communications of the ACM*, **25**(6), 371–384.

Freeman, LC. 1978. Centrality in social networks conceptual clarification. *Social Networks*, **1**, 215–239.

Fricker, T, and Elsner, JB. 2015. Kinetic energy of tornadoes in the United States. *PLoS ONE*, **10**(7), e0131090.

Friedman, DG. 1984. Natural hazard risk assessment for an insurance program. *The Geneva Papers on Risk and Insurance*, **9**(30), 57–128.

Friedman, J, and Shover, CL. 2023. Charting the fourth wave: Geographic, temporal, race/ethnicity and demographic trends in polysubstance fentanyl overdose deaths in the United States, 2010–2021. *Addiction*, **118**(12), 2477–2485.

Friedman, RM. 1999. Constituting the polar front, 1919–1920. Pages 29–39 of: Shapiro, MA, and Grønås, S (eds), *The Life Cycles of Extratropical Cyclones*. Boston, MA: American Meteorological Society.

Frigg, R. 2010. Models and fiction. *Synthese*, **172**(2), 251–268.

Fritz, HM, Mohammed, F, and Yoo, J. 2009. Lituya Bay landslide impact generated mega-tsunami 50th anniversary. *Pure and Applied Geophysics*, **166**, 153–175.

Fuchs, S, Heiss, K, and Hübl, J. 2007. Towards an empirical vulnerability function for use in debris flow risk assessment. *Natural Hazards and Earth System Sciences*, **7**, 495–506.

Fuentes, MA, and Kuperman, MN. 1999. Cellular automata and epidemiological models with spatial dependence. *Physica A*, **267**, 471–486.

Fujita, TT. 1971. Proposed characterization of tornadoes and hurricanes by area and intensity. *SMRP Research Paper*, 1–42.

Fujita, TT. 1973. Tornadoes around the world. *Weatherwise*, **26**(2), 56–83.

Funtowicz, SO, and Ravetz, JR. 1994. The worth of a songbird: Ecological economics as a post-normal science. *Ecological Economics*, **10**, 197–207.

Gage, KL, Burkot, TR, Eisen, RJ, and Hayes, EB. 2008. Climate and vectorborne diseases. *American Journal of Preventive Medicine*, **35**(5), 436–450.

Galanis, P, Sycheva, A, Mimra, W, and Stojadinović, B. 2018. A framework to evaluate the benefit of seismic upgrading. *Earthquake Spectra*, **34**(2), 527–548.

Gall, M, Borden, KA, and Cutter, SL. 2009. When do losses count? Six fallacies of natural hazards loss data. *Bulletin of the American Meteorological Society*, **90**(6), 799–810.

Gall, M, Borden, KA, Emrich, CT, and Cutter, SL. 2011. The unsustainable trend of natural hazard losses in the United States. *Sustainability*, **3**, 2157–2181.

Galway, JG. 1985. J. P. Finley: The first severe storms forecaster. *Bulletin of the American Meteorological Society*, **66**(11), 1389–1395.

Gamba, P. 2014. *Global Exposure Database: Scientific Features*. Pavia: GEM Technical Report 2014-10 V1.0.0, 46 pp., GEM Foundation.

Gardner, M. 1970. Mathematical games: The fantastic combinations of John Corway's new solitaire game 'life'. *Scientific American*, **223**(9), 120–123.

Geller, RJ. 1996. Debate on evaluation of the VAN method: Editor's introduction. *Geophysical Research Letters*, **23**(11), 1291–1293.

Geller, RJ. 1997. Earthquake prediction: A critical review. *Geophysical Journal International*, **131**, 425–450.

Gelman, A, Simpson, D, and Betancourt, M. 2017. The prior can often only be understood in the context of the likelihood. *Entropy*, **19**, 555.

GEM (Global Earthquake Model). 2022. *The OpenQuake-engine User Manual. Global Earthquake Model (GEM) OpenQuake Manual for Engine version devel. (3.18.0)*. Online; accessed 30 September 2023.

Gentile, R, Cremen, G, Galasso, C, Jenkins, LT, Manandhar, V, Mentes, et al. 2022. Scoring, selecting, and developing physical impact models for multi-hazard risk assessment. *International Journal of Disaster Risk Reduction*, **82**, 103365.

Gentry, RC. 1969. Project STORMFURY. *Bulletin of the American Meteorological Society*, **50**(6), 404–410.

Gericke, OJ, and Smithers, JC. 2014. Review of methods used to estimate catchment response time for the purpose of peak discharge estimation. *Hydrological Sciences Journal*, **59**(11), 1935–1971.

Gerstenberger, MC, Wiemer, S, Jones, LM, and Reasenberg, PA. 2005. Real-time forecasts of tomorrow's earthquakes in California. *Nature*, **435**, 328–331.

GFDRR (Global Facility for Disaster Reduction and Recovery). 2023. GED4all exposure taxonomy – Risk Data Library Standard. Online; accessed 13 March 2023.

Ghosh, J, and Padgett, JE. 2010. Aging considerations in the development of time-dependent seismic fragility curve. *Journal of Structural Engineering*, **136**, 1497–1511.

Giardini, D. 2009. Geothermal quake risks must be faced. *Nature*, **462**, 848–849.

Gilbert, GK. 1884. Finley's tornado predictions. *American Meteorological Journal*, **1**, 166–172.

Gilboa, I. 2009. *Theory of Decision under Uncertainty*. New York, NY: Cambridge University Press.

Gill, JC, and Malamud, BD. 2014. Reviewing and visualizing the interactions of natural hazards. *Reviews of Geophysics*, **52**, 680–722.

Gilleland, E, and Katz, RW. 2016. extRemes 2.0: An extreme value analysis package in R. *Journal of Statistical Software*, **72**(8).

GISTEMP Team. 2023. GISS Surface Temperature Analysis (GISTEMP), version 4. NASA Goddard Institute for Space Studies. Online; accessed 27 June 2023.

Glasstone, S, and Dolan, PJ. 1977. *Effects of Nuclear Weapons*. Washington, DC: US Department of Defense and the Energy Research and Development Administration.

Gleditsch, NP, Wallensteen, P, Eriksson, M, Sollenberg, M, and Strand, H. 2002. Armed conflict 1946–2001: A new dataset. *Journal of Peace Research*, **39**(5), 615–637.

Gleick, PH. 2014. Water, drought, climate change, and conflict in Syria. *Water, Climate, and Society*, **6**, 331–340.

Goldblatt, R, Jones, N, and Mannix, J. 2020. Assessing OpenStreetMap completeness for management of natural disaster by means of remote sensing: A case study of three small island states (Haiti, Dominica and St. Lucia). *Remote Sensing*, **12**, 118.

Good, EJ, Ghent, DJ, Bulgin, CE, and Remedios, JJ. 2017. A spatiotemporal analysis of the relationship between near-surface air temperature and satellite land surface temperatures using 17 years of data from the ATSR series. *Journal of Geophysical Research: Atmospheres*, **122**, 9185–9210.

Grasman, RPPP, van der Maas, HLJ, and Wagenmakers, E-J. 2009. Fitting the cusp catastrophe in R: A cusp package primer. *Journal of Statistical Software*, **32**(8).

Graves, R, Jordan, TH, Callaghan, S, Deelman, E, Field, E, et al. 2011. CyberShake: A physics-based seismic hazard model for Southern California. *Pure and Applied Geophysics*, **168**, 367–381.

Gray, WM. 1998. The formation of tropical cyclones. *Meteorology and Atmospheric Physics*, **67**, 37–69.

Grechuk, B, and Zabarankin, M. 2014. Risk averse decision making under catastrophic risk. *European Journal of Operational Research*, **239**, 166–176.

Greenberg, DA, Murty, TS, and Ruffman, A. 1993. A numerical model for the Halifax harbor tsunami due to the 1917 explosion. *Marine Geodesy*, **16**(2), 153–167.

Grimmett, GR, and Stirzaker, DR. 2001. *Probability and Random Processes*. Oxford: Oxford University Press.

Grossi, P, and Kunreuther, H. 2005a. *Catastrophe Modeling: A New Approach to Managing Risk*. New York, NY: Springer.

Grossi, P, and Kunreuther, H. 2005b. Introduction: Needs, stakeholders, and government initiatives. Chapter 1, pages 3–21 of: Grossi, P, and Kunreuther, H (eds), *Catastrophe Modeling: A New Approach to Managing Risk*. New York, NY: Springer.

Grossi, P, and Muir-Wood, R. 2006. *The 1906 San Francisco Earthquake and Fire: Perspectives on a Modern Super Cat*. Newark, CA: Risk Management Solutions, Inc.

Grossi, P, Kunreuther, H, and Windeler, D. 2005. An introduction to catastrophe models and insurance. Chapter 2, pages 23–42 of: Grossi, P, and Kunreuther, H (eds), *Catastrophe Modeling: A New Approach to Managing Risk*. New York, NY: Springer.

Grossman, G, and Rodenhuis, D. 1975. The effects of release of latent heat on the vorticity of a tropical storm over land. *Monthly Weather Review*, **103**(6), 486–496.

Grove, WM. 2005. Clinical versus statistical prediction: The contribution of Paul E. Meehl. *Journal of Clinical Psychology*, **61**(10), 1233–1243.

Grünthal, G. 1998. *European Macroseismic Scale 1998*. Luxembourg: Centre Européen de Géodynamique et de Séismologie.

Guan, D, Wang, D, Hallegatte, S, Davis, SJ, Huo, J, et al. 2020. Global supply-chain effects of COVID-19 control measures. *Nature Human Behaviour*, **4**, 577–587.

Guarino, N, Oberle, D, and Staab, S. 2009. What is an ontology? Pages 1–17 of: Staab, S, and Studer, R (eds), *Handbook on Ontologies*. Berlin and Heidelberg: Springer.

Gudmundsson, MT. 2015. Hazards from lahars and jökulhlaups. Chapter 56, pages 971–984 of: *The Encyclopedia of Volcanoes (2nd ed.)*. London and San Diego, CA: Elsevier Inc.

Guha-Sapir, D, Below, R, and Hoyois, PH. 2022. EM-DAT: The CRED/OFDA International Disaster Database – Université Catholique de Louvain – Brussels – Belgium. Online; accessed 30 January 2022.

Guidolin, M, Chen, AS, Ghimire, B, Keedwell, EC, Djordjevic, S, and Savic, DA. 2016. A weighted cellular automata 2D inundation model for rapid flood analysis. *Environmental Modelling & Software*, **84**, 378–394.

Gunasekera, R, Ishizawa, O, Aubrecht, C, Blankespoor, B, Murray, S, et al. 2015. Developing an adaptive global exposure model to support the generation of country disaster risk profiles. *Earth-Science Reviews*, **150**, 594–608.

Gupta, R, Goodman, B, Patel, N, Hosfelt, R, Sajeev, S, et al. 2019. Creating xBD: A dataset for assessing building damage from satellite imagery. Proceedings of the IEEE/CVF Conference on Computer Vision and Pattern Recognition (CVPR) Workshops, Long Beach, CA, 16–20 June 2019.

Gutenberg, B, and Richter, CF. 1944. Frequency of earthquakes in California. *Bulletin of the Seismological Society of America*, **34**, 184–188.

Haarsma, RJ, Hazeleger, W, Severijns, C, de Vries, H, Sterl, A, et al. 2013. More hurricanes to hit western Europe due to global warming. *Geophysical Research Lettters*, **40**, 1783–1788.

Haklay, M, and Weber, P. 2008. OpenstreetMap: User-generated street maps. *IEEE Pervasive Computing*, **7**(4), 12–18.

Hall, SS. 2011. At fault? *Nature*, **477**, 264–269.

Hallegatte, S. 2008. An adaptive regional input-output model and its application to the assessment of the economic cost of Katrina. *Risk Analysis*, **28**(3), 779–799.

Halley, E. 1693. An estimate of the degrees of the mortality of mankind, drawn from curious tables of the births and funerals at the city of Breslaw; with an attempt to ascertain the price of annuities upon lives. *Philosophical Transactions of the Royal Society of London*, **17**(196), 596–610.

Halley, E. 1705. A synopsis of the astronomy of comets. *Royal Society of London. Philosophical Transactions*, **24**, 1–24.

Hanks, TC, and Cornell, CA. 1994. Probabilistic Seismic Hazard Analysis: A Beginner's Guide. In: *Proceedings of the Fifth Symposium on Current Issues Related to Nuclear Power Plant Structures, Equipment and Piping*, North Carolina State University, Raleigh, NC.

Hanks, TC, and Kanamori, H. 1979. A moment magnitude scale. *Journal of Geophysical Research*, **84**(B5), 2348–2350.

Hao, S. 2010. I-35W bridge collapse. *Journal of Bridge Engineering*, **15**(5), 608–614.

Hardebeck, JL, Felzer, KR, and Michael, AJ. 2008. Improved tests reveal that the accelerating moment release hypothesis is statistically insignificant. *Journal of Geophysical Research*, **113**, B08310.

Hardy, A. 1988. Urban famine or urban crisis? Typhus in the Victorian City. *Medical History*, **32**(4), 401–425.

Harris, I. 2005. Generalised Pareto methods for wind extremes. Useful tool or mathematical mirage? *Journal of Wind Engineering and Industrial Aerodynamics*, **93**, 341–360.

Hauksson, E, Yang, W, and Shearer, PM. 2012. Waveform relocated earthquake catalog for Southern California (1981 to June 2011). *Bulletin of the Seismological Society of America*, **102**, 2239–2244.

Hayes, B. 1993. Balanced on a pencil point. *American Scientist*, **81**, 2–8.

He, X, and Sheffield, J. 2020. Lagged compound occurrence of droughts and pluvials globally over the past seven decades. *Geophysical Research Letters*, **47**, e2020GL087924.

Heal, G, and Millner, A. 2013. *Uncertainty and Decision in Climate Change Economics*. National Bureau of Economic Research (NBER) Working Paper No.18929 Cambridge, MA: NBER.

Heidarzadeh, M, Gusman, AR, Ishibe, T, Sabeti, R, and Sepic, J. 2022. Estimating the eruption-induced water displacement source of the 15 January 2022 Tonga volcanic tsunami from tsunami spectra and numerical modelling. *Ocean Engineering*, **261**, 112165.

Helbing, D, and Kuehnert, C. 2003. Assessing interaction networks with applications to catastrophe dynamics and disaster management. *Physica A*, **328**, 584–606.

Helbing, D, Farkas, I, and Vicsek, T. 2000. Simulating dynamical features of escape panic. *Nature*, **407**, 487–490.

Helbing, D, Farkas, I, and Vicsek, T. 2002. Crowd disasters and simulation of panic situations. Chapter 11, pages 331–350 of: Bunde, A, Kropp, J, and Schellnhuber, HJ (eds), *The Science of Disasters*. Berlin, Heidelberg: Springer.

Helbing, D, Buzna, L, Johansson, A, and Werner, T. 2005. Self-organized pedestrian crowd dynamics: Experiments, simulations, and design solutions. *Transportation Science*, **39**(1), 1–24.

Held, IM, and Soden, BJ. 2006. Robust responses of the hydrological cycle to global warming. *Journal of Climate*, **19**, 5686–5699.

Helmbold, RL, and Rehm, AS. 1995. 'The influence of the numerical strength of engaged forces in their casualties', by M Osipov. *Naval Research Logistics*, **42**, 435–490.

Hergarten, S, and Neugebauer, HJ. 1998. Self–organized criticality in a landslide model. *Geophysical Research Letters*, **25**(6), 801–804.

Herzog, S. 2011. Revisiting the Estonian cyber attacks: Digital threats and multinational responses. *Journal of Strategic Security*, **4**(2), 49–60.

Hethcote, HW. 2000. The mathematics of infectious diseases. *SIAM Review*, **42**(4), 599–653.

Heukelbach, J, Alencar, CH, Kelvin, AA, de Oliveira, WK, and de Góes Cavalcanti, LP. 2016. Zika virus outbreak in Brazil. *The Journal of Infection in Developing Countries*, **10**(2), 116–120.

Hicke, JA, Johnson, MC, Hayes, JL, and Preisler, HK. 2012. Effects of bark beetle-caused tree mortality on wildfire. *Forest Ecology and Management*, **271**, 81–90.

Hill, BM. 1975. A simple general approach to inference about the tail of a distribution. *The Annals of Statistics*, **3**(5), 1163–1174.

Hill, DP, Pollitz, F, and Newhall, C. 2002. Earthquake-volcano interactions. *Physics Today*, **55**(11), 41–47.

Hillier, J. 2017. The perils in brief. Chapter 3, pages 187–296 of: Mitchell-Wallace, K, Jones, M, Hillier, J, and Foote, M (eds), *Natural Catastrophe Risk Management and Modelling, A Practitioner's Guide*. Chichester: Wiley Blackwell.

Hills, JG, and Goda, MP. 1993. The fragmentation of small asteroids in the atmosphere. *The Astronomical Journal*, **105**(3), 1114–1144.

Hirschberg, S, Burgherr, P, Spickerman, G, and Dones, R. 2004. Severe accidents in the energy sector: Comparative perspective. *Journal of Hazardous Materials*, **111**, 57–65.

Hobley, DEJ, Adams, JM, Nudurupati, SS, Hutton, EWH, Gasparini, NM, et al. 2017. Creative computing with Landlab: An open-source toolkit for building, coupling, and exploring two-dimensional numerical models of Earth-surface dynamics. *Earth Surface Dynamcis*, **5**, 21–46.

Hodges, KI. 1995. Feature tracking on the unit sphere. *Monthly Weather Review*, **123**, 3458–3465.

Hofherr, T, and Kunz, M. 2010. Extreme wind climatology of winter storms in Germany. *Climate Research*, **41**, 105–123.

Hohl, R, Schiesser, H-H, and Knepper, I. 2002. The use of weather radars to estimate hail damage to automobiles: An exploratory study in Switzerland. *Atmospheric Research*, **61**, 215–238.

Holland, AP, Riordan, AJ, and Franklin, EC. 2006. A simple model for simulating tornado damage in forests. *Journal of Applied Meteorology and Climatology*, **45**, 1597–1611.

Holland, G. 2008. A revised hurricane pressure–wind model. *Monthly Weather Review*, **136**, 3432–3445.

Holland, GJ. 1980. An analytic model of the wind and pressure profiles in hurricanes. *Monthly Weather Review*, **108**, 1212–1218.

Holliday, JR, Chen, C-c, Tiampo, KF, Rundle, JB, Turcotte, DL, and Donnellan, A. 2007. A RELM earthquake forecast based on pattern informatics. *Seismological Research Letters*, **78**(1), 87–93.

Homer, D, and Li, M. 2017. Notes on using property catastrophe model results. *Casualty Actuarial Society E-Forum, Spring issue*, **2**.

Hope, JR, and Neumann, CJ. 1970. An operational technique for relating the movement of existing tropical cyclones to past tracks. *Monthly Weather Review*, **98**(12), 925–933.

Horton, R. 2023. Offline: Israel–Gaza—What comes next? *The Lancet*, **402**(10412), P1511.

Hosono, K, Miyakawa, D, Uchino, T, Hazama, M, Ono, A, et al. 2016. Natural disasters, damage to banks, and firm investment. *International Economic Review*, **57**(4), 1335–1370.

Hou, WK, Lee, TM-C, Liang, L, Li, TW, Liu, H, et al. 2021. Civil unrest, COVID-19 stressors, anxiety, and depression in the acute phase of the pandemic: a population-based study in Hong Kong. *Social Psychiatry and Psychiatric Epidemiology*, **56**, 1499–1508.

Hsiang, SM, Burke, M, and Miguel, E. 2013. Quantifying the influence of climate on human conflict. *Science*, **341**, 1235367.

Huang, Q. 2004. Seismicity pattern changes prior to large earthquakes – an approach of the RTL algorithm. *Terrestrial, Atmospheric and Oceanic Sciences Journal*, **15**(3), 469–491.

Huang, Q, Sobolev, GA, and Nagao, T. 2001. Characteristics of the seismic quiescence and activation patterns before the $M = 7.2$ Kobe earthquake, January 17, 1995. *Tectonophysics*, **337**, 99–116.

Huang, Z, Dai, K, Wang, J, and Wu, HF. 2016. Investigations of structural damage caused by the fertilizer plant explosion at West, Texas. I: Air-blast incident overpressure. *Journal of Performance of Constructed Facilities*, **30**(4), 04015064.

Huber, C, Finelli, L, and Stevens, W. 2018. The economic and social burden of the 2014 Ebola outbreak in West Africa. *The Journal of Infectious Diseases*, **218**(suppl. 5), S698–S704.

Huizinga, J, de Moel, H, and Szewczyk, W. 2017. *Global Flood Depth-Damage Functions. Methodology and the Database with Guidelines*. JRC Technical Reports. EUR 28552 EN.

Huynen, MMTE, Martens, P, Schram, D, Weijenberg, MP, and Kunst, AE. 2001. The impact of heat waves and cold spells on mortality rates in the Dutch population. *Environmental Health Perspectives*, **109**(5), 463–470.

Hänsel, MC, Drupp, MA, Johansson, DJA, Nesje, F, Azar, C, Freeman, MC, et al. 2020. Climate economics support for the UN climate targets. *Nature Climate Change*, **10**, 781–789.

Hédervári, P. 1963. On the energy and magnitude of volcanic eruptions. *Bulletin Volcanologique*, **25**, 373–385.

Ilachinski, A. 2000. Irreducible Semi-Autonomous Adaptive Combat (ISAAC): An artificial-life approach to land combat. *Military Operations Research*, **5**(3), 29–46.

IPCC (Intergovernmental Panel on Climate Change). 2021. *Climate Change 2021: The Physical Science Basis. Contribution of Working Group I to the Sixth Assessment Report of the Intergovernmental Panel on Climate Change*. Cambridge, UK, and New York, NY: Cambridge University Press.

Irish, JL, Resio, DT, and Ratcliff, JJ. 2008. The influence of storm size on hurricane surge. *Journal of Physical Oceanography*, **38**, 2003–2013.

ISC (International Seismological Centre). 2023. ISC-GEM Earthquake Catalogue. Online; accessed 12 February 2023.

Isimite, J, and Rubini, P. 2016. A dynamic HAZOP case study using the Texas City refinery explosion. *Journal of Loss Prevention in the Process Industries*, **40**, 496–501.

Issermann, M, Chang, F-J, and Jia, H. 2020. Efficient urban inundation model for live flood forecasting with cellular automata and motion cost fields. *Water*, **12**, 1997.

Itami, RM. 1994. Simulating spatial dynamics: cellular automata theory. *Landscape and Urban Planning*, **30**, 27–47.

Iverson, RM. 2000. Landslide triggering by rain infiltration. *Water Resources Research*, **36**(7), 1897–1910.

Jaboyedoff, M, Michoud, C, Derron, M-H, Voumard, J, Leibundgut, G, et al. 2016. Human-induced landslides: Toward the analysis of anthropogenic changes of the slope environment. Pages 217–232 of: Aversa, S, Cascini, L, Picarelli, L, and Scavia, C (eds), *Landslides and Engineered Slopes. Experience,Theory and Practice*. Boca Raton, FL: CRC Press.

Jafarimanesh, A, Mignan, A, and Danciu, L. 2018. Origin of the power-law exponent in the landslide frequency-size distribution. *Natural Hazards and Earth System Sciences Discussions [preprint]*.

Jain, P, Pasman, HJ, Waldram, SP, Rogers, WJ, and Mannan, MS. 2017. Did we learn about risk control since Seveso? Yes, we surely did, but is it enough? An historical brief and problem analysis. *Journal of Loss Prevention in the Process Industries*, **49**, 5–17.

Jaiswal, KS, and Wald, DJ. 2008. *Creating a Global Building Inventory for Earthquake Loss Assessment and Risk Management*. US Geological Survey Open-File Report 2008–1160.

Jakob, M, Stein, D, and Ulmi, M. 2011. Vulnerability of buildings to debris flow impact. *Natural Hazards*, **60**, 241–261.

Jamali, B, Bach, PM, Cunningham, L, and Deletic, A. 2019. A Cellular Automata fast flood evaluation (CA-ffé) model. *Water Resources Research*, **55**, 4936–4953.

Jarvinen, BR, and Neumann, CJ. 1979. Statistical forecasts of tropical cyclone intensity for the North Atlantic Basin. In: *NOAA Technical Memorandum NWS NHC-10*. Fort Worth, TX: National Hurricane Center.

Jasanoff, S. 1998. The political science of risk perception. *Reliability Engineering and System Safety*, **59**, 91–99.

Jelesnianski, CP, Chen, J, and Shaffer, WA. 1992. *SLOSH: Sea, Lake, and Overland Surges from Hurricanes*. Miami, FL: NOAA Technical Report NWS 48, NOAA/AOML Library.

Jiang, Y, Fu, P, and Weng, Q. 2015. Downscaling GOES land surface temperature for assessing heat wave health risks. *IEEE Geoscience and Remote Sensing Letters*, **12**(8), 1605–1609.

Jibson, RW. 1993. Predicting earthquake-induced landslide displacements using Newmark's sliding block analysis. *Transportation Research Record*, **1411**, 9–17.

Jibson, RW. 2007. Regression models for estimating coseismic landslide displacement. *Engineering Geology*, **91**, 209–218.

Johnston, FH, Borchers-Arriagada, N, Morgan, GG, Jalaludin, B, Palmer, AJ, et al. 2020. Unprecedented health costs of smoke-related $PM_{2.5}$ from the 2019–20 Australian megafires. *Nature Sustainability*, **4**, 42–47.

Jones, BA, Grace, D, Kock, R, Alonso, S, Rushton, J, et al. 2013. Zoonosis emergence linked to agricultural intensification and environmental change. *PNAS*, **110**(21), 8399–8404.

Jones, L, and Molnar, P. 1976. Frequency of foreshocks. *Nature*, **262**, 677–679.

Jones, M, Mitchell-Wallace, K, Foote, M, and Hillier, J. 2017. Fundamentals. In: Mitchell-Wallace, K, Jones, M, Hillier, J, and Foote, M (eds), *Natural Catastrophe Risk Management and Modelling, A Practitioner's Guide*. Chichester: Wiley Blackwell.

Jonkman, SN. 2005. Global perspectives on loss of human life caused by floods. *Natural Hazards*, **34**, 151–175.

Jonkman, SN, and Kelman, I. 2005. Deaths during the 1953 North Sea storm surge. *Solutions to Coastal Disasters Conference*, 749–758.

Jowarder, MI. 2022. Aftermath of Uvalde shooting: An account of psychological trauma of survivor children. *Journal of Loss and Trauma*, **28**(7), 588–598.

Jung, K. 2021. Extreme data breach losses: An alternative approach to estimating probable maximum loss for data breach risk. *North American Actuarial Journal*, **25**(4), 580–603.

Kaas, R, Goovaerts, M, Dhaene, J, and Denuit, M. 2008. *Modern Actuarial Risk Theory Using R*. Berlin, Heidelberg: Springer.

Kahneman, D. 2011. *Thinking, Fast and Slow*. New York City, NY: Farrar, Straus and Giroux.

Kahneman, D, and Tversky, A. 1979. Prospect theory: An analysis of decision under risk. *Econometrica*, **47**(2), 263–291.

Kaktins, U, Todd, CD, Wojno, S, and Coleman, N. 2013. Revisiting the timing and events leading to and causing the Johnstown Flood of 1889. *Pennsylvania History A Journal of Mid-Atlantic Studies*, **80**(3), 335–363.

Kam, PM, Aznar-Siguan, G, Schewe, J, Milano, L, Ginnetti, J, et al. 2021. Global warming and population change both heighten future risk of human displacement due to river floods. *Environmental Research Letters*, **16**, 044026.

Kanamori, H. 1981. The nature of seismicity patterns before large earthquakes. Pages 1–19 of: Simpson, DW, and Richards, PG (eds), *Earthquake Prediction: An International Review*. Maurice Ewing Series, Vol. 4. Washington, D.C.: American Geophysical Union.

Kaplan, J, and DeMaria, M. 1995. A simple empirical model for predicting the decay of tropical cyclone winds after landfall. *Journal of Applied Meteorology*, **34**, 2499–2512.

Kappos, A, Panagopoulos, G, Panagiotopoulos, C, and Penelis, G. 2006. A hybrid method for the vulnerability assessment of R/C and URM buildings. *Bulletin of Earthquake Engineering*, **4**(4), 391–413.

Kapucu, N, and Van Wart, M. 2006. The evolving role of the public sector in managing catastrophic disasters. *Administration & Society*, **38**(3), 279–308.

Kareiva, P, and Carranza, V. 2018. Existential risk due to ecosystem collapse: Nature strikes back. *Futures*, **102**, 39–50.

Karstens, CD, Gallus, WA, Jr, Lee, BD, and Finley, CA. 2013. Analysis of tornado-induced tree fall using aerial photography from the Joplin, Missouri, and Tuscaloosa–Birmingham, Alabama, tornadoes of 2011. *Journal of Applied Meteorology and Climatology*, **52**, 1049–1068.

Katsiyannis, A, Whitford, DK, and Ennis, RP. 2018. Historical examination of United States intentional mass school shootings in the 20th and 21st centuries: Implications for students, schools, and society. *Journal of Child and Family Studies*, **27**, 2562–2573.

Katz, RW, and Garcia, RR. 1981. Statistical relationships between hailfall and damage to wheat. *Agricultural Meteorology*, **24**, 29–43.

Katz, RW, Parlange, MB, and Naveau, P. 2002. Statistics of extremes in hydrology. *Advances in Water Resources*, **25**, 1287–1304.

Kean, JW, Staley, DM, Lancaster, JT, Rengers, FK, Swanson, BJ, et al. 2019. Inundation, flow dynamics, and damage in the 9 January 2018 Montecito debris-flow event, California, USA: Opportunities and challenges for post-wildfire risk assessment. *Geosphere*, **15**(4), 1140–1163.

Keeley, JE, Fotheringham, CJ, and Moritz, MA. 2004. Lessons from the October 2003 wildfires in Southern California. *Journal of Forestry*, **102**(7), 26–31.

Keen, S. 2021. The appallingly bad neoclassical economics of climate change. *Globalizations*, **18**(7), 1149–1177.

Keilis-Borok, V. 2002. Earthquake prediction: State-of-the-art and emerging possibilities. *Annual Review of Earth and Planetary Sciences*, **30**, 1–33.

Keith, DW. 2000. Geoengineering the climate: History and prospect. *Annual Review of Energy and the Environment*, **25**(1), 245–284.

Kelsey, D. 1993. Choice under partial uncertainty. *International Economic Review*, **34**(2), 297–308.

Kemp, L, Xu, C, Depledge, J, Ebi, KL, Gibbins, G, et al. 2022. Climate endgame: Exploring catastrophic climate change scenarios. *PNAS*, **119**(34), e2108146119.

Kempner, L, Jr. 2007. *Substation Structure Design Guide*. ASCE Manuals and Reports on Engineering Practice No. 113.

Kennedy, MC, and O'Hagan, A. 2001. Bayesian calibration of computer models. *Journal of the Royal Statistical Society Series B*, **63**(3), 425–464.

Kermack, WO, and McKendrick, AG. 1927. A contribution to the mathematical theory of epidemics. *Proceedings of the Royal Society of London A*, **115**, 700–721.

Kermanshah, A, and Derrible, S. 2017. Robustness of road systems to extreme flooding: Using elements of GIS, travel demand, and network science. *Natural Hazards*, **86**, 151–164.

Kerr, GD, Young, RW, Cullings, HM, and Christy, RF. 2005. Bomb parameters. Chapter 1, pages 42–61 of: Young, RW, and Kerr, GD (eds), *Reassessment of the Atomic Bomb Radiation Dosimetry for Hiroshima and Nagasaki – Dosimetry System 2002*. Hiroshima: The Radiation Effects Research Foundation.

Khan, FI, and Abbasi, SA. 1998a. Models for domino effect analysis in chemical process industries. *Process Safety Progress*, **17**(2), 107–123.

Khan, FI, and Abbasi, SA. 1998b. Techniques and methodologies for risk analysis in chemical process industries. *Journal of Loss Prevention in the Process Industries*, **11**, 261–277.

Khan, FI, and Abbasi, SA. 1999. Major accidents in process industries and an analysis of causes and consequences. *Journal of Loss Prevention in the Process Industries*, **12**, 361–378.

Khan, FI, and Amyotte, PR. 2007. Modeling of BP Texas City refinery incident. *Journal of Loss Prevention in the Process Industries*, **20**, 387–395.

Khanduri, AC, and Morrow, GC. 2003. Vulnerability of buildings to windstorms and insurance loss estimation. *Journal of Wind Engineering and Industrial Aerodynamics*, **91**, 455–467.

Kilburn, CRJ, and Petley, DN. 2003. Forecasting giant, catastrophic slope collapse: Lessons from Vajont, Northern Italy. *Geomorphology*, **54**, 21–32.

Kim, K-H, Ree, J-H, Kim, YH, Kim, S, Kang, SY, and Seo, W. 2018. Assessing whether the 2017 M_w 5.4 Pohang earthquake in South Korea was an induced event. *Science*, **360**, 1007–1009.

King, G. 1983. The accommodation of large strains in the upper lithosphere of the Earth and other solids by self-similar fault systems: The geometrical origin of b-value. *Pure and Applied Geophysics*, **121**(5/6), 761–815.

King, G, and Zeng, L. 2001. Logistic regression in rare events data. *Political Analysis*, **9**(2), 137–163.

King, GCP, Stein, RS, and Lin, J. 1994. Static stress changes and the triggering of earthquakes. *Bulletin of the Seismological Society of America*, **84**(3), 935–953.

King, S, and Bracy, NL. 2019. School security in the post-Columbine era: Trends, consequences, and future directions. *Journal of Contemporary Criminal Justice*, **35**(3), 1–22.

Kirschbaum, D, Stanley, T, and Zhou, Y. 2015. Spatial and temporal analysis of a global landslide catalog. *Geomorphology*, **249**, 4–15.

Kirschbaum, DB, Adler, R, Hong, Y, Hill, S, and Lerner-Lam, A. 2010. A global landslide catalog for hazard applications: Method, results, and limitations. *Natural Hazards*, **52**, 561–575.

Kleindorfer, P, Grossi, P, and Kunreuther, H. 2005. The impact of mitigation on homeowners and insurers: An analysis of model cities. Chapter 8, pages 167–188 of: Grossi, P, and Kunreuther, H (eds), *Catastrophe Modeling: A New Approach to Managing Risk*. New York, NY: Springer.

Klomp, J. 2014. Financial fragility and natural disasters: An empirical analysis. *Journal of Financial Stability*, **13**, 180–192.

Klose, AK, Wunderling, N, Winkelmann, R, and Donges, JF. 2021. What do we mean, 'tipping cascade'? *Environmental Research Letters*, **16**, 125011.

Klugman, SA, Panjer, HH, and Willmot, GE. 2012. *Loss Models. From Data to Decisions (4th ed.)*. Hoboken, NJ: John Wiley & Sons, Inc.

Knaff, JA, DeMaria, M, Sampson, CR, and Gross, JM. 2003. Statistical, 5-day tropical cyclone intensity forecasts derived from climatology and persistence. *Weather and Forecasting*, **18**, 80–92.

Knapp, KR, Kruz, MC, Levinson, DH, Diamond, HJ, and Neumann, CJ. 2010. The International Best Track Archive for Climate Stewardship (IBTrACS). *Bulletin of the American Meteorological Society*, **91**(3), 363–376.

Knapp, KR, Diamond, HJ, Kossin, JP, Kruk, MC, and Schreck, CJIII. 2018. International Best Track Archive for Climate Stewardship (IBTrACS) Project, Version 4. *NOAA National Centers for Environmental Information*.

Knight, FH. 1921. *Risk, Uncertainty and Profit*. Boston, MA, and New York, NY: Houghton Mifflin Company.

Knight, NC, and Heymsfield, AJ. 1983. Measurement and interpretation of hailstone density and terminal velocity. *Journal of the Atmospheric Sciences*, **40**(6), 1510–1516.

Knutson, T, Camargo, SJ, Chan, JCL, Emanuel, K, Ho, C-H, et al. 2020. Tropical cyclones and climate change assessment. *Bulletin of the American Meteorological Society*, **101**(3), E303–E322.

Knutson, TR, Sirutis, JJ, Zhao, M, Tuleya, RE, Bender, M, et al. 2015. Global projections of intense tropical cyclone activity for the late twenty-first century from dynamical downscaling of CMIP5/RCP4.5 scenarios. *Journal of Climate*, **28**(18), 7203–7224.

Kocin, PJ, Weiss, AD, and Wagner, JJ. 1988. The Great Arctic outbreak and East Coast blizzard of February 1899. *Weather and Forecasting*, **3**(4), 305–318.

Koenig, R, and Bauriedel, C. 2009. Generating settlement structures: A method for urban planning and analysis supported by cellular automata. *Environment and Planning B: Planning and Design*, **36**, 602–624.

Komendantova, N, Mrzyglocki, R, Mignan, A, Khazai, B, Wenzel, F, Patt, A, and Fleming, K. 2014. Multi-hazard and multi-risk decision-support tools as a part of participatory risk governance: Feedback from civil protection stakeholders. *International Journal of Disaster Risk Reduction*, **8**, 50–67.

Kong, Q, Allen, RM, Schreier, L, and Kwon, Y-W. 2016. MyShake: A smartphone seismic network for earthquake early warning and beyond. *Science Advances*, **2**(2), e1501055.

Koshimura, S, Namegaya, Y, and Yanagisawa, H. 2009. Tsunami fragility – A new measure to identify tsunami damage. *Journal of Disaster Research*, **4**(6), 479–488.

Kousky, C, and Light, SE. 2019. Insuring Nature. *Duke Law Journal*, **69**, 323.

Krause, A, Kloster, S, Wilkenskjeld, S, and Paeth, H. 2014. The sensitivity of global wildfires to simulated past, present, and future lightning frequency. *Journal of Geophysical Research: Biogeosciences*, **119**, 312–322.

Krausmann, E, Cozzani, V, Salzano, E, and Renni, E. 2011. Industrial accidents triggered by natural hazards: An emerging risk issue. *Natural Hazards and Earth System Sciences*, **11**, 921–929.

Kreps, R. 1990. Reinsurer risk loads from marginal surplus requirements. *PCAS*, **LXXVII**, 196–203.

Kress, M. 2012. Modeling armed conflicts. *Science*, **336**, 865–869.

Kress, M. 2020. Lanchester models for irregular warfare. *Mathematics*, **8**, 737.

Kulkarni, RB, Youngs, RR, and Coppersmith, KJ. 1984. Assessment of confidence intervals for results of seismic hazard analysis. *Proceedings of the Eighth World Conference on Earthquake Engineering*, 263–270.

Kumar, R, and Gardoni, P. 2014. Effect of seismic degradation on the fragility of reinforced concrete bridges. *Engineering Structures*, **79**, 267–275.

Kundzewicz, ZW, and Takeuchi, K. 1999. Flood protection and management: Quo vadimus? *Hydrological Sciences Journal*, **44**(3), 417–432.

Kunreuther, H, Michel-Kerjan, E, and Porter, B. 2005. Extending catastrophe modeling to terrorism. Chapter 10, pages 209–233 of: Grossi, P, and Kunreuther, H (eds), *Catastrophe Modeling: A New Approach to Managing Risk*. New York, NY: Springer.

Kuzak, D, and Larsen, T. 2005. Use of catastrophe models in insurance rate making. Chapter 5, pages 97–118 of: Grossi, P, and Kunreuther, H (eds), *Catastrophe Modeling: A New Approach to Managing Risk*. New York, NY: Springer.

Labat, D, Goddéris, Y, Probst, JL, and Guyot, JL. 2004. Evidence for global runoff increase related to climate warming. *Advances in Water Resource*, **27**, 631–642.

Lacina, B, Gleditsch, NP, and Russett, B. 2006. The declining risk of death in battle. *International Studies Quarterly*, **50**, 673–680.

Lacirignola, M, and Blanc, I. 2013. Environmental analysis of practical design options for enhanced geothermal systems (EGS) through life-cycle assessment. *Renewable Energy*, **50**, 901–914.

LaFree, G, and Dugan, L. 2007. Introducing the Global Terrorism Database. *Terrorism and Political Violence*, **19**(2), 181–204.

Lagomarsino, S, and Cattari, S. 2013. Seismic vulnerability of existing buildings. Chapter 1, pages 1–62 of: Gueguen, P (ed), *Seismic Vulnerability of Structures*. Hoboken, NJ: John Wiley & Sons, Inc.

Lagomarsino, S, and Cattari, S. 2014. Fragility functions of masonry buildings. Chapter 5, pages 111–156 of: Pitilakis, K, Crowley, H, and Kaynia, AM (eds), *SYNER-G: Typology Definition and Fragility Functions for Physical Elements at Seismic Risk*. Dordrecht: Springer.

Lagomarsino, S, and Giovinazzi, S. 2006. Macroseismic and mechanical models for the vulnerability and damage assessment of current buildings. *Bulletin of Earthquake Engineering*, **4**, 415–443.

Lagomarsino, S, Cattari, S, and Ottonelli, D. 2021. The heuristic vulnerability model: Fragility curves for masonry buildings. *Bulletin of Earthquake Engineering*, **19**, 3129–3163.

Lallemant, D, and Kiremidjian, A. 2015. A beta distribution model for characterizing earthquake damage state distribution. *Earthquake Spectra*, **31**(3), 1337–1352.

Lallemant, D, Kiremidjian, A, and Burton, H. 2015. Statistical procedures for developing earthquake damage fragility curves. *Earthquake Engineering & Structural Dynamics*, **44**(9), 1373–1389.

Lalonde, D. 2005. Risk financing. Chapter 7, pages 135–164 of: Grossi, P, and Kunreuther, H (eds), *Catastrophe Modeling: A New Approach to Managing Risk*. New York, NY: Springer.

Lanchester, FW. 1916. *Aircraft in Warfare: The Dawn of the Fourth Arm*. London: Constable and Company Limited.

La Paz, L. 1951. Injuries from falling meteorites. *Popular Astronomy*, **59**, 433–438.

Lapola, DM, Pinho, P, Quesada, CA, Strassburg, BBN, Rammig, A, et al. 2018. Limiting the high impacts of Amazon forest dieback with no-regrets science and policy action. *PNAS*, **115**(46), 11,671–11,679.

Launie, JJ. 1969. The supply function of urban property insurance. *The Journal of Risk and Insurance*, **36**(3), 269–283.

Laurent, P, Mouillot, F, Yue, C, Ciais, P, Moreno, MV, and Nogueira, JMP. 2018a. FRY, a global database of fire patch functional traits derived from space-borne burned area products. *Scientific Data*, **5**, 180132.

Laurent, P, Mouillot, F, Yue, C, Moreno Dominguez, MV, Ciais, P, and Nogueira, JMP. 2018b. List of fire patch properties computed and associated NetCDF maps from the MCD64A1 Collection 6 (2000–2016) and the MERIS fire_cci v4.1 (2005–2011) BA products. *OSU OREME (Dataset)*.

Lautrup, B. 2005. Tsunami physics. Online; accessed 25 November 2023.

Le Feuvre, M, and Wieczorek, MA. 2008. Nonuniform cratering of the terrestrial planets. *Icarus*, **197**, 291–306.

Lee, J-J, Ham, T-Y, and Jeong, W-S. 2022. Economic risk assessment of ammonium nitrate explosions at the Busan Port by determining the building damage using a 3D explosion simulation. *Heliyon*, **8**, e12285.

Lee, JA, and Tchakerian, VP. 1995. Magnitude and frequency of blowing dust on the southern High Plains of the United States, 1947–1989. *Annals of the Association of American Geographers*, **85**, 684–693.

Lee, K-K, Ellsworth, WL, Giardini, D, Townend, J, Ge, S, et al. 2019. Managing injection-induced seismic risks. *Science*, **364**(6442), 730–732.

Legates, DR, Lins, HF, and McCabe, GJ. 2005. Comments on 'Evidence for global runoff increase related to climate warming' by Labat et al. *Advances in Water Resources*, **28**, 1310–1315.

Lehmann, EL. 1993. The Fisher, Neyman–Pearson theories of testing hypotheses: One theory or two? *Journal of the American Statistical Association*, **88**(424), 1242–1249.

Lenssen, N, Schmidt, G, Hansen, J, Menne, M, Persin, A, et al. 2019. Improvements in the GISTEMP uncertainty model. *Journal of Geophysical Research: Atmospheres*, **124**(12), 6307–6326.

Lenton, TM. 2011. Early warning of climate tipping points. *Nature Climate Change*, **1**, 201–209.

Lenton, TM, Held, H, Kriegler, E, Hall, JW, Lucht, W, et al. 2008. Tipping elements in the Earth's climate system. *Proceedings of the National Academy of Sciences*, **105**(6), 1786–1793.

Leonard, M, Westra, S, Phatak, A, Lambert, M, van den Hurk, B, et al. 2014. A compound event framework for understanding extreme impacts. *WIREs Climate Change*, **5**, 113–128.

Lewin, CG, Evans, JV, Goodare, KJ, and Packer, LR. 1989. Calculating devices and actuarial work. *Journal of the Institute of Actuaries*, **116**(2), 215–287.

Li, B, Barker, K, and Sansavini, G. 2018. Measuring community and multi-industry impacts of cascading failures in power systems. *IEEE Systems Journal*, **12**(4), 3585–3596.

Li, G, West, AJ, Densmore, AL, Jin, Z, Parker, RN, and Hilton, RG. 2014. Seismic mountain building: Landslides associated with the 2008 Wenchuan earthquake in the context of a generalized model for earthquake volume balance. *Geochemistry, Geophysics, Geosystems*, **15**, 833–844.

Li, H-N, Liu, Y, Li, C, and Zheng, XW. 2020. Multihazard fragility assessment of steel-concrete composite frame structures with buckling-restrained braces subjected to combined earthquake and wind. *The Structural Design of Tall and Special Buildings*, **29**(11), e1746.

Li, Y, and Ellingwood, BR. 2006. Hurricane damage to residential construction in the US: Importance of uncertainty modeling in risk assessment. *Engineering Structures*, **28**, 1009–1018.

Lichtenstein, S, Slovic, P, Fischhoff, B, Layman, M, and Combs, B. 1978. Judged frequency of lethal events. *Journal of Experimental Psychology: Human Learning and Memory*, **4**(6), 551–578.

Lin, J, and Stein, RS. 2004. Stress triggering in thrust and subduction earthquakes and stress interaction between the southern San Andreas and nearby thrust and strike-slip faults. *Journal of Geophysical Research*, **109**, B02303.

Lin, N, Emanuel, KA, Smith, JA, and Vanmarcke, E. 2010. Risk assessment of hurricane storm surge for New York City. *Journal of Geophysical Research*, **115**, D18121.

Lindenmayer, DB, Kooyman, RM, Taylor, C, Ward, M, and Watson, JEM. 2020. Recent Australian wildfires made worse by logging and associated forest management. *Nature Ecology & Evolution*, **4**, 898–900.

Littell, JS, Peterson, DL, Riley, KL, Liu, Y, and Luce, CH. 2016. A review of the relationships between drought and forest fire in the United States. *Global Change Biology*, **22**, 2353–2369.

Litzenberger, RH, Beaglehole, DR, and Reynolds, CE. 1996. Assessing catastrophe reinsurance-linked securities as a new asset class. *The Journal of Portfolio Management, A Tribute to Fischer Black*, **23**(5), 76–86.

Liu, W, and Zhong, S. 2017. Web malware spread modelling and optimal control strategies. *Scientific Reports*, **7**, 42308.

Lloyd's Market Association. 2013. *Catastrophe Modelling. Guidance for Non-catastrophe Modellers*. London: Lloyd's Market Association.

Loc, HH, Park, E, Chitwatkulsiri, D, Lim, J, Yun, S-H, et al. 2020. Local rainfall or river overflow? Re-evaluating the cause of the Great 2011 Thailand flood. *Journal of Hydrology*, **589**, 125368.

Lomberg, J, and Hora, SC. 1997. Very long term communication intelligence. The case of markers for nuclear waste sites. *Technological Forecasting and Social Change*, **56**, 171–188.

London, J. 1906. *The Story of an Eyewitness*. New York City: Collier Magazine.

Lorenz, EN. 1963. Deterministic nonperiodic flow. *Journal of the Atmospheric Sciences*, **20**, 130–141.

Lotka, AJ. 1910. Contribution to the theory of periodic reaction. *Journal of Physical Chemistry*, **14**(3), 271–274.

Lowrance, WW. 1980. The nature of risk. Pages 5–14 of: Schwing, RC, and WA Albers, Jr (eds), *Societal Risk Assessment: How Safe is Safe Enough?* New York, NY: Springer Science+Business Media, LLC.

Lu, P, Zhang, Z, Li, M, Chen, D, and Yang, H. 2020. Agent-based modeling and simulations of terrorist attacks combined with stampedes. *Knowledge-Based Systems*, **205**, 106291.

Lu, S-M. 2018. A global review of enhanced geothermal system (EGS). *Renewable and Sustainable Energy Reviews*, **81**, 2902–2921.

Ludwig, D, Jones, DD, and Holling, CS. 1978. Qualitative analysis of insect outbreak systems: The spruce budworm and forest. *Journal of Animal Ecology*, **47**(1), 315–332.

Ludwig, D, Hilborn, R, and Walters, C. 1993. Uncertainty, resource exploitation, and conservation: Lessons from history. *Ecological Applications*, **3**(4), 547–549.

Ludwin, RS, Smits, GJ, Carver, D, James, K, Jonientz-Trisler, C, et al. 2007. Folklore and earthquakes: Native American oral traditions from Cascadia compared with written traditions from Japan. *Geological Society, London, Special Publications*, **273**, 67–94.

Lycosthenes, K. 1557. *Prodigiorum ac Ostentorum Chronicon*. Basileae: per Henricum Petri.

Lynch, P. 2008. The origins of computer weather prediction and climate modeling. *Journal of Computational Physics*, **227**, 3431–3444.

Lynett, P, McCann, M, Zhou, Z, Renteria, W, Borrero, J, et al. 2022. Diverse tsunamigenesis triggered by the Hunga Tonga-Hunga Ha'apai eruption. *Nature*, **609**, 728–733.

MacKay, NJ. 2015. When Lanchester met Richardson, the outcome was stalemate: A parable for mathematical models of insurgency. Chapter 6, pages 124–147 of: Forder, RA (ed), *OR, Defence and Security*. London: Palgrave Macmillan.

MacLean, DA. 1984. Effects of spruce budworm outbreaks on the productivity and stability of balsam fir forests. *The Forestry Chronicle*, **60**(5), 273–279.

Mahalingam, A, Coburn, A, Jung, CJ, Yeo, JZ, Cooper, G, and Evan, T. 2018. *Impacts of Severe Natural Catastrophes on Financial Markets*. Cambridge: Cambridge Centre for Risk Studies.

Mailier, PJ, Stephenson, DB, and Ferro, CAT. 2006. Serial clustering of extratropical cyclones. *Monthly Weather Review*, **134**, 2224–2240.

Maillart, T, and Sornette, D. 2010. Heavy-tailed distirbution of cyber-risks. *European Physical Journal B*, **75**, 357–364.

Malamud, BD, and Turcotte, DL. 2000. Cellular-automata models applied to natural hazards. *Computing in Science & Engineering*, **2**(3), 42–51.

Malamud, BD, and Turcotte, DL. 2006. The applicability of power-law frequency statistics to floods. *Journal of Hydrology*, **322**, 168–180.

Malamud, BD, Morein, G, and Turcotte, DL. 1998. Forest fires: An example of self-organized critical behavior. *Science*, **281**, 1840–1842.

Malamud, BD, Turcotte, DL, Guzzetti, F, and Reichenbach, P. 2004. Landslides, earthquakes, and erosion. *Earth and Planetary Science Letters*, **229**, 45–59.

Malmstadt, JC, Elsner, JB, and Jagger, TH. 2010. Risk of strong hurricane winds to Florida cities. *Journal of Applied Meteorology and Climatology*, **49**, 2121–2132.

Malthus, TR. 1798. *An Essay on the Principle of Population*. London: J Johnson.

Mandelbrot, BB. 1974. Intermittent turbulence in self-similar cascades: Divergence of high moments and dimension of the carrier. *Journal of Fluid Mechanics*, **62**(2), 331–358.

Mandelbrot, BB. 1986. Fractals and the rebirth of iteration theory. Pages 151–160 of: Peitgen, H-O, and Richter, PH (eds), *The Beauty of Fractals, Images of Complex Dynamical Systems*. Berlin, Heidelberg: Springer.

Mandelbrot, BB, and Wallis, JR. 1968. Noah, Joseph, and operational hydrology. *Water Resources Research*, **4**(5), 909–918.

Mann, T. 1927. *The Magic Mountain*. London: M Secker.

Marani, M, Katul, GG, Pan, WK, and Parolari, AJ. 2021a. A Global Epidemics Dataset (1500–2020) (V1.0) [Data set]. Zenodo.

Marani, M, Katul, GG, Pan, WK, and Parolari, AJ. 2021b. Intensity and frequency of extreme novel epidemics. *Proceedings of the National Academy of Sciences*, **118**(35), e2105482118.

Markowitz, H. 1952. Portfolio selection. *The Journal of Finance*, **7**(1), 77–91.

Marshall, LR, Maters, EC, Schmidt, A, Timmreck, C, Robock, A, and Toohey, M. 2022. Volcanic effects on climate: Recent advances and future avenues. *Bulletin of Volcanology*, **84**, 54.

Marshall, TP. 1993. Lessons Learned from Analyzing Tornado Damage. *The Tornado: Its Structure, Dynamics, Prediction, and Hazards. Geophysical Monograph*, **79**.

Maslow, AH. 1943. A theory of human motivation. *Psychological Review*, **50**, 370–396.

Masoomi, H, and van de Lindt, JW. 2018. Fatality and injury prediction model for tornadoes. *Natural Hazards Review*, **19**(3), 04018009.

Masse, WB, and Masse, MJ. 2007. Myth and catastrophic reality: Using myth to identify cosmic impacts and massive Plinian eruptions in Holocene South America. *Geological Society, London, Special Publications*, **273**, 177–202.

Matheson, VA, and Baade, RA. 2004. Race and riots: A note on the economic impact of the Rodney King riots. *Urban Studies*, **41**(13), 2691–2696.

Mathias, DL, Wheeler, LF, and Dotson, JL. 2017. A probabilistic asteroid impact risk model: Assessment of sub-300 m impacts. *Icarus*, **289**, 106–119.

Matos, J, Mignan, A, and Schleiss, A. 2015. Vulnerability of large dams considering hazard interactions: Conceptual application of the Generic Multi-Risk framework. *Proceedings of the 13th ICOLD Benchmark Workshop on the Numerical Analysis of Dams*, 285–292.

Matthews, JC. 2016. Disaster resilience of critical water infrastructure systems. *Journal of Structural Engineering*, **142**(8), C6015001.

Matthews, T, Wilby, RL, and Murphy, C. 2019. An emerging tropical cyclone–deadly heat compound hazard. *Nature Climate Change*, **9**, 602–606.

May, RM. 1976. Simple mathematical models with very complicated dynamics. *Nature*, **261**, 459–467.

Mazzocchi, M, Hansstein, F, and Ragona, M. 2010. The 2010 volcanic ash cloud and its financial impact on the European airline industry. *CESifo Forum*, **11**(2), 92–100.

McCaffrey, R. 2008. Global frequency of magnitude 9 earthquakes. *Geology*, **36**, 263–266.

McCallum, BE, Wicklein, SM, Reiser, RG, Busciolano, R, Morrison, J, et al. 2013. Monitoring storm tide and flooding from Hurricane Sandy along the Atlantic coast of the United States, October 2012. *US Geological Survey Open-File Report*, **2013–1043**.

McGarr, A. 1976. Seismic moments and volume changes. *Journal of Geophysical Research*, **81**(8), 1487–1494.

McGarr, A. 2014. Maximum magnitude earthquakes induced by fluid injection. *Journal of Geophysical Research: Solid Earth*, **119**, 1008–1019.

McGuire, RK. 2004. *Seismic Hazard and Risk Analysis*. Oakland, CA: Earthquake Engineering Research Institute.

McKay, DIA, Staal, A, Abrams, JF, Winkelmann, R, Sakschewski, B, et al. 2022. Exceeding 1.5°C global warming could trigger multiple climate tipping points. *Science*, **377**, eabn7950.

McPhail, C, and Wohlstein, RT. 1983. Individual and collective behaviors within gatherings, demonstrations, and riots. *Annual Review of Sociology*, **9**, 579–600.

Meadows, D, Randers, J, and Meadows, D. 2004. *Limits to Growth, The 30-Year Update*. London, Sterling, VA: Earthscan.

Meadows, DH, Meadows, DL, Randers, J, and Behrens, WW. 1972. *The Limits to Growth*. New York, NY: Universe Books.

Meehl, GA, Boer, GJ, Covey, C, Latif, M, and Stouffer, RJ. 2000. The Coupled Model Intercomparison Project (CMIP). *Bulletin of the American Meteorological Society*, **81**(2), 313–318.

Meehl, GA, Washington, WM, Wigley, TML, Arblaster, JM, and Dai, A. 2003. Solar and greenhouse gas forcing and climate response in the twentieth century. *Journal of Climate*, **16**, 426–444.

Meigh, JR, Farquharson, FAK, and Sutcliffe, JV. 1997. A worldwide comparison of regional flood estimation methods and climate. *Hydrological Sciences Journal*, **42**(2), 225–244.

Melchers, RE. 2001. On the ALARP approach to risk management. *Reliability Engineering and System Safety*, **71**, 201–208.

Memari, M, Attary, N, Masoomi, H, Mahmoud, H, van de Lindt, JW, et al. 2018. Minimal building fragility portfolio for damage assessment of communities subjected to tornadoes. *Journal of Structural Engineering*, **144**(7), 04018072.

Mendis, P, Ngo, T, Haritos, N, Hira, A, Samali, B, and Cheung, J. 2007. Wind loading on tall buildings. *Electronic Journal of Structural Engineering*, **7**(1), 41–54.

Merz, B, Kreibich, H, Schwarze, R, and Thieke, A. 2010. Assessment of economic flood damage. *Natural Hazards and Earth System Sciences*, **10**, 1697–1724.

Merz, B, Blöschl, G, Vorogushyn, S, Dottori, F, Aerts, JCJH, et al. 2021. Causes, impacts and patterns of disastrous river floods. *Nature Reviews Earth & Environment*, **2**, 592–609.

Met Office, University of Reading and University of Exeter. 2023. Extreme wind storms catalogue. Online; accessed 14 February 2023.

Metropolis, N. 1987. The beginning of the Monte Carlo method. *Los Alamos Science*, 125–130.

Meyer, PS, Yung, JW, and Ausubel, JH. 1999. A primer on logistic growth and substitution. *Technological Forecasting and Social Change*, **61**, 247–271.

Meyer, R, and Kunreuther, H. 2017. *The Ostrich Paradox: Why We Underprepare for Disasters*. Philadelphia, PA: Wharton Digital Press.

Microsoft. 2023. Global ML building footprints. Online; accessed 09-March-2023.

Mignan, A. 2011. Retrospective on the Accelerating Seismic Release (ASR) hypothesis: Controversy and new horizons. *Tectonophysics*, **505**, 1–16.

Mignan, A. 2012a. Functional shape of the earthquake frequency–magnitude distribution and completeness magnitude. *Journal of Geophysical Research*, **117**, B08302.

Mignan, A. 2012b. Seismicity precursors to large earthquakes unified in a stress accumulation framework. *Geophysical Research Letters*, **39**, L21308.

Mignan, A. 2014. The debate on the prognostic value of earthquake foreshocks: A meta-analysis. *Scientific Reports*, **4**, 4099.

Mignan, A. 2016a. Revisiting the 1894 Omori aftershock dataset with the stretched exponential function. *Seismological Research Letters*, **87**(3), 685–689.

Mignan, A. 2016b. Static behaviour of induced seismicity. *Nonlinear Processes in Geophysics*, **23**, 107–113.

Mignan, A. 2022a. Categorizing and harmonizing natural, technological, and socio-economic perils following the catastrophe modeling paradigm. *International Journal of Environmental Research and Public Health*, **19**, 12780.

Mignan, A. 2022b. A digital template for the Generic Multi-Risk (GenMR) framework: A virtual natural environment. *International Journal of Environmental Research and Public Health*, **19**, 16097.

Mignan, A, and Broccardo, M. 2020. Neural network applications in earthquake prediction (1994–2019): Meta-analytic and statistical insights on their limitations. *Seismological Research Letters*, **91**(4), 2330–2342.

Mignan, A, and Di Giovambattista, R. 2008. Relationship between accelerating seismicity and quiescence, two precursors to large earthquakes. *Geophysical Research Letters*, **35**, L15306.

Mignan, A, and Wang, Z. 2020. Exploring the space of possibilities in cascading disasters with catastrophe dynamics. *International Journal of Environmental Research and Public Health*, **17**, 7317.

Mignan, A, King, G, Bowman, D, Lacassin, R, and Dmowska, R. 2006. Seismic activity in the Sumatra–Java region prior to the December 26, 2004 (M_w = 9.0–9.3) and March 28, 2005 (M_w = 8.7) earthquakes. *Earth and Planetary Science Letters*, **244**, 639–654.

Mignan, A, Grossi, P, and Muir-Wood, R. 2011. Risk assessment of Tunguska-type airbursts. *Natural Hazards*, **56**, 869–880.

Mignan, A, Wiemer, S, and Giardini, D. 2014. The quantification of low-probability–high-consequences events: Part I. A generic multi-risk approach. *Natural Hazards*, **73**, 1999–2022.

Mignan, A, Landtwing, D, Kästli, P, Mena, B, and Wiemer, S. 2015a. Induced seismicity risk analysis of the 2006 Basel, Switzerland, Enhanced Geothermal System project: Influence of uncertainties on risk mitigation. *Geothermics*, **53**, 133–146.

Mignan, A, Danciu, L, and Giardini, D. 2015b. Reassessment of the maximum fault rupture length of strike-slip earthquakes and inference on M_{max} in the Anatolian peninsula, Turkey. *Seismological Research Letters*, **86**(3), 890–900.

Mignan, A, Scolobig, A, and Sauron, A. 2016. Using reasoned imagination to learn about cascading hazards: A pilot study. *Disaster Prevention and Management*, **25**(3), 329–344.

Mignan, A, Broccardo, M, Wiemer, S, and Giardini, D. 2017. Induced seismicity closed-form traffic light system for actuarial decision-making during deep fluid injections. *Scientific Reports*, **7**, 13607.

Mignan, A, Danciu, L, and Giardini, D. 2018. Considering large earthquake clustering in seismic risk analysis. *Natural Hazards*, **91**, S149–S172.

Mignan, A, Broccardo, M, Wiemer, S, and Giardini, D. 2019a. Autonomous decision-making against induced seismicity in deep fluid injections. Pages 369–376 of: Ferrari, A, and Laloui, L (eds), *Energy Geotechnics, SEG 2018*. Springer Series in Geomechanics and Geoengineering.

Mignan, A, Karvounis, D, Broccardo, M, Wiemer, S, and Giardini, D. 2019b. Including seismic risk mitigation measures into the levelized cost of electricity in enhanced geothermal systems for optimal siting. *Applied Energy*, **238**, 831–850.

Mignan, A, Broccardo, M, and Wang, Z. 2021a. Comprehensive survey of seismic hazard at geothermal sites by a meta-analysis of the underground feedback activation parameter a_{fb}. *Energies*, **14**, 7998.

Mignan, A, Ouillon, G, Sornette, D, and Freund, F. 2021b. Global Earthquake Forecasting System (GEFS): The challenges ahead. *The European Physical Journal Special Topics*, **230**, 473–490.

Mignan, A, Spada, M, Burgherr, P, Wang, Z, and Sornette, D. 2022a. Dynamics of severe accidents in the oil & gas energy sector derived from the authoritative ENergy-related Severe Accident Database. *PLoS ONE*, **17**(2), e0263962.

Mignan, A, Mochel, L, and Ducos, G. 2022b. Involving risk reduction practitioners and other experts in the management of super-catastrophes via an online interactive platform. *Frontiers in Earth Sciences*, **10**, 829145.

Mikellidou, CV, Shakou, LM, Boustras, G, and Dimopoulos, C. 2018. Energy critical infrastructures at risk from climate change: A state of the art review. *Safety Science*, **110**, 110–120.

Millennium Ecosystem Assessment. 2005. *Ecosystems and Human Well-being: Synthesis*. Washington, DC: Island Press.

Mills, CA. 1987. The design of concrete structures to resist explosions and weapon effects. *Proceedings of the 1st International Conference on Concrete for Hazard Protections*. Edinburgh, 27–30 September, pp. 61–73.

Mishra, AK, and Singh, VP. 2010. A review of drought concepts. *Journal of Hydrology*, **391**, 202–216.

Mitchell-Wallace, K. 2017. Applications of catastrophe modelling. Chapter 2, pages 47–186 of: Mitchell-Wallace, K, Jones, M, Hillier, J, and Foote, M (eds), *Natural Catastrophe Risk Management and Modelling, A Practitioner's Guide*. Chichester: Wiley Blackwell.

Mitchell-Wallace, K, Jones, M, Hillier, J, and Foote, M. 2017. *Natural Catastrophe Risk Management and Modelling, A Practitioner's Guide*. Chichester: Wiley Blackwell.

Montoya, JM, Pimm, SL, and Solé, RV. 2006. Ecological networks and their fragility. *Nature*, **442**, 259–264.

Mooney, P, and Corcoran, P. 2014. Has OpenStreetMap a role in Digital Earth applications? *International Journal of Digital Earth*, **7**(7), 534–553.

Morgan, TE. 1994. Plague or poetry? Thucydides on the epidemic at Athens. *Transactions of the American Philological Association*, **124**, 197–209.

Morgan-Wall, T. 2023. rayshader: Create maps and visualize data in 2D and 3D. www.rayshader.com, https://github.com/tylermorganwall/rayshader.

Morrison, D. 2006. Asteroid and comet impacts: The ultimate environmental catastrophe. *Philosophical Transactions of the Royal Sociey A*, **364**, 2041–2054.

Motter, AE, and Lai, Y-C. 2002. Cascade-based attacks on complex networks. *Physical Review E*, **66**, 065102(R).

Mueller, B, and Seneviratne, SI. 2013. Hot days induced by precipitation deficits at the global scale. *PNAS*, **109**(31), 12,398–12,403.

Mueller, K, Hough, SE, and Bilham, R. 2004. Analysing the 1811–1812 New Madrid earthquakes with recent instrumentally recorded aftershocks. *Nature*, **429**, 284–288.

Muir-Wood, R, and Mignan, A. 2009. A phenomenological reconstruction of the M_w9 November 1st 1755 earthquake source. Pages 121–146 of: Mendes-Victor, LA, Oliveira, CS, Azevedo, J, and Ribeiro, A (eds), *The 1755 Lisbon Earthquake: Revisited.* Dordrecht: Springer.

Muir Wood, R, Drayton, M, Berger, A, Burgess, P, and Wright, T. 2005. Catastrophe loss modelling of storm-surge flood risk in eastern England. *Philosophical Transactions of the Royal Society*, **363**, 1407–1422.

Mulargia, F. 2001. Retrospective selection bias (or the benefit of hindsight). *Geophysical Journal International*, **146**, 489–496.

Mumby, PJ, Vitolo, R, and Stephenson, DB. 2011. Temporal clustering of tropical cyclones and its ecosystem impacts. *Proceedings of the National Academy of Sciences*, **108**(43), 17,626–17,630.

Munich Re. 2002. *Winter Storms in Europe (II)*. Munich, Germany: Munich Re Group.

Munich Re. 2017. *Natural Catastrophes 2016. Analyses, Assessments, Positions*. Munich Re Topics Geo.

Munz, P, Hudea, I, Imad, J, and Smith, RJ. 2009. When zombies attack!: Mathematical modelling of an outbreak of zombie infection. Chapter 4, pages 133–150 of: Tchuenche, JM, and Chiyaka, C (eds), *Infectious Disease Modelling Research Progress*. New York, NY: Nova Science Publishers, Inc.

Murakami, D, and Yamagata, Y. 2019. Estimation of gridded population and GDP scenarios with spatially explicit statistical downscaling. *Sustainability*, **11**, 2106.

Muriel-Villegas, JE, Alvarez-Uribe, KC, Patiño-Rodríguez, CE, and Villegas, JG. 2016. Analysis of transportation networks subject to natural hazards – Insights from a Colombian case. *Reliability Engineering and System Safety*, **152**, 151–165.

Murnane, RJ. 2006. Catastrophe risk models for wildfires in the wildland–urban interface: What insurers need. *Natural Hazards Review*, **7**, 150–156.

Murphy, AH. 1993. What is a good forecast? An essay on the nature of goodness in weather forecasting. *Weather and Forecasting*, **8**, 281–293.

Murphy, AH. 1996. The Finley affair: A signal event in the history of forecast verification. *Weather and Forecasting*, **11**(1), 3–20.

Murthy, CS, Laxman, B, and Sai, MVR Sesha. 2015. Geospatial analysis of agricultural drought vulnerability using a composite index based on exposure, sensitivity and adaptive capacity. *International Journal of Disaster Risk Reduction*, **12**, 163–171.

Mutiibwa, D, Strachan, S, and Albright, T. 2015. Land surface temperature and surface air temperature in complex terrain. *IEEE Journal of Selected Topics in Applied Earth Observations and Remote Sensing*, **8**(10), 4762–4774.

Myers, N. 1987. The extinction spasm impending: Synergisms at work. *Conservation Biology*, **1**(1), 14–21.

Mörner, N-A. 2007. The Fenris Wolf in the Nordic Asa creed in the light of palaeoseismics. *Geological Society, London, Special Publications*, **273**, 117–119.

Mühlhofer, E, Koks, EE, Kropf, CM, Sansavini, G, and Bresch, DN. 2023. A generalized natural hazard risk modelling framework for infrastructure failure cascades. *Reliability Engineering and System Safety*, **234**, 109194.

Nadim, F, Kjekstad, O, Peduzzi, P, Herold, C, and Jaedicke, C. 2006. Global landslide and avalanche hotspots. *Landslides*, **3**, 159–173.

Nagaya, K, and Nakayachi, K. 2015. Hazard perception and anchoring: A comparison of the three models explaining the anchoring effect. *Journal of Disaster Research*, **10**(4), 678–686.

Nair, J, Wierman, A, and Zwart, B. 2022. *The Fundamentals of Heavy Tails. Properties, Emergence, and Estimation*. Cambridge: Cambridge University Press.

Nandan, S, Kamer, Y, Ouillon, G, Hiemer, S, and Sornette, D. 2021. Global models for short-term earthquake forecasting and predictive skill assessment. *The European Physical Journal Special Topics*, **230**, 425–449.

Nau, RF. 2001. De Finetti was right: Probability does not exist. *Theory and Decision*, **51**, 89–124.

Naughten, KA, Holland, PR, and De Rydt, J. 2023. Unavoidable future increase in West Antarctic ice-shelf melting over the twenty-first century. *Nature Climate Change*, **13**, 1222–1228.

Nauslar, NJ, Abatzoglou, JT, and Marsh, PT. 2018. The 2017 North Bay and Southern California fires: A case study. *Fire*, **1**, 18.

Neal, M. 2014. Preparing for extraterrestrial contact. *Risk Management*, **16**(2), 63–87.

Needham, H, and Keim, BD. 2011. Storm surge: Physical processes and an impact scale. Chapter 20, pages 385–406 of: Lupo, A (ed), *Recent Hurricane Research*. IntechOpen.

Needham, HF, and Keim, BD. 2012. A storm surge database for the US Gulf Coast. *International Journal of Climatology*, **32**(14), 2108–2123.

Néelz, S, and Pender, G. 2013. *Benchmarking the Latest Generation of 2D Hydraulic Modelling Packages*. Bristol: Environment Agency.

Nemtchinov, IV, Svetsov, VV, Kosarev, IB, Golub', AP, Popova, OP, and Shuvalov, VV. 1997. Assessment of kinetic energy of meteoroids detected by satellite-based light sensors. *Icarus*, **130**, 259–274.

Neto, NN, Madnick, S, de Paula, AMG, and Borges, NM. 2021. Developing a global data breach database and the challenges encountered. *ACM Journal of Data and Information Quality*, **13**(1), 3.

Newhall, C, Self, S, and Robock, A. 2018. Anticipating future Volcanic Explosivity Index (VEI) 7 eruptions and their chilling impacts. *Geosphere*, **14**(2), 572–603.

Newhall, CG, and Self, S. 1982. The Volcanic Explosivity Index (VEI): An estimate of explosive magnitude for historical volcanism. *Journal of Geophysical Research*, **87**(C2), 1231–1238.

Newmark, NM. 1965. Effects of earthquakes on dams and embankments. *Geotechnique*, **15**, 139–159.

Neyman, J, and Pearson, ES. 1933. On the problem of the most efficient tests of statistical hypotheses. *Philosophical Transactions of the Royal Society A*, **231**(694–706), 289–337.

Ngai, KM, Burkle, FM, Jr, Hsu, A, and Hsu, EB. 2009. Human stampedes: A systematic review of historical and peer-reviewed sources. *Disaster Medicine and Public Health Preparedness*, **3**(4), 191–195.

NGDC/WDS. 2022. National Geophysical Data Center / World Data Service Global Historical Tsunami Database. *NOAA National Centers for Environmental Information.*

Nicholson, AJ, and Bailey, VA. 1935. The balance of animal populations. – Part I. *Proceedings of the Zoological Society of London*, **105**(3), 551–598.

Nikolkina, I, and Didenkulova, I. 2011. Rogue waves in 2006–2010. *Natural Hazards and Earth System Sciences*, **11**, 2913–2924.

NOAA (National Oceanic and Atmospheric Administration). 2023. Tsunami propagation. Online; accessed 4 March 2023.

NOAA-NCEI. 2023. OAA National Centers for Environmental Information (NCEI) – US billion-dollar weather and climate disasters. Online; accessed 31 January 2023.

NOAA's Storm Prediction Center. 2023. US tornadoes (1950–2021). Online; accessed 14 January 2023.

Nolan, C (Director). 2014. *Interstellar* [Motion picture]. Paramount Pictures.

Nordhaus, WD. 1973. World dynamics: Measurement without data. *The Economic Journal*, **83**(332), 1156–1183.

Nordhaus, WD. 1992. An optimal transition path for controlling greenhouse gases. *Science*, **258**, 1315–1319.

Norio, O, Ye, T, Kajitani, Y, Shi, P, and Tatano, H. 2011. The 2011 eastern Japan great earthquake disaster: Overview and comments. *International Journal of Disaster Risk Science*, **2**(1), 34–42.

Noy, I. 2009. The macroeconomic consequences of disasters. *Journal of Development Economics*, **88**, 221–231.

Nussio, E, Bove, V, and Steele, B. 2019. The consequences of terrorism on migration attitudes across Europe. *Political Geography*, **75**, 102047.

Ogata, Y. 1988. Statistical models for earthquake occurrences and residual analysis for point processes. *Journal of the American Statistical Society*, **83**, 9–27.

Omori, F. 1894. On after-shocks of earthquakes. *Journal of the College of Science, Imperial University of Tokyo*, **7**, 111–200.

Ottemöller, L, and Evers, LG. 2008. Seismo-acoustic analysis of the Buncefield oil depot explosion in the UK, 2005 December 11. *Geophysical Journal International*, **172**, 1123–1134.

Ouyang, M. 2014. Review on modeling and simulation of interdependent critical infrastructure systems. *Reliability Engineering and System Safety*, **121**, 43–60.

Pack, RT, Tarboton, DG, and Goodwin, CN. 1998. *The SINMAP Approach to Terrain Stability Mapping*. 8th Congress of the International Association of Engineering Geology, Vancouver, BC, 21–25 September 1998, pp. 1157–66.

Pagani, M, Monelli, D, Weatherill, G, Danciu, L, Crowley, H, et al. 2014a. OpenQuake Engine: An open hazard (and risk) software for the Global Earthquake Model. *Seismological Research Letters*, **85**(3), 692–702.

Pagani, M, Monelli, D, Weatherill, GA, and Garcia, J. 2014b. The OpenQuake-engine book: Hazard. *Global Earthquake Model (GEM) Technical Report 2014-08*.

Pagani, M, Garcia-Pelaez, J, Gee, R, Johnson, K, Poggi, V, Silva, et al. 2020. The 2018 version of the Global Earthquake Model: Hazard component. *Earthquake Spectra*, **36**(S1), 226–251.

Pagano, T, and Sorooshian, S. 2002. Hydrologic cycle. Pages 450–464 of: MacCracken, MC, and Perry, JS (eds), *Encyclopedia of Global Environmental Change*. Hoboken, NJ: John Wiley & Sons.

Palmer, WC. 1965. Meteorological drought. *US Department of Commerce, Research Paper No. 45*.

Pan, S-Y, Gao, M, Shah, KJ, Zheng, J, Pei, S-L, and Chiang, P-C. 2019. Establishment of enhanced geothermal energy utilization plans: Barriers and strategies. *Renewable Energy*, **132**, 19–32.

Papathoma-Köhle, M, Zischg, A, Fuchs, S, Glade, T, and Keiler, M. 2015. Loss estimation for landslides in mountain areas – An integrated toolbox for vulnerability assessment and damage documentation. *Environmental Modelling & Software*, **63**, 156–169.

Park, J, Bazzurro, P, and Baker, JW. 2007. Modeling spatial correlation of ground motion intensity measures for regional seismic hazard and portfolio loss estimation. Pages 1–8 of: Kanda, J, Takada, T, and Furuta, H (eds), *Applications of Statistics and Probability in Civil Engineering*. Boca Raton, FL: CRC Press.

Parsons, T. 2005. Significance of stress transfer in time-dependent earthquake probability calculations. *Journal of Geophysical Research*, **110**, B05S02.

Parsons, T, Toda, S, Stein, RS, Barka, A, and Dieterich, JH. 2000. Heightened odds of large earthquakes near Istanbul: An interaction-based probability calculation. *Science*, **288**, 661–665.

Parsons, T, Geist, EL, Console, R, and Carluccio, R. 2018. Characteristic earthquake magnitude frequency distributions on faults calculated from consensus data in California. *Journal of Geophysical Research: Solid Earth*, **123**, 10,761–10,784.

Patterson, DT, Westbrook, JK, Joyce, RJV, Lingren, PD, and Rogasik, J. 1999. Weeds, insects, and diseases. *Climatic Change*, **43**, 711–727.

Paté-Cornell, E. 2012. On 'black swans' and 'perfect storms': Risk analysis and management when statistics are not enough. *Risk Analysis*, **32**(11), 1823–1833.

Paul, BK, and Stimers, M. 2012. Exploring probable reasons for record fatalities: The case of 2011 Joplin, Missouri, tornado. *Natural Hazards*, **64**, 1511–1526.

Pei, H, Cui, P, Yin, J, Zhu, H, Chen, X, et al. 2011. Monitoring and warning of landslides and debris flows using an optical fiber sensor technology. *Journal of Mountain Science*, **8**, 728–738.

Peitgen, H-O, and Richter, PH. 1986. *The Beauty of Fractals. Images of Complex Dynamical Systems*. Berlin, Heidelberg: Springer.

Pelinovsky, E, and Poplavsky, A. 1996. Simplified model of tsunami generation by submarine landslides. *Physics and Chemistry of the Earth*, **21**(12), 13–17.

Perez, C. 2010. Technological revolutions and techno-economic paradigms. *Cambridge Journal of Economics*, **34**, 185–202.

Perkins, SE, and Alexander, LV. 2013. On the measurement of heat waves. *Journal of Climate*, **26**, 4500–4517.

Pernin, P. 1971. The Great Peshtigo Fire: An eyewitness account. *The Wisconsin Magazine of History*, **54**(4), 246–272.

Perrow, C. 2001. Accidents, normal. Pages 33–38 of: Smelser, NJ, and Bates, PB (eds), *International Encyclopedia of the Social & Behavioral Sciences*. Amsterdam: Elsevier.

Pescaroli, G, and Alexander, D. 2018. Understanding compound, interconnected, interacting, and cascading risks: A holistic framework. *Risk Analysis*, **38**(11), 2245–2257.

Petley, D. 2012. Global patterns of loss of life from landslides. *Geology*, **40**(10), 927–930.

Petley, DN, Dunning, SA, and Rosser, NJ. 2005. The analysis of global landslide risk through the creation of a database of worldwide landslide fatalities. Pages 367–374 of: Hungr, O, Fell, R, Counture, R, and Ebergardt, E (eds), *Landslide Risk Management*. Boca Raton, FL: CRC Press.

Petrova, EG, and Krausmann, E. 2011. From natural hazards to technological disasters. *Natural Hazards and Earth System Sciences*, **11**, 3063–3065.

Pettersson, T, and Wallensteen, P. 2015. Armed conflicts, 1946–2014. *Journal of Peace Research*, **52**(4), 536–550.

Philips, M, and Markham, Á. 2014. Ebola: A failure of international collective action. *The Lancet*, **384**, 1181.

Phua, J, Weng, L, Ling, L, Egi, M, Lim, C-M, et al. 2020. Intensive care management of coronavirus disease 2019 (COVID-19): Challenges and recommendations. *Lancet Respiratory Medicine*, **8**, 506–517.

Piccardi, L, and Masse, WB. 2007. *Myth and Geology*. London: Geological Society.

Picou, JS. 2009. Katrina as a natech disaster: Toxic contamination and long-term risks for residents of New Orleans. *Journal of Applied Social Science*, **3**(2), 39–55.

Pisaniello, JD, Tingey-Holyoak, JL, Wishart, MJ, Lyon, KN, and García, EB. 2023. Funding dam safety regulation: An international comparative analysis and example application in Australia. *International Journal of Water Resources Development*, **39**(4), 637–662.

Pitilakis, K, Crowley, H, and Kaynia, AM. 2014. *SYNER-G: Typology Definition and Fragility Functions for Physical Elements at Seismic Risk*. Dordrecht: Springer.

Pitt, MA, Myung, IJ, and Zhang, S. 2002. Toward a method of selecting among computational models of cognition. *Psychological Review*, **109**(3), 472–491.

Poe, EA. 1839. The Conversation of Eiros and Charmion. *Burton's Gentleman's Magazine*.

Polacek, A. 2018. Catastrophe bonds: A primer and retrospective. *Chicago Fed Letter*, No. 405. doi: 10.21033/cfl-2018-405.

Popova, OP, Jenniskens, P, Emel'yanenko, V, Kartashova, A, Biryukov, E, et al. 2013. Chelyabinsk airburst, damage assessment, meteorite recovery, and characterization. *Science*, **342**, 1069–1073.

Porter, K. 2021. A beginner's guide to fragility, vulnerability, and risk. In: Beer, M, Kougioumtzoglou, IA, Patelli, E, and Au, IS-K (eds), *Encyclopedia of Earthquake Engineering*. Berlin, Heidelberg: Springer.

Porter, K, Kennedy, R, and Bachman, R. 2007. Creating fragility functions for performance-based earthquake engineering. *Earthquake Spectra*, **23**(2), 471–489.

Powell, MD, and Reinhold, TA. 2007. Tropical cyclone destructive potential by Integrated Kinetic Energy. *Bulletin of the American Meteorological Society*, **88**(4), 513–526.

Powers, JG, Klemp, JB, Skamarock, WC, Davis, CA, Dudhia, J, et al. 2017. The weather research and forecasting model, overview, system efforts, and future directions. *American Meteorological Society*, **98**(8), 1717–1737.

Press, WH, and Teukolsky, SA. 1992. Adaptive stepsize Runge–Kutta integration. *Computers in Physics and IEEE Computational Science & Engineering*, **6**, 188–191.

Przybylinski, RW. 1995. The bow echo: Observations, numerical simulations, and severe weather detection methods. *Weather and Forecasting*, **10**(2), 203–218.

Punge, HJ, Bedka, KM, Kunz, M, and Werner, A. 2014. A new physically based stochastic event catalog for hail in Europe. *Natural Hazards*, **73**, 1625–1645.

Pyle, DM. 1989. The thickness, volume and grainsize of tephra fall deposits. *Bulletin of Volcanology*, **51**, 1–15.

Pyle, DM. 1995. Mass and energy budgets of explosive volcanic eruptions. *Geophysical Research Letters*, **22**(5), 563–566.

Qina, R, Khakzad, N, and Zhu, J. 2020. An overview of the impact of Hurricane Harvey on chemical and process facilities in Texas. *International Journal of Disaster Risk Reduction*, **45**, 101453.

Qiu, J. 2020. Pandemic risk: Impact, modeling, and transfer. *Risk Management and Insurance Review*, **23**(4), 293–304.

Rahmandad, H, and Sterman, J. 2008. Heterogeneity and network structure in the dynamics of diffusion: Comparing agent-based and differential equation models. *Management Science*, **54**(5), 998–1014.

Ramirez, JA, Lichter, M, Coulthard, TJ, and Skinner, C. 2016. Hyper-resolution mapping of regional storm surge and tide flooding: Comparison of static and dynamic models. *Natural Hazards*, **82**, 571–590.

Ramiro, JMS, and Aísa, PAB. 1998. *Risk Analysis and Reduction in the Chemical Process Industry*. Dordrecht: Springer.

Ramsdell, JV, Jr, and Rishel, JP. 2007. *Tornado Climatology of the Contiguous United States*. Richland, WA: Pacific Northwest National Laboratory.

Randall, DA, Wood, RA, Bony, S, Colman, R, Fichefet, T, et al. 2007. Climate models and their evaluation. Chapter 8, pages 589–662 of: Solomon, S, Qin, D, Manning, M, Chen, Z, Marquis, M, et al. (eds), *Climate Change 2007: The Physical Science Basis. Contribution of Working Group I to the Fourth Assessment Report of the Intergovernmental Panel on Climate Change*. Cambridge: Cambridge University Press.

Randers, J, Golüke, U, Wenstøp, F, and Wenstøp, S. 2016. A user-friendly earth system model of low complexity: The ESCIMO system dynamics model of global warming towards 2100. *Earth System Dynamics*, **7**, 831–850.

Rappaport, EN. 2014. Fatalities in the United States from Atlantic tropical cyclones: New data and interpretation. *Bulletin of the American Meteorological Society*, **95**(3), 341–346.

Rasche, L, and Taylor, RAJ. 2019. EPIC-GILSYM: Modelling crop-pest insect interactions and management with a novel coupled crop-insect model. *Journal of Applied Ecology*, **56**, 2045–2056.

Rawtani, D, Gupta, G, Khatri, N, Rao, PK, and Hussain, CM. 2022. Environmental damages due to war in Ukraine: A perspective. *Science of the Total Environment*, **850**, 157932.

Reasenberg, PA, and Jones, LM. 1989. Earthquake hazard after a mainshock in California. *Science*, **243**(4895), 1173–1176.

Reasenberg, PA, and Jones, LM. 1994. Earthquake aftershocks: Update. *Science*, **265**(5176), 1251–1252.

Reed, BC. 2018. A toy model for the yield of a tamped fission bomb. *American Journal of Physics*, **86**(2), 105–109.

Reese, S, Bradley, BA, Bind, J, Smart, G, Power, W, and Sturman, J. 2011. Empirical building fragilities from observed damage in the 2009 South Pacific tsunami. *Earth-Science Reviews*, **107**, 156–173.

Reichert, A, and Shyu, Y-W. 2003. Derivative activities and the risk of international banks: A market index and VaR approach. *International Review of Financial Analysis*, **12**, 489–511.

Reiter, D. 1999. Military strategy and the outbreak of international conflict: Quantitative empirical tests, 1903–1992. *The Journal of Conflict Resolution*, **43**(3), 366–387.

Reitman, NG, Briggs, RW, Barnhart, WD, Jobe, JAT, DuRoss, CB, et al. 2023. Fault rupture mapping of the 6 February 2023 Kahramanmaraş, Türkiye, earthquake sequence from satellite data: US Geological Survey data release. Online; accessed 30 November 2023.

Rengers, FK, McGuire, LA, Oakley, NS, Kean, JW, Staley, DM, and Tang, H. 2020. Landslides after wildfire: Initiation, magnitude, and mobility. *Landslides*, **17**, 2631–2641.

Renn, O. 2008. *Risk Governance, Coping with Uncertainty in a Complex World.* London, Sterling, VA: Earthscan.

Rentschler, J, Avner, P, Marconcini, M, Su, R, Strano, E, et al. 2023. Global evidence of rapid urban growth in flood zones since 1985. *Nature*, **622**, 87–95.

Reynolds, M, Kropff, M, Crossa, J, Koo, J, Kruseman, G, et al. 2018. Role of modelling in international crop research: Overview and some case studies. *Agronomy*, **8**, 291.

Rhodes, RAW. 2007. Understanding governance: Ten years on. *Organization Studies*, **28**(8), 1243–1264.

Richardson, LF. 1948. Variation of the frequency of fatal quarrels with magnitude. *Journal of the American Statistical Association*, **43**, 523–546.

Rigby, SE, Lodge, TJ, Alotaibi, S, Barr, AD, Clarke, SD, et al. 2020. Preliminary yield estimation of the 2020 Beirut explosion using video footage from social media. *Shock Waves*, **30**, 671–675.

Rinaldi, SM, Peerenboom, JP, and Kelly, TK. 2001. Identifying, understanding, and analyzing critical infrastructure interdependencies. *IEEE Control Systems Magazine*, **21**(6), 11–25.

Risk Management Solutions. 2007. *Analytical Methods for Modelling Pandemic Flu*. Online; accessed 12 March 2022.

Ritchie, H, Spooner, F, and Roser, M. 2018. Causes of death. Our World in Data. https://ourworldindata.org/causes-of-death.

Ritchie, PDL, Clarke, JJ, Cox, PM, and Huntingford, C. 2021. Overshooting tipping point thresholds in a changing climate. *Nature*, **592**, 517–523.

Roberts, DC, and Turcotte, DL. 1998. Fractality and self-organized criticality of wars. *Fractals*, **6**(4), 351–357.

Roberts, JF, Champion, AJ, Dawkins, LC, Hodges, KI, Shaffrey, LC, et al. 2014. The XWS open access catalogue of extreme European windstorms from 1979 to 2012. *Natural Hazards and Earth System Sciences*, **14**, 2487–2501.

Robinson, TR. 1849. Description of an improved anemometer for registering the direction of the wind, and the space which it traverses in given intervals of time. *The Transactions of the Royal Irish Academy*, **22**, 155–178.

Röbke, BR, and Vött, A. 2017. The tsunami phenomenon. *Progress in Oceanography*, **159**, 296–322.

Rocha, JC, Peterson, G, Bodin, O, and Levin, S. 2018. Cascading regime shifts within and across scales. *Science*, **362**, 1379–1383.

Romanic, D, Refan, M, Wu, C-H, and Michel, G. 2016. Oklahoma tornado risk and variability: A statistical model. *International Journal of Disaster Risk Reduction*, **16**, 19–32.

Romero, JJ. 2012. Blackouts illuminate India's power problems. *IEEE Spectrum*, **49**(10), 11–12.

Romão, X, Pereira, N, Castro, JM, Crowley, H, Silva, V, et al. 2021. *European Building Vulnerability Data Repository*. Online; accessed 15 March 2023.

Rose, A, and Huyck, CK. 2016. Improving catastrophe modeling for business interruption insurance needs. *Risk Analysis*, **36**(10), 1896–1915.

Rose, AZ, and Blomberg, SB. 2010. Total economic consequences of terrorist attacks: Insights from 9/11. *Peace Economics, Peace Science and Public Policy*, **16**(1).

Roser, M, Appel, C, and Ritchie, H. 2013. Human height. Our World in Data. https://ourworldindata.org/human-height.

Rosier, SHR, Reese, R, Donges, JF, De Rydt, J, Gudmundsson, GH, and Winkelmann, R. 2021. The tipping points and early warning indicators for Pine Island Glacier, West Antarctica. *The Cryosphere*, **15**, 1501–1516.

Rosser, JB, Jr. 2007. The rise and fall of catastrophe theory applications in economics: Was the baby thrown out with the bathwater? *Journal of Economic Dynamics & Control*, **31**, 3255–3280.

Rothermel, RC. 1972. *A Mathematical Model for Predicting Fire Spread in Wildland Fuels*. Ogden, UT: Intermountain Forest and Range Experiment Station, Forest Service, US Department of Agriculture.

Roussi, A. 2020. The battle to contain gigantic locust swarms. *Nature*, **579**, 330.

Royal, A, and Walls, M. 2019. Flood risk perceptions and insurance choice: Do decisions in the floodplain reflect overoptimism? *Risk Analysis*, **39**(5), 957–1191.

Rumpf, J, Weindl, H, Höppe, P, Rauch, E, and Schmidt, V. 2007. Stochastic modelling of tropical cyclone tracks. *Mathematical Methods of Operations Research*, **66**, 475–490.

Russell, LR. 1971. Probability distributions for hurricane effects. *Journal of the Waterways, Harbors and Coastal Engineering Division*, **97**(1), 139–154.

Rutty, GN, Cary, N, and Lawler, W. 2017. Death in crowds. Chapter 3, pages 43–57 of: Rutty, Guy N. (ed), *Essentials of Autopsy Practice*. London: Springer.

Sachtjen, ML, Carreras, BA, and Lynch, VE. 2000. Disturbances in a power transmission system. *Physical Review E*, **61**(5), 4877–4882.

Sadek, S, Dabaghi, M, O'Donnell, TM, Zimmaro, P, Hashash, YMA, and Stewart, JP. 2022. Impacts of 2020 Beirut explosion on port infrastructure and nearby buildings. *Natural Hazards Review*, **23**(2), 04022008.

Sakoda, JM. 1971. The checkerboard model of social interaction. *The Journal of Mathematical Sociology*, **1**(1), 119–132.

Salzano, E, and Cozzani, V. 2005. The analysis of domino accidents triggered by vapor cloud explosions. *Reliability Engineering and System Safety*, **90**, 271–284.

Samuelson, PA. 1977. St. Petersburg paradoxes: Defanged, dissected, and historically described. *Journal of Economic Literature*, **15**(1), 24–55.

Sánchez, JL, Fraile, R, de la Madrid, JL, de la Fuente, MT, Rodríguez, P, and Castro, A. 1996. Crop damage: The hail size factor. *Journal of Applied Meteorology and Climatology*, **35**(9), 1535–1541.

Sargiacomo, M, Servalli, S, Potito, S, D'Andreamatteo, A, and Gitto, A. 2021. Accounting for natural disasters from a historical perspective: A literature review and research agenda. *Accounting History*, **26**(2), 179–204.

Sarkees, MR, and Schafer, P. 2000. The correlates of war data on war: An update to 1997. *Conflict Management and Peace Science*, **18**(1), 123–144.

Sauer, TC, Brown, JS, Boehm, PD, Aurand, DV, Michel, J, and Hayes, MO. 1993. Hydrocarbon source identification and weathering characterization of intertidal and subtidal sediments along the Saudi Arabian coast after the Gulf War oil spill. *Marine Pollution Bulletin*, **27**, 117–134.

Savage, LJ. 1954. *The Foundations of Statistics*. John Wiley & Sons, Inc.

Scawthorn, C, O'Rourke, TD, and Blackburn, FT. 2019. The 1906 San Francisco earthquake and fire—Enduring lessons for fire protection and water supply. *Earthquake Spectra*, **22**(S2), S135–S158.

Schaefer, JT, Kelly, DL, and Abbey, RF. 1986. A minimum assumption tornado-hazard probability model. *Journal of Climate and Applied Meteorology*, **25**, 1934–1945.

Schelling, TC. 1971. Dynamic models of segregation. *Journal of Mathematical Sociology*, **1**, 143–186.

Schiavina, M, Melchiorri, M, Pesaresi, M, Politis, P, Freire, S, et al. 2022. *GHSL Data Package 2022*. Luxembourg: Publications Office of the European Union, JRC 129516.

Schiesser, HH. 1990. Hailfall: The relationship between radar measurements and crop damage. *Atmospheric Research*, **25**, 559–582.

Schiff, JL. 2007. *Cellular Automata, A Discrete View of the World*. Hoboken, NJ: John Wiley & Sons, Inc.

Schmidt, J, Matcham, I, Reese, S, King, A, Bell, R, et al. 2011. Quantitative multi-risk analysis for natural hazards: a framework for multi-risk modelling. *Natural Hazards*, **58**, 1169–1192.

Schneider, PJ, and Schauer, BA. 2006. HAZUS—Its development and its future. *Natural Hazards Review*, **7**(2), 40–44.

Scholz, CH, Sykes, LR, and Aggarwal, YP. 1973. Earthquake prediction: A physical basis. *Science*, **181**(4102), 803–810.

Schorlemmer, D, Zechar, JD, Werner, MJ, Field, EH, Jackson, DD, Jordan, TH, and The REALM Working Group. 2010. First results of the regional earthquake likelihood models experiment. *Pure and Applied Geophysics*, **167**, 859–876.

Schorlemmer, D, Werner, MJ, Marzocchi, W, Jordan, TH, Ogata, Y, et al. 2018. The collaboratory for the study of earthquake predictability: Achievements and priorities. *Seismological Research Letters*, **89**(4), 1305–1313.

Schwarz, J, and Maiwald, H. 2008. Damage and loss prediction model based on the vulnerability of building types. Page 74 of: *4th International Symposium on Flood Defence: Managing Flood Risk, Reliability and Vulnerability*. Toronto, ON, 6–8 May 2008: Institute of Catastrophic Loss Reduction.

Scolobig, A, Komendantova, N, and Mignan, A. 2017. Mainstreaming multi-risk approaches into policy. *Geosciences*, **7**, 129.

Scrimshaw, NS. 1987. The phenomenon of famine. *Annual Review of Nutrition*, **7**, 1–21.

Sealy, KS, and Strobl, E. 2017. A hurricane loss risk assessment of coastal properties in the Caribbean: Evidence from the Bahamas. *Ocean & Coastal Management*, **149**, 42–51.

Self, S. 2006. The effects and consequences of very large explosive volcanic eruptions. *Philosophical Transactions of the Royal Society A*, **364**, 2073–2097.

Sellitto, P, Podglajen, A, Belhadji, R, Boichu, M, Carboni, E, et al. 2022. The unexpected radiative impact of the Hunga Tonga eruption of 15th January 2022. *Communications Earth & Environment*, **3**, 288.

Sen, R, and Borle, S. 2015. Estimating the contextual risk of data breach: An empirical approach. *Journal of Management Information Systems*, **32**(2), 314–341.

Seneviratne, SI, Corti, T, Davin, EL, Hirschi, M, Jaeger, EB, et al. 2010. Investigating soil moisture–climate interactions in a changing climate: A review. *Earth-Science Reviews*, **99**, 125–161.

Seneviratne, SI, Nicholls, N, Easterling, D, Goodess, CM, Kanae, S, et al. 2012. Changes in climate extremes and their impacts on the natural physical environment. Chapter 3, pages 109–230 of: Field, CB, Barros, V, Stocker, TF, Qin, D, Dokken, DJ, et al. (eds), *Managing the Risks of Extreme Events and Disasters to Advance Climate Change Adaptation*. Cambridge: Cambridge University Press.

Sharma, A, McCloskey, B, Hui, DS, Rambia, A, Zumla, A, et al. 2023. Global mass gathering events and deaths due to crowd surge, stampedes, crush and physical injuries – Lessons from the Seoul Halloween and other disasters. *Travel Medicine and Infectious Disease*, **52**, 102524.

Shelley, M. 1826. *The Last Man*. London: Henry Colburn.

Sherwood, SC, and Huber, M. 2010. An adaptability limit to climate change due to heat stress. *Proceedings of the National Academy of Sciences*, **107**(21), 9552–9555.

Sheu, J-B, and Kuo, H-T. 2020. Dual speculative hoarding: A wholesaler–retailer channel behavioral phenomenon behind potential natural hazard threats. *International Journal of Disaster Risk Reduction*, **44**, 101430.

Shoemaker, EM. 1983. Asteroid and comet bombardment of the Earth. *Annual Review of Earth and Planetary Sciences*, **11**, 461–494.

Shultz, JM, Walsh, L, Garfin, DR, Wilson, FE, and Neria, Y. 2014. The 2010 Deepwater Horizon oil spill: The trauma signature of an ecological disaster. *The Journal of Behavioral Health Services & Research*, **42**(1), 58–76.

Sills, GL, Vroman, ND, Wahl, RE, and Schwanz, NT. 2008. Overview of New Orleans levee failures: Lessons learned and their impact on national levee design and assessment. *Journal of Geotechnical and Geoenvironmental Engineering*, **134**(5), 556–565.

Silva, V, Henshaw, P, Huyck, C, and O'Hara, M. 2018. *GED4ALL. Global Exposure Database for Multi-Hazard Risk Analysis*. Pavia: D5 - Final Report. GEM Technical Report 2018- 05, GEM Foundation.

Simiu, E, and Heckert, NA. 1996. Extreme wind distribution tails: A 'peaks over threshold' approach. *Journal of Structural Engineering*, **122**, 539–547.

Simmons, KM, Sutter, D, and Pielke, R. 2013. Normalized tornado damage in the United States: 1950–2011. *Environmental Hazards*, **12**(2), 132–147.

Simon, SI. 1981. The dilemma of war and military exclusion clauses in insurance contracts. *American Business Law Journal*, **19**(1), 31–45.

Simpson, NP, Mach, KJ, Constable, A, Hess, J, Hogarth, R, et al. 2021. A framework for complex climate change risk assessment. *One Earth*, **4**, 489–501.

Simpson, RH. 1974. The hurricane disaster–potential scale. *Weatherwise*, **27**, 169–186.

Smil, V. 2008. *Global Catastrophes and Trends, The Next Fifty Years*. Cambridge, MA, London, UK: The MIT Press.

Smith, D. 2017. Pandemic risk modelling. Pages 463–495 of: Pompella, M, and Scordis, NA (eds), *The Palgrave Handbook of Unconventional Risk Transfer*. London: Palgrave Macmillan.

Smith, JA. 1987. Regional flood frequency analysis using extreme order statistics of the annual peak record. *Water Resources Research*, **23**, 1657–1666.

Smolka, A. 2006. Natural disasters and the challenge of extreme events: Risk management from an insurance perspective. *Philosophical Transactions of the Royal Society*, **364**, 2147–2165.

Sobolev, GA. 2007. On applicability of the RTL prognostic algorithms and energy estimation to Sakhalin seismicity. *Journal of Volcanology and Seismology*, **1**(3), 64–78.

Soffiantini, G. 2020. Food insecurity and political instability during the Arab Spring. *Global Food Security*, **26**, 100400.

Somerville, PG, Smith, NF, Graves, RW, and Abrahamson, NA. 1997. Modification of empirical strong ground motion attenuation relations to include the amplitude and duration effects of rupture directivity. *Seismological Research Letters*, **68**, 199–222.

Song, SG, Beroza, GC, and Segall, P. 2008. A unified source model for the 1906 San Francisco earthquake. *Bulletin of the Seismological Society of America*, **98**(2), 823–831.

Sornette, D. 2006. *Critical Phenomena in Natural Sciences*. Berlin and Heidelberg: Springer.

Sornette, D, Malevergne, Y, and Muzy, J-F. 2004. Volatility fingerprints of large shocks: endogenous versus exogenous. Pages 91–102 of: Takayasu, H (ed), *The Application of Econophysics*. Tokyo: Springer.

Spence, RJS, Kelman, I, Baxter, PJ, Zuccaro, G, and Petrazzuoli, S. 2005. Residential building and occupant vulnerability to tephra fall. *Natural Hazards and Earth System Science*, **5**(4), 477–494.

Spencer Jones, H. 1947. Edmond Halley and his times. *Popular Astronomy*, **55**, 2–17.

Standohar-Alfano, CD, and van de Lindt, JW. 2015. Empirically based probabilistic tornado hazard analysis of the United States using 1973–2011 data. *Natural Hazards Review*, **16**(1), 04014013.

Starr, C. 1980. Introductory remarks. Pages 2–4 of: Schwing, RC, and WA Albers, Jr (eds), *Societal Risk Assessment, How Safe is Safe Enough?* New York, NY: Springer Science+Business Media, LLC.

Starr, JR. 1991. Water wars. *Foreign Policy*, **Spring**(82), 17–36.

Stathopoulos, T. 2003. Wind loads on low buildings: In the wake of Alan Davenport's contributions. *Journal of Wind Engineering and Industrial Aerodynamics*, **91**, 1565–1585.

Stearns, PN. 2008. Texas and Virginia: A bloodied window into changes in American public life. *Journal of Social History*, **42**(2), 299–318.

Steffen, W, Rockström, J, Richardson, K, Lenton, TM, Folke, C, et al. 2018. Trajectories of the Earth system in the Anthropocene. *Proceedings of the National Academy of Sciences*, **115**(33), 8252–8259.

Steinbeck, J. 1939. *The Grapes of Wrath*. The Viking Press-James Lloyd.

Steinbock, D. 2023. *What Led to the Gaza–Israel Catastrophe? The Nightmare after 50 Years of Failed Military Policies*. Online; accessed 11 November 2023.

Steinhauser, G, Brandl, A, and Johnson, TE. 2014. Comparison of the Chernobyl and Fukushima nuclear accidents: A review of the environmental impacts. *Science of the Total Environment*, **470–471**, 800–817.

Stewart, MG, Netherton, MD, and Rosowsky, DV. 2006. Terrorism risks and blast damage to built infrastructure. *Natural Hazards Review*, **7**(3), 114–122.

Stirling, M, Goded, T, Berryman, K, and Litchfield, N. 2013. Selection of earthquake scaling relationships for seismic-hazard analysis. *Bulletin of the Seismological Society of America*, **103**(6), 2993–3011.

Stockton, FR. 1882. *The Lady, or the Tiger?* New York City: The Century Magazine.

Storchak, DA, Di Giacomo, D, Bondar, I, Engdahl, ER, Harris, J, et al. 2013. Public release of the ISC-GEM Global Instrumental Earthquake Catalogue (1900–2009). *Seismological Research Letters*, **84**, 810–815.

Storchak, DA, Di Giacomo, D, Engdahl, ER, Harris, J, Bondár, I, et al. 2015. The ISC-GEM Global Instrumental Earthquake Catalogue (1900–2009): Introduction. *Physics of the Earth and Planetary Interiors*, **239**, 48–63.

Strogatz, SH. 2018. *Nonlinear Dynamics and Chaos. With Applications to Physics, Biology, Chemistry, and Engineering*. Boca Raton, FL: Taylor & Francis.

Stull, R. 2011. Wet-bulb temperature from relative humidity and air temperature. *Journal of Applied Meteorology and Climatology*, **50**, 2267–2269.

Styron, R, and Pagani, M. 2020. The GEM Global Active Faults Database. *Earthquake Spectra*, **36**, 160–180.

Sullivan, AL. 2009. Wildland surface fire spread modelling, 1990–2007. 1: Physical and quasi-physical models. *International Journal of Wildland Fire*, **18**, 349–368.

Sun, W, Bocchini, P, and Davison, BD. 2020. Applications of artificial intelligence for disaster management. *Natural Hazards*, **103**, 2631–2689.

Swiss Re. 2013. *A History of Insurance*. Zurich: Swiss Reinsurance Company Ltd.

Swiss Re Institute. 2020. *Sigma No 2/2020. Natural Catastrophes in Times of Economic Accumulation and Climate Change*. Zurich: Swiss Re.

Swiss Re Institute. 2021. *Sigma No 1/2021. Natural Catastrophes in 2020: Secondary Perils in the Spotlight, but Don't Forget Primary-Peril Risks*. Zurich: Swiss Re.

Swiss Re Institute. 2022. *Sigma No 1/2022. Natural Catastrophes in 2021: The Floodgates Are Open*. Zurich: Swiss Re.

Swiss Re Institute. 2023. *Sigma No 1/2023. Natural Catastrophes and Inflation in 2022: A Perfect Storm*. Zurich: Swiss Re.

SwissRe sigma. 2012. *Natural Catastrophes and Man-made Disasters in 2011*. Zurich: Swiss Re.

Syphard, AD, and Keeley, JE. 2015. Location, timing and extent of wildfire vary by cause of ignition. *International Journal of Wildland Fire*, **24**, 37–47.

Szewranski, S, Chruscinski, J, Kazak, J, Swiader, M, Tokarczyk-Dorociak, K, and Zmuda, R. 2018. Pluvial Flood Risk Assessment tool (PFRA) for rainwater management and adaptation to climate change in newly urbanised areas. *Water*, **10**, 386.

Szinicz, L. 2005. History of chemical and biological warfare agents. *Toxicology*, **214**, 167–181.

Tainaka, K-I. 1994. Intrinsic uncertainty in ecological catastrophe. *Journal of Theoretical Biology*, **166**, 91–99.

Tainter, JA. 1988. *Collapse of Complex Societies*. Cambridge: Cambridge University Press.

Talbott, EO, Youk, AO, McHugh-Pemu, KP, and Zborowski, JV. 2003. Long-term follow-up of the residents of the Three Mile Island accident area: 1979–1998. *Environmental Health Perspectives*, **111**(3), 341–348.

Taleb, NN. 2007. *The Black Swan: The Impact of the Highly Improbable*. New York City, NY: Random House.

Tang, MJ, and Baker, QA. 1999. A new set of blast curves from vapor cloud explosion. *Process Safety Progress*, **18**(4), 235–240.

Tang, P, Xia, Q, and Wang, Y. 2019. Addressing cascading effects of earthquakes in urban areas from network perspective to improve disaster mitigation. *International Journal of Disaster Risk Reduction*, **35**, 101065.

Tarbotton, C, Dall'Osso, F, Dominey-Howes, D, and Goff, J. 2015. The use of empirical vulnerability functions to assess the response of buildings to tsunami impact: Comparative review and summary of best practice. *Earth-Science Reviews*, **142**, 120–134.

Taszarek, M, Allen, JT, Groenemeijer, P, Edwards, R, Brooks, HE, Chmielewski, V, and Enno, S-E. 2020. Severe convective storms across Europe and the United States. Part I: Climatology of lightning, large hail, severe wind, and tornadoes. *Journal of Climate*, **33**(23), 10,239–10,261.

Taylor, KE, Stouffer, RJ, and Meehl, GA. 2012. An overview of CMIP5 and the experiment design. *Bulletin of the American Meteorological Society*, **93**(4), 485–498.

Taylor, P, and Carter, J. 2020. *Oasis Financial Module*. Online; accessed 28 August 2023.

Thom, R. 1975. *Structural Stability and Morphogenesis, An Outline of a General Theory of Models*. Reading, MA: WA Benjamin, Inc.

Thompson, FML. 1968. The Second Agricultural Revolution, 1815–1880. *The Economic History Review*, **21**(1), 62–77.

Thornton, HG. 1922. On the development of a standardised agar medium for counting soil bacteria, with especial regard to the repression of spreading colonies. *Annals of Applied Biology*, **9**(3–4), 241–274.

Thurner, S, Hanel, R, and Klimek, P. 2018. *Introduction to the Theory of Complex Systems*. Oxford: Oxford University Press.

Tilling, RI, and Lipman, PW. 1993. Lessons in reducing volcano risk. *Nature*, **364**, 277–280.

Tin, D, Hertelendy, AJ, Hart, A, and Ciottone, GR. 2021. 50 years of mass-fatality terrorist attacks: A retrospective study of target demographics, modalities, and injury patterns to better inform future counter-terrorism medicine preparedness and response. *Prehospital and Disaster Medicine*, **36**(5), 531–535.

Toda, S, Stein, RS, Sevilgen, V, and Lin, J. 2011. *Coulomb 3.3 Graphic-Rich Deformation and Stress-Change Software for Earthquake, Tectonic, and Volcano Research and Teaching – User Guide*. US Geological Survey open-file report 2011-1060.

Tomlinson, CJ, Chapman, L, Thornes, JE, and Baker, CJ. 2011. Including the urban heat island in spatial heat health risk assessment strategies: A case study for Birmingham, UK. *International Journal of Health Geographics*, **10**, 42.

Tonnang, HEZ, Nedorezov, LV, Ochanda, H, Owino, J, and Löhr, B. 2009. Assessing the impact of biological contorl of *Plutella xylostella* through the application of Lotka–Volterra model. *Ecological Modelling*, **220**, 60–70.

Tonnang, HEZ, Hervé, BDB, Biber-Freudenberger, L, Salifu, D, Subramanian, S, et al. 2017. Advances in crop insect modelling methods – Towards a whole system approach. *Ecological Modelling*, **354**, 88–103.

Toreti, A. 2014. Gridded Agro-Meteorological Data in Europe. European Commission, Joint Research Centre (JRC) [Dataset]. Online; accessed 5 March 2023.

Totschnig, R, Sedlacek, W, and Fuchs, S. 2011. A quantitative vulnerability function for fluvial sediment transport. *Natural Hazards*, **58**, 681–703.

Touzé-Peiffer, L, Barberousse, A, and Le Treut, H. 2020. The Coupled Model Intercomparison Project: History, uses, and structural effects on climate research. *WIREs Climate Change*, **11**, e648.

Trautman, LJ, and Ormerod, PC. 2017. Corporate directors' and officers' cybersecurity standard of care: The Yahoo data breach. *American University Law Review*, **66**(5), 1231–1291.

Trenberth, KE. 2011. Attribution of climate variations and trends to human influences and natural variability. *WIREs Climate Change*, **2**, 925–930.

Trifunac, MD, and Todorovska, MI. 2002. A note on differences in tsunami source parameters for submarine slides and earthquakes. *Soil Dynamics and Earthquake Engineering*, **22**, 143–155.

Trump, DJ. 1987. *The Art of the Deal*. New York City, NY: Random House.

Trunfio, GA. 2004. Predicting wildfire spreading through a hexagonal cellular automata model. Pages 385–394 of: Sloot, PMA, Chopard, B, and Hoekstra, AG (eds), *ACRI 2004, LNCS 3305*. Heidelberg: Springer.

Trunfio, GA, D'Ambrosio, D, Rongo, R, Spataro, W, and Di Gregorio, S. 2011. A new algorithm for simulating wildfire spread through cellular automata. *ACM Transactions on Modeling and Computer Simulation*, **22**(1), 6:1–6:26.

Turchin, P. 2008. Arise 'cliodynamics'. *Nature*, **454**, 34–35.

Turco, M, Jerez, S, Augusto, S, Tarín-Carrasco, P, Ratola, N, et al. 2019. Climate drivers of the 2017 devastating fires in Portugal. *Scientific Reports*, **9**, 13886.

Turner, GM. 2008. A comparison of *The Limits to Growth* with 30 years of reality. *Global Environmental Change*, **18**, 397–411.

Turrini, E, and Ghosh, S. 2010. A pragmatic, experiential definition of computer crimes. Chapter 1, pages 3–23 of: Ghosh, S, and Turrini, E (eds), *Cybercrimes: A Multidisciplinary Analysis*. Berlin and Heidelberg: Springer.

Tversky, A, and Kahneman, D. 1992. Advances in prospect theory: Cumulative representation of uncertainty. *Journal of Risk and Uncertainty*, **5**, 297–323.

UK Environment Agency. 2021. 2D benchmarking: Evaluating the latest generation of the hydraulic models for FCRM. Online; accessed 18 October 2023.

Ulam, S. 1962. On some mathematical problems connected with patterns of growth in figures. *Proceedings of Symposia in Applied Mathematics*, **14**, 215–224.

Ullah, S, Abdrakhmatov, K, Sadykova, A, Ibragimov, R, Ishuk, A, et al. 2015. EMCA Central Asia seismic source model. V. 1.2. GFZ Data Services. Online; accessed 30 November 2023.

Ulvrová, M, Paris, R, Kelfoun, K, and Nomikou, P. 2014. Numerical simulations of tsunamis generated by underwater volcanic explosions at Karymskoye lake (Kamchatka, Russia) and Kolumbo volcano (Aegean Sea, Greece). *Natural Hazards and Earth System Sciences*, **14**, 401–412.

UNDRR (United Nations Office for Disaster Risk Reduction), and ISC (International Science Council). 2020. *Hazard Definition & Classification Review. Technical Report*. Geneva: United Nations.

UNISDR (United Nations International Strategy for Disaster Reduction). 2009. *2009 UNISDR Terminology on Disaster Risk Reduction*. Geneva: UNISDR.

United Nations. 2015. *Sendai Framework for Disaster Risk Reduction 2015–2030*. United Nations – Headquarters United Nations Office for Disaster Risk Reduction.

US Bureau of Labor Statistics. 2023. Shelter in U.S. city average, all urban consumers, not seasonally adjusted. Online; accessed 24 October 2023.

Utsu, T, Ogata, Y, and Matsu'ura, RS. 1995. The centenary of the Omori formula for a decay law of aftershock activity. *Journal of Physics of the Earth*, **43**, 1–33.

van Aalst, MK. 2006. The impacts of climate change on the risk of natural disasters. *Disasters*, **30**(1), 5–18.

Van Den Eeckhaut, M, Poesen, J, Govers, G, Verstraeten, G, and Demoulin, A. 2007a. Characteristics of the size distribution of recent and historical landslides in a populated hilly region. *Earth and Planetary Science Letters*, **256**, 588–603.

Van Den Eeckhaut, M, Poesen, J, Dusar, M, Martens, V, and Duchateau, PH. 2007b. Sinkhole formation above underground limestone quarries: A case study in South Limburg (Belgium). *Geomorphology*, **91**, 19–37.

van der Elst, NJ, Page, MT, Weiser, DA, Goebel, THW, and Hosseini, SM. 2016. Induced earthquake magnitudes are as large as (statistically) expected. *Journal of Geophysical Research Solid Earth*, **121**, 4575–4590.

Van Luik, A, Patterson, R, Shafer, D, and Klein, T. 2013. Memory saves lives: Intergenerational warnings effectiveness. Page 13556 of: *WM2013 Conference*, 24–28 February 2013, Phoenix, AZ.

van Stiphout, T, Wiemer, S, and Marzocchi, W. 2010. Are short-term evacuations warranted? Case of the 2009 L'Aquila earthquake. *Geophysical Research Letters*, **37**, L06306.

van Wijk, MT, Tittonell, P, Rufino, MC, Herrero, M, Pacini, C, et al. 2009. Identifying key entry-points for strategic management of smallholder farming systems in sub-Saharan Africa using the dynamic farm-scale simulation model NUANCES-FARMSIM. *Agricultural Systems*, **102**, 89–101.

Varghese, B, and Rau-Chaplin, A. 2013. Accounting for secondary uncertainty: Efficient computation of portfolio risk measures on multi and many core architectures. *WHPCF '13: Proceedings of the 6th Workshop on High Performance Computational Finance*. New York, NY: Association for Computing Machinery.

Varotsos, P, Alexopoulos, K, and Nomicos, K. 1981. Seismic electric currents. *Proceedings of the Academy of Athens*, **56**, 277–286.

Verburg, PH, Dearing, JA, Dyke, JG, van der Leeuw, S, Seitzinger, S, et al. 2016. Methods and approaches to modelling the Anthropocene. *Global Environmental Change*, **39**, 328–340.

Verhulst, P-F. 1838. Notice sur la loi que la population poursuit dans son accroissement. *Correspondance Mathématique et Physique*, **10**, 113–121.

Vicente-Serrano, SM, Lopez-Moreno, J-I, Beguería, S, Lorenzo-Lacruz, J, Sanchez-Lorenzo, A, et al. 2014. Evidence of increasing drought severity caused by temperature rise in southern Europe. *Environmental Research Letters*, **9**, 044001.

Vickery, PJ, and Twisdale, LA. 1995. Wind-field and filling models for hurricane wind-speed predictions. *Journal of Structural Engineering*, **121**, 1700–1709.

Vickery, PJ, Skerlj, PF, and Twisdale, LA. 2000. Simulation of hurricane risk in the US using empirical track model. *Journal of Structural Engineering*, **126**(10), 1222–1237.

Vickery, PJ, Skerlj, PF, Lin, J, Twisdale, LA, Jr, Young, MA, and Lavelle, FM. 2006. HAZUS-MH hurricane model methodology. II: Damage and loss estimation. *Natural Hazards Review*, **7**, 94–103.

Vijayalakshmi, A, and Meena, M. 2020. Crop destructor-locust – A critical review. *Journal of Xidian University*, **14**(7), 1798–1804.

Virtanen, P, Gommers, R, Oliphant, TE, Haberland, M, Reddy, T, et al. 2020. SciPy 1.0: Fundamental algorithms for scientific computing in Python. *Nature Methods*, **17**(3), 261–272.

Vitolo, R, Stephenson, DB, Cook, IM, and Mitchell-Wallace, K. 2009. Serial clustering of intense European storms. *Meteorologische Zeitschrift*, **18**(4), 411–424.

Vittori, E, Fulloni, S, and Piccardi, L. 2007. Environment and natural hazards in Roman and Medieval texts: Presentation of the CLEMENS database project. *Geological Society, London, Special Publications*, **273**, 51–59.

Voltaire. 1756. *Poème sur le désastre de Lisbonne*. Geneva: Cramer.

Voltaire. 1759. *Candide ou l'Optimisme*. Geneva: Cramer.

Volterra, V. 1928. Variations and fluctuations of the number of individuals in animal species living together. *ICES Journal of Marine Science*, **3**(1), 3–51.

von Hippel, FN. 2011. The radiological and psychological consequences of the Fukushima Daiichi accident. *Bulletin of the Atomic Scientists*, **67**(5), 27–36.

von Neumann, J. 1966. *Theory of Self-Reproducing Automata*. University of Illinois Press.

von Neumann, J, and Morgenstern, O. 1944. *Theory of Games and Economic Behavior*. Princeton University Press.

von Rebeur-Paschwitz, E. 1889. The Earthquake of Tokio, April 18, 1889. *Nature*, **40**, 294–295.

Vreugdenhil, CB. 1994. *Numerical Methods for Shallow-Water Flow*. Dordrecht: Springer.

Wagenaar, D, de Jong, J, and Bouwer, LM. 2017. Multi-variable flood damage modelling with limited data using supervised learning approaches. *Natural Hazards and Earth System Sciences*, **17**, 1683–1696.

Wald, DJ, Jaiswal, KS, Marano, KD, Bausch, DB, and Hearne, MG. 2010. *PAGER — Rapid Assessment of an Earthquake's Impact*. US Geological Survey Fact Sheet 2010–3036.

Walker, E, Mitchell, D, and Seviour, W. 2020. The numerous approaches to tracking extratropical cyclones and the challenges they present. *Weather*, **75**(11), 336–341.

Walker, GR. 2011. Modelling the vulnerability of buildings to wind — a review. *Canadian Journal of Civil Engineering*, **38**, 1031–1039.

Walker, KF, Cohen, AL, Walker, SH, Allen, KM, Baines, DL, and Thornton, JG. 2014. The dangers of the day of birth. *BJOG*, **121**(6), 714–718.

Walker, T, Xu, Y, Gramlich, D, and Zhao, Y. 2022. The impact of natural disasters on the performance and solvency of US banks. *International Journal of Managerial Finance*, **19**(1), 136–154.

Walsh, JE, Phillips, AS, Portis, DH, and Chapman, WL. 2001. Extreme cold outbreaks in the United States and Europe, 1948–99. *Journal of Climate*, **14**, 2642–2658.

Walsh, K, and Pittock, AB. 1998. Potential changes in tropical storms, hurricanes, and extreme rainfall events as a result of climate change. *Climatic Change*, **39**, 199–213.

Walsh, KJE, Camargo, SJ, Vecchi, GA, Daloz, AS, Elsner, J, et al. 2015. Hurricanes and climate: The US CLIVAR Working Group on hurricanes. *Bulletin of the American Meteorological Society*, **96**(6), 997–1017.

Wang, K, Chen, Q-F, Sun, S, and Wang, A. 2006. Predicting the 1975 Haicheng earthquake. *Bulletin of the Seismological Society of America*, **96**(3), 757–795.

Wang, R, Dearing, JA, Langdon, PG, Zhang, E, Yang, X, et al. 2012. Flickering gives early warning signals of a critical transition to a eutrophic lake state. *Nature*, **492**, 419–422.

Wang, Z. 2009. Comment on 'Sigma: Issues, insights, and challenges' by F. O. Strasser, N. A. Abrahamson, and J. J. Bommer. *Seismological Research Letters*, **80**(3), 491–493.

Wang, Z, Broccardo, M, Mignan, A, and Sornette, D. 2020. The dynamics of entropy in the COVID-19 outbreaks. *Nonlinear Dynamics*, **101**, 1847–1869.

Ward, SN, and Asphaug, E. 2000. Asteroid impact tsunami: A probabilistic hazard assessment. *Icarus*, **145**, 64–78.

Ward, SN, and Day, S. 2001. Cumbre Vieja Volcano – Potential collapse and tsunami at La Palma, Canary Islands. *Geophysical Research Letters*, **28**(17), 3397–3400.

Ward, SN, and Day, S. 2011. The 1963 landslide and flood at Vaiont Reservoir Italy. A tsunami ball simulation. *Italian Journal of Science*, **130**(1), 16–26.

Watson, JT, Gayer, M, and Connolly, MA. 2007. Epidemics after natural disasters. *Emerging Infectious Diseases*, **13**(1), 1–5.

Webb, JDC, Elsom, DM, and Meaden, GT. 2009. Severe hailstorms in Britain and Ireland, a climatological survey and hazard assessment. *Atmospheric Research*, **93**, 587–606.

Weinstein, ND. 1984. Why it won't happen to me: Perceptions of risk factors and susceptibility. *Health Psychology*, **3**(5), 431–457.

Wells, DL, and Coppersmith, KJ. 1994. New empirical relationships among magnitude, rupture length, rupture width, rupture area, and surface displacement. *Bulletin of the Seismological Society of America*, **84**(4), 974–1002.

Westerling, AL, Hidalgo, HG, Cayan, DR, and Swetnam, TW. 2006. Warming and earlier spring increase western US forest wildfire activity. *Science*, **313**, 940–943.

Wheatley, S, Hofmann, A, and Sornette, D. 2020. Addressing insurance of data breach cyber risks in the catastrophe framework. *The Geneva Papers on Risk and Insurance – Issues and Practice*, **46**, 3–78.

White, JA, and Foxall, W. 2016. Assessing induced seismicity risk at CO_2 storage projects: Recent progress and remaining challenges. *International Journal of Greenhouse Gas Control*, **49**, 413–424.

White, R, Engelen, G, and Uljee, I. 1997. The use of constrained cellular automata for high-resolution modelling of urban land-use dynamics. *Environment and Planning B: Urban Analytics and City Science*, **24**(3), 323–343.

White, SH, del Rey, AM, and Sánchez, GR. 2007. Modeling epidemics using cellular automata. *Applied Mathematics and Computation*, **186**, 193–202.

Whitman, S, Good, G, Donoghue, ER, Benbow, N, Shou, W, and Mou, S. 1997. Mortality in Chicago attributed to the July 1995 heat wave. *American Journal of Public Health*, **87**(9), 1515–1518.

Whitney, H. 1955. On singularities of mappings of Euclidean spaces. I. Mappings of the plane into the plane. *Annals of Mathematics*, **62**(3), 374–410.

Wieczorek, GF, Larsen, MC, Eaton, LS, Morgan, BA, and Blair, JL. 2001. *Debris-Flow and Flooding Hazards Associated with the December 1999 Storm in Coastal Venezuela and Strategies for Mitigation*. USGS Publications, Open-File Report 2001-144.

Wikipedia. 2023. Mediterranean tropical-like cyclone. Online; accessed 11 November 2023.

Wilhelmson, RB, Jewett, BF, Shaw, C, Wicker, LJ, Arrott, M, et al. 1990. A study of the evolution of a numerically modeled severe storm. *The International Journal of Supercomputer Applications*, **4**(2), 20–36.

Willcock, S, Cooper, GS, Addy, J, and Dearing, JA. 2023. Earlier collapse of Anthropocene ecosystems driven by multiple faster and noisier drivers. *Nature Sustainability*, **6**, 1331–1342.

Williams, ER, and Lhermitte, RM. 1983. Radar tests of the precipitation hypothesis for thunderstorm electrification. *Journal of Geophysical Research*, **88**(C15), 10,984–10,992.

Willoughby, HE, and Rahn, ME. 2004. Parametric representation of the primary hurricane vortex. Part I: Observations and evaluation of the Holland (1980) model. *Monthly Weather Review*, **132**, 3033–3048.

Willoughby, HE, Darling, RWR, and Rahn, ME. 2006. Parametric representation of the primary hurricane vortex. Part II: A new family of sectionally continuous profiles. *Monthly Weather Review*, **134**, 1102–1120.

Wing, OEJ, Pinter, N, Bates, PD, and Kousky, C. 2020. New insights into US flood vulnerability revealed from flood insurance big data. *Nature Communications*, **11**, 1444.

Winkler, J, Dueñas-Osorio, L, Stein, R, and Subramanian, D. 2010. Performance assessment of topologically diverse power systems subjected to hurricane events. *Reliability Engineering and System Safety*, **95**, 323–336.

Winkler, J, Dueñas-Osorio, L, Stein, R, and Subramanian, D. 2011. Interface network models for complex urban infrastructure systems. *Journal of Infrastructure Systems*, **17**, 138–150.

Wittneben, BBF. 2012. The impact of the Fukushima nuclear accident on European energy policy. *Environmental Science & Policy*, **15**, 1–3.

Woessner, J, Danciu, L, Giardini, D, Crowley, H, Cotton, F, et al. 2015. The 2013 European Seismic Hazard Model: Key components and results. *Bulletin of Earthquake Engineering*, **13**, 3553–3596.

Wolfram, S. 1983. Statistical mechanics of cellular automata. *Reviews of Modern Physics*, **55**(3), 601–644.

Wolfram, S. 2002. *A New Kind of Science*. Champaign, IL: Wolfram Media.

Woo, G. 1999. *The Mathematics of Natural Catastrophes*. London: Imperial College Press.

Woo, G. 2002. Natural catastrophe probable maximum loss. *British Actuarial Journal*, **8**(5), 943–959.

Woo, G. 2011. *Calculating Catastrophe*. London: Imperial College Press.

Woo, G. 2015. An actuarial perspective on pandemics. *Longevity Bulletin: The Pandemics Edition*, **1**, 3–5.

Woo, G, and Mignan, A. 2018. Counterfactual analysis of runaway earthquakes. *Seismological Research Letters*, **89**(6), 2266–2273.

Wood, HO, and Neumann, F. 1931. Modified Mercalli intensity scale of 1931. *Bulletin of the Seismological Society of America*, **21**(4), 277–283.

Woodward, RT, and Bishop, RC. 1997. How to decide when experts disagree: Uncertainty-based choice rules in environmental policy. *Land Economics*, **73**(4), 492–507.

Wright, N. 2022. Seeing and interpreting visions of the next age in Interstellar. *Journal of Religion & Film*, **26**(1), 1–44.

Wu, H-C, Murphy, H, Greer, A, and Clay, L. 2024. Evacuate or social distance? Modeling the influence of threat perceptions on hurricane evacuation in a dual-threat environment. *Risk Analysis*, **44**, 724–737.

Wu, Y-M, Chang, C-H, Zhao, L, Teng, T-L, and Nakamura, M. 2008. A comprehensive relocation of earthquakes in Taiwan from 1991 to 2005. *Bulletin of the Seismological Society of America*, **98**, 1471–1481.

Wyss, M, and Habermann, RE. 1988. Precursory seismic quiescence. *Pure and Applied Geophysics*, **126**(2-4), 319–332.

Xie, W, Li, N, Li, C, Wu, J-d, Hu, A, and Hao, X. 2014. Quantifying cascading effects triggered by disrupted transportation due to the Great 2008 Chinese Ice Storm: Implications for disaster risk management. *Natural Hazards*, **70**, 337–352.

Xu, C, Xu, X, and Shyu, JBH. 2015. Database and spatial distribution of landslides triggered by the Lushan, China Mw 6.6 earthquake of 20 April 2013. *Geomorphology*, **248**, 77–92.

Xu, Y, Zhang, L, and Jia, J. 2008. Lessons from catastrophic dam failures in August 1975 in Zhumadian, China. Pages 162–169 of: Reddy, KR, Khire, MV, and Alshawabkeh, AN (eds), *GeoCongress 2008: Geosustainability and Geohazard Mitigation* Reston, VA: Geo-Institute of The American Society of Civil Engineers.

Yang, Y, Saatchi, SS, Xu, L, Yu, Y, Choi, S, et al. 2018. Post-drought decline of the Amazon carbon sink. *Nature Communications*, **9**, 3172.

Yin, Y, Zheng, W, Li, X, Sun, P, and Li, B. 2011. Catastrophic landslides associated with the M8.0 Wenchuan earthquake. *Bulletin of Engineering Geology and the Environment*, **70**, 15–32.

Youngs, RR, and Coppersmith, KJ. 1985. Implications of fault slip rates and earthquake recurrence models to probabilistic seismic hazard estimates. *Bulletin of the Seismological Society of America*, **75**(4), 939–964.

Youngs, RR, Chiou, S-J, Silva, WJ, and Humphrey, JR. 1997. Strong ground motion attenuation relationships for subduction zone earthquakes. *Seismological Research Letters*, **68**, 58–73.

Zack, M. 2015. Rebuilding mathematically: A study of the rebuilding of Lisbon and London. *Nexus Network Journal*, **17**, 571–586.

Zahler, RS, and Sussmann, HJ. 1977. Claims and accomplishments of applied catastrophe theory. *Nature*, **269**, 759–763.

Zeeman, EC. 1976. Catastrophe theory. *Scientific American*, **234**, 65–70, 75–83.

Zhao, C, Liu, B, Piao, S, Wang, X, Lobell, DB, et al. 2017. Temperature increase reduces global yields of major crops in four independent estimates. *Proceedings of the National Academy of Sciences*, **114**(35), 9326–9331.

Zhao, Q, Guo, Y, Ye, T, Gasparrini, A, Tong, S, et al. 2021. Global, regional, and national burden of mortality associated with non-optimal ambient temperatures from 2000 to 2019: A three-stage modelling study. *Lancet Planet Health*, **5**, e415–425.

Zhou, Q, Zhang, Y, Chang, K, and Brovelli, MA. 2022. Assessing OSM building completeness for almost 13,000 cities globally. *International Journal of Digital Earth*, **15**(1), 2400–2421.

Zhou, Y, Nelson, K, Mohr, KI, Huffman, GJ, Levy, R, and Grecu, M. 2019. A spatial-temporal extreme precipitation database from GPM IMERG. *Journal of Geophysical Research: Atmospheres*, **124**, 10,344–10,363.

Zuccaro, G, Cacace, F, Spence, RJS, and Baxter, PJ. 2008. Impact of explosive eruption scenarios at Vesuvius. *Journal of Volcanology and Geothermal Research*, **178**, 416–453.

Zucchini, W. 2000. An introduction to model selection. *Journal of Mathematical Psychology*, **44**, 41–61.

Index

accident, 7, 265
accuracy, 251
adjacency matrix, 167, 198, 209, 212
agent-based model (ABM), 174–178, 182, 186, 189
  definition, 165
airburst. *See* asteroid impact
armed conflict, 9, 56, 69, 165, 224, 225
  definition and examples, 55
  impact of global warming, 220
  intensity (system dynamics), 188
  size distribution, 89
asteroid impact, 10, 11, 16, 24, 30, 75, 80
  definition and examples, 42
  intensity footprint, 99
  size distribution, 91
  source and size, 69
  Torino scale, 135
  vulnerability curve, 140
average annual loss (AAL), 147, 197, 239

banking system failure
  cyber-heist, 50
  definition and examples, 52
Bayes' theorem, 95
Bayesian inference, 259
Bernoulli trial, 258
beta distribution, 238, 272
bias
  in catastrophe data, 9
  in perception, 264
binomial distribution, 132
black swan, 88
blackout, 36, 43, 49, 212
  cascading failure (topological) model, 209–211
  cascading failure algorithm, 210
  definition and examples, 47
  intensity footprint (cellular automaton), 167
  size distribution, 91
building
  accelerated fatigue, 205
  ageing, 203
  capacity curve, 138, 205
  construction type, 122
  resonance, 122
business interruption, 14, 40, 48, 49, 54, 119, 132, 178, 212, 222, 223, 229
  definition and examples, 51
  impact of global warming, 220
  intensity footprint, 104

cascading effects, 213
  compound event, 212–229
  domino effects, 206–212
  due to global warming, 218
  tipping cascade, 220, 225
  trophic cascade, 221
casualty, 5, 8, 9
  by building damage state, 142, 262
  matrix, 141
  by overpressure, 141
  by rising water, 142
CAT model, 13, 16
  cellular automata and ABMs in CAT model, 165
  for crops as exposure, 188
  event interactions in CAT model, 202
  as forecast, 245
  global warming in CAT model, 220
  history, 20
  sandbox, 151–153
  system dynamics in CAT model, 182, 188
  time dependency in CAT model, 203
catastrophe, 4, 8
  bond, 240
  in literature, 19
  model. *See* CAT model
  object, 67
  ontology, 60
  super-cat, 193, 209, 211, 214
cellular automaton, 162–164, 167–174, 178
  definition, 162
  forest-fire model, 173
  road network, 125
  sandpile model, 164
  urban growth, 126
  von Neumann and Moore neighbourhoods, 162
centrality measures, 199
  betweenness (as flow proxy), 211
civilizational collapse. *See* societal collapse
climate change. *See* global warming
coastal flood. *See* storm surge; tsunami
coefficient of determination, 137
cold wave, definition and examples, 36

comet impact. *See* asteroid impact
complementary cumulative distribution function (CCDF), 64, 88, 89, 95
complex Earth system, 178, 193, 212, 232
complex system, 161
  punctuated equilibrium, 177
  regime shift, 212, 222, 223
  tipping point, 214, 219
compound event. *See* cascading effects
confusion matrix, 251
consequence function, 131, 141, 262
convective storm, 76, 86
  definition and examples, 33
  source, 82
criminality, 7, 50
critical infrastructure failure. *See* peril, technological
crop
  damage by hail, 143
  loss, 25, 34, 36, 37, 41, 49, 134, 184, 219, 223
cumulative distribution function (CDF), 80
  empirical CDF (eCDF), 80
    pseudocode, 91
cumulative log-normal distribution, 130–132, 140, 144
cyberattack, 9, 55
  definition and examples, 49
  size distribution, 91

dam flood, 169
  definition and examples, 46
damage
  function, 15
  secondary uncertainty, 140
  state (DS), 130–132, 135, 136, 138, 140
deterministic vs. probabilistic, 64
disaster. *See* catastrophe
dispersion
  index of dispersion, 196
  statistics, 196
distance
  Chebyshev, or chessboard, 163
  Euclidean, 98
  Manhattan, 163
distribution function. *See* cumulative distribution function (CDF)
diversification, 241
domino effects. *See* cascading effects
downscaling, 128
drought, 9, 10, 17, 19, 35, 38, 49, 56, 124, 216
  definition and examples, 36
  impact of global warming, 222
dynamical system. *See* system dynamics

earthquake, 9, 18–20, 26, 28, 30, 43, 48, 49, 73, 77, 93, 122, 130, 132, 136, 138, 140, 172, 204, 209, 257
  clustering (by stress transfer), 200
  damage amplification, 205
  definition and examples, 22
  forecasting, 248–253
  hazard secondary uncertainty, 112
  intensity footprint, 98
  intensity prediction equation, 261
  MMI scale, 133
  PSHA (entire pipeline), 106–114
  size distribution, 92
  source and size, 73
  stochastic rupture model, 108
  vulnerability curve, 141
economic
  disaster, 51
  loss. *See* loss
ecosystem collapse, 221–223, 270
  Amazon rainforest dieback, 221
  lake eutrophication, 222
energy
  blast energy, 69
  of combustion, 86
  of explosion, 69
  of flood, 72
  kinetic energy, 69, 77, 83
  potential energy, 78
  seismic energy, 75
  thermal energy, 73
energy supply failure, 43
  e.g., blackout. *See* blackout
epidemic, 9, 49, 53, 56, 124, 165, 211
  definition and examples, 39
  intensity (system dynamics), 179
  intensity footprint (ABM), 182
  size distribution, 96
error function, 87
event
  chain of events, 197, 207
  freak event, 10, 42
  loss, 146
  loss table (ELT), 62, 147, 149, 151, 197
    definition, 146
    standard deviation, 238
  mega-event, 9, 193
  meta-event, 10, 193, 194, 216
  speculative event, 10
  table, 61, 107
  tail event, 9, 193
exceedance probability (EP) curve, 149, 240, 262
  aggregate EP (AEP) curve, 149, 196
  occurrence EP (OEP) curve, 149, 196
existential risk, 9, 225
explosion, 24, 43, 85, 189
  in chain of events, 207
  definition and examples, 45
  intensity footprint, 99
  source and size (including nuclear explosion), 69
  vulnerability curve, 140

exponential
  decay, 82
  distribution, 95
  function, 92
  growth, 184
exposure
  characteristics, 120–124
  as crop, 124
  databases, 124
  as ecosystem services, 223
  module, 14, 119–128
  occupancy class, 120
  as population, 124
  primary and secondary modifiers, 120
  simulation, 126
  time-dependency, 203, 242
extratropical cyclone, 29
  definition and examples, 32
  size distribution, 97
  source, 82
  statistical clustering, 196
  vulnerability curve, 145
extreme value distributions, 94–97

famine, 9, 17, 37, 41, 51, 228, 229
  definition and examples, 55
  impact of global warming, 220
fatality. *See* casualty
financial crisis, definition and examples, 52
financial module. *See* loss, module
fire, 18, 19, 24, 34, 42, 43, 48, 49, 53, 70, 173, 174, 176, 189
  definition and examples, 45
  in chain of events, 207
  source and size, 85
flood. *See also* coastal flood; fluvial flood; pluvial flood
  impact of global warming, 220
  intensity footprint (cellular automaton), 168
  vulnerability curve, 142
fluvial flood, 29, 142, 169
  damage claims, 136
  definition and examples, 28
  size distribution, 93, 96
  source and size, 71
force, 138, 175
forecasting, 235, 245–253
fragility curve, 15, 129, 132, 138–141, 203, 205
  definition, 130
Fréchet distribution, 94, 97

Gaussian. *See* normal distribution
generalized extreme value (GEV) distribution, 94–96
generalized Pareto distribution (GPD), 90, 95–97
geomagnetic storm, definition and examples, 43
global warming, 4, 5, 10, 105, 179, 193, 203, 213–220, 222, 225, 245, 270
  consequences for perils, 218
  ESCIMO climate model, 214
gravitational acceleration, 78
Gumbel distribution, 95, 96

hailstorm
  definition and examples, 34
  source and size, 83
  TORRO H scale, 134
  vulnerability curve, 142
hazard, 5
  curve, 64, 113
  as energy transfer, 66
  event source and size, 67
  intensity, 63
  intensity footprint, 97–106
  map, 65, 113
  modelling pipeline (earthquake), 106–114
  module, 13
  secondary uncertainty, 64
heatwave, 37, 124, 216
  definition and examples, 35
  impact of global warming, 220
  intensity footprint, 105
Heaviside step function, 65
heavy-tailed distribution, 86, 87
hyperbolic tangent function, 133

impact function. *See* damage, function
infectious disease. *See* epidemic
insurance, 5, 13–15, 18, 19, 21, 27, 54, 119, 136, 137, 143, 189, 194, 236–240, 254, 256, 263, 267, 268
  diversification, 237
  financial structure, 238
  -linked securities (CAT bond), 240
  pricing, 239
  survival constraint, 240
intensity. *See* hazard, intensity
inventory. *See* exposure
inverse transform sampling, 80

landslide, 9, 24, 28, 30, 36, 47, 49, 79, 135, 204, 209, 211
  definition and examples, 26
  impact of global warming, 220
  intensity footprint (cellular automaton), 170
  size distribution, 93
  source and size, 83
  triggering by rainstorm or earthquake, 171
  vulnerability curve, 143
law of large numbers, 264
levee failure. *See* dam flood
life table, 18

lightning, 9, 34, 37, 38, 48, 83, 173, 209
  definition and examples, 34
  source, 85
likelihood. *See* probability
log-normal distribution, 202
logistic
  function, 80, 142
  growth, 184, 228
loss, 3, 5, 8
  amplification, 193
  aversion, 267
  footprint, 145
  ground-up loss, 15, 119, 145, 239
  insured loss, 239
  intangible loss, 223
  module, 15, 145–151
  return period, 149
  secondary uncertainty, 146
  standard deviation, 237
  table, 146, 237

Markov
  chain, property, 197
  model, 247
  transition matrix, 198, 207, 247
mass school shooting, definition and examples, 54
maximum likelihood estimation (MLE), 89
mean damage ratio (MDR), 118, 131, 136, 140, 145, 204, 238
  definition, 129
method of moments, 140
micromort, 263
minimax, maximin, maximax, 270
Monte Carlo method, 20, 76, 146, 210, 260
  definition, 80
myth (of catastrophe), 16

Natech event, 44
negative binomial distribution, 195
normal distribution, 87
numerical integration, Euler's method, 179, 181

ordinary differential equation (ODE), 161
  cusp catastrophe, 186
  epidemic (SIR) model, 181
  Lanchester equations, 188
  logistic growth, 184
  Lotka–Volterra model, 184, 222
  master equation, 199
  pseudocode, 181
outbreak. *See* epidemic

pandemic. *See* epidemic
Pareto distribution, 88, 95
partial differential equation (PDE), 168
perfect storm, 10
peril, 4
  biological, 39–41
    impact of global warming, 219
  climatological, 35–39
    impact of global warming, 218
  extraterrestrial, 41–43
  geophysical, 22–27
    impact of global warming, 219
  hydrological, 27–30
    impact of global warming, 219
  meteorological, 31–35
    impact of global warming, 218
  socio-economic, 50–56
    impact of global warming, 219
  technological, 43–50
    chemical, 45
    domino effects, 206
    impact of global warming, 219
    nuclear, 45
    risk mitigation, 244
pest infestation, 38, 56
  definition and examples, 41
  intensity (system dynamics), 183
  intensity footprint (ABM), 186
pluvial flood, 36, 142, 169, 170
  definition and examples, 29
Poisson distribution, 62, 64, 146, 149, 195, 197, 238, 259
pollution, 8, 26, 38, 57, 165, 222
pound per square inch, 99
power law, 87–94, 164, 166, 167, 173, 193
probability, 5, 7, 61, 64, 66, 87, 113, 130, 194, 202, 237
  definition, 11
  density function (PDF), 64, 80, 88, 260
    definition, 88
  empirical/frequency-based probability, 12, 264
  objective probability, 11
  subjective probability, 259
probable maximum loss (PML), 151
process error. *See* uncertainty, aleatory
pseudocode
  ELT to YLT, 146
  empirical CDF, 91
  epidemic (SIR) model, 181
  hazard module, 154
  risk modules, 155
public service failure, definition and examples, 53

rainstorm, 26, 27, 71, 93, 170, 171
  impact of global warming, 220
  size distribution, 97
random
  field, 112
  number/sampling, 80, 108, 121, 125, 209, 238
randomness. *See* uncertainty, aleatory
return level plot, 95

return period, 62
riot. *See* social unrest
risk, 5, 11
  ambiguity aversion, 270
  assessment, 6
  aversion, 266, 268
  -based capital (RBC), 240
  cession, 236
  governance, 7, 235, 253–271
  individual risk, 262
  load, 239, 241
  management, 6
  matrix, 5
  metrics, 147–151, 235, 237
  parameter risk. *See* uncertainty, epistemic
  perception, 264
  process risk. *See* uncertainty, aleatory
  reduction, 7, 242–245
    in exposure module, 243
    in hazard module, 244
    in vulnerability module, 243
  seeking, 268
  transfer, 6, 236–242
  triangle, 13

self-organized criticality (SOC), 87, 164, 168, 173
self-similar, 87
sigmoid function, 130
simulation table. *See* year event table (YET)
social unrest, 51, 53
  definition and examples, 53
  intensity footprint (ABM), 176
societal collapse, 10, 224–229
  resource depletion, 10, 224, 225
  World2 and World3 models, 225
socio-political peril. *See* peril, socio-economic
Solvency II, 240
stakeholder, 235, 254
stampede, 244
  definition and examples, 53
  intensity footprint (ABM), 174
statistical moment, 88
stochastic, 61, 62, 65, 76, 80, 82, 83, 110–112, 121, 196, 201
  event set, 61, 86, 107, 110, 151, 183, 220
  uncertainty. *See* uncertainty, aleatory
storm. *See* convective storm; extratropical cyclone; tornado; tropical cyclone
storm surge, 31, 134, 142, 169, 204, 246
  definition and examples, 29
  impact of global warming, 220
  intensity footprint, 106
  source and size, 75
stress, 138, 193, 201
system dynamics, 178–188
  catastrophe theory, 187
  causal loop diagram, 179
  compartmental (SIR) model, 179
  cusp catastrophe, 186, 188, 213
  ESCIMO climate model, 214
  hysteresis, 187, 223
  master equation, 199
  pseudocode, 181
  socio-ecological model, 221
  stocks, flows, feedback loops, 178
  World models, 225

tail
  conditional expectation (TCE), 151
  fattening, 197
  value at risk (TVaR), 151
terrorist attack, 53, 69, 71, 174, 175, 265
  damage of blast, 140
  definition and examples, 54
  size distribution, 89
thunderstorm. *See* lightning
TNT equivalent, 69
tornado, 31, 210
  damage footprint, 137, 337
  definition and examples, 34
  Enhanced Fujita (EF) scale, 135
  forecasting, 251
  intensity footprint, 100
  source and size, 76
  vulnerability curve, 145
toxic spill, 43, 170
  in chain of events, 207
  definition and examples, 45
transport system failure, 209
  definition and examples, 48
tropical cyclone, 20, 29, 33, 132, 209
  definition and examples, 31
  forecasting, 246–248
  impact of global warming, 220
  intensity footprint, 102
  Saffir–Simpson scale, 134
  size distribution, 93, 97
  source and size, 81
  statistical clustering, 196
  stochastic path model, 248
  vulnerability curve, 145
tsunami, 10, 17, 18, 24, 26, 27, 42, 45, 47, 142, 169, 204
  definition and examples, 30
  size distribution, 93
  source and size, 77

uncertainty, 11
  aleatory, 11, 64, 130
  epistemic, 11, 258–263
    logic tree, 260
    principle of indifference, 264
  Knightian, 12
  ontological, 12

primary, 15, 259
secondary, 16, 238, 259
uniform distribution, 80
utility, 266
cumulative prospect theory (CPT), 267
expected utility theory (EUT), 267

value at risk (VaR), 150, 240
volcanic eruption, 10, 16, 24, 28, 30, 79, 204, 218
definition and examples, 25
impact on earthquake damage, 204
intensity footprint, 104
size distribution, 93
source and size, 72
volcanic explosivity index (VEI), 72
vulnerability curve, 144
vulnerability, 6, 119, 122
amplification, 204
curve, 129, 136, 140
definition, 131
module, 15, 129–145
surface, 204

war. *See* armed conflict
water system failure, 71, 170, 209
definition and examples, 49
weather event. *See* flood; storm
Weibull distribution, 94, 96, 143
wildfire, 26, 34, 36, 41, 84, 216, 218, 222
definition and examples, 37
intensity footprint (cellular automaton), 173
size distribution, 94
windstorm. *See* extratropical cyclone

year event table (YET), 62, 202
year loss table (YLT), 62, 148, 151, 195, 197
definition, 146
from ELT (pseudocode), 146
standard deviation, 238

For EU product safety concerns, contact us at Calle de José Abascal, 56–1°,
28003 Madrid, Spain or eugpsr@cambridge.org.

www.ingramcontent.com/pod-product-compliance
Lightning Source LLC
LaVergne TN
LVHW080017210926
848481LV00011B/3044